Mechanics and Architectural Design

Mechanics and Architectural Design

Proceedings of 2016 International Conference

2016 International Conference on Mechanics and Architectural Design
Suzhou, Jiangsu, China, 14 – 15 May 2016

Editors

Shi-Hong Zhang
Beijing University of Civil Engineering & Architecture, China

Peng-Sheng Wei
National Sun Yat-sen University, Taiwan

World Scientific

NEW JERSEY • LONDON • SINGAPORE • BEIJING • SHANGHAI • HONG KONG • TAIPEI • CHENNAI • TOKYO

Published by

World Scientific Publishing Co. Pte. Ltd.
5 Toh Tuck Link, Singapore 596224
USA office: 27 Warren Street, Suite 401-402, Hackensack, NJ 07601
UK office: 57 Shelton Street, Covent Garden, London WC2H 9HE

British Library Cataloguing-in-Publication Data
A catalogue record for this book is available from the British Library.

MECHANICS AND ARCHITECTURAL DESIGN
Proceedings of 2016 International Conference

ISBN 978-981-3149-01-4

Printed in Singapore

Contents

Committees

Editor

Prof. Peng-Sheng Wei, Professor, National Sun Yat-Sen University, Taiwan (China)

Co-editor

Prof. Shi-Hong Zhang, Beijing University of Civil Engineering & Architecture, China

Technical Program Committee

Prof. J. A. Esfahani, Ferdowsi University of Mashhad, Wikipedia

Dr. Mais Nijim, Texas A&M- Kingsville, The United States of America

Prof. Zhao-Qun Du, Donghua University, China

Dr. Wan-Bo Yu, Dalian University, China

Dr. Yong-Hui An, Dalian University of technology, China

Dr. Lin-Gai Luo, CNRS(French National Scientific Research Centre), China

Dr. Yu-Chao Wang, Sun Yat-sen University, China

Dr. Meng Guo, Harbin Institute of Technology, China

Dr. Fu-Jun Xu, Donghua University, China

Prof. Xuan-Ming Ding, Chongqing University, China

Preface

MAD 2016 was held in Suzhou, China from May 14–15, 2016. This conference, an excellent international academic forum, calls for wide interest of researchers and practitioners among different universities and research institutions on the world. Any paper with a broad topic related to fields of Mechanics and Architectural Design is welcomed. We strongly desire to provide the conference a perfect platform to share resources, exchange opinions and inspire further studies.

MAD 2016 Proceeding has collected the most up-to-date, comprehensive, and worldwide state-of-art knowledge on Mechanics and Architectural Design. All the accepted papers have been submitted for a strict peer-review by 2-4 expert referees, considering originality, significance and clarity. The conference program is extremely rich, profound, and featuring high-impact and breakthrough contributions. We sincerely hope that the conference not only shows the participants a broad overview of the latest research results on related fields, but also provides them with a significant platform for academic connection and exchange.

The final conference program consists of 90 papers divided into 2 sessions. The proceedings will be published in a volume by World Scientific Publishing Company.

We would like to express our sincere gratitude to all the TPC members and organizers for their hard work, spending precious time and endeavor preparation for the conference. Our deepest thanks are also to the volunteers and staffs for their long-hour's work and generosity devoted to the conference. The last but not the least, we would like to thank the authors, speakers and participants for their great contributions to the success of MAD2016.

MAD 2016 Organizing Committee

Session 1

Mechanics

Influence of gap on aerodynamic performance of wind turbine airfoil with discrete trailing edge flaps

Heng-Fan Li*, Ya-Lei Jia, Yu-Jin Zhao, Zhong-He Han

Key Lab of Condition Monitoring and Control for Power Plant Equipment
(North China Electric Power University)
Ministry of Education,
Baoding 071003, China
**E-mail: lhfdt100@163.com*

To study the influence of gap on aerodynamics of wind turbine airfoil, the wind turbine airfoil S809 is used as the research object, the structure of S809 airfoil model with discrete trailing edge flaps is designed in the article, and then discrete trailing edge flaps geometry model and integrated trailing edge flaps model are established. The gaps between the airfoil main body and trailing edge flaps is optimized, and the gap is selected as 1‰ chord . To calculate aerodynamic performance of airfoil with 10% chord length trailing edge flaps, computational fluid dynamics method is adopted. The flow field around airfoil is analyzed in theory, and the calculated results show that pressure coefficient distribution curves and pressure contour of airfoil model with discrete trailing edge flaps is consistent with that of the integrated airfoil, there are some airflow flowing through the gap from pressure surface to suction surface. The lifting coefficient of airfoil model with discrete trailing edge flaps is 1% lower than that of airfoil model with integrated trailing edge flaps , the gap between the airfoil main body and trailing edge flaps have little impact on airfoil aerodynamic performance, the influence of the gap can be neglected in the aerodynamics performance analysis.

Keywords: Airfoil; Discrete Trailing Edge Flaps; Lift-To-Drag Ratio; Computational Fluid Dynamics.

1. Introduction

It is well known that both extreme and fatigue loads can cause cumulative fatigue damage in wind turbine components, and eventually leading to failures [1]. The former is often induced by drastic wind velocity and direction changes, while the later is frequently originated from wind shear, turbulence, tower shadow, wake and yawed flow. If active aerodynamic control method of wind turbine blades was adopted, the loadings on the wind turbine and the cost per kWh would be decreased significantly [2]. So, active aerodynamic load control has become a very hot research direction in the field of wind turbine and shows great promise for reducing wind turbine fatigue damage.

There are many kinds of actuator of intelligent blade, including deformable trailing edge flaps, the discrete trailing edge flaps and sliding microtab[3]. Li Chuanfeng et al proposed that deformable trailing edge flaps can enhanced lift coefficient of airfoil and decrease the fluctuate load on wind turbine [4]. Van Dam studied sliding microtab and show that sliding microtab can improve pressure distribution of airfoil, and increased the lift coefficient and lift-to-drag ratio of airfoil[5]. Moving trailing edge flaps is studied by Lee [6], Lackner and Kuik studied the effect of discrete trailing edge flaps on 5MW upwind wind turbine.

The structure of the deformable trailing edge flaps are complex and it is difficult to achieve variable angle controlling. The structure of the sliding microtab is simple and the response speed is rapid, but the manufacture cost of it is high. The advantages of the discrete trailing edge flaps is that its structure is simple and its response speed is faster, furthermore, the manufacturing cost of it is relatively low.

With discrete trailing edge flaps , the lift coefficient and the lift-to-drag ratio of airfoil is increased, further more, it is easy to adjust deflect angle of flap, the flap can also increase the output power of wind turbine at low wind speed, the flap can assist pitch system and it can achieve faster response times than pitch control system. But the gaps between the flaps and the main body will appear in manufacturing [7], and the performance of the aerodynamic of the airfoil will be influenced [8]. The studies on the structure design and aerodynamic characteristic of the discrete trailing edge flaps at home and abroad is not sufficient. This blocked the further development of this method. In order to fully understand the aerodynamic performance of the airfoil with discrete trailing edge flaps, it is necessary study the influence of the gaps on the aerodynamics of the airfoil.

The S809 airfoil is selected as research object, the gap structure of the airfoil with 10% chord length trailing edge flap are optimized. The influence of the gap between the flaps and the main body on the aerodynamic performance of the trailing edge flaps at different attack angle is analyzed. The CFD analysis was carried out using commercial software Fluent 14.0. The post processing of the results was done using ANSYS CFD POST. This work is the preparation for more works on the study of the trailing edge flaps.

2. Model of the trailing Edge Flaps

2.1. *Geometry model*

S809 airfoil was selected as the reference airfoil, and chord length is 1000 mm. S809 airfoil is a specialized wind turbine airfoil, which is developed earlier and comparatively more mature. The experimental data of this airfoil is rich and reliable. So selecting S809 airfoil as study object.

Flap was attached to S809 airfoil at 10% chord length from trailing edge (Fig. 1). The gap between airfoil main body and flap was optimized, the gap was equally distributed and is 1mm wide. Figure 2a is the trailing edge flap model with gap, and Figure 2b is the model without gap.

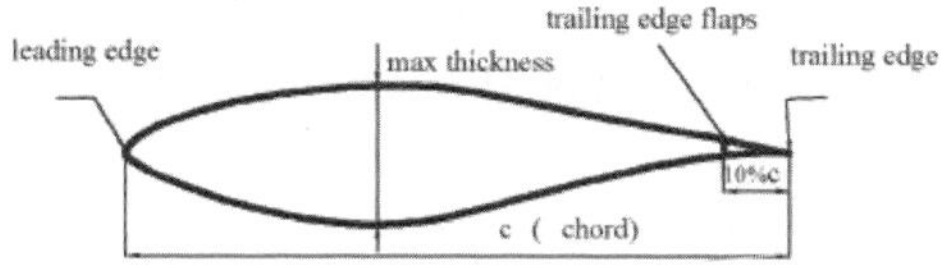

Fig. 1. Airfoil S809.

a Discrete trailing edge flaps model

b Integrated trailing edge flaps model

Fig. 2. Model of airfoil S809 with trailing edge flap.

2.2. *Computational grids*

The computational domain was composed of a semicircle grid and a rectangular grid. The flap model located at a place near the center of the semicircle and is oriented horizontally. The total computational domain is meshed with 2D structural mesh and C grid topology is applied in this domain. In order to avoid the disturbs of the inlet and outlet boundaries, the distance between entrance section and the leading edge of the airfoil is defined as 19 times chord length. The distance between the outlet section of computational domain and the trailing edge of the airfoil is 19 times chord length, and the upper and lower boundary is 20 times chord length respectively. The left semicircle boundary is defined as velocity-inlet and the right boundary is defined as outlet. The surface of the airfoil is defined as non-slip wall. The first layer grid near the surface of model meet the demand that value of y^+ is near 1. The grid structure of the computational domain is shown in Fig. 3.

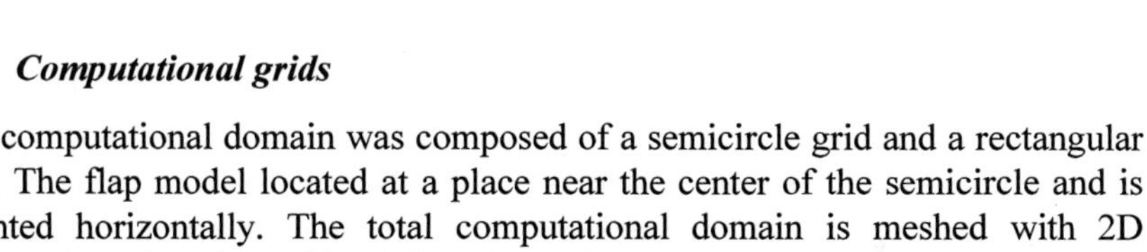

a The grid structures of the whole airfoil

b The grid structure of S809 with discrete trailing edge flaps

c The grid structure of S809 with integrated trailing edge flaps

Fig. 3. Comparison of grid structure.

2.3. *Related Parameters of Computation*

Reynolds number Re=2×106, Number of iterations is set as 5000, and the attack angles of the airfoil without flap are 0°, 1.02°, 5.13°, 9.22°, 14.24°, 20.15° respectively. The aerodynamic performance of the model with discrete trailing edge flaps and model with integrated trailing edge flaps at different attack angle was calculated, and the flaps is set as 10% chord length.

3. Results and Discussions

3.1. *The influence of the gap on the lift coefficient*

The lift coefficient of the fixed 10° deflection angle flap model with discrete trailing edge flaps and with integrated trailing edge flaps are shown in Fig. 4. 'S809 Exp data' means the experimental data , 'with gap data' means the result of the flap model with discrete trailing edge flaps, and 'without gap data' means the result of the flap model with integrated trailing edge flaps. From the Fig. 4, we can see that the lift coefficient of the two flap model is all higher than that of the airfoil without flap. The maximum increment of the lift coefficient of the airfoil with flap is 70% when attack angle is 0. The lift coefficient curve of the two flap model is similar. The maximum difference is 1% when the attack angle is 0°. With the increase of the attack angle, the difference of the lift coefficient between the two model decreases. When the attack angle is 8°, the lift coefficient and the lift coefficient of the two flap model is essentially the same. The influence of the gap can be ignored when analyze the aerodynamic performance of the airfoil. This means that flap model with discrete trailing edge flaps can be replaced by flap model with integrated trailing edge flaps to analyze the aerodynamic performance of the airfoil.

3.2. *The influence of the gap on the drag coefficient*

Figure 5 shows that the drag coefficient of the airfoil model with flap is bigger than that of the airfoil without flap when the attack angle is smaller than 10°. When the attack angle is 10°, the drag coefficient difference between the airfoil model with flap and the model without flap are very small, and then with increase of attack angle the drag coefficient of airfoil model without flap surpassed that of airfoil model with flap. The drag coefficient of the flap model with gap is bigger than that of the flap model without gap, but the difference is small,the difference is about 2% when attack angle is 0, and with increase of attack angle, the drag coefficient difference of the two flap model decreases, when the attack angle is bigger than 10°,the drag coefficient difference is approximately 0 (Fig. 5).

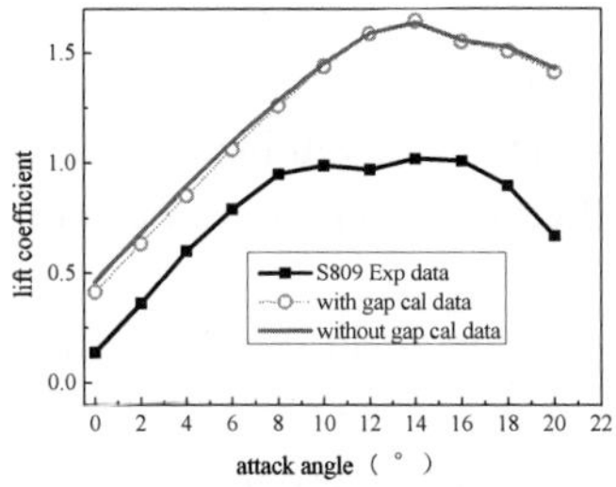

Fig. 4. Curve of lift coefficients at 10° deflecting angle.

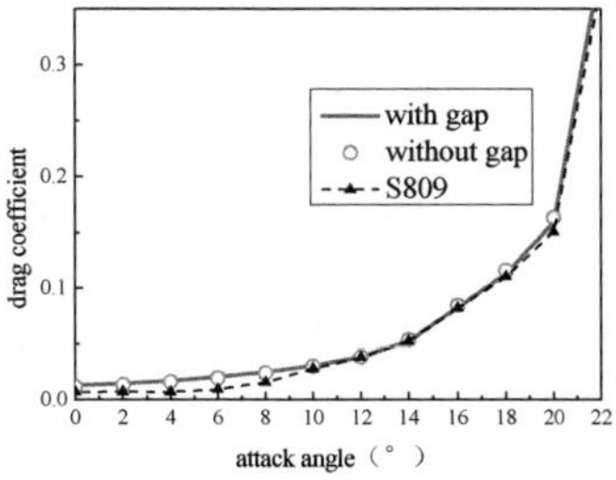

Fig. 5. curve of drag coefficients at 10° deflecting angle.

4. Conclusion

The S809 airfoil model with 10% chord length flaps is studied, All the results are shown below:

(1) Trailing edge flaps can increase the camber of the airfoil, and then increased the lift coefficient of the airfoil.

(2) Gap changes the balance between the pressure surface and suction surface of the airfoil, this make that the lift coefficient of flap model with gap decrease, and drag coefficient increase.

(3) When the gap between the flap and the airfoil body is 1‰c , it has little influence on aerodynamics of flap model. When the attack angle is 0, the lift coefficient of the discrete trailing edge flaps model is 1% lower than that of integrated trailing edge flaps model. With the increase of the attack angle, the difference of the lift coefficient between the two models decreases. When the attack angle is bigger than 8°, the difference is smaller than 0.5% and this means the gap has little influence on the lift of the airfoil.

(4) The gap between the airfoil main body and flaps is reasonably designed and the air of the lower surface of the airfoil flow out along the tangent of the upper surface. So it has little influence on flow field near the airfoil and then it will has little influence on the lift and drag coefficient of airfoil.

Acknowledgment

This work is supported by Fundamental Research Funds for the Central Universities (No. 2015MS107).

References

1. Y. Li, J. Wang, J. Hua, Experimental investigations on the effects of divergent trailing edge and Gurney flaps on a supercritical airfoil. *Aerospace Science and Technology* **11**, 91-99 (2007).
2. M. Chandrasekhara, Optimum Gurney flap height determination for "lost-lift" recovery in compressible dynamic stall control. *Aerospace Science and Technology* **14**, 551-556 (2010).
3. T. K. Barlas, G. Van Kuik, Review of state of the art in smart rotor control research for wind turbines. *Progress in Aerospace Sciences* **46**, 1-27 (2010).
4. C. Li, Y. Xu, J. Xu. Influence of Slot on Wind Turbine Airfoil With Trailing Edge Flaps. *Journal engineering thermophysics* **32**, 1851-1854 (2011).
5. C. P. van Dam, D. Y. Nakafuji, C. Bauer, K. Standish, D. Chao, in *Micromachining and Microfabrication.* (International Society for Optics and Photonics, 2003), pp. 28-39.
6. S. J. Johnson, J. P. Baker, C. Van Dam, D. Berg, An overview of active load control techniques for wind turbines with an emphasis on microtabs. *Wind Energy* **13**, 239-253 (2010).
7. M. A. Lackner, G. van Kuik, A comparison of smart rotor control approaches using trailing edge flaps and individual pitch control. *Wind Energy* **13**, 117-134 (2010).
8. A. I. Jose, J. D. Baeder, in Proceedings of 47th AIAA aerospace sciences meeting including the new horizons forum and aerospace exposition, Orlando, Florida. (2009), pp. 5-8.

Dynamic testing and analysis of Poisson's ratio constants of timber

Gang-Gang Wang[1], Xian Zhang[2], Zi-Zhen Gao[2], Yun-Lu Wang[2], Cao Yu, Zheng Wang[2,*]

[1]*College of Mechanical and Electronical Engineering, Nanjing Forestry University, Nanjing, 210037, China*

[2]*College of Material Science and Engineering, Nanjing Forestry University, Nanjing, 210037, China*

[*]*E-mail: wangzheng63258@163.com*

www.njfu.edu.cn

The objective of this work was to propose a method for dynamic testing of Poisson's ratio of dimension lumber. Based on the first-order bending mode shape of cantilever slab, Sitka Spruce was selected as research object using instruments including four-channel dynamic resistance strain gauge and CRAS vibration and a set of collection-analysis system for dynamic signal. Free vibration of cantilever specimen of Sitka Spruce and mild steel can be stimulated by knocking, and the fundamental vibration should be reserved by filtering processing. Additionally, decaying curve of oscillatory waves for transverse and longitudinal strain of fundamental vibration should also be recorded and displayed. Obtained from the positive (negative) peak in oscillatory wave curve for transverse strain, Poisson's ratios along grain and across grain on radial section and that across grain on transverse section of Sitka Spruce specimens were measured. Then measured results were applied to analyse the accuracy of calculated modulus of elasticity obtained by plugging the first-order bending frequency measured by cantilever slab into equation of cantilever beam, and this method is proved to be robust and accurate. The method based on first-order bending mode shape of cantilever slab to measure dimension lumber Poisson's ratio is efficient and simple, with good repeatability and high accuracy. Besides, the same method was also applied to test Poisson's ratio of mild steel to validate the correctness of the method on wood. It is observed that the positive (negative) peak in oscillatory wave curve for transverse strain is corresponding to that for longitudinal strain, meaning that the phase difference, between oscillatory wave curves for transverse strain and longitudinal strain, is 180°, or that transverse strain and longitudinal strain are reversed. The average value should be taken after calculating the ratios between peak-to-peak values read from the first channel and the second channel in oscillatory wave curve. The results show that the Poisson's ratio along grain of Sitka Spruce specimen is one order higher than that across grain, which indicates the anisotropy of wood. The test results validated that the elastic modulus of the lumber can be accurately derived by substituting the first-order fixed frequency, measured with the cantilever plate of Sitka Spruce specimen, into the cantilever theoretical equation. This research can provide reference to practical measurement and assessment of mechanical properties of wood and steel.

Keywords: Lumber; Cantilever Slab; Bending; Vibration; Along Grain; Across Grain; Poisson's ratio.

1. Introduction

Elastic modulus and Poisson's ratio are the two important parameters in representing mechanical behavior of a material. Poisson's ratio of a material, the absolute value of ratio between transverse strain and longitudinal strain when pulling (pressing) in axial direction, is an elastic constant reflecting the transverse deformation of material [1-3]. There are a lot of ways to measure the Poisson's ratio of a material, including mechanical method with two pairs of extensometer; acoustic methods, likebrillouin.

Scattering (SBS), surface acoustic wave (SAW) and acoustic micrology (AM); optical methods, such as elastic wave theory and elastic synchronous vibration theory; as well as electrometric methods and so on. At present, however, studies on Poisson's ratio of lumber are still lacking, especially on the dynamic testing for Poisson's ratio of viscous-elastic materials [4]. In 1995, Chan, et al. [5] predicted the Poisson's ratio of some viscous-elastic materials using finite element method and boundary element method. The results showed that finite element method could accurately predict the Poisson's ratio of compressible and incompressible viscous-elastic materials. In 1999, Quan Tiehan, et al. [6] measured the viscous-elastic Poisson's ratio of non-metallic material using dual-laser lever and electrical extensometer for transverse deformation. A year later, Renanlt, et al. [7] calculated , the Poisson's ratio of metal film with X-ray diffractometer, based on the continuum theory. In 2004, Wang et al. [8] measured the elastic modulus E and the Poisson's ratio μ of bamboo materials using strain electrometric method and Visual Basic (VB) program.

With different load properties of a viscous-elastic material, Poisson's ratio can be divided into two types: static viscous-elastic Poisson's ratio and dynamic complex Poisson's ratio, which respectively represent the response of transverse strain to longitudinal strain under static tensile load and dynamic tensile load. Owing to the viscoelasticity of a material, a test specimen may become relaxed or creeping under an external applied load, and the response of transverse strain should lag behind longitudinal strain. Therefore, the formula for viscous-elastic Poisson's ratio is much more complex than that for linear-elastic material [4] [9-12]. Also tension (compression) in axial direction is one-way stress, while plate in bending belongs to dual-direction stress. In 1996, Ma [13] studied the elastic constant of unidirectional composite material plate in dynamic (static) determination [1]. In addition, he proposed a method for Poisson's ratio measurement with plate bending vibration, which has been successfully applied to typical unidirectional composite materials. Meanwhile, he analyzed the distribution of ratio between transverse strain and longitudinal strain, in length direction on plate surface, when typical one-way composite plate is in fundamental (first-order) bending vibration. If the ratio between transverse and longitudinal strain equals to Poisson's ratio, the position on the plate is almost in one-way stressed state. Poisson's ratio can be obtained by measuring the oscillatory wave curves of transverse strain and longitudinal strain at this location with crossing strain rosette attached on it and cantilever slab in (first-order) bending vibration [13].

Given that Sitka Spruce (P·sitchensis) is generally used in manufacturing furniture, ordinary architecture, boxboard, pulp and paper and castanets (of piano and guitar), this work tested the Poisson's ratio along grain and across grain on radial section of quarter-sawed lumber as well as the Poisson's ratio of grain specimen on transverse section by means of dynamic mechanics. In addition, the accuracy was analyzed with testing results of elastic modulus, calculated by substituting the first-order bending frequency, measured with cantilever plate specimen, into cantilever formula. The work provides an alternative method for measuring mechanical properties of lumber in wooden architecture, furniture, wood processing and other industries.

2. Materials and Methods

2.1. *Species*

Sitka Spruce made in U.S.A, with straight grain, has gravity of 0.37-0.42g/cm^3. It is the main material of piano castanet, known for its good resonance in acoustics.

2.2. *Instrument and tools, clamps and measures*

2.2.1. *One set of resistance strain gauge*

The system is a crossing strain rosette, composed of one unit of YD-28A four-channel dynamic resistance strain gauge, two bridge boxes (two channels for measuring Poisson's ratio) and B×120-5AA resistance strain gage (sensitivity ratio 2.08+1%).

The working mechanism of this measuring instrument is to turn mechanical quantity into electric quantity in utilization of properties of metal materials. Strain gage, the transition element for strain measurement, is made by winding extremely fine metal resistance wire, or printing and corroding the metal foil, and then firmly stuck on specimen with a fast-curing 502 glue. The resistance value of strain gage may change, namely turning mechanic quantity (deformation) into electric quantity (change of resistance value), when the length of specimen varies under exogenous action. At last, the resistance value was converted into corresponding strain value ε.

2.2.2. *CRAS vibration and a set of collection-analysis system for dynamic signal*

This system includes AZ802 signal conditioning box, AZ308R signal collecting box as well as collection and analysis software of Ad Cras signal.

2.2.3. *Tools, clamps and measures*

One monkey wrench, bench screw, rubber hammer, Guangzhuo 701 outside micrometer (0~25), Kanghong steel tap (0~5), and HK-30 wood moisture content (MC) meter.

2.3. *Specimen dimension and orientation*

2.3.1. *Quarter-sawn timber (along grain): specimen dimension 280mm× 60mm×12.2mm, 5 pieces, cantilever extending to 240mm (aspect ratio is 4);*

2.3.2. *Transverse plate (across grain):specimen dimension 280mm× 60mm×12.2mm, 5 pieces, cantilever extending to 240mm (aspect ratio is 4);*

2.3.3. *Quarter-sawn timber (across grain): transversely cutting a specimen in 123mm×36mm×8mm from the quarter-sawn lumber in 500mm×123mm×12.2mm as the specimen of quarter-sawn lumber with grain, cantilever extending to 108mm (aspect ratio is 3), five pieces.*

The moisture content (MC) of above specimens after kiln drying and equalization is 9-11%

2.4. *Schematic diagram for Poisson's ratio definition*

Fig. 1 shows the orientation ordinates of this specimen, in which L, T and R separately represent longitudinal direction, tangent direction and radial direction of annual ring. The subscript of Poisson's ratio μ is composed of two numbers, the first referring to the stretching direction and orientation of longitudinal strain gage, while the second to the orientation of transverse strain gage. For instance, points to the absolute ratio between the transverse and longitudinal strain generated by L directional stretching. The Poisson's ratio always equals to the absolute value of ratio between transverse strain and longitudinal strain when sticking the strain gage according to the coordinates and figures in this work.

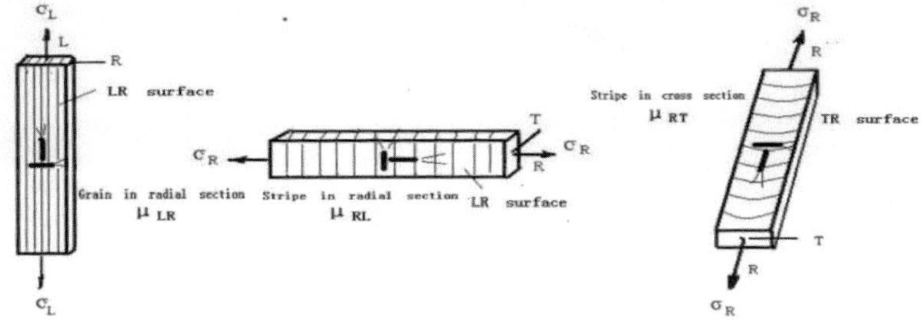

Fig. 1. Schematic diagram for Poisson's ratio (partial) definition.

2.5. *Test set up*

This test was conducted under an indoor temperature of 18°C, humidity of 45%, no high-intensity magnetic field interference, using a voltage of AC 220V anda frequency of 50Hz.

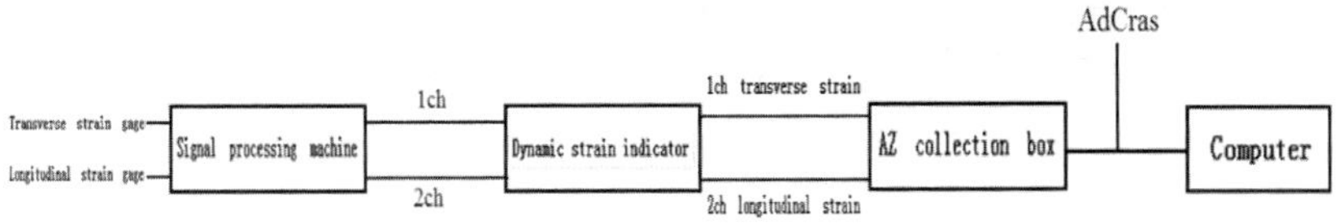

Fig. 2. Block diagram for test.

In Fig. 2, longitudinal strain and transverse strain should be respectively measured along the longitudinal and transverse strain gages of Sitka Spruce specimen. Signal processing machine has functions of signal amplification and filtering. The dynamic strain meter uses two channels to separately measure the transverse and longitudinal strains by adopting the bridge circuit connection of 1/4 bridge. The output signal from dynamic strain meter will digitalize the analog signal with A/D chip in AZ collection box, and then draws the oscillatory wave curve with software.

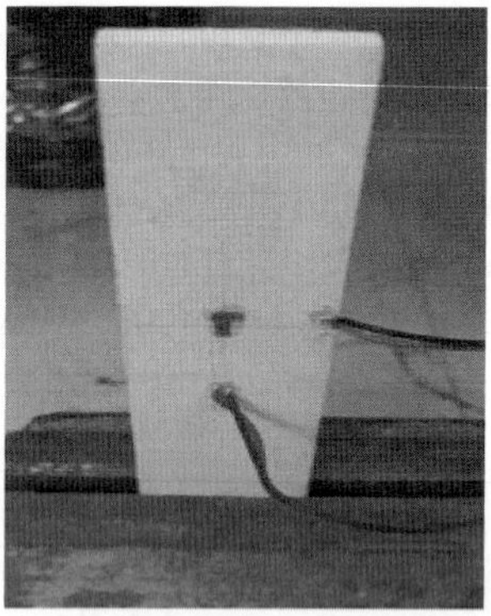

a) Specimen across grain on transverse section

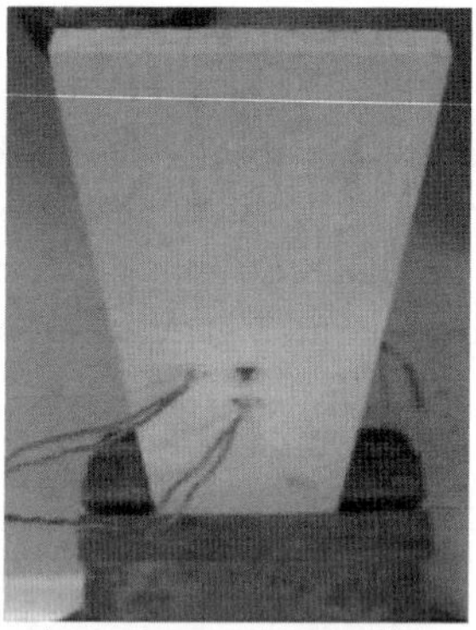

b) Specimen along grain on radial section

Fig. 3. Test site of a specimen.

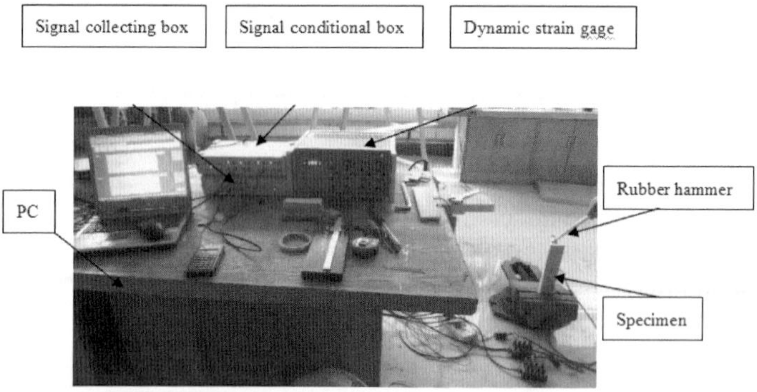

Fig. 4. Full view of test site.

Free vibration of cantilever specimen can be stimulated by knocking, and the fundamental vibration should be reserved by filtering processing. Additionally, decaying curve of oscillatory waves for transverse and longitudinal strain of fundamental vibration should also be recorded and displayed. It is observed that the positive (negative) peak in oscillatory wave curve for transverse strain is corresponding to that for longitudinal strain, meaning that the phase difference, between oscillatory wave curves for transverse strain and longitudinal strain, is 180°, or that transverse strain and longitudinal strain are reversed. Poisson's ratio can be obtained from the ratio between transverse strain peak and longitudinal strain peak at the same time. In order to get a good repeatability of Poisson's ratio, the knocking strength should ensure that longitudinal and transverse wave curves will not be clipped.

For specimens along grain on radial section and across grain on transverse section (aspect ratio is 4), the specimen should be clamped for 4cm, and the crossing strain rosette be stuck at 12cm (that is the half of span L, 0.5L) away from the fixed end. For specimen with stripe at radial section (aspect ratio is 3), the specimen should be clamped for 1.5cm, and the crossing strain rosette be stuck at 6.48cm (0.6L) away from the fixed end; Instruments should be connected as shown in Fig. 2. The sensitivity of strain gage should be set as and standardized for confirmation; Data collection and processing program (Ad Cras) should be called for test establishment;

Parameter setting: With collection triggered, sampling frequency should be determined based on specimen dimension. For example, the sampling frequency along grain at radial section in this test was 5120Hz, that across grain on radial section 1280Hz, and that for stripe at cross section 256Hz. One (1) should be selected as the average time. In the case that the setting of filtering frequency was very significant, it should be ensured that bending vibration at fundamental frequency only occurred in specimen knocking. Due to the bad filtering effects of general dynamic strain gage, signal processing machine were connected with the gage in block diagram;

An oscillography was conducted to check if the circuit was properly connected, and the parameter setting was correct;

The specimen was stimulated in free vibration at fundamental frequency after being knocked. After that, data were collected to check the oscillatory wave curve; All test data were saved in disks for further analysis..

3. Results and Discussion

3.1. *Wave form analysis*

Fig. 5 show the oscillatory curve of specimens with grain at radial section. These results demonstrated the anisotropic nature of dimension lumber. As shown in Fig. 6

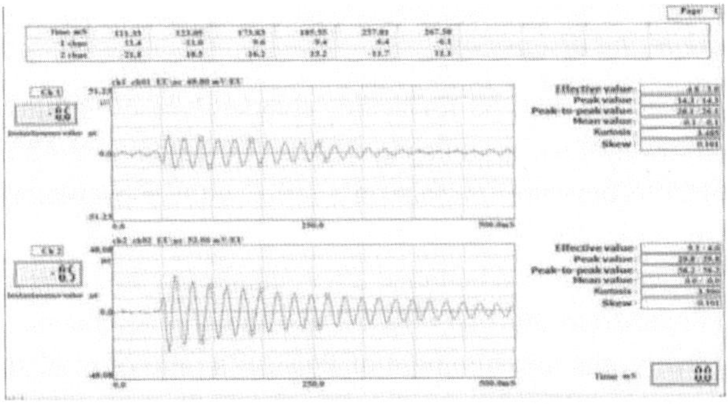

Fig. 5. Oscillatory curves of a specimen with stripe at cross section.

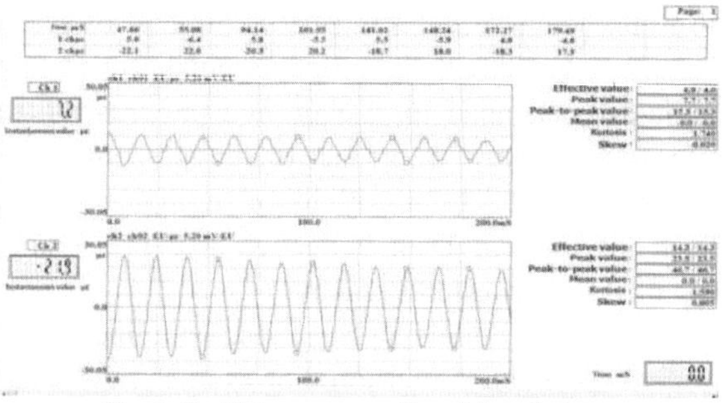

Fig. 6. Oscillatory curves of a mild steel plate.

3.2. *Test results*

Peak-to-peak values of the first channel and the second channel can be read from the oscillatory curve, and the average value (taking the absolute value) of the ratio, between peak-to-peak values of the first channel and the second channel, can be adopted as the Poisson's ratio. The results are shown in Table 1.

Table 1. Measurements of Poisson's ratios, and for Sitka Spruce.

Specimen No.	Along grain on radial section μLR	Across grain on radial section μRL	Across grain on transverse section μRT
1#	0.38	0.045	0.57
2#	0.37	0.017	0.58
3#	0.35	0.029	0.62
4#	0.37	0.047	0.56
5#	0.36	0.038	0.63
Average	0.37	0.035	0.59
Standard deviation	0.011	0.012	0.031
Coefficient of variation (CV)	3.1%	34.3%	5.3%

3.3. *Verification and discussion*

In order to validate the method for testing lumber Poisson's ratio in this work, the strain rosette was stuck in midspan of 180mm×45mm×3mm mild steel plate (aspect ratio 4:1) to measure its Poisson's ratio, with its oscillatory wave curve shown in Fig. 6.

The average value should be taken after calculating the ratios between peak-to-peak values read from the first channel and the second channel in oscillatory wave curve. Finally, the measurement of Poisson's ratio of mild steel should be (the standard value is 0.25-0.28).

It can be measured that =0.37, and =0.035;=0.37 and =0.029. This means the Poisson's ratio of grain at radial section measured by dynamic method in this work is the same as that in Reference, while there is an error in Poisson's ratio of stripe for the small specimen dimension and transverse strain;

The ratio, between, Poisson's ratio of grain at radial section, and μRL, Poisson's ratio of stripe at radial section, is 10.6, which means the Poisson's ratio of grain at radial section is one order higher than that of stripe at radial section.

According to the Formula, the elastic modulus of grain at radial section is one order higher than of stripe at radial section, which illustrates the anisotropy of lumber;

When applying the rigidity coefficient of one-way slab to the grain of radial section [13], then, and . When the length-width ratio of cantilever plate L/b is 4, the relationship of with plate length L, thickness h, density and the first-order bending frequency can be represented as follows:

$$Q_{11} = \frac{E_1}{1-\mu_{21}\mu_{12}} = 39.11\frac{\rho L^4 f_1^2}{h^2}, \quad (L/b=4) \tag{1}$$

The formula for elasticity modulus calculated with cantilever equation is: (1)

Elastic modulus, calculated by substituting measured and into expression, was about 0.7% larger than elastic modulus calculated with cantilever formula, which means the elastic modulus calculated with cantilever formula (1) is 0.7% smaller than the actual one.

4. Conclusions

The dynamic method for Poisson's ratio measurement with the first-order bending mode shape of cantilever plate is proved to be robust and very accurate;

Poisson's ratio of grain at radial section of Sitka Spruce is one order higher than that of stripe at cross section, which demonstrates the anisotropic nature of the dimension lumber;The test results validated that the elastic modulus of the lumber can be accurately derived by substituting the first-order fixed frequency, measured with the cantilever plate of Sitka Spruce specimen, into the cantilever theoretical equation.

Acknowledgment

This work is supported by Special Scientific Funding Program for North Jiangsu, 2015 (BN2015143) and Undergraduate Training Program for Innovation of Jiangsu Province, 2015 (201510298111X).

References

1. Xu Zhilun, Elastic Mechanics, Second Edition (First Volume) [M]. People's Education Press, 1982: 334-339.
2. Xu Zhilun, Elastic Mechanics, Third Edition (xwce). People's Education Press, 1990: 12-17.
3. Liu Hongwen. Mechanics of Material, Second Edition (First Volume) [M]. Higher Education Press.1983:287-293.
4. Shan Guifang, YangWei, Peng Jianmin, et al. Advances in Test Methods for Poisson’s Ratio of Materials [J]. Materials Leader, 2006, 20(3): 15-19.
5. Chan Y W, Oyadiji S O, Wright J R, et al. Predicting the Vibration Characteristics of Elements Incorporating Incompressible Viscoclastic Materials[J]. Proceeding of SPIE the International Society for Optical Engineering,1995, 2445:293.
6. Quan Tiehan. Double Laser Lever/Electric Measure Lateral Displacement Gauge [J]. Instrument Technique and Sensor, 1999, 4: 36
7. Renault P O, Badawi K F, Coudeau, et al. Experimental Method for Measuring the Poisson’s Ratio in Thin Films and Multilayers Using a Tensile Machine Set up on X-ray Goniometer [J].EPJ Appl PHYS, 2000, 10(2):91.
8. Wang Zheng, Wang Zhiqiang, Application of Strain Electric Method and VB Program in Measuring Bamboo Material's Elasticity Modulus E and Poisson Ration μ[J]. Wood Processing Machinery, 2004, 15(1):16-17).
9. Zhao Bohua. Study on Viscoelastic Poisson’s Ratio with Solid Propellants [J]. Journal of Beijing Institute of Technology, 1994,1:87.
10. Vogel D, Grosser, Schubert A, et al. MicroDAC Strain Measurement for Electronics Packaging Structures [J]. Optics and Lasers in Engineering, 2001, 36(2):195.
11. Hao Songlin, Initial Viscoelastic Poisson Ratio of High-Filled Composite Propellants [J]. Journal of Solid Rocket Technology, 1996, 19(4):46.
12. He Manjun, High Polymer Physics[M]. Fudan University Press, 1990.
13. Ma Gongxun, Dynamic (Static) Measurement of the Elastic Constants in Uniaxially-Reinforced Composite Sheet [J]. Acta Material CompositalSimica, 1996, 13(2):117-123.

A simple analytical model of the rotary friction welding of thin-wall tube

L. B. Yang[1,2,*], X. M. Qin[1], R. C. Reed[2], P. Bowen[2]
[1]*School of Mechanical Engineering, Beijing Institute of Technology, ZhuHai, ZhuHai, Guangdong, 519088, P.R. China*
[2]*School of Metallurgy and Materials, University of Birmingham, Edgbaston, Birmingham, B15 2TT, UK*
[*]*E-mail: 1394126856@qq.com*

Rotary friction welding is routinely used in joining the latest grades of super alloys. There is still uncertainty of the formation of the burr (internal flash) near welding interface despite the simulation work done in this field. In this paper, a simple model was developed to describe the process of rotary friction welding of tubes from the viewpoint of the fluid mechanics. It explains that the extrusion of the material during the last seconds of the welding is a direct result of the quick stop of the rotating part. The model also indicates that the axial shortening rate in rotary friction welding is not only influenced by the material flow rate in side directions, but also by the change rate of the rotation speed. The simple formula from this model can predict the axial shortening (upset) in the friction welding process..

Keywords: Rotary Friction Welding; Axial Shortening; Navier-Stokes Equations.

1. Background

1.1. *Introduction of rotary friction welding*

Rotary friction welding, including inertia welding and direct drive friction welding, is widely used in industry to supply various welds of high integrity. During the process of rotary friction welding, one of the parts to be welded is first accelerated to a high angular velocity; welding begins when a stationary part is pushed against the rotating one under pressure. A schematic illustration of the process parameters of rotary friction welding are shown in Fig. 1. It is believed that a joint can thus be produced without incurring the gross melting of the material; as a result, the cracking and distortion of the welds during solidification are considerably reduced. This character of the rotary friction welding makes it suitable to join the latest grades of high-strength superalloys, which are crucial for producing modern high efficiency, low-carbon emission jet engines, while the conventional fusion welding methods often cause the problem of cracking.

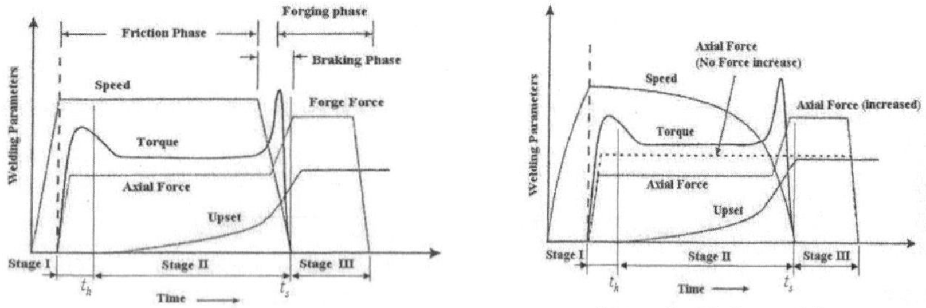

(a) direct drive friction welding (b) inertia friction welding

Fig. 1. Schematic illustration of the parameters of rotary friction welding process.

1.2. *Heat affected zone in friction welding*

During friction welding, the material near the welding interfaces undergoes drastic temperature changes and severe plastic deformation. This region is mentioned as the heat affected zone (HAZ). Usually the HAZ can be divided into several regions, as shown in Fig. 2a [1] .

The white zone, also called contact zone, or plasticized area, is the area which experiences the maximum temperature and plastic strain [2]. The metals in this zone rub with each other and fragments of metals transfer from one rubbing surface to the other. Some researchers hold that a thixotropic deformation behaviour is facilitated. The material in this zone has a very fine grain size due to severe plastic deformation and full dynamic recrystallisation. Some authors also hold that there is partial melting of materials in this region [1, 3, 4] .

It is not difficult to understand that the flow behaviour of the metal in this plasticized area can have strong impact on the quality of the friction welds. There is nevertheless little modelling work focussed in this area despite the metallographic studies invested [5,6]. In friction welding, flashes are caused by the severe plastic deformation of metals near the welding interface. The shape of flash is usually similar to those shown in Fig. 2. There are often some materials extruded out along the weld line, particularly when the rotary part is about to coming to a stop. This part of metal is stated as internal flash in D'Alvise's PhD thesis [1], which is also mentioned as burr according to British standard [7]. A picture of flashes and burr in friction welding of superalloys is illustrated in Fig. 2. Although lots of models have been built to describe the procedure, to the author's knowledge, there is hardly any clear explanation why this burr happens during friction welding. In this study, a model for it is detailed.

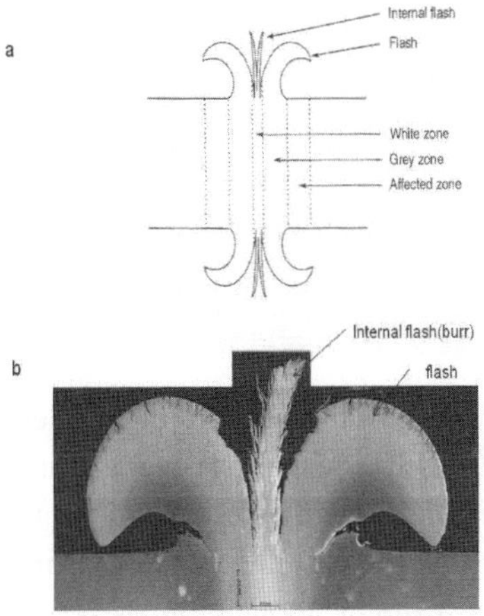

Fig. 2. Different regions in HAZ of friction welds, (**a**) a schematic illustration [1] (**b**) a photo of flashes[8] .

2. Analytical Model for Plasticized Layer in Friction Welding

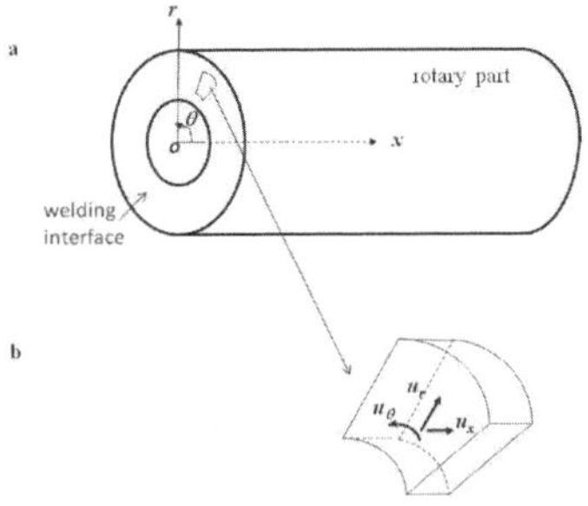

Fig. 3. Model used in the analysis: (**a**) a round tube with cylindrical coordinate system indicated (**b**) an element of the thin layer with velocity components u_r, u_x and u_θ.

In friction welding, parts to be welded with circular cross section are frequently encountered. Thus it may be reasonable to assume that two round tubes of identical geometry, material to be welded. Here the material of the tube used is mild steel. The thin layer of the plasticized metal near the weld line, which is mainly the white zone shown in Fig. 2, is what we are most interested in and this can be treated as a layer of a viscous fluid due to the high temperature there. If we want to describe the behaviour of this soft layer, a cylindrical coordinate system shown in Fig. 3 has to be set up first. Only one part - the rotary part - is illustrated due to the mirror symmetry along weld line. Then the

Navier-Stokes equation used to describe the movement of the fluids can be written as follows [9][10].

$$\begin{cases} \rho\left\{\frac{\partial u_r}{\partial t}+u_r\frac{\partial u_r}{\partial r}+\frac{u_\theta}{r}\frac{\partial u_r}{\partial \theta}+u_x\frac{\partial u_r}{\partial x}-\frac{u_\theta^2}{r}\right\}=-\frac{\partial P}{\partial r}+\mu\left\{\frac{1}{r}\frac{\partial}{\partial r}\left(r\frac{\partial u_r}{\partial r}\right)+\frac{1}{r^2}\frac{\partial^2 u_r}{\partial \theta^2}+\frac{\partial^2 u_r}{\partial x^2}-\frac{u_r}{r^2}-\frac{2}{r^2}\frac{\partial u_\theta}{\partial \theta}\right\}+\rho g_r \\ \rho\left\{\frac{\partial u_\theta}{\partial t}+u_r\frac{\partial u_\theta}{\partial r}+\frac{u_\theta}{r}\frac{\partial u_\theta}{\partial \theta}+u_x\frac{\partial u_\theta}{\partial x}+\frac{u_r u_\theta}{r}\right\}=-\frac{\partial P}{r\partial \theta}+\mu\left\{\frac{1}{r}\frac{\partial}{\partial r}\left(r\frac{\partial u_\theta}{\partial r}\right)+\frac{1}{r^2}\frac{\partial^2 u_\theta}{\partial \theta^2}+\frac{\partial^2 u_\theta}{\partial x^2}-\frac{u_\theta}{r^2}+\frac{2}{r^2}\frac{\partial u_r}{\partial \theta}\right\}+\rho g_\theta \\ \rho\left\{\frac{\partial u_x}{\partial t}+u_r\frac{\partial u_x}{\partial r}+\frac{u_\theta}{r}\frac{\partial u_x}{\partial \theta}+u_x\frac{\partial u_x}{\partial x}\right\}=-\frac{\partial P}{\partial x}+\mu\left\{\frac{1}{r}\frac{\partial}{\partial r}\left(r\frac{\partial u_x}{\partial r}\right)+\frac{1}{r^2}\frac{\partial^2 u_x}{\partial \theta^2}+\frac{\partial^2 u_x}{\partial x^2}\right\}+\rho g_x \end{cases} \quad (1)$$

Where u_r, u_x and u_θ represent the velocity components in the radial (r), axial (x) and circumferential (θ) directions respectively. μ is the viscosity of the thin layer. ρ is the density. P is the pressure exerted on the layer. g_r, g_x and g_θ denote the external body force components in the r, x and θ directions respectively. The Navier-Stokes equation is a direct application of Newton's second law on the movement of the fluid. The three expressions describe the principle of the conservation of momentum in a fluid along the radial, circumferential and axial directions respectively. At the same time, the velocity fields of the fluid should obey the continuity equation

$$\frac{1}{r}\frac{\partial}{\partial r}(ru_r)+\frac{1}{r}\frac{\partial u_\theta}{\partial \theta}+\frac{\partial u_x}{\partial x}=0 \quad (2)$$

The Navier-Stokes equations seem formidable and difficult to solve. However, in the case of rotary friction welding of round tubes, these equations can be greatly simplified when coping with the thin layer at the mating interfaces.

Since what we are interested in is the thin layer of the plasticized metal in the weld line, the effect of body forces on it usually is very small and can be neglected. That means that the three body force components g_r, g_x and g_θ can be taken as zero. Another assumption is that there is no gradient of velocity and pressure in the circumferential direction as the thin layer is in a circular shape. Thus,

$$\frac{\partial u_r}{\partial \theta}=\frac{\partial u_x}{\partial \theta}=\frac{\partial u_\theta}{\partial \theta}=\frac{\partial P}{\partial \theta}=0 \quad (3)$$

Then the second expression in equation (1), which describes the momentum balance in circular direction, can be written as

$$\rho\left\{\frac{\partial u_\theta}{\partial t}+u_r\frac{\partial u_\theta}{\partial r}+u_x\frac{\partial u_\theta}{\partial x}+\frac{u_r u_\theta}{r}\right\}=\mu\left\{\frac{1}{r}\frac{\partial}{\partial r}\left(r\frac{\partial u_\theta}{\partial r}\right)+\frac{\partial^2 u_\theta}{\partial x^2}-\frac{u_\theta}{r^2}\right\} \quad (4)$$

As the tangential velocity can be expressed as $u_\theta = r\omega$, where r, ω represent radius and rotation speed of the thin layer respectively, one can replace u_θ with $r\omega$ in equation (4), so that

$$\rho\left\{\frac{\partial(r\omega)}{\partial t}+u_r\frac{\partial(r\omega)}{\partial r}+u_x\frac{\partial(r\omega)}{\partial x}+\frac{u_r(r\omega)}{r}\right\}=\mu\left\{\frac{1}{r}\frac{\partial}{\partial r}\left(r\frac{\partial(r\omega)}{\partial r}\right)+\frac{\partial^2(r\omega)}{\partial x^2}-\frac{(r\omega)}{r^2}\right\} \quad (5)$$

We can assume the rotation speed of the thin layer decreases linearly along the direction of its thickness, that is, $\omega = \omega_0\, x/h$, where the ω_0 is the nominal angular speed of the rotating part, its typical variation is like the speed curve in Fig.1. The symbol x is the local coordinate in the plasticized layer along axial direction; it is set a value of zero on the thin layer's border to the static part and a value of h at the border to the rotating part. Then we have $\partial^2\omega/\partial x^2 = 0$, Hence the right side of the equation (5) equals zero, from the left side we can further get:

$$r\frac{\partial\omega}{\partial t}+2u_r\omega+ru_x\frac{\partial\omega}{\partial x}=0$$

So we have $u_r = -\frac{r}{2\omega}\left(\frac{\partial\omega}{\partial t}+u_x\frac{\partial\omega}{\partial x}\right)$

Replace ω with ω_0 , u_r can be further written as

$$u_r = -\frac{r}{2\omega_0}\frac{\partial\omega_0}{\partial t}-\frac{r}{2x}u_x \quad (6)$$

The continuity equation can also be written as follows after using expression (3)

$$\frac{\partial u_r}{\partial r}+\frac{u_r}{r}+\frac{\partial u_x}{\partial x}=0 \quad (7)$$

Substituting the expression for u_r in equation (7) with the expression in equation (6), we finally have

$$\dot{\varepsilon}_x=\frac{\partial u_x}{\partial x}=\frac{1}{\omega_0}\frac{\partial\omega_0}{\partial t}+\frac{u_x}{x} \quad (8)$$

Equation (6) and equation (8) are the formulae we are interested in.

3. Discussion and Results

3.1. *Formation of the internal flash*

Equation (6) and equation (8) are the results of the direct application of the Navier-Stokes equations on the circular thin viscous layer of the metal near the weld line. They define the necessary requirement for the velocity fields to satisfy the principle of momentum conservation.

From equation (6), we find that the radial flow speed u_r, *i.e.*, the speed of the welding metal flowing in the flash direction, is not only related with the rate of axial shortening (upsetting rate) u_x, but also strongly influenced by the rotation speed ω. If we look at the first item on the right side of the equation (6), which represents the effect of angular velocity, its absolute value is expected to increase very quickly at the last seconds of the stage II in Fig. 1, because the value of the rotation speed ω_0 is nearing zero.

Further, the rate of the deceleration of the rotation $\frac{\partial \omega_0}{\partial t}$ is found high at this moment from the speed curve of the rotary friction welding. From equation (8), the upsetting speed u_x also increases when the rotation is coming to a stop. This increase of u_x leads to a further increase of the radial speed u_r in equation (6). So from these two formulae, we can see that there is an unavoidable rapid increase of the radial flow speed of the welding metal due to the cessation of rotation in the process. This rapid increase of the radial flow speed u_r leads to the rapid squeezing of the soft material near the weld line at the last seconds of the steady state stage, which is also referred as the extrusion phenomenon in inertia welding. This extrusion behaviour of material is the cause of the burr shown in Fig. 2. Thus the formation of the burr is a direct result of the balance of the momentum due to the rapid stop of rotation.

From equation (6) we also find that the radial flow speed u_r increases with an increase in the radius *r*, which implies that the extrusion of the burr is more likely to happen during the friction welding of larger parts.

3.2. *Prediction of axial shortening*

Equation (8) shows that the axial strain rate $\dot{\varepsilon}_x$ is also influenced by the rotation speed. This formula can be used to estimate the upset (axial shortening) of rotary friction welding. We can take the axial velocity at the border of the thin layer to rotating part as the axial shortening rate of the whole welding part. That means the *x* in equation (8) equals *h*. Using the deceleration curve shown in Fig. 1, the whole upset of inertia welding can be worked out through integration.

As the effect of the pressure is not included in equation (8), to make this formula have general applicability, it can be further modified as follows,

$$\dot{S} = C_1 \frac{1}{\omega_0}\frac{d\omega_0}{dt} + C_2 \sigma_p \tag{9}$$

where $\dot{S}$, ω_0, σ_p, *t* denote axial shortening rate, angular velocity, nominal axial compression stress and time, respectively. C_1, C_2 are material constants.

The first item at the right of equation (9) represents the effect of angular velocity of the rotating part, while the second part mainly the effect of pressure. For mild steel EN 3B, the following parameters are tentatively adopted,

$$\dot{S} = 495\frac{1}{\omega_0}\frac{d\omega_0}{dt} + 0.13\sigma_p \quad (10)$$

where the unit of $\dot{S}$, ω_0, σ_p t are mm/s, rad/s, MPa and *s* respectively.

According to equation (10), at the start of stage II (equilibrium stage), the value of the first item is usually small since the rotational speed changes little, the axial shortening is mainly dominated by the pressure of the process, however, the value of the first item increases significantly due to the lower down of the rotation speed at the end of this stage, which accounts for the rapid increase of the upset during the last seconds of this period in experimental data. The predictions by equation (10) are in broadly agreement with the experimental data reported by Duffin et al [11].

4. Conclusions

In this study, the model of the plasticized layer indicates that the formation of the burr is a direct result of the balance of the momentum due to the rapid change of angular velocity. The model also shows that the axial shortening rate is not only influenced by the material flow rate in side directions, but also by the change in the rate of the rotation speed in inertia welding of tubes. The simple formula derived to predict the axial shortening is agreeable with the experimental data, which could be used to estimate the upset of the rotary friction welding.

Acknowledgments

The first author (L.Yang) is grateful for the financial aid from School of Metallurgy and Materials, University of Birmingham.

This work is partially supported by the financial aid from Beijing Institute of Technology, Zhuhai.

References

1. Laurent D'Alvise. *Development of a finite element model for the simulation of the inertia friction welding process between dissimilar materials*. PhD thesis, L' Ecole Nationale Superieure Des Mines De Paris, (2002).

2. M. Maalekian, Friction welding - critical assessment of literature. *Sci. Technol. Weld. Joining* 12 (2007) 738–759.
3. O.T. Midling and O. Grong. A process model for friction welding of Al-Mg-Si alloys and Al-SiC metal matrix composites. i. HAZ temperature and strain rate distribution. *Acta Metall. Mater.* 42(5) (1994) 1595 – 1609.
4. M.Soucail, A. Moal, L.Naze, E. Massoni, C.Levaillant, and Y. Bienvenu. Microstructural study and numerical simulation of inertia friction welding of astroloy. In *Superalloy 1992, TMS,* (*Warrendale, PA*, 1992).
5. M. Preuss, J.W.L. Pang, P.J. Withers, and G.J. Baxter. Inertia welding nickel-based superalloy. i. metallurgical characterization. *Metall. Mater. Trans. A*, 33A (2002) 3215 – 3225.
6. Z.W. Huang, H.Y. Li, M. Preuss, M. Karadge, P. Bowen, S. Bray, and G. Baxter. Inertia friction welding dissimilar nickel-based superalloys alloy 720Li to IN718. *Metall. Mater. Trans. A*, 38A (2007) 1608–1620.
7. Bs en iso 15620:2000. British standard, (2000).
8. Friedrich H. Daus. Unpublished research work, University of Birmingham, (2007).
9. Oscar Pinkus and Beno Sternlicht. *Theory of Hydrodynamic Lubrication.* (McGraw-Hill Inc., 1961).
10. Donald F.Young, Bruce R.Munson, and Theodore H. Okiishi. *A brief introduction to fluid mechanics*. (John Wiley and Sons Inc., 2001).
11. F. D. Duffin and A. S. Bahrani, Frictional behaviour of mild steel in friction welding, Wear, 26 (1973) 53-74.

Local working string mechanical behavior above subsea wellhead

Bao-Kui Gao*, Jing-Wei Ren, Liang Gao
Department of Petroleum Engineering, China University of Petroleum Beijing, Beijing, 102249, China
**E-mail: gaobaokui@126.com*
www.cup.edu.cn

During deep water well drilling and completion, subsea well head, blowout preventer stacks (BOPS) and drilling riser compose the wellbore from mud line to platform. The riser is connected to BOPS with a flexible joint which usually has angular deflection because of riser deformation. Subsea wellhead, BOPS, flexible joint and riser form a complex local constraint to inner working string. In this paper, working string deformation above subsea wellhead was modeled, and effects of flexible joint position, angular deflection and axial force on string mechanical behavior were investigated. According to calculation, working string has two potential risks: yielding from axial stress and collapsing from pressing upon flexible joint. The angular deflection of flexible joint is the dominating parameter to working string safety. An interesting finding is that a slender string is not safer than a thick one.

Keywords: Deep Water; Wellhead; Riser; Flexible Joint; String.

1. Foreword

The wellhead usually locates at mud line when developing deep water reservoirs. Above wellhead are blowout preventer stacks (BOPS) and drilling riser, and all of them composing the wellbore from mud line to platform [1, 2]. The BOPS is strong and does not permit any lateral movement, while the riser is relatively long and thin and sways from current. Flexible joint is used between BOPS and riser to protect riser from breaking [3]. The maximum angular deflection of flexible joint is 7° to 12° [4].

Such a local wellbore structure brings potential trouble to inner working strings. Casing string and tubing string are prone to getting stuck or even collapse near flexible joint. Although such accidents have reported, studies are mainly made on large scale contact between riser and inner string [5~8], and local mechanical behavior of working string has not been studied seriously up to now.

From mud line to platform, inner string has to adjust its deformation constantly according to the movement of riser [9,10]. While in practice, the riser is in tensile state and its dynamic characters are inconspicuous, so the riser can be modeled as a quasi-steady state. Inner working string near wellhead, constrained by BOPs and riser, can be treated as a steady state. In this paper, local working string mechanical behavior model was established and calculated under the structure characters as well as operation process [11,12]. Numerical method was used to solve the differential equation. Lateral deformation and maximum bending stress were discussed and finally, suggestions were offered for working string safety near flexible joint.

2. Working String Models

2.1. *Local wellbore description*

Local wellbore and working string structure can be simplified as that in Figure 1. The sinuosity in Figure 1a) is flexible joint, above it is riser and below it is BOPS and wellhead. The riser is unsteady [13]. In order to describe its restraint to working string, local riser lateral deformation is expressed as sine curve and only a small part of one period will be used.

Erect planar rectangular coordinates as that in Figure 1a), then in a limited length above origin of X coordinate wellbore lateral displacement can be expressed as

$$\begin{cases} Y_r(X) = 0, & 0 \le X \le H \\ Y_r(X) = A\sin\dfrac{2\pi(X-H)}{L}, & X \ge H \end{cases}. \tag{1}$$

Where H is distance from flexible joint to origin of coordinates, m; A and L is amplitude and wavelength of riser wave respectively, m.

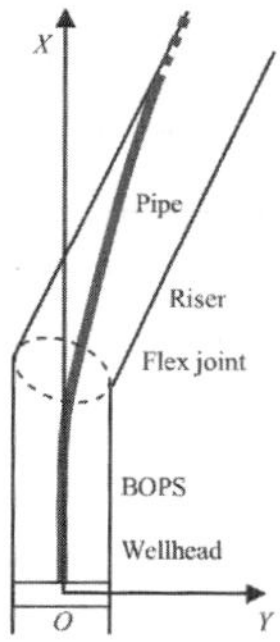

a) Local well structure

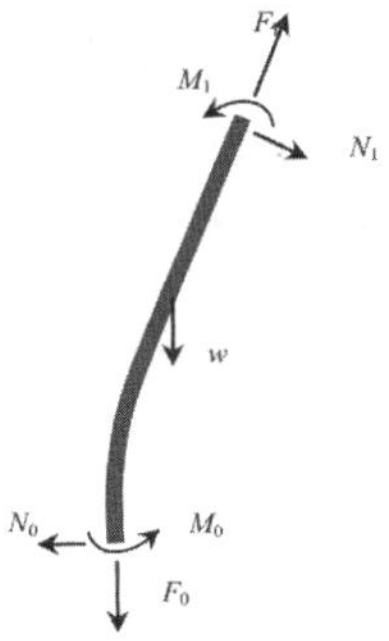

b) Loads on string section

Fig. 1. Local string mechanical model.

2.2. *Loads on inner working string section*

Working string is a slender pipe; its lower end is treated as fixed and after a suspending section it becomes continuous contact with riser from tangent point. Intercepting string section from origin to tangent point, loads on the separate section are shown in Figure 1b). Where F_i, N_i and M_i (i=0, 1) is axial load (kN), lateral load (kN) and bending moment (kN·m) respectively, subscript 0 and 1 denotes loads at fixed end and at tangent point respectively; w is distributed load from pipe gravity, kN/m. All these loads act in the X-Y plane.

2.3. *String displacement differential equation*

Riser lateral displacement is still in small deformation category. Riser diameter is far larger than that of working string, and therefore riser deformation doesn't be changed by inner pipe.

Take into account load conditions, working string lateral displacement equation can be established as

$$EIY''' - (F_0 + wX)Y' = -N_0. \tag{2}$$

Where EI is the tubular bending stiffness, kN·m^2.

2.4. *Restraint conditions*

At fixed end, known parameters include: displacement $Y(0)=0$ angular deflection $Y'(0)=0$ and axial force $F(0)=0$.

Normally, working string axial force is tension and contacts with wellbore at upper side, as shown in Figure 1. The tangent point is not a determinate position, so in order to make use of it, a supposed position, X_T, is given as a variable. Then boundary conditions at tangent point can be expressed as follows:

Lateral displacement

$$Y(X_T) = b - a + A\sin\frac{2\pi(X_T - H)}{L}. \tag{3}$$

Where b is working string outside radius, m; a is riser internal radius, m.

Angular deflection

$$Y'(X_T) = A\frac{2\pi}{L}\cos\frac{2\pi(X_T - H)}{L}. \tag{4}$$

Bending moment

$$M(X_T) = -EI \cdot A\left(\frac{2\pi}{L}\right)^2 \sin\frac{2\pi(X_T - H)}{L}. \tag{5}$$

Where M is tubing string bending moment, kN·m.

2.5. *Solution of differential equation*

Working string deformation is controlled by gravity, axial force and riser displacement. It is a static indeterminacy problem and the differential equation is transcendental. So in this paper finite difference method is used to get the solution of differential equation. During searching for the numerical solution, some restraint conditions cannot be satisfied precisely, in this investigation the priority is ranked as: lateral displacement → angular deflection → bending moment → shearing force.

3. Working String Mechanical Behavior

In the following calculation, main parameters of riser and working string are listed in Table 1.

Table 1. Dimensions of riser and working string.

Type	Outside diameter(mm)	Internal diameter(mm)
Riser	539.75	509.0
3.5″ tubing	88.9	76.0
4.5″ tubing	114.3	88.9
7.0″ casing	177.8	159.4

3.1. *Working string deformation and bending stress*

Let riser wavelength equal 200m, wave amplitude equal 2.8m, then joint angular deflection will be 5°. Let distance from joint to fixed end equal 3m; let axial force vary from 0 to 500kN. Performing calculation based on these data working string mechanical behavior can be obtained.

(1) Error analysis

To this special restraint problem, the only mechanical parameter that can be proved by theoretical value is bending moment at tangent point. String bending moment above tangent point is decided by riser deformation. Comparing numerical solution with exact solution, precision of numerical solution can be obtained.

By intensive comparison, maximum relative errors are all within 4%, so the precision is acceptable.

As a whole, small axial force corresponds to larger error, and the reverse is true. This phenomenon may come from the assumption that the pipe contacts riser continuously above tangent point. When axial force is small, bending rigidity is the dominant factor to pipe deformation, and the numerical solution is too sensitive to element length. This kind of error is difficult to be eliminated. Another error source is the step length during solving the transcendental equation. Trial and error method is used to find the tangent point, and an unlucky step increment may miss the exact solution. This kind of error can be reduced.

(2) Position of tangent point

The relationship between tangent point position and axial force are drawn in Figure 2, three curves correspond to three types of pipe. The horizontal ordinate is pipe axial force at fixed end, and the vertical ordinate is the distance from tangent point to fixed end.

Features of tangent point position are as follows

- Tangent point rises with axial force.
- Small size pipe has higher tangent point than large size pipe.
- Tangent point position of small size pipe changes faster than large size pipe at lower axial force region. The change tends to be mitigated when axial force is large enough.

(3) Maximum bending stress

The relationships between maximum bending stresses and string axial forces are shown in Figure 3. Three curves correspond to three types of pipe. Maximum bending stresses take place at fixed end.

Features of maximum bending stress are as follows

- Maximum bending stress increases along with axial force.
- When axial force closes to zero, large size string has greater bending stress; when axial force exceeds 100kN, small size string has larger bending stress.

When axial force exceeds 200kN, maximum bending stress will soon approach material yield limit. In engineering, a 200kN axial force is very easy to be exceeded, so bending strength is a failure risk.

It is remarkable that slender pipe bears larger bending stress. This phenomenon is somewhat a deviation from our perceptual knowledge and attention should be paid to avoid mistakes.

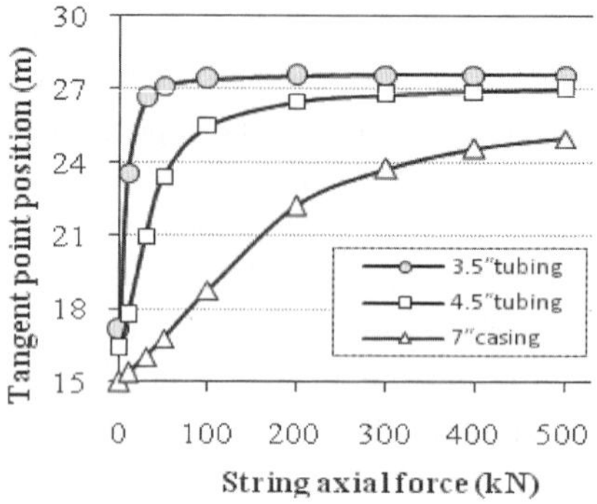

Fig. 2. Tangent point positions.

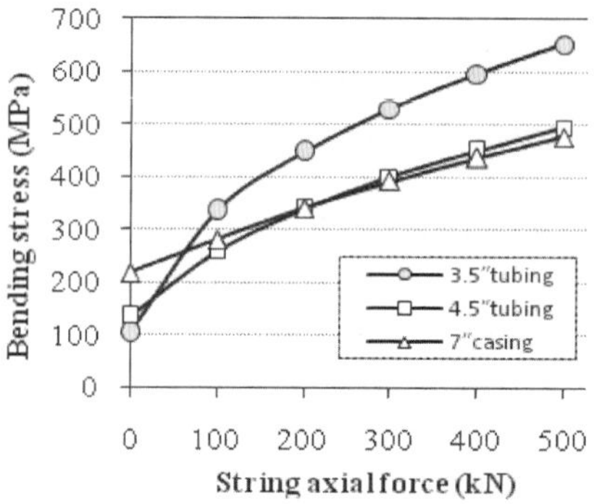

Fig. 3. Maximum bending stresses.

3.2. *Influence of flex joint position and deflection*

In this section only 4.5″ tubing is considered.

(1) Maximum bending stress

Let axial force be 300kN, maximum bending stresses versus joint position and joint deflection are drawn in Figure 4, where the two lines, H=2m and H=4m, denote the distance from joint to fixed end.

Figure 4 shows that

(a) Joint position has less influence on bending stress.

(b) Joint deflection has more obvious influence on bending stress.

Thus it can be seen, control joint deflection is crucial to limit maximum bending stress.

(2) Joint contact with tubing

Comprehensive investigations show that tubing will come into contact with flexible joint in some circumstances. Contact can be regarded as a danger. Take initial contact as critical point, relationship among axial force, joint position and joint deflection is drawn in Figure 5.

Curves in Figure 5 indicate that

(a) Contact takes place earlier under large axial force, but the influence is not prominent.

(b) Joint position and deflection form an exponential curve which is more crucial to contact than axial force.

Structures of wellhead and BOPS are unlikely to be changed, so to restrict flex joint deflection is also necessary for preventing contact.

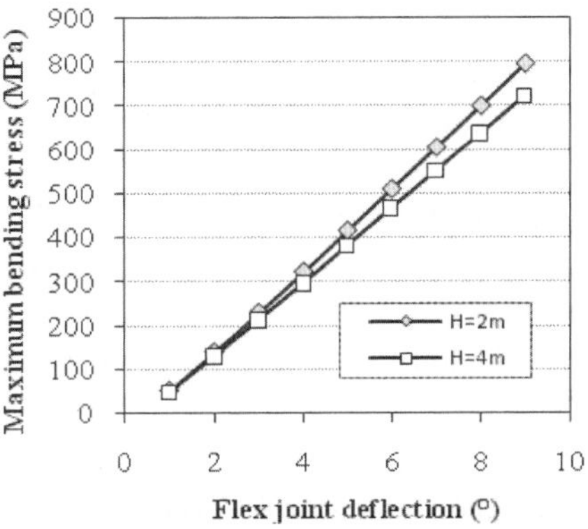

Fig. 4. Joint parameters to bending stress.

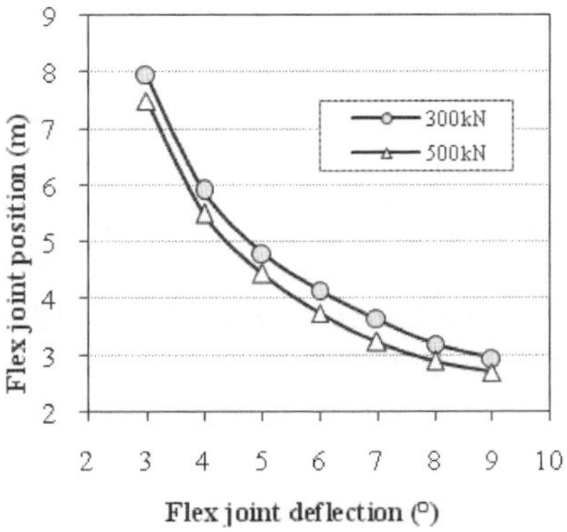

Fig. 5. Critical contact conditions.

4. Conclusions

In the vicinity of subsea wellhead and riser flexible joint, inner working string bears two kinds of risk: stress strength at fixed end and collapse at flexible joint. In order to reduce these risks, local structure including wellhead, BOPS and riser flexible joint should be optimized and string axial force should be limited in a rational range.

The angular deflection of flexible joint is the dominating parameter to working string mechanical behavior. It is induced by marine circumstance and can only be confined through exerting proper tension to riser.

Once the operations have started, axial force becomes the one and only parameter that can be controlled.

Under such a special local constraints a slender string is not safer than a thick one. Mechanical investigation must be made on all kinds of working string.

Acknowledgments

This work is supported by "National Basic Research Program of China" (973 Program, 2015CB251205).

References

1. Denison, E.B., Kolpak, M.M., Garrett, D.L., A Comprehensive Approach to Deepwater Marine Riser Management, *SPE 13339*, (1985).
2. I. Cummins, E. Neme, W. Todd, Deepwater Riser Management, *IADC/SPE 59104*, (2000).

3. International Organization for Standardization, Petroleum and natural gas industries: Drilling and production equipment (part 1): Design and operation of marine drilling riser equipment, *ISO 13624-1-2009* (Geneva, 2009).
4. Chang Yuanjiang, Duan Menglan, Design and Operation of Marine Drilling Risers: A Case History of Deepwater Drilling in the South China Sea, *Natural Gas Industry*, 34(5): 106-111 (2014).
5. Bueno R C S, Morooka C K, Analysis Method for Contact Forces between Drillstring-Well-Riser, *SPE 28723*, (1994).
6. Wendler C, Maia C T, Planning and Conducting Well Tests in Deep and Ultra-deep Water to Mitigate Potential Risks and Justify Expense for the Operator, *OTC 21915*, (2011).
7. Bavidge M. Husky Liwan Deepwater Subsea Control System, *OTC 23960*, (2013).
8. Chang Y J, Chen G M, Zhang L, et al, Calculation and Analysis of Operability Envelopes for Deepwater Drilling Riser System, *OTC 24778*, (2014).
9. Qi Meisheng, Wen Jihong, Chen Guoming, Chang Yuanjiang, Stiffness Analysis of Marine Riser Flexible Joint Elastomer, *CHINA PETROLEUM MACHINERY*, 42(8): 71-74 (2014).
10. Xu Liangbin, Zhou Jianliang, Wang Rongyao, et al, Analysis of Deep Water Drilling Platform Evacuation from Imminent Typhoons with Riser Hanging-off in the South China Sea, *China Offshore Oil and Gas*, 27(3): 101-107 (2015).
11. H.J.J. van den Boom, J. N. Dekker, A. M. van Elsacker, Dynamic Aspects of Offshore Riser and Mooring Concepts, *SPE 16622*, (1988).
12. Robert J. Stomp, Graham J. Fraser, Stephen C. Actis, Luke F. Eaton, and Kerry C. Freedman, Deepwater DST Planning and Operations from a DP Vessel, *SPE 90557*, (2004).
13. CHEN Guoming, LIU Xiuquan, CHANG Yuanjiang, XU Lianabin, Advances in Technology of Deepwater Drilling Riser and Wellhead, *Journal of China University of Petroleum*, 37(5): 129-139 (2013).

Experimental study on flow characteristics in bending channel

Dan Wang, Jie Gu*

College of Marine Sciences, Shanghai Ocean University,
Shanghai, China
**E-mail: jgu@shou.edu .cn*
www.shou.edu.cn

In this paper, the flow characteristics in bending channel are studied using physical model. The water level and velocity along and cross bending channel are measured by the automatic water level measurement system and the velocity measurement system, respectively. The influences of inflow discharges and water levels at the outlet on the flow characteristics in the bending channel are summarized. The results show that the water level on the concave bank rises and the water level on the convex bank drops, when the flow moves from the straight segment to the curve segment. The maximum of flow velocity appears in the convex bank firstly, the main stream line by the convex bank then gradually moves towards the concave bank. The change of inflow discharge or water level at the outlet affects water level and flow patterns in the bending channel.

Keywords: U-Shaped Bend Channel; Flow Characteristics; Inflow Discharge; Water Level and Velocity.

1. Introduction

Natural river generally exist the curve, it can be regarded as the basic unit of river [1]. When the water flow come from the straight segment to curve segment, it is also affected by the centrifugal force when it is under influence of the gravity.Under the combined action of the two, the surface morphology and flow structure in bending reach occur adjustments and changes, forming the motion characteristics of bend flow. The flow characteristics of bending reach are as follows: the transverse gradient of water surface, the vertical gradient of water surface, bend circulation and redistribution of flow velocity [2][3].

Thomson J [4] by test found the spiral flow of bending reach for the first time, and described the corresponding mechanism about the formation of circulation flow and the cause of transverse sediment transport. Since then, many experts and scholars at home and abroad through field observation, experimental research, theoretical analysis and numerical simulation method, researched and analyzed the flow characteristics of bending reach [5][6]. Based on "potential flow theory", Milovic [7] et al derived the first theoretical law--"the law of area"; Zhang Hongwu [8][9] et al in the movable bed model test of muddy water found that, flow characteristics of bending reach and sediment concentration in water have a close relationship. But when the water is in the high sediment state, this effect is not too obvious; Yangtze River Channel Design Research Institute [10][11] by experiment on curved channel generalized model found that, the line of water surface on each section from entrance to exit of bending reach is nonlinear, and the degree of inclination is not same.

Because there are many factors relate to the change of flow characteristics in bending reach, it is also an important research topic in river hydraulics [12]. Therefore, this paper used the laboratory physical model, according to different operating condition, focused on the impact of water discharge and water level on the flow characteristics in bending reach.

2. Research Methods and Experimental Arrangements

In this paper, using the laboratory U-shaped water tank, the automatic water level measurement system and the velocity measurement system, under different water discharge or water level conditions, using the tail gate control water level, to analyze the influence of water discharge or water level on the flow characteristics in bending reach, and gives the change regulation of water level and flow velocity under corresponding conditions.

Figure 1 is the layout of observation points and observation sections. A1 to E2 are ultrasonic probe of water level. Among them, A1, B1, C1, D1 and E1 were arranged on one side of the convex bank, A2, B2, C2, D2 and E2 were arranged on one side of the concave bank. The distance between each pair of probes is 0.25m, the distance away from the wall at about 0.125m. The data of flow velocity measured by Particle Image Velocimetry (PIV). In this study, there are five observation sections of flow velocity(A to E), each section is parallel with the bottom of water channel, and the distance away from the bottom of water channel is about 0.15m.

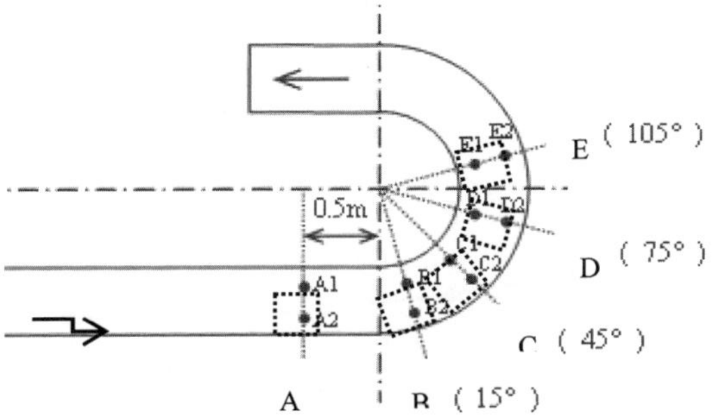

Fig. 1. Layout of observation points and observation sections.

3. Test Scheme

The first test scheme: the inlet water discharge was set to different numerical values, and in turn adjusted the tail gate, to make the water level of tail gate maintain at around 0.30m (see Table 1); the second test scheme: the inlet water discharge was set at $60\,m^3/h$, and adjusted the tail gate, to make the water level of tail gate controlled in several different numerical values (see Table 2).

Table 1. The first test scheme.

Group Number	Water Discharge(m^3/h)	Water level(m)
1	40	0.30
2	50	0.30
3	60	0.30
4	70	0.30

Table 1. The second test scheme.

Group Number	Water Discharge(m^3/h)	Water level(m)
1	60	0.27
2	60	0.30
3	60	0.33
4	60	0.36

4. Experimental Result and Discussion

4.1. *Flow characteristics in bending reach*

4.1.1. *The change of water level*

Figure 2 is the water level of each observation point when water discharge is 60m3/h and water level is 0.30m. It can be seen from the figure, when the water flow come from the straight segment to curve segment, the water level on the concave bank is rising obviously, then the water level decreases gradually along the way; and the water level on the convex bank is always dropping.

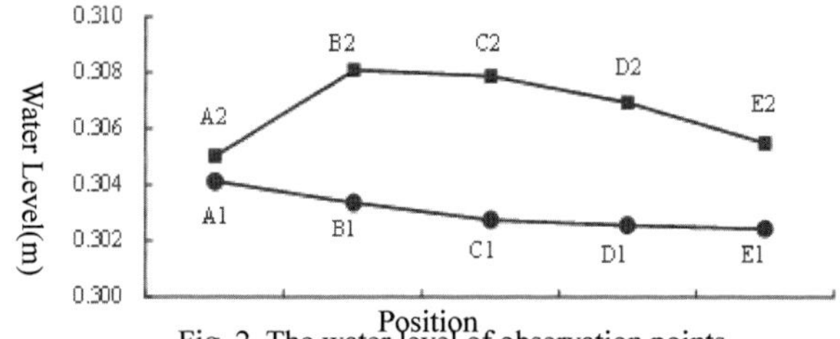

Fig. 2. The water level of observation points.

4.1.2. *The change of flow velocity*

Figure 3 is the flow velocity contour map of each observation section, when the water discharge is 60m3/h and the water level is 0.30m. It can be seen from the figure, when the water flow in straight segment, the flow velocity between concave bank and convex bank is well-distributed; when the water flow get into curve segment, flow velocity on convex bank increases obviously, flow velocity on concave bank decreases obviously. But along with the water flow advances in bending reach, flow velocity on convex bank gradually decreases, flow velocity on concave bank increases unceasingly, the main stream line by the convex bank gradually towards the concave bank. Finally, flow velocity on concave bank greater than convex bank.

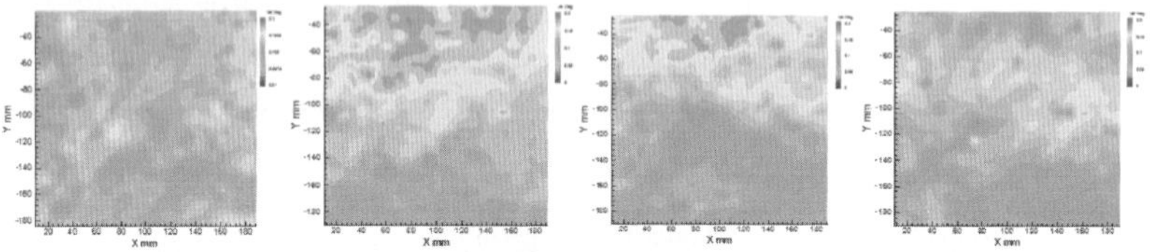

(a) section A (b) section B (c) section C (d) section D

(e) Observation section E

Fig. 3. The flow velocity contour map of each observation section.

4.2. *Effect of water discharge on bend flow*

4.2.1. *The change of water level*

Figure 4 is the water level difference of corresponding observation points on five cross section when the water discharge are different and the water level is same. It can be seen from the figure, when the water level is constant, with the increase of water discharge, water level difference between concave bank and convex bank on different cross section all increase. And under various operating condition, water level difference between concave bank and convex bank along the way all performance first increase and then decrease.

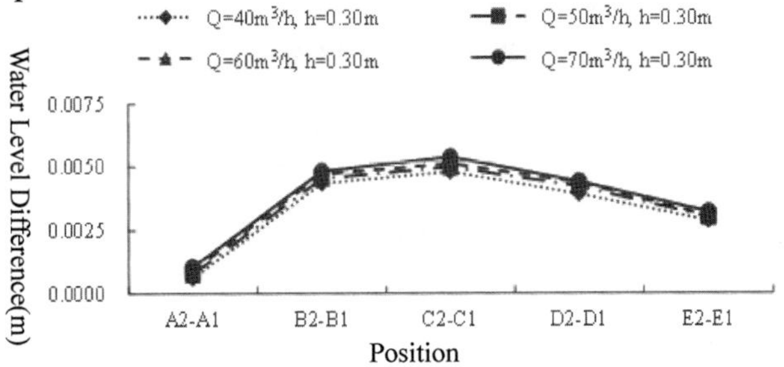

Fig. 4. Water level difference between concave bank and convex bank

4.2.2. *The change of flow velocity*

Figure 5 is the flow velocity contour map of observation section B when the water discharge are different and the water level is same. It can be seen from the figure, when the water level is constant, with the increase of water discharge, the flow velocity of entire section performance increases.

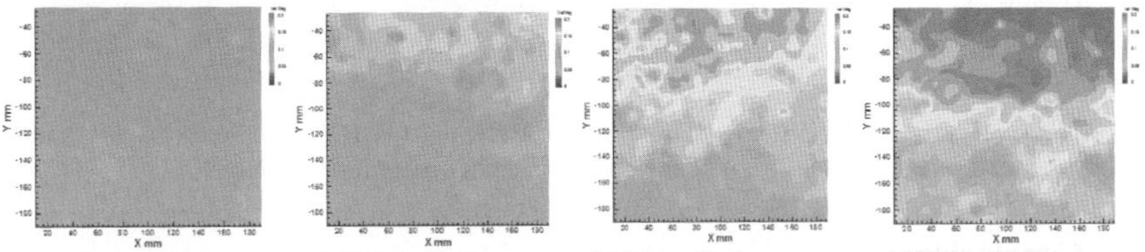

(d)Water discharge is $40m^3/h$ (c)Water discharge is $50m^3/h$ (b)Water discharge is $60m^3/h$ (a)Water discharge is $70m^3/h$

Fig. 5. The flow velocity contour map of observation section B

4.3. *Effect of water level on bend flow*

4.3.1. *The change of water level*

Figure 6 is the water level difference of corresponding observation points on five cross section when the water discharge is same and the water level are different. It can be seen from the figure, when the water discharge is constant, with the increase of water level, water level difference between concave bank and convex bank on different cross section all decrease. And under various operating condition, water level difference between concave bank and convex bank along the way all performance first increase and then decrease.

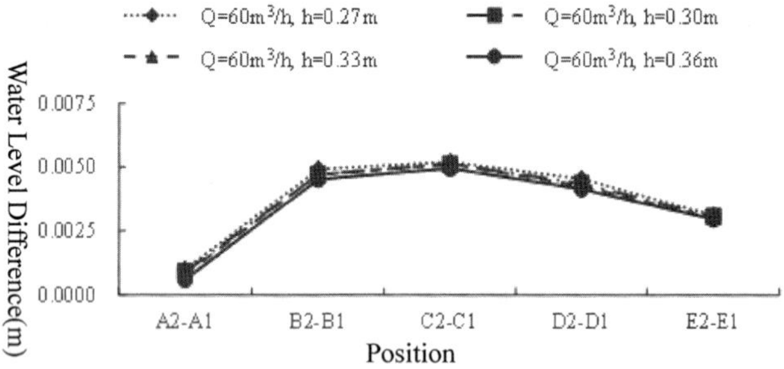

Fig. 6. Water level difference between concave bank and convex bank

4.3.2. *The change of flow velocity*

Figure 7 is the flow velocity contour map of observation section B when the water discharge is same and the water level are different. It can be seen from the figure, when the water discharge is constant, with the increase of water level, the flow velocity of entire section performance decreases.

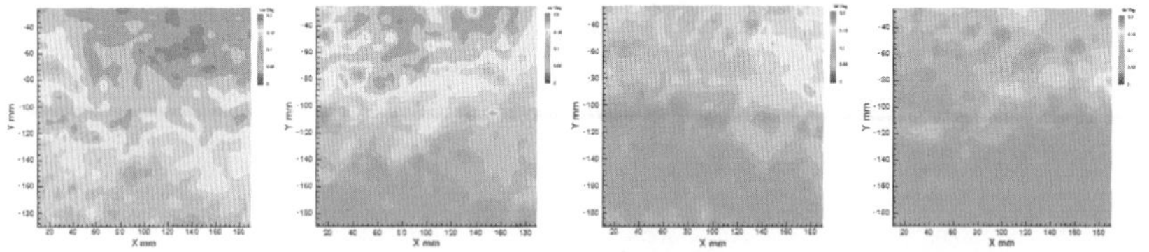

a. Water level is 0.27m   b. Water level is 0.30m   c. Water level is 0.33m   d. Water level is 0.36m

Fig. 7. The flow velocity contour map of observation section B

5. Conclusions

In this paper, using physical model test to study the effects of water discharge and water level on the flow characteristics of bending reach. The experimental results show that:

(1) When the water flow come from the straight segment to curve segment, the water level on the concave bank increased firstly and decreased gradually, while the water level on the convex bank always been in decline.

(2) After the water flow come to curve segment, maximum of flow velocity appears in the convex bank firstly, then the main stream line by the convex bank gradually towards the concave bank.

(3) When the water level is constant, with the increase of water discharge, the water level difference and the flow velocity between concave bank and convex bank all increase; when the water discharge is constant, with the increase

of water level, the water level difference and the flow velocity between concave bank and convex bank all decrease.

References

1. Wang Wei, Xu Wei-lin, Cai Jin-de. The theory and Application of water and sediment movement in bending reach[M]. Chengdu: Chengdu University of Science and Technology press,1994. (In Chinese).
2. Dong Yao-hua. Essential features and numerical modelling of bend flow[J]. Journal of Yangtze River Scientific Research Institute, 1996, 13(1): 25-29. (In Chinese).
3. Rui De-fan. Numerical model for circumfluence and movement of sediment in continuous meandering river[D]. Chengdu: Sichuan University, 2005. (In Chinese).
4. Thomson J. On the Origin and Winding of Rivers in Alluvial Plains. Proc. Royal Society of London, 1876, 25(3).
5. Diao Ming-jun, Yang Hai-bo. A summary of the study and progress on hydraulics of curved flow[J]. Journal of Southwest University for Nationalities, 2007,33(3): 596-601. (In Chinese).
6. Wang Ping-yi. Curved river dynamics[M].Chengdu: Chengdu University of Science and Technology press,1995. (In Chinese).
7. Rozovskii I L. Flow of Water in Bends of Open Channels. Isracl Program for Scientific Translation, 1965.
8. Zhang Hong-wu. Stochastic modeling of roughness on alluvial river bed[J]. Journal of Wuhan Institute of water conservancy and electric power, 1986(3): 92-99. (In Chinese).
9. Zhang Hong-wu, Lv Xin. Bend hydraulics [M]. Beijing: Water Conservancy and Electric Power Press, 1993. (In Chinese).
10. Liu Huai-hai. The research report about experiment on curved channel generalized model[R].Wuhan: Yangtze River Channel Design Research Institute, 2008. (In Chinese) .
11. Tang Ren-jie, Hu Xu-yue. Research situation of water flow in bend reach[J]. Journal of Waterway and Harbor, 2009, 30(2): 108-112.(In Chinese).
12. Liu Yue-qin. Development of research on essential characteristics of meandering rivers[J]. Pearl River, 2003(2):1-4. (In Chinese).

The dynamics simulation analysis on turbocharger turbine-shaft system

Fei Cao*, Shou-Ren Wang†, Gao-Qi Wang‡, Ya-Dong Li§
School of mechanical engineering, University of Jinan,
Jinan, Shandong Province, China
*E-mail: *cccfff000@163.com; †sherman0158@tom.com; ‡me_wanggq@ujn.edu.cn;*
§liyadongdl@163.com
www.ujn.edu.cn

The study object in this paper is turbine-shaft system of a turbocharger. The turbine is made of k418 alloy and the rotating shaft is made of 42CrMo. The 3D model is created by using SolidWorks software and the finite element dynamic analysis is done by ANSYS. It gets the turbine-shaft system's static modal and dynamic modal which considers the influence of the gyroscopic effect and prestress. Then it get the critical speed of the system through plotting the campbell diagram. This is very important reference information for designing the turbine . This article also analyzes the stress of the system under some specific rotate speeds and predicts the limit rotor speed of the turbine which is made of k418 alloy.

Keywords: Turbine; K41; Finite; The Modal; Critical Speed; The Limit Rotor Speed.

1. Introduction

The demand for automobile dynamic performance is getting higher and higher with the development of auto industry, so the Turbo technology has been widely used in automobile engine. Turbo technology was first mentioned in the end of the 19th century and used in diesel engines. And it was used in gasoline engines Turbocharger is the machine that use the exhaust which comes from engine rapidly to impact turbine blades and the turbine driven coaxial impeller. At the same time the impeller blades will compress air into the engine cylinder, so it can increase the air-fuel ratio in the engine cylinder. Air-fuel ratio's increasing in Engine cylinder makes fuel burn more fully. That's to say the same weight of fuel can spurt greater power, saving fuel and reducing environmental pollution. As the core parts of turbocharger, turbine sustains the high-speed impact of high temperature exhaust, leading to turbine blades produce huge internal stress. It is worth noting that turbine-shaft system will produce uneven whirl in the process of high-speed rotating and when the whirl frequency is same as the system natural frequency it will cause resonance. Resonance is pretty harmful, because it can cause big deformation and dynamic stress of the mechanical structure, even leading to the destruction of the institutions, endangering the personal safety. Reasonable and reliable structure natural frequency and vibration mode is very important for the design of the structure under dynamic load [2]. In this paper, by using the finite element analysis software ANSYS, it gets the critical speeds of the system and this is convenient for designers to improve its critical speed by optimizing the turbine rotor structure, as well as it also helps people to avoid the turbocharger's critical speeds in daily use, driving safely.

Stress on impeller blades is increasing with the speed of turbo getting higher and higher, and mainly is centrifugal stress. Analyzing the mechanical properties of the turbine-shaft system under specific speed is very important to study turbine's working performance and reliability.

2. Modeling and Meshing

2.1. *Modeling*

As shown in the figure, this study system includes a shaft and the turbine. The turbine is made of k418 alloy and the shaft is made of 42CrMo. The turbine has nine blades and the structure of the blades is very complex. The size of turbine and shaft was measured by using measuring instrument and the results are as follows: diameter of the turbine is 40 mm, blade thickness is 1 mm, and the shaft length is 84 mm.

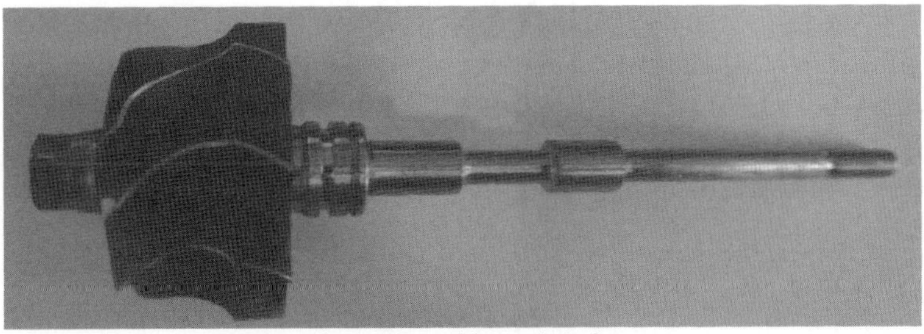

Fig. 1. The turbine and shaft.

In order to facilitate finite element analysis, it simplify the entity structure, and then creat 3D model by using Solidworks software, ignoring the small fillet and thread parts, as shown as in figure 2.

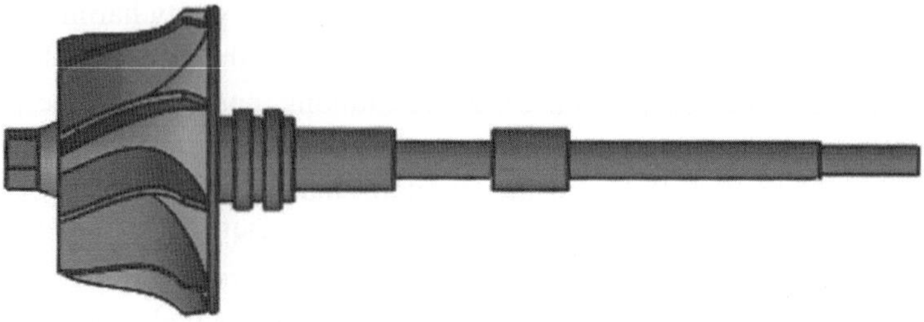

Fig. 2. The 3D model.

The 3D model is imported into ANSYS, selecting solid 186 element type and defining material properties. Turbine and shaft material parameters are listed in the table 1.

Table 1. The parameters of material.

	Material	Density[g/cm^3]	Modulus of elasticity[GPa]	Poisson's ratio	Yield limit [MPa]
Turbine	K418	7.8	176	0.3	≈850
shaft	42CrMo	7.85	212	0.28	>930

2.2. *Meshing*

Then the model was meshed with refining the blade root and margin by using tetrahedron free meshing method. The model after meshing is shown in the figure 3.

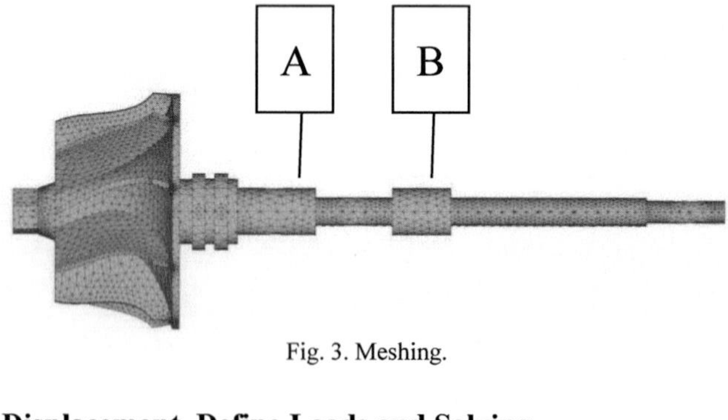

Fig. 3. Meshing.

3. Apply Displacement, Define Loads and Solving

As shown as in the figure 3, the bearing are installed in the part A and B of shaft in working status, so the movement of the two parts is constrainted and the rotation angular velocity around Y axis is set up .

3.1. *Modal analysis*

The system's natural frequency and vibration are acquired through modal analysis, that are the inherent mechanical properties of structure.

Modal analysis is a kind of method to study structure dynamic characteristics effectively started in the modern times, letting people master the speed range of the structure's resonance frequencies and predict structure's actual vibration response reasonably. The model after meshing has 41883 nodes and 125649 elements. Each node has three degrees of freedom namely the entire structure has 125649 degrees of freedom. The dynamic equilibrium equation of each node expressed as follows[3].

$$M\ddot{S} + C\dot{S} + KS = F \tag{1}$$

The meshing of letters are follows: M means the mass matrix, C represents the damping matrix, K means the stiffness matrix, F is the external force, S represents the node displacement, q represents the node speed, w is the node acceleration. Considering the structure has 125649 degrees of freedom, M and K are all125649 order phalanx so 125649 order natural frequencies can be obtained, but the former 6 order natural frequencies solved are enough to meet the needs of the research in this paper [4].

Table 2. The former 6 steps static inherent frequencies.

Steps	1	2	3	4	5	6
Frequency[Hz]	2013.3	2013.4	3765.6	3768	5348	10004
Vibration type	Curve	Curve	Curve	Curve	Twist and Blade twisting	Blade twisting

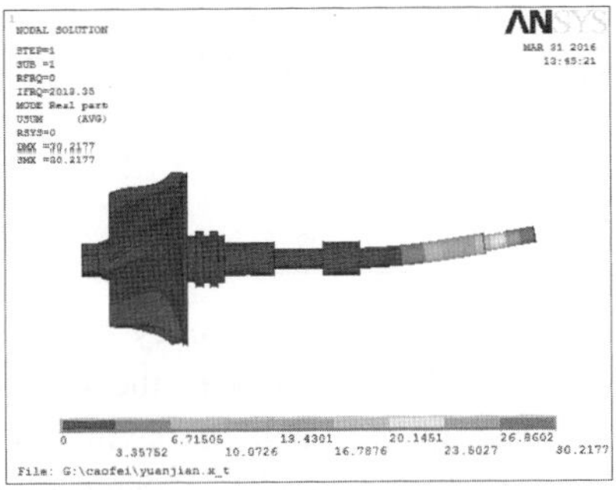

Step=1

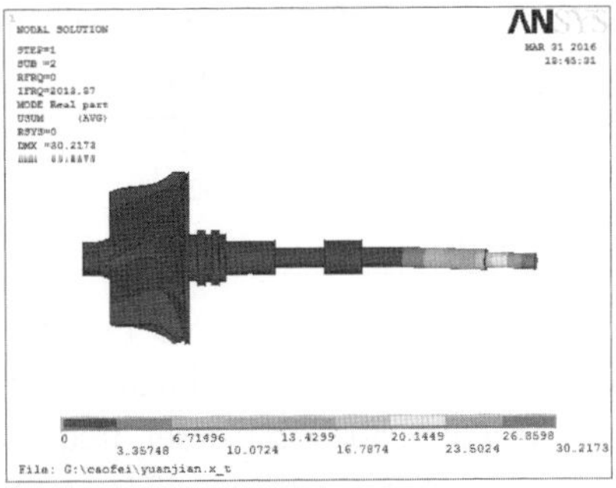

Step=2

Fig. 4. The former 6 vibration mode.

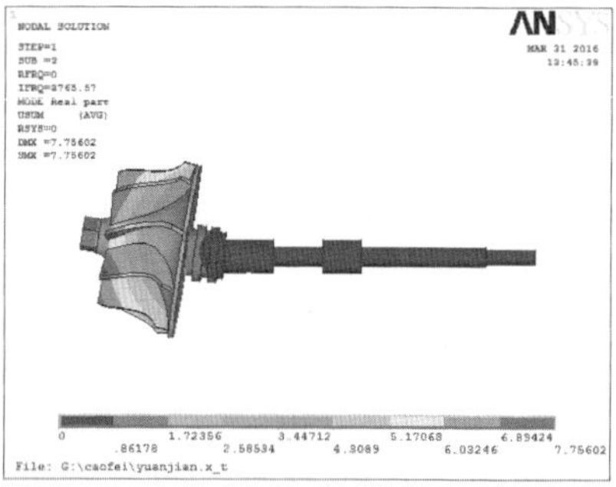

Step=3

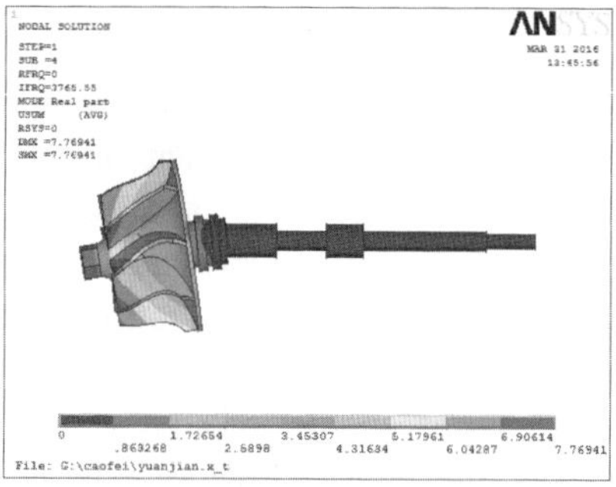

Step=4

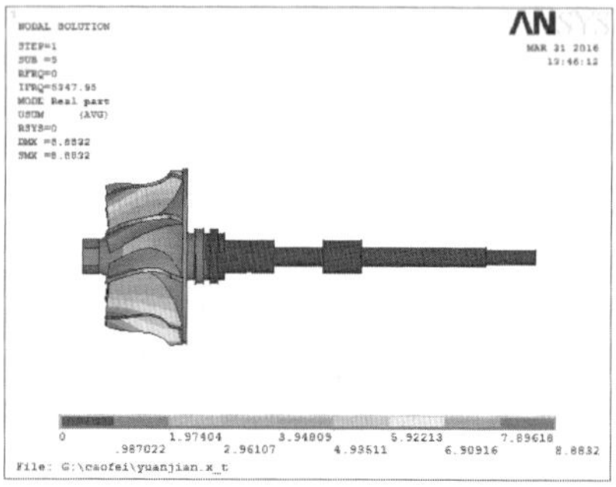

Step =5

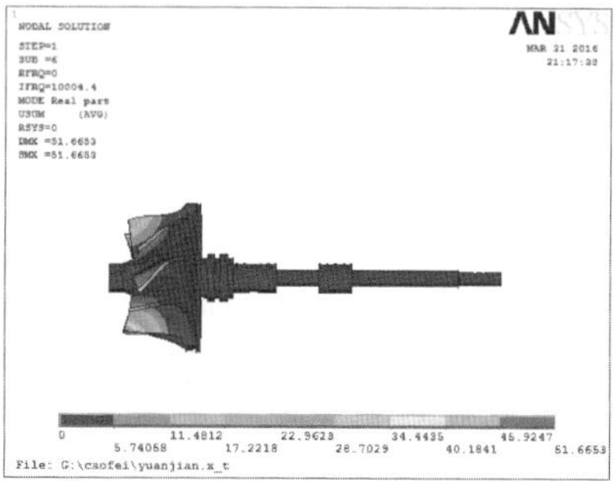

Step=6

Fig. 4. *(Continued)*

From the figure 4 we can know that the turbine-shaft system become deformed big. Analyzing the modal considering the rotating prestress and Coriolis effect and then compare the solution with the static natural frequency. From the comparison we can find the influence of rotating prestress and Coriolis on system's inherent frequency.

Table 3. The natural frequencies under different rotational speed.

Rotational speed [r/min]	40000	80000	120000
Step 1 [Hz]	1352.2	686.82	20.54
Step 6 [Hz]	10416	10362	10273

Table 3 shows that the turbine-shaft system's natural frequency declines with the increase of rotational speed due to centrifugal force and the vortex. Among them, the first order frequency decrease quickly and the sixth order frequency changes slowly. The reason for this change maybe is that system stiffness reduces under the comprehensive influence of prestress and vortex.

According to the change of turbine shaft-system's inherent frequency with speed increasing, this paper plots the Campbell diagram by using ANSYS software, as shown in the figure 5. Campbell diagram is an effective means to monitor system's vibration response amplitude. It's X axis is speed and Y axis is inherent frequency. A slope of 1 ray is drew from the origin and these intersection points of the slope and inherent frequency lines are the turbine-shaft system's critical speeds [5].

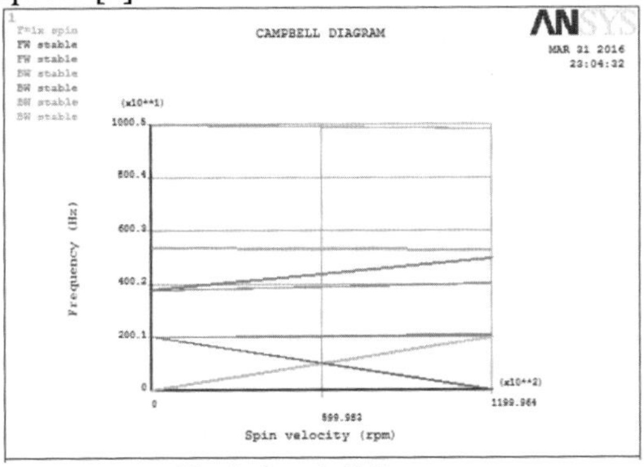

Fig. 5. Campbell diagram.

From figure 5 we know the system will appear resonance at the speed 63000 r/min and 120000 r/min, so people should through this speed quickly and avoid driving at the critical speed.

3.2. *Stress analysis under specific speed*

It analyzes the change of turbine stress along with the rotating speed increasing and predict the limit rotor speed. Speed values are divided into five levels: 40000 r/min, 80000 r/min, 120000 r/min, 160000 r/min, 200000 r/min. The maximum stress under various speed are shown in figure 6.

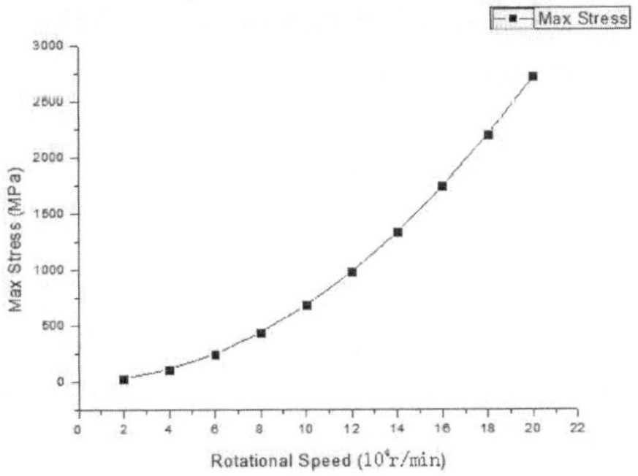

Fig. 6. The maximum stress under different rotational speed.

It can be seen from the figure 6 that the maximum stress value at the speed of 100000 r/min and 120000 r/min are close to the yield limit of K418 material .So it selects the speed of 110000 r/min to be analyzed and the stress nephogram is shown in figure 7.

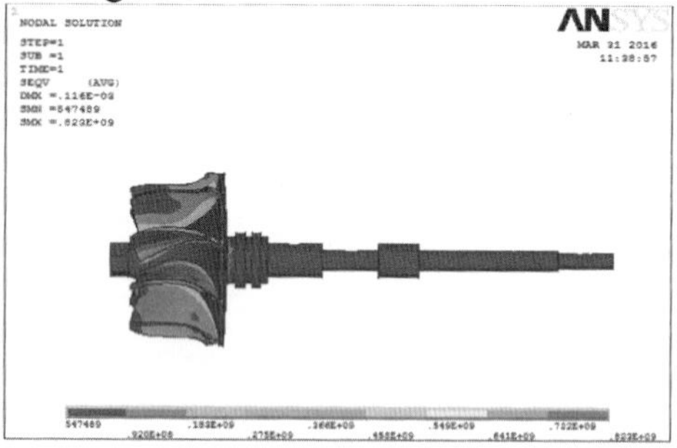

Fig. 7. The stress nephogram under 110000r/min.

The figure 7 shows when the turbine-shaft system works on the speed of 110000r/min, the maximum stress is very close to the material's yield strength which is 850MPa[6],and the maximum stress appears in the blade root. Blade will occur plastic deformation when the speed value is more than 110000r/min, therefore it is regarded as the reference limit speed of k418 turbine .

4. Conclusion

(1) The former 6 orders natural frequency and resonance of the system type are gotten by analyzing the dynamic characteristics of the turbine-shaft system and these date are very important for optimizing structure and improving the natural frequency.

(2) The critical speed of system was obtained through plotting Campbell diagram thus users can avoid it in daily driving.

(3) It analyzes the stress of k418 turbine under different rotational speed in this paper, finding that when rotating speed reach to 110000 r/min, the maximum stress is close to yield strength of k418 alloy, so the speed of 110000 r/min is regarded as the referential limit speed.

Acknowledgments

This work is supported by the National Natural Science Foundation of P.R. China (ID: 51372101) and Taishan scholar engineering special funding. (2016.01.01-2020.12.30).

References

1. Ugur Kesgin. Effect of turbocharging system on the performance of a natural gas engine. Energy Conversion and Management. 16 (2005) 11-32.
2. Yanzhu Liu, Wenliang Chen, Liqun Chen, *Vibration mechanics*(Higher Education Press, 1998). (In Chinese).
3. Liu G.R. Quek S.S. *The Finite Element Method* (Hunan University Press, 2004).
4. Qingwu Wang, Fang Zuo, Renxi Hu, *ANSYS 10.0 Advanced Mechanical Design Application*s (Mechanical Industry Press, 2006). (In Chinese).
5. Gang Tong, Yongli Qiao, Yanting Ai, *Coupling resonance characteristics of the aero engine turbine rotor blade – analysis*, Journal of aeronautics and astronautics university. 5 (2010) 6-10. (In Chinese).
6. Wenming Yuan, Xin Tang, Faxin Liu, Rong Zhang, *The influence of fine grain casting on the tensile properties of 418 alloy*, Journal of aviation materials. 16 (1996) 39-43. (In Chinese).

Session 2

Civil, Architectural Engineering

Analysis of the effects of beam and column semi-rigid connection on the mechanical properties of assembled pallet rack

Deng-Chao Niu, Zhuan Wang*, Shan Mi
School of Mechanical Engineering, University Of Science and Technology Beijing, Beijing, China
**Email: wzpaper@126.com*

This work presents a general analysis of assembled pallet rack, evaluating the influence of beam and column semi-rigid connection on the mechanical properties by finite element analysis software, based on SAP84. The numerical results show that the connection of beam and column has significant influence on the rack's stability. The results are analyzed and get some useful conclusions.

Keyword: Assembled Pallet Rack; Semi-Rigid Connection; Stability of Rack; Finite Element Analysis

1. Introduction

In the modern system of storage rack, the high utilization, access convenient of pallet rack is popular in warehouses, so the safety performance is very important.

Presently, analysis of rack's behavior, the finite element method is widely applied. J. Bonada [1] performed analysis ultimate load values of cold-formed pallet rack with the finite element method and the generalised beam theory. Ch. Affolter [2] investigated the failure reason collapse of a high storage rack. The first-order modal of the linear buckling analysis as the initial condition of the nonlinear static analysis is the most reasonable. For assembled rack, the connection way of components, especially the connection way of beam and column has great effect on global stability. At present most of research work, the rigid connection of beam and column is majority. Actually, beam and column connection is semi-rigid connection by bolts. Nadia Baldassino [3] investigated the connection way of beam end. Put forward the beam end connection will provide a lateral stiffness, it cannot be ignored, therefore, on a model of rack, the beam end semi-rigid connection is considered to be the most reasonable. Arlene M.S. Freitas [4] investigated behavior of steel storage drive-in racks, evaluating the influence of each of their components with respect to global stability. Keshav K. Sangle [5] performed analysis of cold-formed pallet rack structures, semi-rigid connections and different column section have significant influence on the rack's global stability.

In different coefficient of beam and column semi-rigid connection conditions, the finite element analysis is carried out on the pallet rack by finite element analysis software, based on SAP84, and results provide reference for rack structural optimization.

2. FEM Calculation Model

2.1. *Structure model*

In this paper, taking assembled pallet rack as research object. Column, beam, diagonal bracing and coupling bar are main components of rack. The overall grid structure model is shown in figure 1.

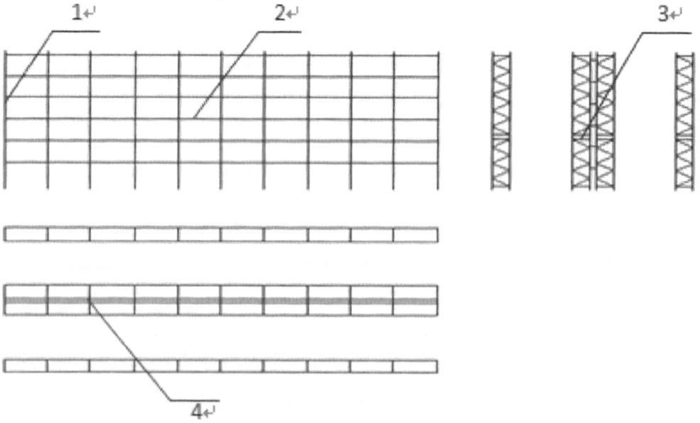

1——column; 2——beam; 3——diagonal bracing; 4——coupling bar

Fig. 1. overall grid structure model.

2.2. *FEM model simplification*

FEM model is establisher by finite element analysis software, based on SAP84, the main components of rack FEM model are simplified as frame element. Meshing structure model bases on its nature grid, the load and constraint condition on rack are simplified to the corresponding node. In the coordinate system, X represents the down-aisle direction, Y represents the cross-aisle direction, Z represents the rack height direction.

In SAP84, the connection way of 3-D frame element is changed by setting the coefficients of semi-rigid connection. When the coefficient of semi-rigid connection for 1, said as the rigid connection between elements; When the semi-rigid connection coefficient is 0, said as hinged connection between elements; When the coefficient of semi-rigid connection between 0 and 1, said as the semi-rigid connection between elements. In this paper, setting column and ground, column and beam connection is semi-rigid connection, as the rigid connection between other elements. Set the coefficient of column and ground semi-rigid connection is 0.5, changing the coefficient of column and beam semi-rigid connection, under different coefficient of semi-rigid connection, analysis the performance of assembled pallet rack. In SAP84, all nodes of 3-D frame element have six degrees of freedom, including three translational degrees of freedom and three rotational degrees of freedom. The number 0 indicates that the degree of freedom is release and the number 1 is constraint. In this FEA model, R = 1, 1, 1, 1, 1, 1 indicates DOF of column with ground nodes, R = 0, 0, 0, 0, 0, 0 indicates DOF of other frame nodes.

2.3. *Calculation loading*

In this paper, the method of ultimate bearing capacity was taken as the calculation method. The types of load and partial coefficient are shown in table 1.

Table 1. Types of load and partial coefficient.

Type of load	Explanation	Partial coefficient
Constant load	Weight of the rack	1.2
Live load	Weight of goods unit	1.4
Vertical impact load	50% of the maximum static load of goods unit	1.4
Horizontal load	1% of the sum of total Constant load and Live load	1.4

2.4. *Calculation conditions*

In this paper, four calculation conditions are studied. Types of load include constant load, live load, impact load and horizontal load. Calculation conditions and load combinations are shown in table 2.

Table 2. Calculation conditions and load combinations.

Calculation conditions	Load combinations
Condition 1	Constant load and live load
Condition 2	Constant load, live load and vertical impact load
Condition 3	Constant load, live load and horizontal load in X direction
Condition 4	Constant load, live load and horizontal load in Y direction

3. Simulation Experiment

Based on the FEA rack model above, in different calculation conditions, with the coefficient of beam and column semi-rigid connection is the independent variable, 11 simulation experiment schemes are designed, to analyze the effect of beam and column connection on behavior of pallet rack, maximum stress of column and maximum displacement of node is analyzed as the main performance parameters. Simulation experiment schemes are shown in table 3.

Table 3. Simulation experiment schemes.

conditions / Analysis indicators / Semi-rigid coefficients	condition 1	condition 2	condition 3	condition 4
0,0.01,0.05,0.1, 0.15,0.2,0.3,0.5, 0.7,0.9,1	□ The maximum stress of column elements □ The maximum displacement of nodes	The same as the condition 1	□ The maximum stress of column elements □ The maximum displacement of nodes □ The bending moment and axial force of the maximum stress element	The same as the condition 1

4. Results Analysis

4.1. *Result analysis of condition 3*

Loads of condition 3 includes Constant load, live load and horizontal load in X direction, and horizontal load in X direction has a great influence on the stability of rack, so the experimental results of condition 3 is analyzed independently. Under condition 3, maximum stress, displacement, axial force and bending moment results are shown in table 4, figure 2, figure 3 and figure 4.

As can be seen from the experimental data:

(1) The X-displacement of nodes is sensitive to the coefficient of beam and column semi-rigid connection, with the decrease of the semi-rigid coefficient, X-displacement increases gradually, fluctuation is large, Y-displacement and Z-displacement have little change.

(2) The stress of columns is sensitive to the coefficient of beam and column semi-rigid connection, with the decrease of the semi-rigid coefficient, the stress of columns increases gradually, fluctuation is large.

(3) The bending moment of columns is sensitive to the coefficient of beam and column semi-rigid connection, the axial force is insensitivity. With the decrease of the semi-rigid coefficient, the axial force of columns decreases gradually, fluctuation is small, but the bending moment of columns increases gradually and fluctuation is large, so the bending moment of columns increase is the mainly reason for the stress of columns increase.

Table 4. Displacement and stress of condition 3.

Semi-rigid coefficient	Displacement/mm			Stress/MPa	Axial force/N	Bending moment/(N*mm)
	X	Y	Z			
1	9.96	0.23	3.72	148.37	84466.55	469584.75
0.9	11.02	0.23	3.71	149.05	84236.33	490067.45
0.7	13.95	0.23	3.69	150.88	83886.59	537265.08
0.5	18.92	0.23	3.68	153.78	83660.15	602248.77
0.3	29.45	0.23	3.67	159.17	83541.76	714054.66
0.2	41.13	0.23	3.67	164.27	83521.73	817196.55
0.15	51.60	0.23	3.67	168.31	83521.69	898304.13
0.1	69.94	0.23	3.67	174.53	83530.00	1023000.00
0.05	111.91	0.23	3.67	186.59	83540.00	1265000.00
0.01	236.33	0.23	3.67	216.04	83560.00	1856000.00
0	338.91	0.23	3.67	238.34	83560.00	2304000.00
Fluctuation	3302.71%	0	1.36%	60.64%	1.13%	390.65%

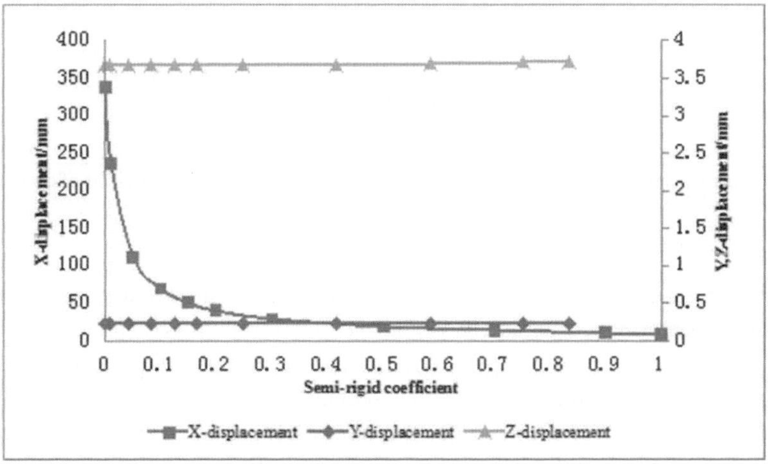

Fig. 2. Relationship between X, Y, Z displacement and semi-rigid coefficients of condition 3.

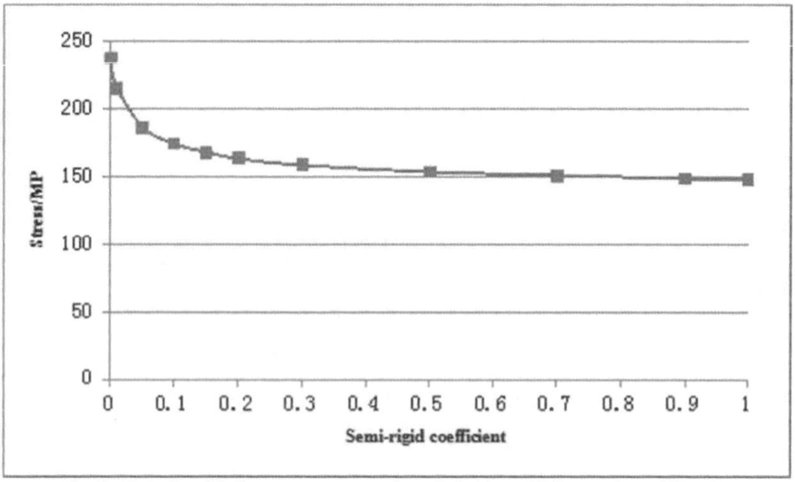

Fig. 3. Relationship between stress and semi-rigid coefficients of condition 3.

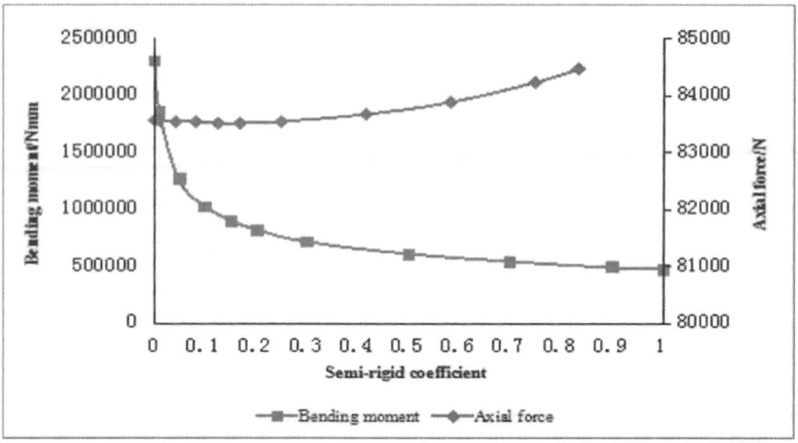

Fig. 4. Relationship between bending moment, axial force and semi-rigid coefficients of condition 3.

4.2. *Result analysis of condition 1, 2, 4*

Maximum stress of column and maximum displacement of node is analyzed as the main performance parameters, and calculate percentage fluctuations. The results under different semi-rigid coefficient are shown in table 5, table 6.

As can be seen from the experimental data:

(1) Result of condition 1, 2, the stress of columns is sensitive to the coefficient of beam and column semi-rigid connection, with the increase of the semi-rigid coefficient, the stress of columns increases gradually, fluctuation is large. Result of condition 4, the stress of columns is insensitive.

(2) The X, Y, Z displacement of nodes is insensitive to the coefficient of beam and column semi-rigid connection. With the decrease of the semi-rigid coefficient, it has nearly no effect on the X, Y, Z displacement of nodes. When the coefficient of beam and column semi-rigid connection is 0, which means the column-to-ground connection is fully articulated, the maximum X-displacement of condition 1,4 is 0, the maximum X-displacement of condition 2 is almost 0.

Table 5. Stress of condition 1, 2, 4.

Semi-rigid coefficient	Condition 1 stress/MPa	Condition 2 stress/MPa	Condition 4 stress/MPa
1	138.53	181.89	142.23
0.9	133.80	175.06	142.12
0.7	124.86	157.01	142.00
0.5	124.56	132.11	141.85
0.3	124.21	124.21	141.65
0.2	124.02	124.02	141.53
0.15	123.93	123.93	141.46
0.1	123.84	123.84	141.40
0.05	123.74	123.74	141.32
0.01	123.67	123.67	141.26
0	123.66	123.66	141.25
Fluctuation	12.02%	47.09%	0.70%

Table 6. Displacement of condition 1, 2, 4.

Semi-rigid coefficient	condition 1			condition 2			condition 4		
	X	Y	Z	X	Y	Z	X	Y	Z
1	1.17	0.23	3.72	1.59	2.52	3.75	1.17	2.68	3.84
0.9	1.14	0.23	3.71	1.53	2.52	3.74	1.14	2.68	3.83
0.7	1.02	0.23	3.69	1.38	2.52	3.72	1.03	2.68	3.81
0.5	0.85	0.23	3.68	1.15	2.51	3.71	0.85	2.69	3.80
0.3	0.60	0.23	3.67	0.84	2.51	3.70	0.60	2.69	3.80
0.2	0.43	0.23	3.67	0.64	2.50	3.70	0.43	2.69	3.80
0.15	0.34	0.23	3.67	0.52	2.50	3.70	0.34	2.69	3.80
0.1	0.24	0.23	3.67	0.39	2.50	3.70	0.24	2.69	3.80
0.05	0.12	0.23	3.67	0.25	2.50	3.71	0.12	2.69	3.80
0.01	0.03	0.23	3.67	0.11	2.50	3.71	0.03	2.69	3.80
0	0.00	0.23	3.67	0.00	2.50	3.71	0.00	2.69	3.80
Fluctuation		0	1.34%		0.79%	1.07%		0.37%	1.04%

5. Conclusion

This paper analyzed the assembled pallet rack in different the coefficients of beam and column semi-rigid connection by using the finite element analysis software, and get the results of displacement and stress under different semi-rigid coefficients are analyzed.

The results of research show that, when rack was applied the lateral force, the displacement of nodes and the stress of columns is sensitive to the coefficient of beam and column semi-rigid connection. With the decrease of the semi-rigid coefficient, the displacement of nodes and the stress of columns increases gradually, which means the greater of the semi-rigid coefficient, rack is more stable and safety. When rack was not applied the lateral force, the stress of columns is sensitive to the coefficient of beam and column semi-rigid connection. With the decrease of the semi-rigid coefficient, the displacement of nodes and the stress of columns decreases gradually, which means the smaller of the semi-rigid coefficient, rack is more stable and safety.

References

1. Bonada J, Casafont M, Roure F, et al. Selection of the initial geometrical imperfection in nonlinear FE analysis of cold-formed steel rack columns. Thin-Walled Structures 2012; 51 (2012) 99–111.
2. Affolter Ch, Piskoty G, Wullschleger L,et al. Collapse of a high storage rack. Engineering Failure Analysis 2009; 16 (2009) 1846–1855.
3. Baldassino Nadia, Bernuzzi Claudio. Analysis and behaviour of steel storage pallet racks. Thin-Walled Structures 2000; 37 (2000) 277–304.
4. Freitas Arlene MS, Souza Flavio T, Freitas Marcilio SR. Analysis and behavior of steel storaged rive-in racks. Thin-Walled Structures 2010; 48 (2010) 110–117.
5. Sangle Keshav K, Bajoria Kamal M, Talicotti Rajshekar S. Elastic stability analysis of cold-formed pallet rack structures with semi-rigid connections. Journal of Constructional Steel Research 2012; 71 (2012) 245–262.
6. GB/T28576-2012 Calculation of industrial rack design, Beijing: Standards Press of China, 2012. (in Chinese).

Strength analysis of six-pillar hydraulic support base

Jiang-Yu Wu, Mei-Mei Feng*, Kun Tang

State Key Laboratory for Geomechanics and Deep Underground Engineering, School of Mechanics & Civil Engineering, China University of Mining & Technology, Xuzhou, Jiangsu 221116, China

**E-mail: wujiangyu@cumt.edu.cn, fengmeimei@cumt.edu.cn*

The support base is an important bearing part of the hydraulic support in backfilling and coal mining face, which not only transmits the pressure from roof to seam floor but also assures strength, rigidity and seam floor specific pressure. In order to evaluating the reliability of hydraulic support, improving the structure and reducing the cost, we established the three-dimensional numerical model of six-pillar hydraulic support to conduct a statics simulation and conduct the mechanics characteristic and strength analysis. We analyzed the stress distribution characteristics of support part in two dangerous working conditions, and obtained the equivalent stress and deformation distribution of support base, discussed the strength characteristic and influence factor of the danger zone. By analyzing and comparing, in addition to determining the dangerous working conditions and the most dangerous point, we not only studied the strength characteristics and influencing factors of the base structure of the danger zone, but also analyzed the causes and effects of the stress concentration, which can provide reference for further structural optimization.

Keywords: Six-Pillar Hydraulic Support; Support Base; Strength Characteristics; Stress Concentration.

1. Introduction

For a long time, the world's coal mining industry has been exploring the development of mining technology filling, and experienced a hardship and tortuous development path. Some filling techniques have been studied successively, such as cement filling, cream filling, high water material filling, wind filling, grouting and polishing gangue separation filling techniques [1-4]. Now, the researchers have been performing various degrees of improvement and perfection on these technologies to meet the needs of large-scale industrial applications. For example, the author's development team affiliated with China University of Mining & Technology in collaboration with a number of coal companies and coal machinery enterprises, developed solid backfilling and fully-mechanized coal mining integration technology with full independent intellectual property rights to enable the filling coal technology research made a crucial breakthrough [5-11]. This technology is one of the important methods of green coal mining, which can effectively solve the environmental issues caused by coal mining under buildings, railways and water-bodies mining. In addition, it has been applied successfully in more than a dozen mines, such as Pingdingshan, Yanzhou, Jining and so on. Hydraulic support in backfilling and coal mining face is one of the main devices to guarantee the safe and efficient production in fully-mechanized coal mining face and the one-time investment on the support accounts filling total capital of mechanized mining equipment for over 70%. At the same time, the hydraulic support ensured that the necessary fill space, filling channels and filling power for the backfilling mining to realize the purpose of mining operations in parallel with backfilling under the cover of the same bracket.

Hydraulic support design based primarily on the interaction system of hydraulic support and surrounding rock, which together constituted by the "roof - Filling bracket - the floor, the coal wall - stand", and the mechanics model is establish to analyze the maximum concentrate load of support and the forces of each column when the ideal backfilling reached[12]. Literature [13] introduced the structure of a new type of backfill supports with six props and its characteristic, and the process of moving support is depicted and some advice is given for methods of moving supports in different conditions. Literature [14] proposed the hydraulic support design conception of mechanized mining with solid backfilling, the mechanic model of underground pressure that is used to analyze the roof deformation induced by the initiative loading of support is established. According to the Flamant's solution of a concentrated force act on infinite half plane, the analytical solution that can calculate the displacement curve of roof deformation induced by the active force provided by supporter are obtained. Literature [15] constructed a mechanics of the top beam and three rows of columns of a new type hydraulic support to analyze the relationship between the forces and the three rows of columns between the top force beams. Moreover, the Pro/E software is used to establish the three-dimensional model for kinematics simulation analysis of support and simulate the movement process of the main components, draw the motion profile from all the major components in the work process, and check the parameters of support.

The base, as a very important bearing part, must be satisfied with the rigidity, the strength and the floor specific pressure while the base transmits the pressure from the roof to the floor. So it is great significance to study its force characteristics and strength analysis to access the reliability of the filling hydraulic support, improve its structure and reduce its cost. Therefore, the finite element analysis of the base structure occupies an important position in the whole process of the filling hydraulic support design. In this article, the Pro/E three-dimensional modeling software was used to establish the ZZC8000/20/38 six-pillar hydraulic support model, which is imported into the finite element analysis software ANSYS Workbench to analyze the base strength under special working conditions and obtain the hazardous operating condition and dangerous points to prepare for the optimization design of the base structure.

2. Three-Dimensional Model

According to the design specifications of ZZC8000/20/38 six-pillar hydraulic support, Pro/E three-dimensional modeling software was used to design the main parts including the base, front and rear top beams, front and rear spatial linkage, column and so on. The support must be assembled and the interferential detection must be conducted after completing all parts of hydraulic support three-dimensional solid model before establishing the finite element model. There were three kinds of assemble design types in Pro/E assemble design modules, which were bottom-up, top-down and hybrid[15], and here we adopted the bottom-up assembly design. Interference of phenomenon occurs between the spatial linkage and the column, the guard plate and the column of hydraulic support by the position and size of interference indicated in the interference checking. After several attempts to modify and the full ZZC8000 / 20/38 three-dimensional solid model of the filling hydraulic support in coal mining is eventually received as shown in Figure 1.

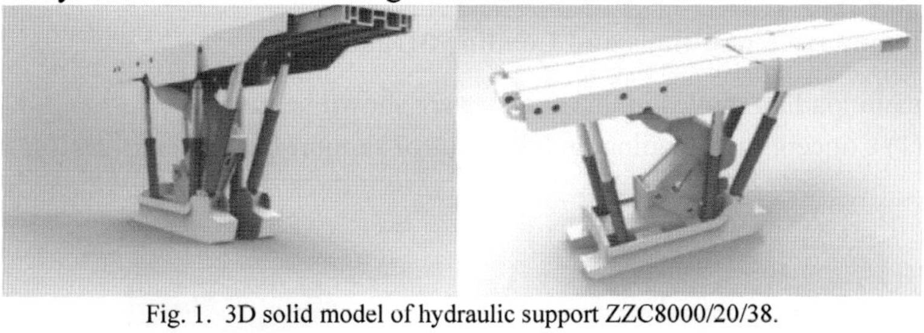

Fig. 1. 3D solid model of hydraulic support ZZC8000/20/38.

Table 1. Mechanical properties of Q550.

Density ρ/g·cm^{-3}	Elastic Modulus E/MPa	Poisson's ratio μ	Yield Strength σ_s/MPa	Tensile strength σ_b/MPa	Elongation δ_5/%
7.85	2.06e+5	0.28	550	630	>16

The body material of the ZZC8000/20/38 hydraulic support discussed in this paper is Q550 steel plate, which was a high-strength structural steel. The Q550 steel plate model adopted the linear elastic isotropic material, and the basic mechanical properties are as shown in Table 1.

However, the default element used in statics analysis ANSYS Workbench was Solid186 hexahedral element, which was a high order three-dimensional 20-node hexahedral solid structure of element. The structure of this element is so complicated that high quality hexahedral meshes can't be drawn, because the element degeneration in the whole hydraulic support division. So we need to modify the default element type for Solid187 element to avoid generating accuracy problems. Here, we selected the higher-order three-dimensional 10-node tetrahedral Solid187 element, which had a secondary displacement mode for better simulating the irregular model.

Both the element shape and the model meshing type are need to specify in the meshing, here, the tetrahedral elements and the free meshing are selected. Solid model could be divided for the finite element mesh generated finite element model as shown in Figure 2

Fig. 2. Finite element model.

3. Boundary Conditions and Loading Means

The main loading modes in this analysis are shown below.

(1) Concentrated loading on both ends of support base

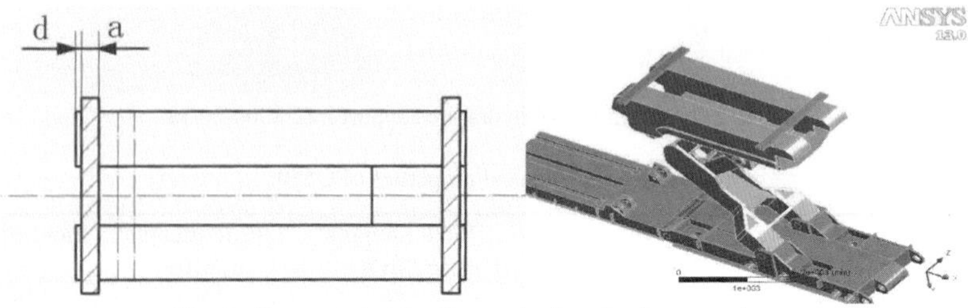

Fig. 3. Concentrated loading on both ends of support base.

(2) Torsion loading of support base

This condition of the torsion loading of the support base is caused by the uneven floor.

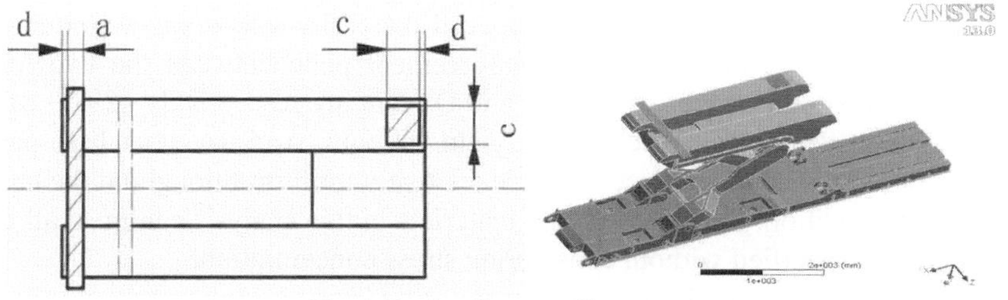

Fig. 4. Torsion diagram of basement.

In the loading diagrams above, a = 150 mm; b = 200 mm; c = 300 mm; d ≥ 100 mm; and the thickness of pad was 50 mm. According to "General technical conditions", the test loading method adopts loading within the column, the test pressure varies with the change of the manner of the exerting pressure upon support. The experimental test pressure P and height H are shown in Table 2 where the F_r is rated operational pressure.

Table 2. The loading requirements.

loading modes	*H*/mm	*P*/kN
Concentrated loading on both ends of support base	3200	1.2 F_r
Torsion loading of support base	3200	1.2 F_r

4. Results Analysis

4.1. *Strength characteristics of support base under Concentrated loading on both ends of support base*

The stress contours of the whole support and the support base under concentrated loading on both ends of support base of the six-column hydraulic support respectively show in Figure 5 and Figure 6. The corresponding stress state of the support base in high-stress areas shows in Table 3. we can find that: (1) In this condition, the larger regional of stress gradients variation are mainly distributed in four reinforcement rib plates of the support base, while the stress of the top beam and linkage is distributed evenly and symmetrically; (2) The stress distribution of support base was more complicate, the high-stress areas were respectively at point C and point A with the value of 575.92 MPa and 527.08 MPa, which are near the yield strength and dangerous, however both the two stresses can be ignored because of the stress concentration caused by fixedly connecting with the base and pad and the hinged constraint of the reinforcement plate and rear linkage. In addition, the stress concentration appeared around B point of welding position of the main reinforcement and transverse ribs, which are between the front axle and linkage. And there also was a small amount of stress concentration at D point of reinforcement plate between the two long pads. The maximum values of B point and D point are respectively 429.39 MPa and 185.16 MPa without reaching the yield strength. And the other base parts were exposed to a smaller force which is relatively uniform throughout the base. (3) In this condition, the structure and materials in this analysis can meet all the requirements specified without considering stress concentration.

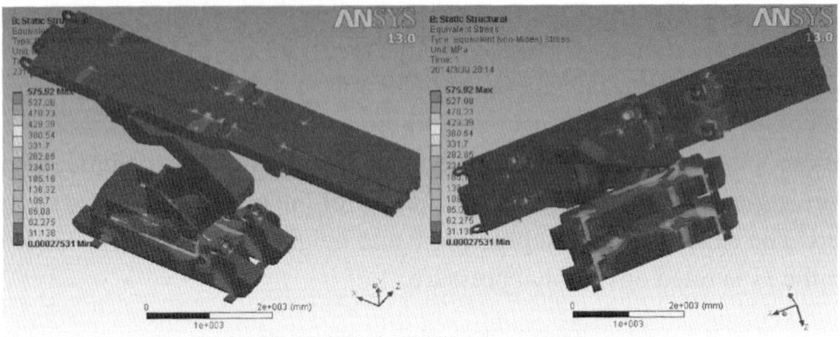

Fig. 5. Stress contours of the whole support under concentrated loading on both ends of support base.

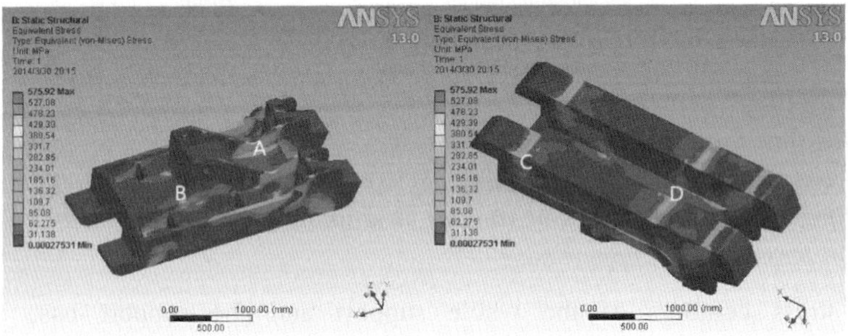

Fig. 6. Stress contours of support base

Table 3. The high-stress areas of support base under concentrated loading on both ends of support base.

Locations	σ_{eqv}/MPa	σ_s/MPa	$\sigma_{eqv}/\sigma_s \times 100\%$	Dangerous situation
A	527.08	550	95.8	No
B	429.39	550	78.1	No
C	575.92	550	104	Yes
D	185.16	550	33.7	No

The displacement contours of the support under concentrated loading on both ends of support base show in Figure 7. It can be seen that the maximum displacement of the support is at E point of the rear column net in the support base about 1.3 mm, and there was small amount of deformation displacement at F point of reinforcement plate between two pads. According to hydraulic support top-down order (top beam→linkage→base), the displacement of various structure deformation showed a gradual increasing trend and the deformation of the base displacement has a uniform and symmetrical distribution, which basically satisfied with "General specification for powered supportMT312-2000".

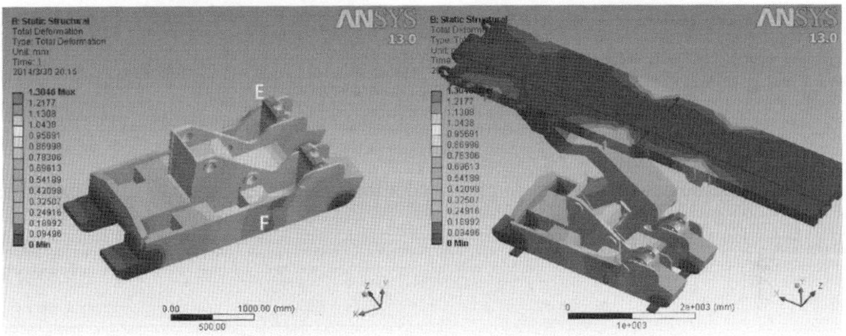

(a) Support base (b) Whole support

Fig. 7. Displacement contours of the support under concentrated loading on both ends of support base.

4.2. *Strength characteristics of support base under torsion loading conditions of support base*

Figure 8 and Figure 9 respectively show the stress contours of the whole support and the support base of the six-column hydraulic support under torsion loading conditions of support base, and the corresponding stress state of the support base in high-stress areas was shown in Table 4. It can be seen that: (1) In this condition, the larger regional of stress gradients variation mainly distributed in four reinforcement rib plates of the support base and the linkages, and the stress distribution of the base was more complicated, and the stress distribution of top beam mainly in column nets; (2) The high-stress areas of the support base mainly include two regional ranges. In the side of pad, there was greater stress concentration at point A in reinforcement plate near the hinged hole of the base and rear linkage with the value of 766.58 MPa. In addition, the larger stress concentration so appeared at B point of the base contacted with the long pad with the value of 699.56 MPa. (3) As is known, the stresses at point A and point B in high-stress areas of support base had exceeded the maximum yield strength of the material. So, we must take some measures to enhance the support base safe. Reselecting the material and redesigning the structure of the support base structure are the most common methods. However, reselecting the high-strength material is not easy to realize for some reasons, so it was necessary for us to make the structural optimization on the support base to ensure the structure safe in this working condition, which will be introduce in next paper.

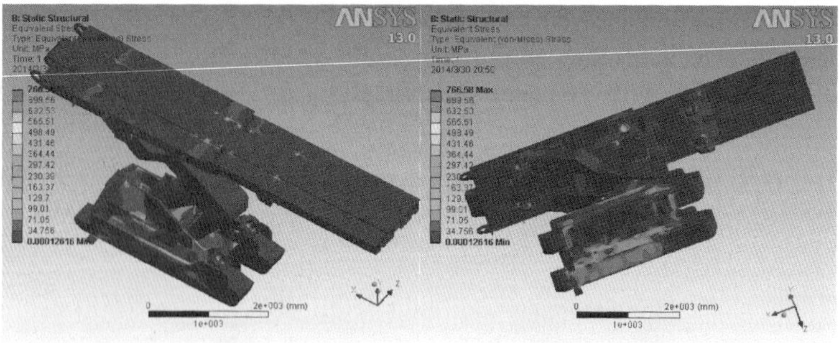

Fig. 8. Stress contours of the whole support under torsion loading conditions of support base.

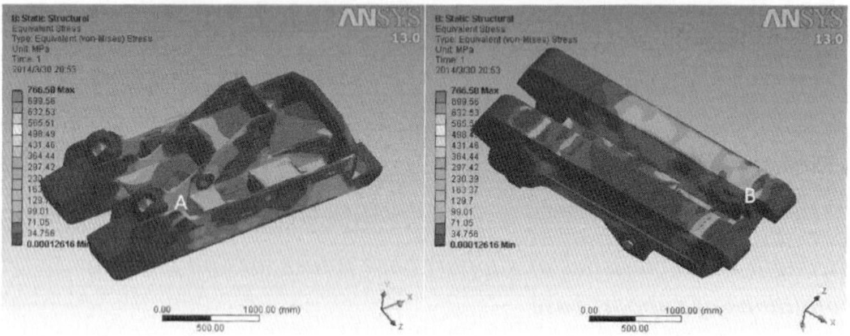

Fig. 9. Stress contours of the support base under torsion loading conditions of support base.

Table 4. The high-stress areas of support base under torsion conditions of support base.

Locations	σ_{eqv}/MPa	σ_s/MPa	$\sigma_{eqv}/\sigma_s \times 100\%$	Dangerous situation
A	766.58	550	139	Yes
B	699.56	550	127	Yes

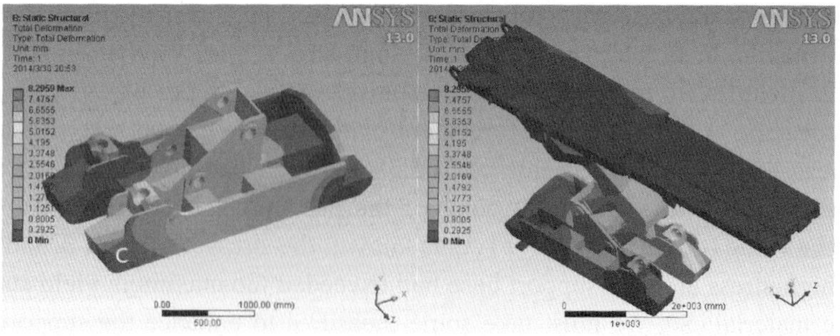

(a) Support base (b) Whole support

Fig. 10. Displacement contours of the support under torsion conditions of support base.

The displacement contours of the support under torsion conditions of support base show in Figure 10. We can see that the support deformation mostly distributed on the linkages and the support base, and the displacement distributed asymmetrical on the support base that was peak at C point of the other end of end base of pad with the maximum value 8.3 mm. According to hydraulic support top-down order (top beam→linkage→base), the displacement of various structure deformation showed a gradual increasing trend and the deformation of the base displacement has a uniform and symmetrical distribution, which basically satisfied with "General specification for powered support MT312-2000".

4.3. *Comparison of Strength characteristics of support base in five working conditions*

Table 5 was the maximum equivalent stress and the corresponding location of hydraulic support base in five working conditions. We defined the two working conditions with the concentrated loads on both ends of the base and torsion loading of support base as Case 1, Case 2. By analyzing and comparing, no matter in what conditions, the reinforcement ribs were relatively dangerous near the rear of base and the hinged hole of linkage, and the greater stress concentration of peak approximately 766.58 MPa had exceeded the maximum yield strength of the material in torsion loading of support base (Case 2). Although the main structural components were welded with plastic material low carbon steel Q550B (carbon content less than 0.25%) which did not affect their overall strength under static load stress concentration, hydraulic support in the working face often in periodic dynamic loads needed to improve their material and structural design.

Table 5. The maximum stress of the hydraulic support base in five working conditions (MPa).

Conditions	Locations	σ_{eqv} /MPa	σ_s /MPa	σ_{eqv}/σ_s ×100%	Dangerous situation
Case 1	The position of contacting with base plate and two pads	575.92	550	104	Yes
Case 2	The reinforcement plates near the hinged hole of rear linkage	766.58	550	139	Yes

5. Conclusion

(1) Pro/E three-dimensional modeling software were used to design the main parts of ZZC8000/20/38 six-pillar hydraulic support and the assembly was finished bottom-up according to the bottom-up assembly design. Then the three-dimensional numerical model of six-pillar hydraulic support in backfilling fully mechanized coal mining was established to conduct a statics simulation on the

whole support under five working conditions based on "MT312-2000 hydraulic support general technical conditions", and the corresponding stress distribution characteristics of support base was analyzed.

(2) By analyzing and comparing on the stress distribution characteristics of support base in two working conditions, we determined the dangerous working conditions and the most dangerous points, which were the ribs of reinforcement near the base and rear linkage under torsion loading conditions of support base (Case 2). So it was necessary for us to make the structural optimization on the support base to ensure the structure safe in this working condition. The stress and deformation of the support in others working conditions were within the allowable safety limits, which satisfied the requirements of "MT312-2000 hydraulic support general technical conditions".

(3) There is no industry standard of filling and coal mining hydraulic support currently, therefore we only utilized the current generic "MT312-2000 hydraulic support general technical conditions" to adopt the method of equivalent treatment to achieve the loading on the support, which can largely reflect the real situations. However, there were slightly disparities between the indication above loading methods and actual working conditions, even couldn't accurately simulate the periodical compression of support and the stress in impact loading conditions. But this is the focus of the author's future researching work.

References

1. Karfakis M G, Bowman C H, Topuz E. Characterization of coal-mine refuse as backfilling material[J]. Geotechnical and Geological Engineering, v 14, n 2, p 129-150, May 1996.
2. Senyur M G. Fabric of coal-mine refuse as backfilling material and its relation to grain-size distribution parameters[J]. Journal of The South African Institute of Mining and Metallurgy, v 98, n 1, p 39-47, Jan-Feb 1998.
3. Quamruzzaman C, Mostofa A K M G, Howladar M, et al. Mine backfill design study for Barapukuria coal mine, Dinajpur, Bangladesh[J]. Journal of Mines, Metals and Fuels, v 57, n 7, p 163-167+173, July 2009.
4. Kovalenko Y F, Sidorin Y V, Ustinov K B. Deformation of a coal seam with a system of isolated gas-filled fissures[J]. Journal of Mining Science, v 48, n 1, p 27-38, January 2012.
5. MIAO Xie-xing, ZHANG Ji-xiong, FENG Mei-mei. Waste-filling in fully-mechanized coal mining and its application[J]. Journal of China University of Mining and Technology, v 18, n 4, p 479-482, December 2008.

6. Zhang Ji-Xiong, Miao Xie-Xing, Guo Guang-Li. Development status of backfilling technology using raw waste in coal mining[J]. Journal of Mining and Safety Engineering, 2009, 26(4):3 95-401.
7. Feng Ju, Zhang Ji-xiong, Huang Yan-li, et al. Waste filling technology under condition of complicated geological condition working face[J]. Source: Procedia Earth and Planetary Science, v 1, n 1, p 1220-1227, September 2009, Proceedings of the International Conference on Mining Science and Technology, ICMST 2009.
8. Miao Xie-Xing, ZhangJi-Xiong, Guo Guang-Li. Study on waste-filling method and technology in fully-mechanized coal mining[J]. Journal of the China Coal Society, v 35, n 1, p 1-6, January 2010 Language: Chinese.
9. Huang Yan-li, Zhang Ji-xiong, Zhang Qiang, et al. Backfilling technology of substituting waste and fly ash for coal underground in China coal mining area[J]. Environmental Engineering and Management Journal, 2011, 10(6): 769-775.
10. Zhang Qiang, Zhang Ji-xiong, Wu Xiao-gang, et al. Roof gap rationality research of backfilling-coal mining hydraulic support[J]. Journal of the China Coal Society, v 38, n 8, p 1325-1330, August 2013 Language: Chinese.
11. Miao Xie-xing, Zhang Ji-xiong. Key technologies of integration of coal mining-gangue washing-backfilling and coal mining[J]. Journal of the China Coal Society, v 39, n 8, p 1424-1433, August 1, 2014 Language: Chinese.
12. MIAO Xie-xing. Progress of fully mechanized mining with solid backfilling technology[J]. Journal of the China Coal Society, v 37, n 8, p 1247-1255, August 2012 Language: Chinese.
13. Xu Jun-ming, Tan Fu-qing, Ju Feng, et al. Principle and structure of backfilling supports with six props[J]. China Mining Magazine, 2011, 20(4): 101-104.
14. Miao, Xie-Xing. Principle of underground pressure control in fully-mechanized coal mining with solid filling and force analysis of mining support[J]. Journal of China University of Mining and Technology, v 39, n 6, p 795-801, November 2010 Language: Chinese.
15. Zhou Yue-jin, Zhang Ji-xiong, Nie Shou-jiang, et al. Mechanical analysis and kinematic simulation of a hydraulic support used in backfilling and the coal mining face[J]. Journal of China University of Mining and Technology, v 41, n 3, p 366-370, May 2012 Language: Chinese.

Reasonable thickness of metro station with single arch large-span in upper-soft and lower-hard stratum

Zi-Guang Zhang, Ke-Guo Sun , Wen-Ge Qiu
Key Laboratory of Transportation Tunnel Engineering, Ministry of Education, School of Civil Engineering, Southwest Jiaotong University,
Chengdu 610031,China
E-mail:shuibingcx@sina.com

Based on the Junfeng Road station of Qingdao metro engineering, this paper studies the reasonable thickness of overburden rock stratum above the mining tunnel of single arch long-span station by theoretical analysis, numerical calculation and feedback analysis on field monitoring. Results show that stress release and transfer caused by tunnel excavation in upper-soft and lower-hard stratum distribute mainly in the surrounding rock stratum and have less impact to soft stratum. With the same thickness of overburden, the safety margin drops with the decreasing of surrounding rock strength and increasing of thickness of overburden rock; the thickness of overburden rock in Junfeng Road station tunnel is reasonable, but the excavation method of upper, middle and lower bench which contains 9 excavation parts is conservative, and the excavated parts and support measurements should be optimized to realize informationalized design and construction.

Keywords: Mining Tunnel; Upper-Soft and Lower-Hard Stratum; Single Arch Long-Span Station; Thickness of Overburden Rock Stratum.

1. Introduction

Due to the advantages of good integrality, clear effect of stress, open space, relatively low price, good ventilation, benefit to structure waterproofing and so on, single arch long-span metro station is widely adopted in the cities of Yokohama, Paris, St Peterburg, Milan, Qingdao, Dalian, and Chongqing [1]~[4]. In the upper-soft and lower-hard stratum, the shallow buried tunnels constructed in the hard stratum makes important theoretical and practical significance on the stability of surrounding rock, its bearing capacity, choosing tunneling method, reducing the support measures and construction risk, saving construction cost. However, domestically and internationally, the understanding and standard have not been unified about the thickness of overburden rock.

Zheng Yingren [2]and Zhang Xianfeng [5] pointed out that the new rock stratum which is 5~7 m above the vault of long span tunnel can bear itself completely. Recently, there is no specific theoretical formula illustrating the thickness of overburden rock. Many attentions have been paid to the foundation of cavern [6], submarine tunnel [7] and mined out area [8], but researches for the thickness of overburden rock above the metro tunnel have been rarely reported.Therefore, this paper makes deep analysis on the reasonable thickness of overburden rock above the undercut tunnel of single arch long-span station by the means of theoretical analysis, numerical calculation and field monitoring based on the Junfeng Road station. Conclusions are of important reference value and guide meanings for the similar engineering.

2. Project Overviewing

The length of Junfeng Road station is 179.50 m, the excavation width and height are 20.79 m and 18.37 m. The main structure of underground two-story mining metro station is in horseshoe-shape with single arch long-span and the method of excavation is lateral straight wall Crossed (CRD) method that upper, middle and lower benches which contains 9 excavation parts [1], The depth is 7.38~15.63 m and the thickness of overburden rock is 4.78~9.78 m, as shown in Fig. 1.

Junfeng Road metro station is located at the junction of Jingkou Road and Junfeng Road, the direction of station is from northwest to southeast along the Jingkou Road. The traffic on this road is comparatively heavy. The nearest building from the main structure of station is about 6.1 meters. The construction terrain undulates relatively, and the biggest elevation difference is 8.25 meters. Landform is erosion monadnock and accumulation slope, the artificial plain pile on the surface of site which is bioclastic gravel layer of quaternary system is between 1.90 and 2.60 meters thick. Coarse grain granite of late Yanshan stage underlies the plain fill, with vein rocks such as lamprophyre and granite porphyry distributed in it locally. The thickness of strongly weathered zone is 0.4 to 4.10 meters, rock is crushing and soft, the measured standard penetration test (SPT) is 59~100, and the basic quality grade of rock mass is V; the moderate weathered zone is 0.2~16.7 meters thick, the average value of uniaxial saturated compressive strength is 24.0MPa, the index of rock mass integrity K_V is about 0.3~0.5, rock mass is relatively crushing and soft in this zone, and the basic quality grade of rock mass is III~IV. The thickness of slightly weathered zone is between 0.40 and 38.40 m, the value of uniaxial saturated compressive strength is 78.8~105.6MPa, the index of rock integrity K_V is over 0.6, the rock of this zone is between relatively broken and completer hard rock, and the basic quality grade of rock mass is II~III.

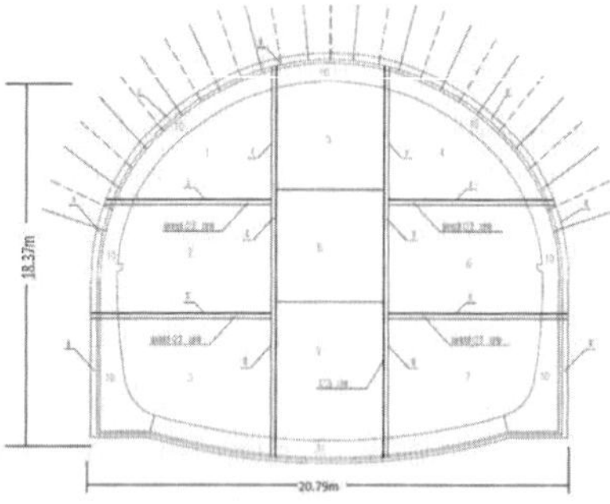

Fig. 1. The schedule of lateral straight wall CRD in the main tunnel.

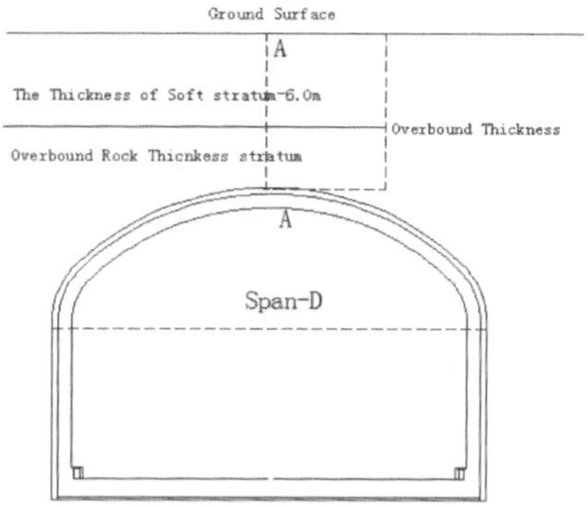

Fig. 2. The analyzed model.

The calculation parameters have been showed in Table 1 according to the statistical results for physical and mechanical properties of the stratums along the metro, related research results [9] and the current code.

Table 1. Mechanical parameters of the rock and soil.

Stratum condition	Elastic modulus E (Gpa)	Poisson ratio u	Bulk density γ (KN/m^3)	Cohesion c (Mpa)	Internal friction angle φ (°)
Strongly weathered zone	0.05	0.28	22.5	0.042	31
Moderate weathered zone	5.0	0.25	24.5	0.6	35
Slightly weathered zone	22.0	0.22	26.0	1.6	43

3. Theoretical Analysis

According to the principle of stress arch of surrounding rock subjecting to the theory of relaxation load, tunnel excavation will not have influence on ground surface when the depth of tunnel is 2.0~2.5 times of the height of vault crown collapse, and the thickness of overburden is considered as the reasonable thickness which meets with engineering safety and economy [5]. The results of researches have shown that empirical formulas on thickness of overburden rock above the shallow-depth tunnel in upper-soft and lower-hard stratum from the Code for Design on Tunnel of Railway and Protodyakonov's theory are helpful and instructive [10], shown as in the formula (1) and (2).

$$H=(2\text{-}2.5)\times 0.45\times 2^{S-1}\left[1+i(D-5)\right] \tag{1}$$

$$H = (2\text{-}2.5)\frac{\frac{D}{2} + \mathrm{h}tg\left(45^\circ - \frac{\varphi}{2}\right)}{f_m} \tag{2}$$

Where H is the thickness of overburden; D is the span of gross tunnel; s is surrounding rock grade; i is factor of tunnel span, and i is 0.1 and 0.2 separately when D is more than 5m and less or equal to 5m; h is tunnel height; φ is internal friction angle and f_m is protodyakonov coefficient.

According to the sizes and mechanical parameters of Junfeng Road metro station, the grade of surrounding rock in moderate weathered zone is III and f_m is 4, meanwhile in slightly weathered zone is II and 6, the thickness of overburden is 2 times of vault crown. The values of reasonable thickness of overburden rock stratum in the underlying moderate and slightly weathered hard stratums obtained by the formula (1) and (2) respectively are (4.68 m, 6.11 m) and (9.36 m, 9.16 m), in accordance with the range of value of engineering. Thus the thickness of overburden designed in Junfeng Road station is reasonable.

4. Analysis on Numerical Calculation Results

4.1. *Analysis model*

In this paper, the layers are assumed as homogeneous continuous elastic-plastic body with equalized strength, qualifying Mohr-Coulomb's criteria. The model is simplified that the overlying stratum is soft and the underlying rock is hard. Land surface and stratums are assumed as horizontally distributed layers. To highlight the upper-soft and lower-hard stratum, referring to actual engineering condition, the calculation parameters of fill of quaternary system and strongly weathered layer are assumed as the ones of strongly weathered soft stratum, and the parameters of hard stratum is respectively the same as the ones of moderate and slightly weathered zone. The generalized model is showed as in Fig. 2.

4.2. *Analysis method and conditions*

The thickness of overburden rock stratum is regarded as plane strain problem and calculated by finite element method, and the DP4 yielding principle is used as the formula in (3). The calculation model ranges of lateral boundaries is confined to over 4 times of tunnel diameter; the left and right boundaries are constrained by horizontal displacement; the lower boundary is immovably confined; the surface of upper boundary is ground surface and free constraint.

Only gravity of rock mass is regarded as initial stress, the load on the ground surface is uniform load with 20 KPa. During the numerical simulation, the excavation of the metro tunnel is one-time and full-face, blasting vibration, support measure while other factors during the construction are not taken into consideration.

$$F = \alpha\ I_1 + \sqrt{J_2}\ \text{-k} \tag{3}$$

Where I_1,J_2 are the first invariant of stress tensor and the second invariant of stress deviator, $\alpha = \dfrac{\sin\varphi}{\sqrt{3\left(3+\sin^2\varphi\right)}}$, and $k = \dfrac{3c\sin\varphi}{\sqrt{3\left(3+\sin^2\varphi\right)}}$.

Based on the characteristics of actual engineering, the following three conditions are calculated numerically in this paper.

Condition 1: the thickness of soft stratum is 6.0 m with underlying moderate weathered stratum.

Condition 2: the thickness of soft stratum is 6.0 m with underlying slightly weathered stratum.

Condition 3: the thickness of soft stratum is 12.0 m with underlying slightly weathered stratum.

4.3. *Analysis on calculation result*

4.3.1. *Analysis on stress characteristics*

The characteristics of A-A stress variation of overburden rock in the range of tunnel vault and ground surface are shown as in Fig. 3. In this graph, compressive stress is positive value and tensile stress is negative. Stress release and transfer of surrounding rock induced by tunnel excavation affect stronger in hard stratum and influence weaker on soft stratum. When stress transfers, the radial stress decreases and tangential stress increases while the position of maximum tangential stress is 1.0 m from the interface of hard and soft stratum. The value of maximum stress increases more and more rapidly with the decrease of thickness of overburden rock and there is tensile stress in the range from 0.5 to 1.0 m in vault wall. Under the consideration of the same thickness of overburden, the maximum value of stress enlarges with the increase of physical and mechanical properties of the rock and the thickness of soft stratum. The safety and stability of single arch long-span and shallow overburden tunnel surrounding rock in upper-soft and lower-hard stratum are affected by bearing rock stratum in vault, and are related to the quality of rock mass and the thickness of soft stratum. The thickness of overburden rock stratum is the core and key factor that determines the safe and stability of surrounding rock.

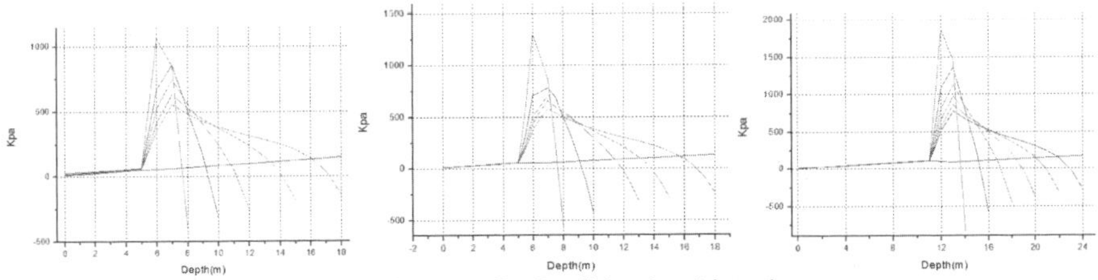

(a) Tangential stress for Condition 1 to 3 in order

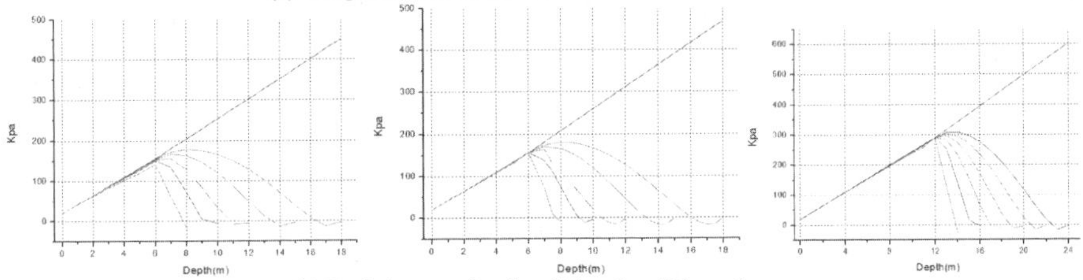

(b) Radial stress for Condition 1 to 3 in order

Fig. 3. The graph of stress distribution.

4.3.2. *The characteristic of surrounding rock stable strength reserve*

In safe coefficient of tunnel surrounding rock, the tunnel is instable when the vertical displacement of tunnel vault increases rapidly [11]. Calculation procedure can be described as following: the stress and strain of initial model is analyzed when tunnel is excavated; c and φ, the factors of rock mass strength are reduced using (4) and then input into the initial model to recalculate the stress and strain. The vertical displacement of tunnel vault is recorded during the calculation and the reduction factor when the displacement changes abruptly is considered as the safe factor of tunnel surrounding rock.

$$c' = \frac{c}{\omega}, \ \varphi' = arc\left(\frac{\tan\varphi}{\omega}\right) \tag{4}$$

In this formula, ω is the reduction factor.

The typical curves of displacement changing with reduction factor in different conditions are demonstrated as in Fig. 4, and the result of safe factor calculation is illustrated as in Table 2.

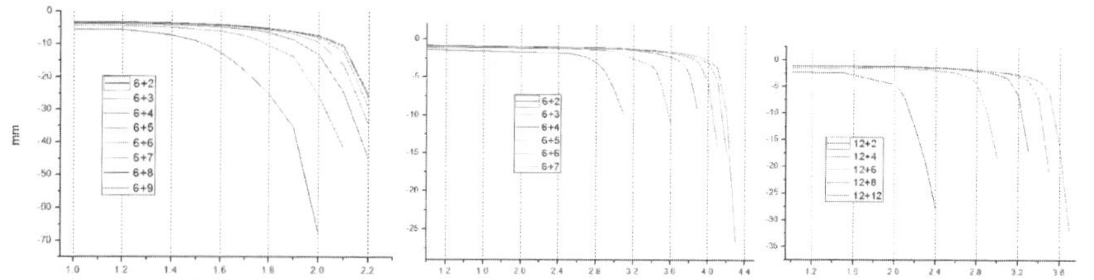

Fig. 4. Curves of displacements vs. strength reduction factor at some key points corresponding to Condition 1 to 3.

Table 2. Safe factor of tunnel surrounding rock.

Overburden Thickness(m)	2	3	4	5	6	7	8	9
Condition 1	1.58	1.78	1.93	1.97	2.02	2.06	2.06	2.07
Condition 2	2.87	3.42	3.75	3.78	3.82	3.95	4.11	4.25
Condition 3	2.03	——	2.80	——	3.04	——	3.32	——

The curve of surrounding rock safe factor changing with overburden thickness is shown as in Fig. 5. It can be conclude that safe factor increases rapidly when overburden thickness is less than 4.0m; the curve changes abruptly when the thickness is about 4.0m; the safe factor increases slowly as the thickness increases from 4.0m; safe factor decreases varying with the rock mass strength decreases and the thickness of soft stratum increases. It is reasonable because the Strength Reduction Finite Element Method is a limited state method. The higher the Shear strength parameters of rock mass c and φ are, the higher the safe factor is; The less the thickness of soft stratum is, the less the additional stress of underlying hard rock mass in the same location around the tunnel is,; the more the Mohr's stress circle close to coordinate origin is and the less the radius is, the higher the safe factor is [12], as in Fig. 6.

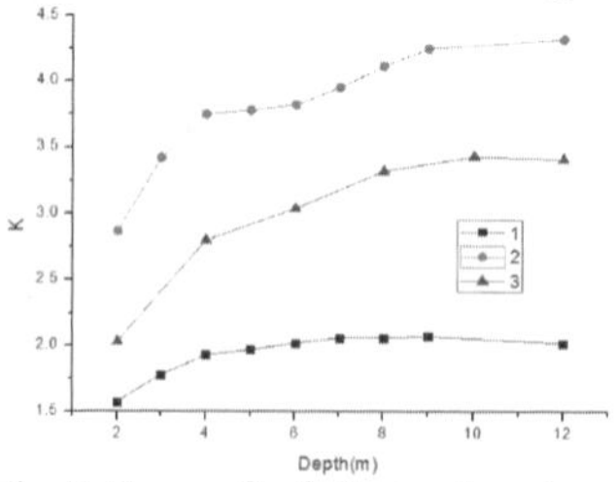

Fig. 5. Curve of safe factor changing with overburden thickness.

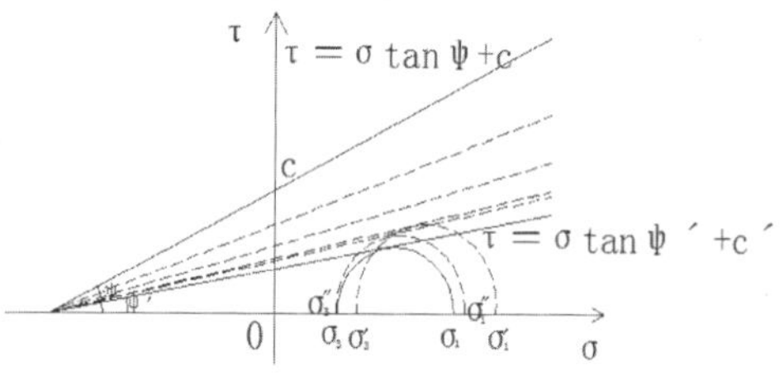

Fig. 6. Safe factor in different stress path.

The reasonable thickness of overburden rock in mined metro station tunnel is considered relatively. Increasing the overburden rock thickness has little influence on stability of tunnel surrounding rock. However, Cosidering convenience for passagers and metro operating costs, thickness should be selected in the range of flat stage after safe factor of tunnel surrounding rock increases rapidly. So the reasonable thickness of overburden rock in Qingdao Junfeng road metro station is in the range from 4.0 to 7.0m.

It is can be seen from calculation results that surrounding rock itself can meet with the stability request with no support during long span tunnel full-face excavation in Upper-soft and Lower-hard Stratum, so the rock bearing ability has high safe reserve and exploit potentiality in Junfeng road station, self-supporting ability of surrounding rock should be fully used and the excavated parts and support measurement should be optimized. Risk management and control measures should be adopted flexibly according to the quality of stratum, and the support can be enhanced in some fracture zones. To ensure engineering safety and achieve informational design and construction, engineering factors should be adjusted corresponding to feedback information from the field monitoring.

5. Analysis of Effects in Engineering Practice

Junfeng Road station tunnel has been constructed completely and delivered. The surrounding rock and environment of tunnel are monitored and measured completely during construction, and monitoring items and distribution of measurement position in main structure of station are as shown in Fig. 7.

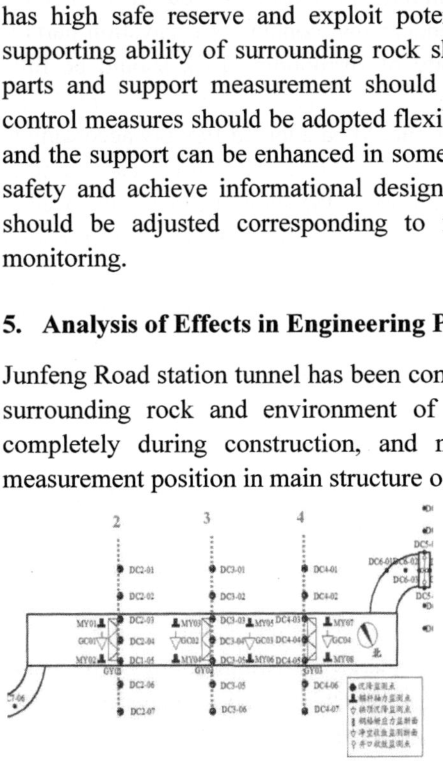

Fig. 7. The distribution of monitor points.

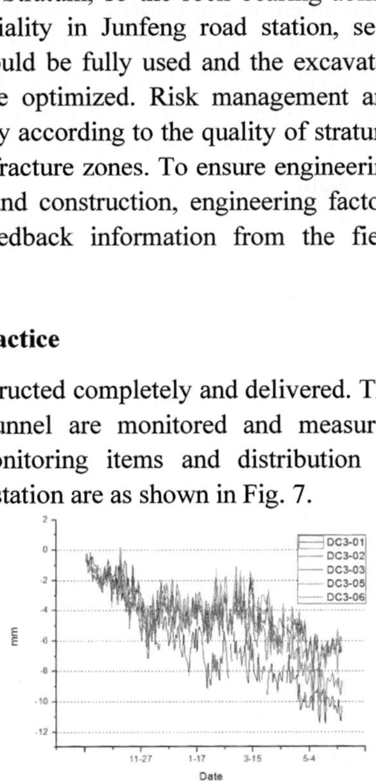

Fig. 8. The time-history curve of ground settlement.

Ground settlement which responses the rule of the whole process of tunnel excavation is analyzed in this paper. The ground settlement time-history curve is wavy because of many-time excavation and engineering blasting, but the curve finally reaches a steady state, as the typical curve shown in Fig. 8. The maximum value of ground settlement in main structure of station is 12.7 mm, and other points are less than 11.0 mm. These values are under the value of deformation allowance (16.0 mm) [13]. It is verified that the thickness of overburden rock of Junfeng Road station tunnel is reasonable.

6. Conclusions

The stress release and transfer caused by tunnel excavation in upper-soft and lower-hard stratum distribute mainly in the hard stratum and influence less on soft stratum. The safety margin decreases with the decreasing of overburden rock strength and increasing of the thickness of soft stratum. The thickness of overburden rock in Junfeng Road station tunnel is reasonable, but the excavation method of upper, middle and lower benches that contains 9 excavation parts is conservative. The self-supporting ability of surrounding rock should be fully used and the excavated parts and support measurement should be optimized. Risk management and control measures should be adopted flexibly according to the quality of stratum. Engineering factors should be adjusted corresponding to feedback information from the field monitoring. the support can be enhanced in fracture zones, to realize the technology of information design and construction.

This paper is mainly based on the stratum parameters of Qingdao area, the conclusions should be a good reference to single arch long-span metro mining tunnel in similar stratums, and its analysis applicability can be extended for other stratums and tunnel cross-section with different values .

References

1. P. Cheng, J. W. Liu, P. Cao. The Optimization Analysis on Construction Scheme of Large Span Metro Station in Hard Rock Stratum. Chinese Journal of Underground Space and Engineering, Vol. 10 (2014), pp. 635-640.
2. Y. R. Zheng, X. Z. Xie, N. J. Zhang. Design and Research on the Xiaoshizi's Station Of Chongqing Underground Light Railway. In tunnel and underground engineering society of China civil engineering society proceedings of the VIII annual session (1994), pp. 239-245.
3. Z. H. Shi, M. Zhang, M. H. Song, et al. Collected Works in Contemporary Civil Engineering and Architectural Science and Technology. (Shaanxi Science and Technology Press, Xi'an, 2006), pp. 216-218.
4. Y. Liang. Study on Setting Principle and Construction Method Selection of Subsurface Excavation Metro Station. Master's thesis. Bijing Jiaotong University (2012), pp. 43-52.
5. X. F. Zhang. Discussion on buried depth of Qingdao metro station by underground tunneling. Journal of Railway Engineering Society, Vol. 10 (1996), pp. 313-316.

6. Y. X. Zhang, G. L. Wang, J. Y. Hu. Stability Analysis Method and Practice of Rock cavern foundation.(Science and Technology Press, Beijing, 2005), pp. 44-47.
7. S. C. Li, B. S. Xu, L. Y. Wei. Channel Tunnel Method for Determining the Minimum Rock Cover Thickness of Drilling and Blasting Method Construction. (Science and Technology Press, Beijing, 2013).
8. M. S. Zhang, W. C. Zhu, Z. S. Hou, X. Q. Guo. Numerical Simulation for Determining the Safe Roof Thickness and Critical Goaf Span. Journal of Mining & Safety Engineering, Vol. 29 (2012), pp. 543-548.
9. Z. Yang, Y.R. Zheng, et al. Analysis on Stability for the Surrounding Rock of Tunnel and Exploring the Strength Parameters. Chinese Journal of Underground Space and Engineering, Vol. 5 (2009), pp. 283-289.
10. P. F. Li, Y. Zhou, et al. Calculation method for surrounding rock pressure and application scopes. China Railway Science, Vol. 34 (2013), pp. 55-60.
11. C. Kong, W. G. Qiu, et al. A Study of the Rock Mass Stability of a Tunnel Group Based on Numerical Limit Analysis. Modern Tunnelling Technology, Vol. 55 (2013), pp. 66-71.
12. Z. R. Zhang, Z. Q. Yang, et al. Study on Point Safety Factors under different stress Paths. Tunnel Construction, Vol. 29 (2009), pp. 626-628;
13. GB50911-2013.Code for Monitoring Measurement of Urban Rail Transit Engineering. (China Building Industry Press, Beijing, 2013), pp. 171-172.

Analysis of the column bearing capacity with lateral load in precipitator casing

Ning Wang, Tong Lu, Kai-Li Xing, Deng-Feng Wang

School of environment and civil engineering, Jiangnan University, Wuxi, Jiangsu, China

E-mail: 1164052835@qq.com

http://www.jiangnan.edu.cn

In consideration of initial geometric imperfections and load disturbance, by using nonlinear finite element method, the deformation of precipitator casing is analyzed, with the load applied in the casing and column . Meanwhile, the stress and deformation of the structure are analyzed.

Keywords: Precipitator Casing; Column; Lateral Load; Load Capacity; Nonlinear Finite Element Method.

1. Introduction

Precipitator is widely used in electric power, metallurgy, chemical industry and building materials industry with important environmental protection equipment. Currently, the research is focus on the collecting technology and the research on the bearing mechanism of the supporting structures is limited.

The most important components in process of precipitator, the retaining structure by - wall the column structure system, structure and load condition are more complicated. This paper studies the shell wall stiffening plate, frame column for H type steel column, two continuous welded connections, forming the whole force. The shell wall panels are mainly subjected to the lateral loads such as negative pressure and wind load, while the shell column is mainly borne by the vertical load of the device, such as the weight of the device and the accumulation of ash and so on.

H steel column is considered as the structure system of the stressed skeleton, while stiffened steel panels plays the role of stressed skin. On the one hand, it shares the load on the column. On the other hand, it provides support for the column to improve its stability. As for the study of the stressed skin effect, based on large number of experiments, Davies and Bryan carried out a theoretical analysis, put forward the calculation method of the shear strength and flexibility of stressed skin diaphragm [1]. It took the flexibility of whole stressed diaphragm as the sum of each component of stressed diaphragm. The strength calculation based on the particular computing hypothesis, the weakest part was determined and the minimum value of the failure load was taken as the critical design load. The research of Wright et al. showed that the connection between diaphragm and the supporting frame was important for the shear performance of the stressed diaphragm [2]. Lucas et al. studied the cold- formed purlin- sheeting calculation model. One was an entire model with purlins and sheeting, the other is a simplified model with only purlins. In the simplified model, the shear and rotational stiffness of the sheeting was simulated by using spring, which connected the edge of the purlins and the sheeting. By setting the reasonable spring parameters, the responding of two kinds of models and experiments got similar results [3, 4]. Lee et al. investigated the bearing capacity of the composite panels composed of gypsum wallboard and cold-formed steel stud, which were subjected to axial pressure. The connection of the gypsum wallboard and column flange was simulated by spring. The results showed that the gypsum wallboard was prone to bending or flexural-torsional buckling, the deformation was localized. The stud was prone to flexural-torsional buckling and the stud interval had little influence on the bearing capacity [5] .

As the research foundation, this paper mainly examines the bearing performance of shell column when shell wall plate is directly beared lateral load. Under the lateral load, the column is used as the supporting member on both sides of the wall plate, and the support is arranged between the two sides. In fact, the column is multi span continuous elastic boundary on two sides of the wall plate. Considering the structure of the stiffened steel plate and the column of H type steel structure is complex and initial imperfections and nonlinear effects, the bearing capacity is very difficult by the analytic method of direct solution column. Therefore ,we can study on the failure mechanism of shell column by means of nonlinear finite element method applying uniform load to structure damage to the wall.

2. The Establishment of the Finite Element Model of the Precipitator Shell

According to the actual project structure, and to improve the efficiency of modeling and calculation, the calculation model of two span wall panels is established..The total height of the column wall is H, the width is W, the stiffener spacing is S, and wall thickness is T. Along the height direction, column is equipped with equidistant transverse strut to provide constraint on the direction of vertical wall for the column. Support spacing is d. Because the wall bottom and hopper is connected, in order to improve the stiffness of the structure, The bottom of the wall plate is provided with a stiffening beam. In order to consider the diversity of structure size, five groups of model are set up. The structure dimensions are shown in Table 1

Table 1. Structure size.

group	Model1	Model2	Model3	Model4	Model5
Column height /mm	10000	17500	14500	12500	10500
Wall width /mm	5000	3000	4000	5300	4750
Stiffening rib spacing /mm	500	1000	500	500	500
Stiffener spacing of H type steel column section size	300×200×8 ×12	250×200×8× 12	300×200 ×8×12	300×200×8	300× 250×8 ×12
Wall thickness /mm	6	5	6	4	4
Angle size	100×63×6	100×6	100×6	100×6	100×6

In this paper, the finite element program ANSYS is used to analyze things. All structural components are simulated by Shell181 element. Considering the influence of material nonlinearity of steel materials, The ideal elastic plastic model is used in the steel material and Von-Mises criteria is used to determine whether the steel will yield. With Q235 steel, The yield strength fy is 235 MPa, the elastic modulus E is 2.06×105 MPa and Poisson's ratio v is 0.3. While conducting the analysis of ultimate bearing capacity of wall panels, considering the geometric nonlinear effects, the arc length method is used to track the structural response path.

Transverse load on the wall plate transfers to the housing pilla. Because of this, column bears main bending moment and loss of stability will occur. The initial imperfection will have a significant impact on the structural stability, while for the engineering structure, the initial defects are inevitable in the process of manufacturing and installation. Therefore, it is needed to consider the influence of defects. Modeling according to the actual structure, after applying the corresponding constraints, the unit line load is applied to the side of the flange of the side of the side wall of the vertical column. As shown in Figure 2, the disturbing load q 'Value:

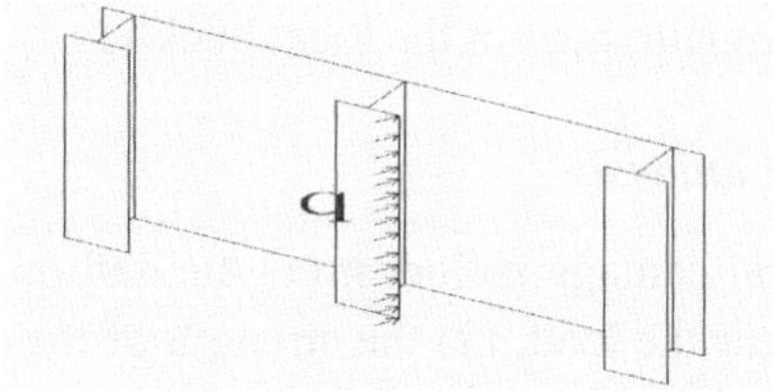

Fig. 1. Application of disturbance load diagram.

3. Nonlinear Analysis of the Shell Under Lateral Load

3.1. *Nonlinear analysis of shell under lateral loads*

Considering that the damage form of the structure is likely to be unstable, according to the modeling method of nonlinear model, the initial geometric imperfections and disturbance loads are applied to the structure, and the five sets of models are obtained.

The stress distribution and deformation distribution of each model are obtained, and the stress magnitude and the Z (direction of vertical wall panel) to the maximum deformation, the maximum deformation and the maximum stress of the structure are shown in Table 2:

Table 2. Structural distortion and stress extreme value.

group	Model1	Model2	Model3	Model4	Model5
Column maximum stress/MPa	235	210.727	112.6	231.56	231.345
Column Z to the deformation extremum/mm	5.83	4.14	2.2	1.94762	1.064
Wall Z to deformation extreme value/mm	60.5	43.16	53.53	112.67	36.2
Maximum structural stress/MPa	235	235	235	235	235

The following conclusions can be drawn from the results of calculation: (1) When the lateral load reaches the extreme value, the deformation of the wall is much larger than that of the column. (2) When the maximum stress reached the yield point, the structural stress reached the yield point, and the yield zone occurred in the cross of the supporting plate and stiffened rib. (3) When the structure is destroyed, the column has little deformation, but the maximum stress of the column is close to or reached the yield strength. (4) The load is directly on the wall, and the column is driven by the wall panels, and the displacement is a change of the whole column, but not the local buckling.

3.2. *Structural failure analysis*

The causes of structural damage includes: (1) the deformation of the wall is too large to continue to bear the load; (2) the strength of the connecting plate of the supporting position of the column is broken; (3) the cross regional strength of the stiffening rib is destroyed.

3.3. *Column internal force analysis*

When the structure is damaged, the maximum internal force of column internal force composition is shown in Table 3. The data in the table can be seen that the effect of axial force cannot be ignored in the column structure. Due to the axial force and side wall column flange of continuous welded connection, wall and column work can be regarded as the extension of wallboard one side of the column flange .So wall can bend moment and shear force together with the column. In terms of column cross section, because the wall plate is shared by the wall plate, stress on one side flange connected with wall plate is low while away from the wall side of the flange force level is relatively high. The column section normal stress value is not zero. There is axial force on the cross section.

While there is no axial force of the shell theory in the traditional answer, which is a great difference between finite element model of the solution and the theoretical solution.

Table 3. Column section internal force finite element solution.

group	Model1	Model2	Model3	Model4	Model5
axial force /N	-65614.99	-107015.2	-36799.81	-186642.1	-123454.3
bending moment /KNm	-79.34749	-60.23344	-37.50958	-33.84479	-44.75717
shear force /N	275979.2	132325.6	67816.5	113155.5	130981.8

The maximum internal force of the column is in the supporting position of the column, and the shear force and bending moment of the column are relatively large.

3.4. *Column stability analysis*

When the transverse load is directly applied on the wall plate utill the structure reach the ultimate bearing capacity, only the maximum stress of the column of the model 1 reaches the yield strength and the remaining models of the column has not yet been destroyed. This shows that when most of the models are destroyed in the wall panels, the column has not yielded ,the stress level is not high and no instability has occurred. In order to further study whether the column in the higher stress level will be unstable, the same model is distributed on the wall of the wall and the load is transformed into the direct acting on the column to conduct nonlinear loading. When the column is damaged, the effect of the load is reached the yield strength, and the maximum stress position occurs at the position of the support.

According to the results of the load on the column, When the load reaches the maximum value, the column has a large plastic development and the largest deformation point is located in the flange of the column across the flange. For the wing edge which is connected with the wall panel, although there is a large deformation in the connection side of the wall plate, even if the load is applied directly to the wall, the structure will not occur instability. Because the deformation is because of the joint and several action of the wall plate, which causes the column to have the whole deformation and the side section is drawn. For the wing edge which is far from the wall, section may occur instability under pressure. But from the point of view of the development of the displacement, the compression flange of the column has not accelerated the development of Z to deformation and no significant decay occurs in the stiffness. Also, the deformation is not large when the load is reached. So the column is not stable.

4. Conclusions

(1) Sructural damage is due to the deformation of the wall is too large or the stiffener and the connecting plate yield caused.

(2) After considering the initial imperfections and disturbance load, the failure mode of the column is the strength failure. The column stress reaches the yield strength, and has a certain bearing capacity, which has a certain ductility.

References

1. Davies J M and Bryan E R, “Manual of stressed skin diaphragm design”, Granada Publishing. London, 1982.

2. Wright Howard D. and Hossain K. M. Anwar, "In-plain shear behavior of profiled steel sheeting", Thin-walled Structures, vol. 29, pp. 79–100, 1997.
3. R. M. Lucas, F. G. A. Al-Bermani, and S. Kitipornchai, "Modelling of cold-formed purlin-sheeting systems-Part 1: Full Model", Thin-Walled Structures, vol. 27, pp. 223–243, 1997.
4. R. M. Lucas, F. G. A. Al-Bermani, and S. Kitipornchai, "Modelling of cold-formed purlin-sheeting systems-Part 2: Simplified Model", Thin-Walled Structures, vol. 27, pp. 263–286, 1997.
5. Youngki Lee and Thomas H. Miller, "Axial strength determination for gypsum-sheathed, cold-formed steel wallboard stud composite panels", Journal of Structural Engineering, ASCE, vol. 127, pp. 608–615, 2001.

Model experiment of displacement response on ice sheet induced by moving air cushion load

Zhi-Yong Ding, Yu-Chen Li*
College of science, Naval University of Engineering,
Wuhan, Hubei, China
**E-mail: gamiliyuchen@163.com*

According to the differential equation of thin elastic vibrating sheet and the similarity theory, the similar relationships between model and prototype are established. A high accuracy non-contact laser displacement measuring system is developed to obtain the surface deformation of the ice-like material. A speed measurement and positioning system is established for the moving air cushion load. At different speed of moving load, a series of experiments are carried out in towing channel which the depth of water is variable. At critical speed the maximum deformation of ice-like material can be caused.

Keywords: Air Cushion Load; Ice Sheet; Model Experiment; Critical Speed.

1. Introduction

The existence of freezing up for three months per year on many regions of Yellow River poses a great threat to those people who live near the river in accordance to the frequent appearance of ice run when the spring comes. To break the ice and clear the river channel in time is an important task. Exploring safe and efficient ways to break the ice has become an urgent problem to be solved immediately.

The deformation of ice sheet when the Air cushion vehicle (ACV) is moving on ice were studied by Kozin V.M.[1,2]. The ice sheet response of a steadily moving point load in three dimensions was calculated by Davys J.W.[3] and Milinazzo F.[4]. Considering the nonlinear effect, the response of ice sheet was obtained by Bonnefoy F.[5] and Parau E.[6] when a load moves on ice close to the critical speed. Experiments of the deformation of real ice sheet caused by moving vehicle were carried out by Takizawa T.[7] and Squire V.A.[8] respectively. Some laboratory tests have been carried out by Whitten and Hinchey[9], involved styrofoam sheet, as an analogue ice.

2. Experiment Principle and Main Parameters

The ice sheet is supported by water in finite uniform depth, the thickness of ice is *h*, the density of ice is ρ_i. The pressure of moving air cushion load on the ice surface is *q*, the load moves along *x*-direction, and its constant speed is *V*. The depth of water is *H*, the density of water is ρ_w. The absolute coordinate system *oxy* is established on the undisturbed water surface. The *z*-direction is vertically upward, the undisturbed water surface is at $z = 0$, the water bottom is at $z= -H$, and the interface of ice and water is at $z = -h'\rho_i$, as shown in Fig. 1.

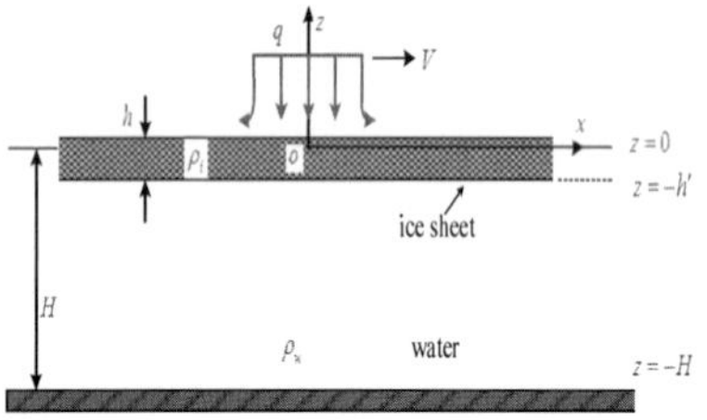

Fig. 1. The coordinate system of air cushion ice sheet and water.

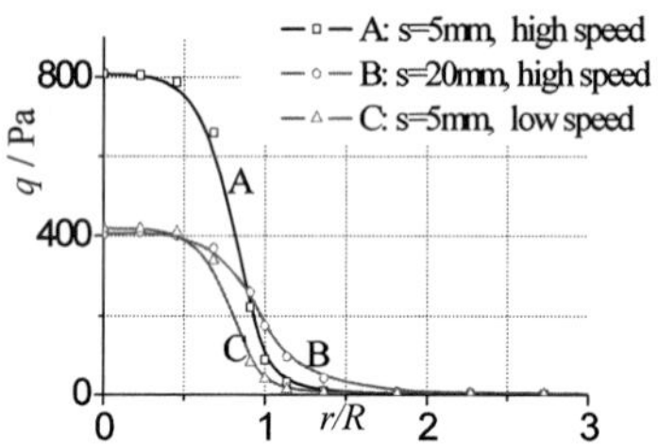

Fig. 2. The pressure distribution along radial direction.

According to the similar relationship between model and prototype, the EPS foam pad with small density is selected as ice-like materials. The strict fulfillment of global similarity is very difficult in model experiment, because of existence of a high requirement to develop the ice-like material for us. Therefore, the selection of the ice-like material can only take the principle of local similarity in the model test.

The relative deformation can be measured by adding load along longitudinal transverse direction, then the elastic modulus of the foam pad can be calculated. The measuring results show that: (1) The thickness of the foam pad h 3.5mm, the density $\rho = 18.35\text{kg/m}^3$. (2) The longitudinal elastic modulus $E = 1.474\times 10^6$ Pa, and the transverse elastic modulus $E = 4.83\times 10^5 \, Pa$. (3) The longitudinal elastic modulus is about three times of transverse elastic modulus, which presents anisotropical characteristics.

Air cushion load is simulated by blower, and the wind speed may be low or high. The vertical distance s from the blower vent to the surface of ice-like material can be adjusted by changing the height of the blower.

Before moving the load, the pressure distribution along the radial direction can be obtained by making the blower continuous work, as shown in Fig. 2, where the horizontal coordinate is the dimensionless distance along radial direction, R is the blower vent radius; the vertical coordinate is the pressure of air cushion load. It indicate that: (1) There is a relatively stable and high pressure area just under the vent. (2) The pressure drops quickly near $r = R$, as shown in curve A, C. (3) The variation of cushion pressure is very small to need not be considered when $r \geq 2R$, as shown in curve B. (4) The surface integral under the curve of pressure points corresponds to the impact force of air cushion.

3. Experimental Instrument

Model test is carried out in the towing channel, its length is 11 m, width is 0.59 m. The adjustable water depth and trailer speed is 0~0.6 m and 0~1.0 m/s respectively.

The depth of water is H and the ice-like material lays on the surface of the water. The blower is towed along the centerline of the channel by the trailer system outside. The laser displacement sensor is fixed on the test section of the channel, with the distance y from measuring point to the centerline of the channel. When the blower is power-on, the surface deformation of ice-like material can be measured by laser displacement sensor. The equipment layout in the model test is shown in Fig. 3, The principle of laser displacement sensor is from item 1 to 6 as shown in Fig. 4.

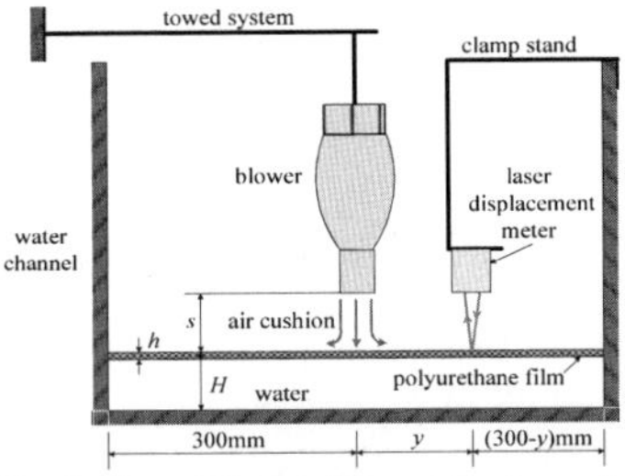

Fig. 3. The equipment layout in channel.

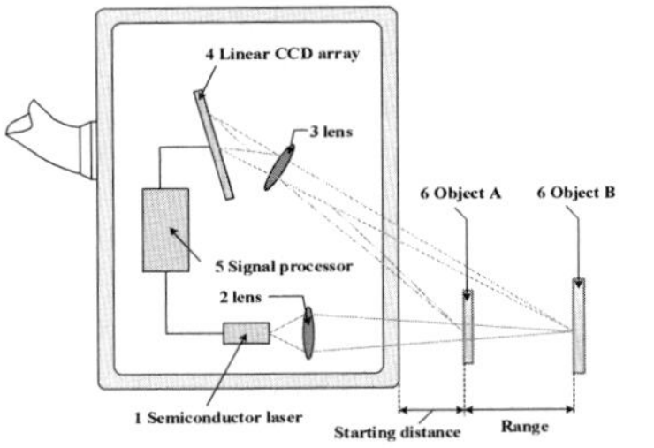

Fig. 4. The principle of laser displacement sensor.

4. Typical Experiment Results and Conclusion

The thickness of foam pad $h = 3.5$mm, the depth of water $H = 52.5$mm, the height of the blower vent $s = 5$mm, the transverse distance to measuring point $y = 60$mm, the distance from the starting position of blower to the measuring point is 7.04 m. To high speed blower, the maximum of wind pressure $p_{\max} = 800\text{Pa}$. Under different speeds of air cushion load, the surface displacement of the ice-like material can be acquired by the displacement measuring system, as shown in Fig. 5 ~ Fig. 7. The horizontal coordinate is time(s), the vertical coordinate is the surface displacement of ice-like material (mm). Where curve A is the displacement, point B is the position of air cushion load, V is the moving speed of air cushion load.

A series of surface displacement curves of ice-like material can be obtained by changing the speed of air cushion load, and the curves of displacement depression peak varying with the speed of air cushion load are drawn, as shown in Fig. 8. Where the horizontal coordinate is the speed of air cushion load (m/s), and the vertical coordinate is the surface displacement depression peak (mm). Curve E and C is corresponding to the air cushion load of high speed and low speed blower respectively.

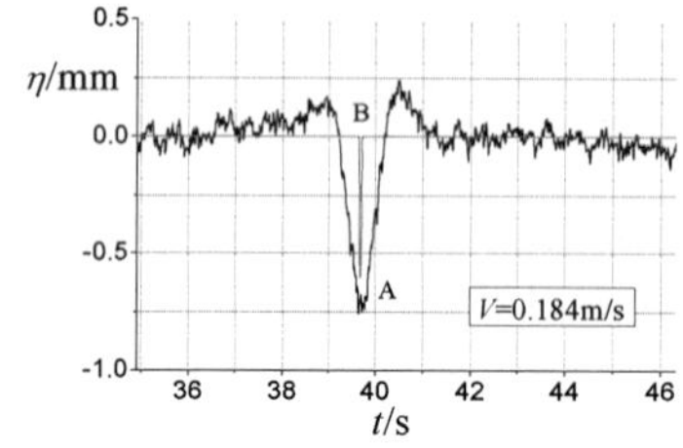

Fig. 5. Displacement curve of ice-like materials when V=0.581m/s.

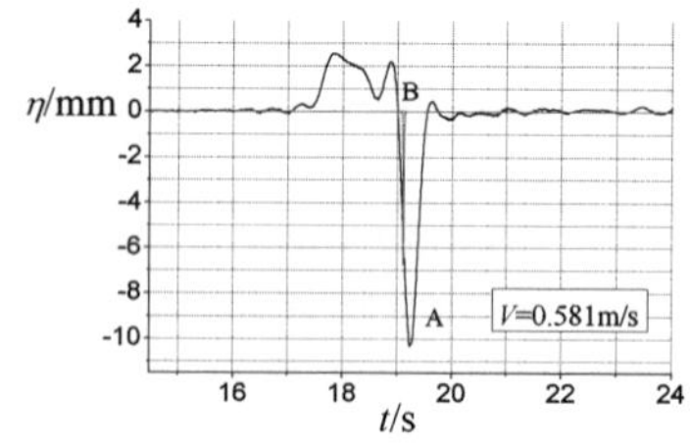

Fig. 6. Displacement curve of ice-like materials when V=0.184m/s.

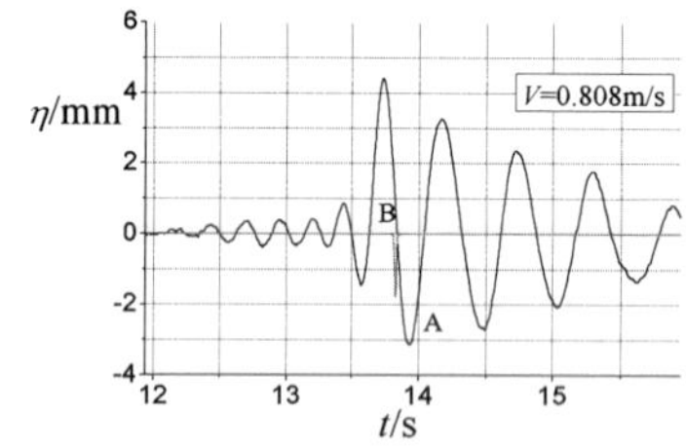

Fig. 7. Displacement curve of ice-like materials when V=0.808m/s.

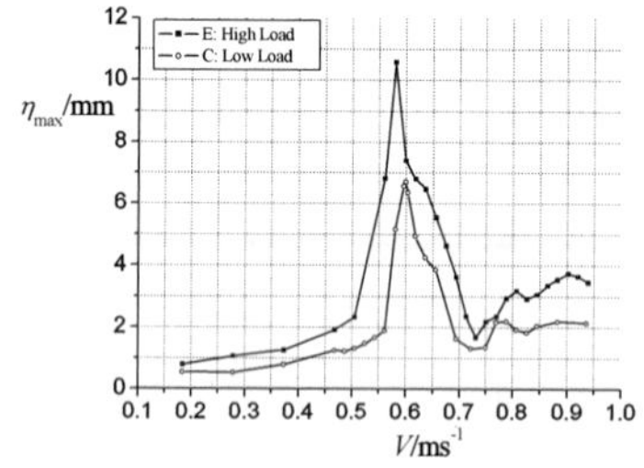

Fig. 8. Depression peak and critical speed influenced by wind speed.

The surface deformation depression peak can reach the maximum at the critical speed. Under above experiment condition, the critical speed is among 0.55 ~ 0.60 m/s, as shown in Fig. 6 and Fig. 8.

At the low speed (V=0.184 m/s), the surface deformation is similar to that caused by static load, the curve is bilateral symmetry, and the position of air cushion load is nearby the surface displacement depression peak, as shown in Fig. 5.

At the critical speed (V=0.581 m/s), there are several upward displacement peaks before the moving load, which is just like the solitary wave in ship waves, and there is a sharp declined surface displacement behind moving load, as shown in Fig. 6.

When the speed of moving load (V=0.808 m/s) is larger than the critical speed, there are a series of increasing displacement peaks before the moving load, which is caused by solid elastic waves with short wavelengths. There are a series of decreasing displacement peaks behind the moving load, which is caused by fluid gravity waves with long wavelengths. Obviously, the surface displacement depression peak at this speed is much smaller than the one at critical speed, as shown in Fig. 7.

If the air cushion load increases, the surface displacement depression peak caused by it will be bigger, but the effect on the critical speed is a little when the air cushion load is changed, as shown in Fig. 8.

5. Conclusion

In this paper, the relationships among the deformation of ice-like material water depth, ice-like material properties, pressure and speed of moving load are studied, the resonance mechanism of ice breaking is revealed, it provides the basis of the ice breaking for ACV.

A critical speed exists in the system of ice-water vibration. When air cushion load moves at critical speed, the energy exerted on ice sheet by air cushion load cannot be emitted away. When ice sheet is exerted by air cushion load continuously, the deformation amplitude of ice sheet will enlarge and the resonance effect of ice-water system will be caused, finally the ice sheet may be broken. It provides the basis of determination the speed of ACV for ice breaking.

The experimental results show that: (1) One or multiple critical speeds exist when air cushion load is moving on the ice–like material, and at these speeds, the surface deformation depression peaks can reach the maximum. (2) At the low speed, the surface deformation caused by moving load is similar to static load. (3) At the critical speed, there is a sharp declined surface displacement behind moving load. (4) When the speed of moving load is larger than the critical speed, the surface displacement depression peak is much smaller than the one at critical speed. (5) If the air cushion load increases, the surface displacement depression peak will increase, but the effect on the critical speeds is a little.

Acknowledgments

This work was supported by the National Natural Science Foundation of China (No. 51479202).

References

1. Kozin V M, Pogorelova A V. Wave resistance of amphibian aircushion vehicles during motion on ice fields. Journal of Applied Mechanical and Technical Physics: 2003, 44(2): 193-197.
2. Zhestkaya V D, Kozin V M. Numerical solution of the problem of the effect of a shock pulse on an ice sheet. Journal of Applied Mechanical and Technical Physics, 2008, 49(2): 285-290.
3. Davys J W, Hosking R J and Sneyd A D. Waves due to a steadily moving source on a floating ice plate. J. Fluid Mech. 1985, 158: 269–287.
4. Milinazzo F, Shinbrot M and Evans N W. A mathematical analysis of the steady response of floating ice to the uniform motion of a rectangular load. J. Fluid Mech. 1995, 287:173–197.
5. Bonnefoy F, Meylan M H and Ferrant P. Nonlinear higher-order spectral solution for a two-dimensional moving load on ice. J. Fluid Mech. 2009, 621: 215–242.
6. Parau E, Dias F. Nonlinear effects in the response of a floating ice plate to a moving load. J. Fluid Mech. 2002, 460: 281–305.
7. Takizawa T. Deflection of a floating sea ice sheet induced by a moving load. Cold Regions Science and Technology, 1988, 11: 171–180.
8. Squire V A, Robinson W H, Langhorne P J, et al. Vehicles and aircraft on floating ice. Nature, 1988, 33: 159–161.
9. Whitten, J. and Hinchey, M.J. Critical speed data for model floating ice roads and runways. Canadian Aemnautics and Space Journal.1988, 34: 151-161.

Kinematic analysis of multi-link mechanical press based on rigid-flexible coupling model

Shu-Sheng Wang, Jing-Lin Yao, Xin-Guo Zhang, Xiang Zheng*

Mechanical Engineering College, Yangzhou University,
Yangzhou, 225127, PR China
R&D department, Magneti Marelli Exhaust Co.,Ltd.,
Hefei, 230601, PR China
The JFMMRI Metalforming Machinery Co., LTD,
Yangzhou, 225127, China
**E-mail:9zhengxiang@163.com*
**Corresponding author*

Set up dynamic rigid-flexible coupling model of multi-link press machine's drive system, through the analysis in finite-element software ANSYS and dynamic analytical software ADAMS, contrast the difference of displacement, velocity and acceleration curve between rigid model and rigid-flexible coupling model, research the influence of transmission components' elastic deformation on the whole press machine. Rigid-flexible coupling analysis, which provide further study method to control precision of multi-link press machine, offer more accurate boundary condition for further stress strain dynamic response analysis.

Keywords: Multi-Link Press; Finite Element Method; Rigid-Flexible Coupling; Dynamic Simulation; ADAMS.

1. Introduction

With the development of the industry and economic, manufacturing industry put forward new requirements to product design and manufacture. Compares to the traditional crank connecting rod press, multi-press can close to the work piece at a lower speed, the pressure and velocity is more uniform, which can increase productivity, reduce noise and prolong the service life of mould.

Multi-rod drive system is one of the key components of multi-link press machine. Usually, we assume the component as a rigid part; in this case, it won't produce deformation. But in some need to consider the deformation of component under special circumstances, we must take some parts of the model into a flexible body to meet the precision requirement. When the component is under the force, we need to study the stress value and its distribution [1], then the components needs to be flexible. This paper combines the finite element analysis software with motion analysis software, analyzed the drive train's rigid motion and the coupled motion. Comparative analysis shows coupled dynamic analysis method which considering the connecting rod's elastic deformation is more close to the actual situation. Through the analysis of the coupled method, provide further design method of the multi-link mechanical press and improve the control precision.

2. Analytical Model of Multi-Link Mechanical Press

In this paper, drive system of press machine is an eight linkage type structure, and in the machine, there are four eight-link structures [2], it can be realized in the form of double acting job. The drive train diagram as shown in Figure 1, adopts eccentric structure instead of crank connecting rod structure.

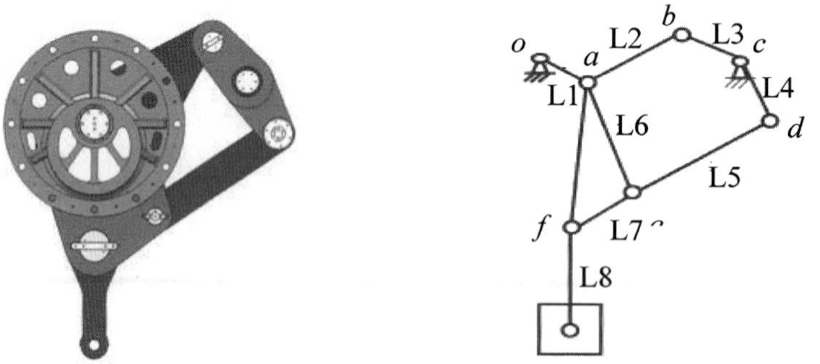

L1-Eccentric; L2-Upper-link; L3-Upper rocker; L4-Lower rocker;
L5-Rocker; L6-Lower link; L7-A tripod; L8-Guide post

Fig. 1. Drawing and diagram of eight connecting rod drive mechanism assembly.

3. The Coupled Modeling Method Based on ADAMS and ANSYS

In the actual movement process, force and moment of transmission bar system are passed through the elastic elements of connecting rod component to the press machine's body, under the action of the force and moment, transmission bar system can produce flexible deformation, the deformation will affect transmission mechanism kinematics and elastic kinematics, also will affect the precision of the stamping and cause vibration, so we need to establish a rigid-flexible couples model [3]. Make dynamic analysis and simulation using flexible multi-link model, the coupled dynamics simulation process of drive mechanism is shown in Figure 2:

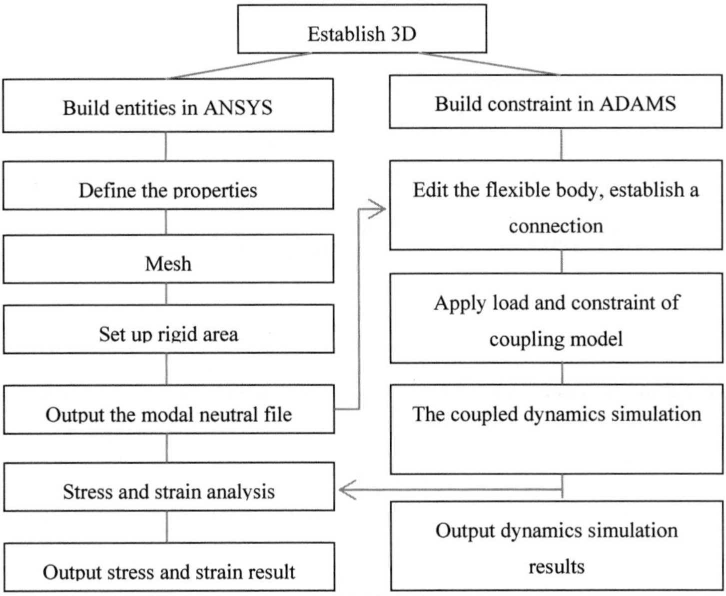

Fig. 2. Joint simulation process of rigid-flexible coupling dynamic system.

Discrete components into a fine grid through finite element analysis software, calculate modal of these components [4], then the calculate modal is saved as a modal neutral file (MNF file) and directly read to establish a flexible body in ADAMS. We cannot directly applied virtual constraint vice or slip plane movement on the flexible body when we are connecting rod component in ADAMS, we have to link object through massless connect parts, and then the constraint imposed on the massless connection objects. The method in this article is to create a node in each axis center of the connecting rod and define material property, build rigid connections which connect the node in the axis center with other nodes around the hole, so the flexible body model is set up [5], it has six order modal. The information of generated modal neutral file contains mass centroid, moment of inertia, frequency and vibration mode. Figure 3 shows the partial rigidity model of upper link, joint point locate in the center of the axis, shaft hole is a rigid area, the purpose of partial rigid is to ensure the movement of flexible part has the same characteristics with the basis movement [6].

According to the finite element method, analyze the modal of components. Based on vibration theory, the low frequency vibration mode plays a major role in the system vibration, the higher order vibration frequency effect smaller, and higher order modes will be fast attenuation under the effect of damping.

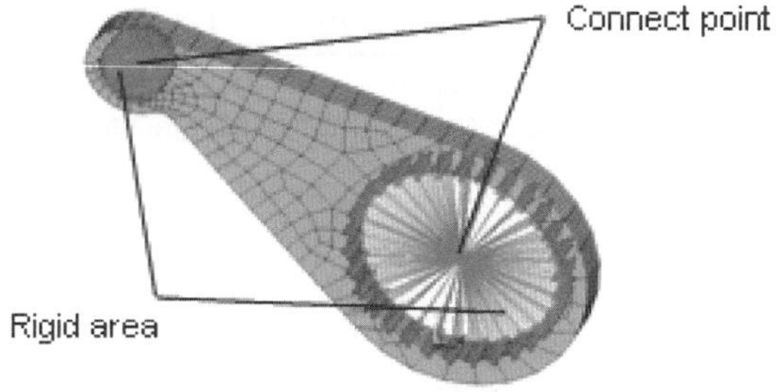

Fig. 3. Partial rigidity model of upper link.

Table 1. 1-6 Order modal of each bar.

	Upper link	Lower link	Guide post
Order 1	0	0	7.778E-006
Order 2	3.92E-004	0	0
Order 3	6.61E-004	4.64E-004	2.37E-004
Order 4	372.51	359.63	291.22
Order 5	407.97	316.94	320.47
Order 6	390.25	322.89	337.69

Import modal neutral file into ADAMS, establish a flexible upper link, lower link, guide bar instead of original rigid model, obtain the coupled dynamics model.

4. Movement Characteristics Analysis of Coupled Multi-Link Structure

First, it is assumed that each transmission component as a rigid member, using principle of rigid-body dynamics, establish multi-rigid-body dynamics model, the displacement [7], velocity and acceleration curve of sliding block is obtained by motion analysis. The results of the analysis are shown in Figure 4.

Same as the front multi-link rigid body dynamics simulation, set up eight connecting rod drive mechanism of the coupled dynamics model, the simulation time is 3.5s (about a cycle). In the simulation, the slider start from bottom dead center position [8], the displacement, velocity and acceleration curve of coupled model as shown in Figure 5.

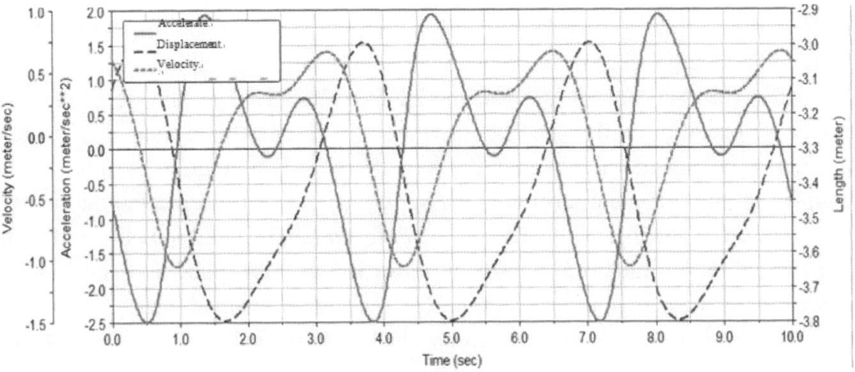

Fig. 4. Displacement, velocity and acceleration curve of rigid model.

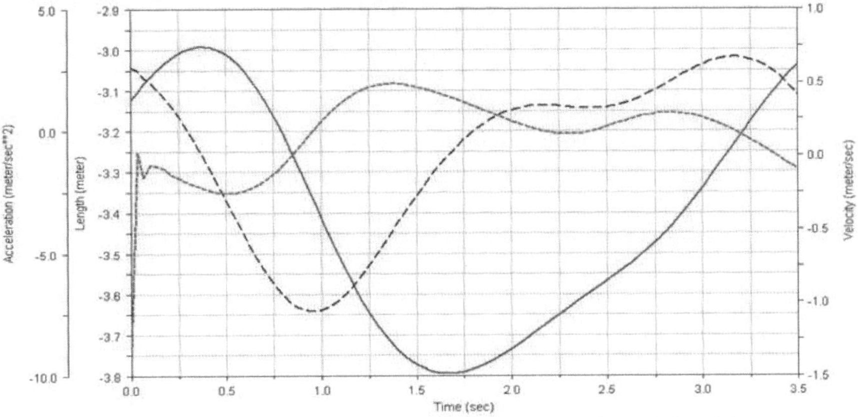

Fig. 5. Displacement, velocity and acceleration curve of coupled model.

Through the contrast of figure4 and figure5 can be seen: the motion characteristic curve of rigid model and the coupled model is basically the same, there is only a tiny difference, by looking in the ADAMS amplifier, slider stroke's displacement of the rigid body model is 798.46 mm, the displacement of coupled model is 800.56 mm, the displacement amplitude in the coupled model increased 2.10 mm. The maximum slider velocity of rigid body model is 0.66 $m{\cdot}s^{-2}$, the maximum speed of coupled model is 0.71 $m{\cdot}s^{-2}$, increased 0.05 $m{\cdot}s^{-2}$. There is a big difference between two acceleration curves, compare figure 4 to figure 5, the general line of acceleration curves are the same, but in the coupled model, the volatility of acceleration is smaller than the rigid one[9], the curve is relatively smooth. But the acceleration mutate at the bottom dead center, this is due to the flexible stamping part release impact load absorbed before. Although the deviation less effect on the slider movement characteristics, but compared with the traditional sense of the multi-rigid-body modeling, coupled analysis approach the real movement of institutions.

After turn the connecting rod model into flexible part, the occurrence of elastic deformation will affect the movement of the machine, thus close to real reaction mechanism of movement. From the above simulation results, in the coupled model, the final motion is the superposition of rigid motion and elastic deformation of flexible body. The elastic deformation of flexible body leading to obvious vibration in startup phase, cause for irregularity speed and acceleration curve of mechanism motion. Summarize the above analysis graph, under the rigid body model, each curve, such as displacement, velocity, acceleration are smooth curves, and presents a cyclical change rule with the change of the crank shaft angle. After the introduction of flexible connecting rod, each curve appeared irregular volatility, especially in the starting stage; a very big change appeared on its peak, but the overall curve is the same with rigid body model one. That means crooked deformation of flexible connecting rod affects the whole body load, make work situation more complicated and more close to the real motion state, the simulation precision is higher, the load data is more reliable, which provide more accurate boundary conditions for the further study of dynamic response analysis.

5. Conclusion

(1)In the former motion analysis of multi-link press drive train, we did not involve the elastic deformation of main components. In fact, high speed and flexibility are the characteristics of the connecting rod, therefore, we can't ignore the influence of elastic deformation, and it will affect the system's dynamic stress and dynamic deformation.

(2)The purpose of rigid-flexible analysis is to get a more accurate load transfer relationship through elastic deformation and coupling between the elastomer.

(3)Replaced the rigid body in dynamics simulation model with flexible body, set up rigid-flex dynamic model, apply the same load and constrain with multi-rigid-body model, research the kinematic and dynamic characteristics through simulation. The results of rigid-flexible model are more close to the real movement.

Acknowledgment

This research was financially supported by the science foundation of Jiangsu province of the People's Republic of China (BY2013063-07).

References

1. X. Gao. Mechanical properties research of six-link press. Machine Design & Research, (04) 2013, 143-146.
2. H.F. Li. Transient analysis of crane boom coupled system. Wuhan university of technology, Huhan, China, 2010.
3. S.S. Yu, S.D. Xu, J.X. Lei. Linkage analysis and the development of virtual prototype technology. Machinery Design & Manufacture, (05) 2004, 118-120.
4. Q.K. Han, Z. Luo. Dynamics analysis control and simulation of multi-body mechanical system, Scientific publishing club, Beijing, 2010.
5. Y. Jiang. Dynamics shafting vibration characteristic analysis of 4105 diesel engine based on the coupled multi-bod. Jangsu University, Zhengjiang, China, 2010.
6. S.J. Shuai. Z.Q. Lv. Dynamic multi-rigid-body system. Beijing institute of technology press, Beijing, China, 1992.
7. Tso P L, Liang K C. A nine-bar linkage for mechanical forming presses [J]. International Journal of Machine Tools & Manufacture, 2002, 42(1): 139-145.
8. Hsieh W H, Tsai C H. On a novel press system with six links for precision deep drawing [J]. Mechanism and Machine Theory, 2011, 46 (02):239-252.
9. X. Zheng, K.R. Fang, C. Wang. Finite element simulation and test for four cylinder diesel engine body. Yangzhou University (Natural Science Edition), (1) 2010,42-46.

An n-th order shear deformation theory for buckling analysis of laminated composite plates

Song Xiang*, Ji Wang, Chun Lu, Wei-Ping Zhao
Liaoning Key Laboratory of General Aviation, Shenyang Aerospace University, Shenyang, 110136, China
*E-mail: *xs74342@sina.com, 458725606@qq.com, 2372178@qq.com, 2535693329@qq.com*
www.sau.edu.cn

In the present paper, an n-th order shear deformation theory is presented to perform the buckling analysis of simply supported laminated composite plates. Critical buckling loads of simply supported laminated composite plates under uniaxial buckling load are calculated by a Navier-type analytical method.

Keywords: N-th Order Shears Deformation Theory; Buckling Analysis; Laminated Composite Plates; Critical Buckling Load.

1. Introduction

The use of composite material for the aerospace and automobile has grown significantly in the past decades, because their superior characteristics can be tailored to meet specific design requirements. With the ever-increasing applications of laminated composites, buckling of such laminates has attracted considerable attention.

Kumar et al. [1] proposed the buckling behaviour of laminated composite skew plates with different boundary conditions subjected to linearly varying in-plane loads. Chen and Qiao [2] presented a semi-analytical solution for the shear buckling of composite laminated plates with all four edges elastically-restrained against rotation. Vescovini and Dozio [3] presented a unified Lévy-type solution procedure for the buckling analysis of both thin and thick composite plates under biaxial loads. Sahoo and Singh [4] proposed a new trigonometric zigzag theory for the analysis of laminated composite and sandwich plates. Huang et al. [5] presented an efficient finite element modelling technique, for the prediction of buckling response of grid stiffened composite panels having different stiffening arrangements. Shiau et al. [6] studied thermal buckling behavior of composite laminated plates by making the use of finite element method. Topal and Uzman [7] dealt with multi-objective optimization of symmetrically angle-ply square laminated plates subjected to biaxial compressive and uniform thermal loads. Barbero et al. [8] presented a mixed, quadrilateral, 3D plate element for static linear and buckling analysis of folded laminated composite structures.

In the present paper, an n-th order shear deformation theory is presented to perform the buckling analysis of simply supported laminated composite plates. Critical buckling loads of simply supported laminated composite plates under uniaxial buckling load are calculated by a Navier-type analytical method.

2. Governing Equations

The displacement field in higher-order shear deformation theories is expressed as follows:

$$
\begin{aligned}
U &= u(x,y) - z\phi_x(x,y) - \frac{1}{n}\left(\frac{2}{h}\right)^{n-1} z^n \left(\phi_x(x,y) + \frac{\partial w(x,y)}{\partial x}\right) \\
V &= v(x,y) - z\phi_y(x,y) - \frac{1}{n}\left(\frac{2}{h}\right)^{n-1} z^n \left(\phi_y(x,y) + \frac{\partial w(x,y)}{\partial y}\right) \\
W &= w(x,y)
\end{aligned}
\tag{1}
$$

Where u, v, w, ϕ_x and ϕ_y are unknown displacement functions; h is plate thickness.

Employing the principle of virtual displacements, we obtain the following Euler–Lagrange equations:

$$
\begin{aligned}
&\frac{\partial N_x}{\partial x} + \frac{\partial N_{xy}}{\partial y} = 0 \\
&\frac{\partial N_y}{\partial y} + \frac{\partial N_{xy}}{\partial x} = 0 \\
&\frac{\partial Q_x}{\partial x} - C_2\frac{\partial R_x}{\partial x} + \frac{\partial Q_y}{\partial y} - C_2\frac{\partial R_y}{\partial y} + C_1\left(\frac{\partial^2 R_x}{\partial x^2} + 2\frac{\partial^2 R_{xy}}{\partial x \partial y} + \frac{\partial^2 R_y}{\partial y^2}\right) \\
&+\bar{N}_{xx}\frac{\partial^2 w}{\partial x^2} + 2\bar{N}_{xy}\frac{\partial^2 w}{\partial x \partial y} + \bar{N}_{yy}\frac{\partial^2 w}{\partial y^2} = 0 \\
&\frac{\partial M_x}{\partial x} - C_1\frac{\partial P_x}{\partial x} + \frac{\partial M_{xy}}{\partial y} - C_1\frac{\partial P_{xy}}{\partial y} - Q_x + C_2 R_x = 0 \\
&\frac{\partial M_{xy}}{\partial x} - C_1\frac{\partial P_{xy}}{\partial x} + \frac{\partial M_y}{\partial y} - C_1\frac{\partial P_y}{\partial y} - Q_y + C_2 R_y = 0
\end{aligned}
\tag{2}
$$

where $C_1 = \frac{1}{n}\left(\frac{2}{h}\right)^{n-1}, C_2 = \left(\frac{2}{h}\right)^{n-1}$, $\bar{N}$ is the load.

$$N_{\alpha\beta} = \int_{-h/2}^{h/2} \sigma_{\alpha\beta} dz, \quad M_{\alpha\beta} = \int_{-h/2}^{h/2} \sigma_{\alpha\beta} z dz, \quad P_{\alpha\beta} = \int_{-h/2}^{h/2} \sigma_{\alpha\beta} z^n dz \tag{3}$$

$$Q_\alpha = \int_{-h/2}^{h/2} \sigma_{\alpha z} dz, \quad R_\alpha = \int_{-h/2}^{h/2} \sigma_{\alpha z} z^{n-1} dz \tag{4}$$

By substituting Eq. (4) into Eq. (3), the governing equations in terms of displacements can be obtained.

3. Solutions Methods

In this paper, the simply supported boundary conditions and governing equations are satisfied by the following displacement functions:

$$u = \sum_{m=1}^{\infty}\sum_{n=1}^{\infty} U_{mn} \cos(\alpha x)\sin(\beta y)$$

$$v = \sum_{m=1}^{\infty}\sum_{n=1}^{\infty} V_{mn} \sin(\alpha x)\cos(\beta y)$$

$$w = \sum_{m=1}^{\infty}\sum_{n=1}^{\infty} W_{mn} \sin(\alpha x)\sin(\beta y) \tag{5}$$

$$\phi_x = \sum_{m=1}^{\infty}\sum_{n=1}^{\infty} X_{mn} \cos(\alpha x)\sin(\beta y)$$

$$\phi_y = \sum_{m=1}^{\infty}\sum_{n=1}^{\infty} Y_{mn} \sin(\alpha x)\cos(\beta y) \tag{6}$$

where $\alpha = \frac{m\pi}{a}$, $\beta = \frac{n\pi}{b}$. Substituting Eq. (5) into the governing equations and collecting the coefficients, the following equation can be obtained:

$$[K - \bar{N}M]\{\Delta\} = \{0\}, \{\Delta\}^T = \{U_{mn}, V_{mn}, W_{mn}, X_{mn}, Y_{mn}\} \tag{7}$$

The critical buckling load $\bar{N}$ can be obtained by solving the eigenvalue equations (7).

4. Numerical Examples

The material properties of each layer are as follows:

$$\frac{E_1}{E_2} = 10, 20, 30, 40; G_{12} = G_{13} = 0.6E_2; G_{23} = 0.5E_2; v_{12} = 0.25; \quad (8)$$

The buckling loads are non-dimensionalized by the following equation:

$$\bar{N} = \frac{\bar{N}_{xx} a^2}{E_2 h^3} \quad (9)$$

Table 1. The critical buckling load of a simply supported laminated composite plates (0°/90°/0°, a/h=10) under uniaxial load.

Theory	E_1/E_2		
	3	10	20
Present n=3	5.3890	9.8320	14.8890
Present n=5	5.3990	9.8770	15.0000
Present n=7	5.4080	9.9180	15.1040
Present n=9	5.4150	9.9470	15.1810

The critical buckling loads of a simply supported square laminated composite plate (0°/90°/0°, a/h=10) under uniaxial load are listed in Table 1. According to the Tables 1, the present results are in good accuracy. it is found that present theory does not require any shear correction factor and allows the user to experiment the best order number to approximate the structural problem under investigation.

Acknowledgment

This work is supported by Liaoning Bai Qian Wan talents Program (No. 2014921048).

References

1. A. Kumar, S.K. Panda, R. Kumar, Buckling behaviour of laminated composite skew plates with various boundary conditions subjected to linearly varying in-plane edge loading, International Journal of Mechanical Sciences, 2015; 100(1): 136-144.
2. Q.Y. Chen, P.Z. Qiao, Shear buckling of rotationally-restrained composite laminated plates, Thin-Walled Structures, 2015; 94:147-154.

3. R. Vescovini, L. Dozio, Exact refined buckling solutions for laminated plates under uniaxial and biaxial loads, Composite Structures, 2015, 127(1): 356-368.
4. R. Sahoo, B.N. Singh, A new trigonometric zigzag theory for buckling and free vibration analysis of laminated composite and sandwich plates, Composite Structures, 2014, 117:316-332.
5. L. Huang, A.H. Sheikh, C.T. Ng, Michael C. Griffith, An efficient finite element model for buckling analysis of grid stiffened laminated composite plates, Composite Structures, 2015, 122 : 41-50.
6. L.C. Shiau, S.Y. Kuo, C.Y. Chen, Thermal buckling behavior of composite laminated plates, Composite Structures, 92 (2010) 508-514.
7. U. Topal, Ü. Uzman, Multiobjective optimization of angle-ply laminated plates for maximum buckling load, Finite Elements in Analysis and Design, 46 (2010) 273-279.
8. E.J. Barbero, A. Madeo, G. Zagari, R. Zinno, G. Zucco, A mixed isostatic 24 dof element for static and buckling analysis of laminated folded plates, Composite Structures, 116 (2014) 223-234.

Post occupancy evaluation theory application in the urban park

Yu Zhang
Art & Design School, Dalian Polytechnic University,
Dalian, Liaoning Province, China
E-mail:zy810112@163.com

The article try to use Post Occupancy Evaluation(POE) research in the field of architectural design theory introduced into urban parks, and discuss the feasibility of its application and the ways to implement in urban parks, thus in this fresh research provided a reference for further research

Keywords: Post Occupancy Evaluation (POE); Urban Park; Possibility Implementation Way.

1. Introduction

Urban Park is a typical type of garden green space, popular throughout the world. In recent years, for the investigation of the urban parks of various types of evaluation has been very common, especially for the park environmental benefit, economic value, evaluation of scenic resources are numerous to list, but it is a pity that in the domestic from the point of view of the consumer survey of urban park use only a few evaluation work. This not only caused the actual work of design quality completely dependent on the designer's experience and personal preferences, but also makes the landscape design in colleges and universities education in teaching and learning are affected, so that is difficult to promote the quality of the design and use.

The purpose of this paper is to use Post occupancy evaluation in urban park, analyze the feasibility of its application and implementation of the way, provide reference basis for the further research and specific case application in urban park and urban park system of evaluation to establish, and look forward to the attention of more people to urban park users.

2. Post Occupancy Evaluation Summarize

Post occupancy evaluation is English POST OCCUPANCY EVALUATION (POE) in translation. Although academia in POE's extension is still controversial, but its basic content is clear. That is, from the user's perspective, the study of evaluation system for designed and being used facilities. Summarize its characteristics, we can describe as follows: POE refers to building or outdoor environment space has been built for many years, in a standardized, systematic program, to collect users data for the evaluation of environmental information, through scientific analysis, to understand their evaluation of the target environment; The complete verification design environment in extent to meet the needs of the group using; Through the summary of reliable information, can provide scientific reference for future similar construction, in order to maximize the comprehensive benefit and quality of design [1].

Its research methods mainly through people's perception ability judged environment quality, specific methods used in the research field is very similar with the other environmental psychology methods. The most important method is the questionnaire, and such as investigation, visit, behavior observation, data collection and the results analysis, etc. In a narrow sense, POE can be as a review of design intent, goals, and built environment between the actual performance of gain and loss, In the broad sense, POE can provide information for management of the construction industry, planning and design work, is a tool to improve the decision-making quality.

POE method grow up in the mid - 1960s of UK, it's the mature things has standardized and market-oriented in the west [2-5], but introduced into China only more than 20 years. First the method is just spread and applied in the evaluation of architectural design field. In 1981, Taiwan architect wan wan Li published an article in the magazine, <the building use review> translation, Zimring and Rigenstein co-author. established the foundation of POE professional language. Later since some architects, planners, managers, and other related fields scholar's participation, POE function and attention of academic and social is increasingly attracting attention, Influence gradually comes to the campus outside space, the residential public space environment, commercial street and square, hospital, etc. but overall the domestic research in this field of practice quantity is less, and mostly confined to academic circles, far less than that in abroad. POE application in urban park has not yet been reported [6-7].

3. Equations Analysis on Feasibility of POE Research Theory Introduced to the Urban Park

The main object of service for the urban park, which are the people masses; The main task of the urban park is to provide a well training and sightseeing space for the users. More concerned to the users, understand their needs and interests, to understand their views and opinions to the environment, are the most important guarantee of success in design, that's why we advocated "people-oriented" idea today. POE's thought is a kind of respect for the user's humanistic thought and the scientific spirit of seeking truth from facts, pay attention to investigation and scientific analysis in the design is a good medicine for the treatment of the current design. POE supervision and feedback mechanism will greatly improve the design quality and make it more close to the demand of the society, also can stand the test of the market more. So use POE in urban park, understand the user's real requirement are particularly important for the sustainable development of urban park and optimization design.

In addition, from the perspective of disciplinary development, POE methods introduced to the field of landscape architecture, a reasonable reference for related disciplines theory and research method of our advantage, promote the theoretical research and the development of landscape, it is also the need of the development of landscape, for landscape subject is a synthesize architecture, art, botany and other multi-disciplinary integrated disciplines. Constantly inject fresh for its reasonable research methods and ideas is also an effective way to maintain its academic vitality [8].

4. Implement Way of Poe Research Theory Applied in the Urban Park

In the domestic architecture field, parts of POE theory and practice have been implemented, this parts are completed more by teachers and students who studied architecture in the university, such as Tongji university, Hunan university, Dalian university of technology, some professors and graduate students and so on, some of these ways can also be applied to urban park in POE studies.

4.1. *Basic research*

In terms of basic research, researchers can declare national research funds, research careful and meticulous study abroad such as evaluation index, technology, technical documents and operating procedures, master the details of the technology. At the same time, To understand the social institution of foreign POE institutions, and research POE theory significance from the viewpoint of government administration, design ethics, and theory research, through achievements in scientific research and published articles to introduce POE.

4.2. *Practice research*

In terms of practice research, It is important to make the government understanding the technology of POE, Co-operation with relevant government departments to set up the project, we can set up from the large urban park which needs POE currently, select pilot, employ foreign experts share together when need. In the implementation of POE, finding problem, summarize experience, do practice for the government to set up a special POE organization. After a period of time to practice, we can make up a POE theory and method system which accords with the situation of China.

In addition, we need a broader way. For example, in the basic research, we should strengthen the popularization of this field, such as in some of architecture, agriculture and forestry universities, landscape architecture and garden major, not only offer the courses in planning, but also study special POE courses at the same time, I think it is very important to make the students learn to use POE technology, can improve the quality of the landscape design talent. Besides we can compose the research team, carry out the actual case studies to improve the students understanding of landscape design integrity [9]. College teachers and students also can undertake part of the social cooperation tasks, help some park finish POE work, for the park's reconstruction, revenue of colleges and universities, the research of scholars, the course practice and practical ability of students, the current situation of the development of students for professional and market understanding and grasping is very beneficial. It is not just a way of thinking, but also a method, a kind of choice [3].

At the same time in practice, we can also list the POE requirements in the design terms of the contract, improve the system gradually. Such as a reasonable duty of design office to replenish and modify after the urban park in use and POE research. These can greatly improve the designer's sense of responsibility and the consciousness of "after-sales service", and not just in form. We should also support some private POE service, make it become a kind of new business, they can provide consulting services for design office, also can accumulate a lot of practical experience, affect the design idea and the way, to promote development of landscape industry [10].

5. Conclusions

In a word, POE provides a new idea and approach for the development of urban park. Exploration of this article is only the start for POE in urban park as a fresh research topic, the groundwork for future related research can be used as bedding. But because of the different urban population, economic level, living habits, the different development model, the factors influencing park usage may be different, more comprehensive and perfect analysis and further research also stay after step in future.

References

1. Clare C.M and Carolyn F, *People Places: design guidelines for urban open space*. (John Wiley & Sons, 1998).
2. J. Whyte, D.M. Gann. Closing the loop.between design and use: post-occupancy evaluation. Building Research & Information, 2001, Volume 29, Number 6 / November 1: 456～459.
3. John Danish, John Cray, Kemohan, and Anne Salmond. Post-occupancy in new Zealand, *Design Studies* 3 (April l982):77～83.
4. Christine. Fitzgerald, Bill. Tramposch. Soft open post-occupancy evaluation at Te Papa, the Museum of New Zealand, *Visitor Studies Today,* United States, Summer 1998 Vol.1: Issue 2.
5. Preiser W F E, Rabinowitz H Z, White E T. Post-Occupancy Evaluation. New York: Van Nostrand Reinhold, 1988: 39- 46.
6. White J, Gann DM. Closing the loop between design and use. *Building Research & Information*, 2001, 29(6) : 456-459.
7. Christine F, Tramposch B. Soft opening post- occupancy evaluation at Te Papa, the Museum of New Zealand. Visitor Studies Today, 1998, (2) : 21-29.
8. Xiaolei Zhu and songxian Wu, Impact on the architectural design and meaning to China of POE. *Architectural Journal* 2002(5), pp. 42-44.
9. Leiqing Xu and Gongxia Yang, *Environmental psychology,* (Shang Hai: Tongji University Press, 2002).
10. Lingling Luo and Wei Lu, International trends of POE research and realistic thinking introduced to China. *Architectural Journal* 2004(8), pp. 82-83.

The deflection formula of 2N + 1 variable cross section symmetric simply supported beam

Li-Qun Zhang*, Li-Jiao Zhang, Zhen-Hai Xu, Dian Gao
Civil Department, Hubei Institute of Architecture and Civil Engineering, Zhangjiakou, Hebeisheng, China
**E-mail:1529019804@qq.com*

The use of variable section cantilever principle of stiffness superposition method and calculation results for solving the deflection of multi-segment (odd segment) becomes symmetrical cross-section simply supported beam when concentrated loads effects at central girder in normal circumstances, to derive the deflection formula of 2N + 1 variable cross section symmetric simply supported beam when concentrated loads effects at central girder in normal circumstances

Keywords: 2N + 1 Variable Cross Section Symmetric Simply Supported Beam; Concentrated Load; Deflection Formula.

1. Introduction

Symmetric Tapered simply supported beam is widely used in the life, but difficult to calculate, and for the deflection formulas of 2N + 1 variable cross section symmetric simply supported beam rarely been mentioned. This paper use the deflection formulas and methods of N segments variable section cantilever beam to derive the deflection formula of 2N + 1 variable cross section symmetric simply supported beam when concentrated loads effects at central girder in normal circumstances.

2. Formula Derivation

The bending moment diagram of simply supported beam under concentrated load at the central [1]and the bending moment diagram of cantilever effected concentrated load at the free end [1] are triangulars. And the bending moment diagram of cantilever effected concentrated load at the free end is the half of the The bending moment diagram of simply supported beam under concentrated load at the central from the shape to look. So using stiffness superposition method [2-3] solve the central deflection equation of simply supported beam.

2.1. *The deflection of three segments of varying cross-sectional symmetry simply supported beam*

For simply supported beam shown in figure 1, its central deflection as follow:

$$\omega = \frac{F^2(L_1 + L_2)}{6EI_1} + \frac{F^2 L_2^3}{6EI_1}\left(\frac{I_1}{I_2} - 1\right)$$

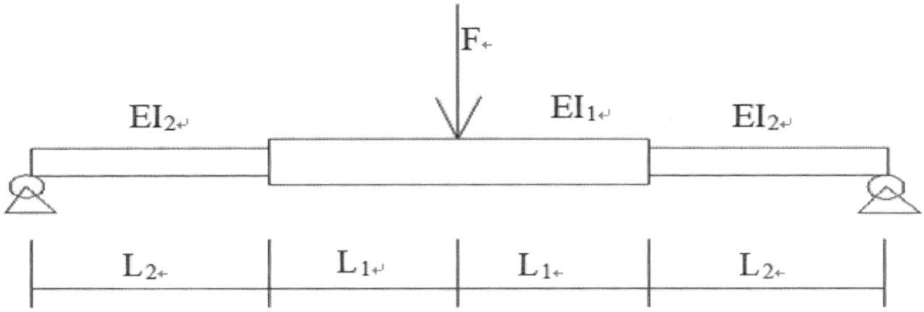

Fig. 1.

2.2. *The deflection of five segments of varying cross-sectional symmetry simply supported beam*

For simply supported beam shown in figure 2, its central deflection as follow:

$$\omega = \frac{F^2(L_1 + L_2 + L_3)^3}{6EI_1} + \frac{F^2(L_2 + L_3)^3}{6EI_1}\left(\frac{I_1}{I_2} - 1\right)$$

$$+ \frac{F^2 {L_3}^3}{6EI_1}\left(\frac{I_1}{I_3} - 1\right)$$

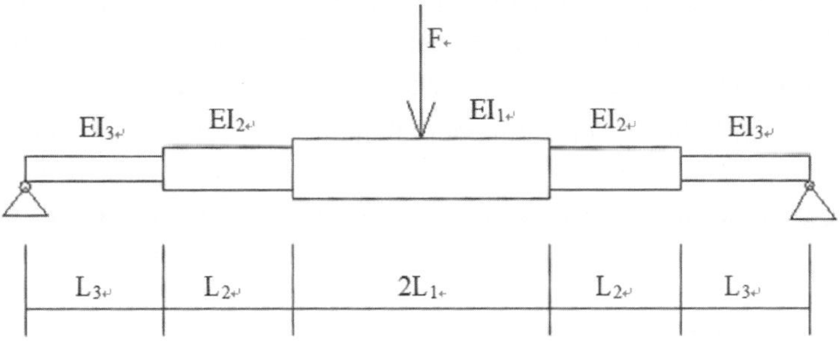

Fig. 2.

2.3. *The deflection of 2N+1 segments of varying cross-sectional symmetry simply supported beam*

For simply supported beam shown in figure 3, its central deflection as follow:

$$\omega = \frac{F^2(L_1+L_2+\cdots+L_n)^3}{6EI_1} + \frac{F^2(L_2+\cdots+L_n)^3}{6EI_1}\left(\frac{I_1}{I_2}-1\right) + \cdots + \frac{F^2L_n}{6EI_1}\left(\frac{I_1}{I_n}-1\right)$$

Summarized by the above-mentioned: The deflection formula of 2N + 1 variable cross section symmetric simply supported beam as follow:

$$\omega = \frac{F^2}{6EI_1}\left(\sum_{m=1}^{n} L_m\right)^3 + \frac{F^2}{6EI_1}\sum_{m=2}^{n}\left[\left(\sum_{i=m}^{n} L_i\right)\left(\frac{I_1}{I_m}-1\right)\right]$$

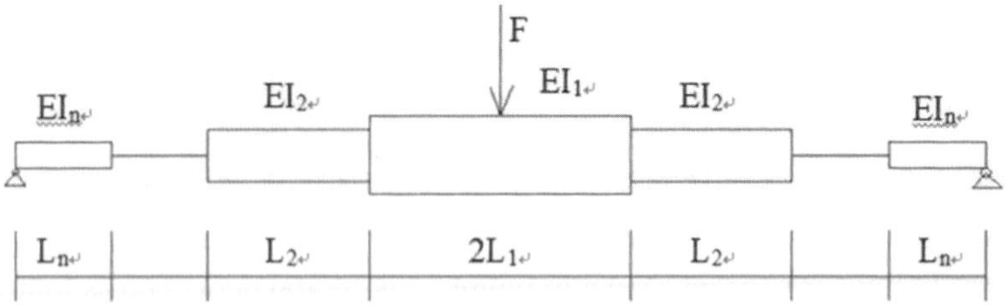

Fig. 3.

3. Finite Element Contrast

The results of above-mentioned research have a certain multiple relationships with the results of literature [2], namely when the variable cross-section case of the half of Symmetric simply supported beam structure is consistent with the cantilever of literature [2]. The span deflection ω of simply supported beam span of applied load P at central and the beam end deflection ω′ of cantilever under load P at free end have a certain relationship as follow: $\omega = \frac{1}{2}\omega'$

Establishing finite element model of simply supported beam and cantilever with ABAQUS shown in figure 4, observe the relationship where load P. ①: Sectional dimension of 3 sections symmetric simply supported beam L_1 segment is 0.4m*0.4m; cross-sectional dimension of L_2 Segment is 0.4m*0.2m; $L_1 = L_2 = 1$m. ②: Sectional dimension of 5 sections symmetric simply supported beam L_1 segment is 0.4m*0.4m; cross-sectional dimension of L_2 Segment is0.4m*0.3m; cross-sectional dimension of L_3 Segment is 0.4m*0.2m; $L_1 = L_2 = L_3 = 1$ m. ③: Sectional dimension of 7 sections symmetric simply supported beam L_1 segment is 0.4m*0.4m; cross-sectional dimension of L_2 Segment is0.4m*0.3m; cross-sectional dimension of L_3 Segment is0.4m*0.2m; cross-sectional dimension of L_4 Segment is 0.4m*0.1m; $L_1 = L_2 = L_3 = L_4 = 1$ m. Elastic modulus $E = 2.1 \times 10^{11}$Pa. cantilever structure is the half supported beam . Structures were applied loadP=50000N. Deflection Table 1 as follow.

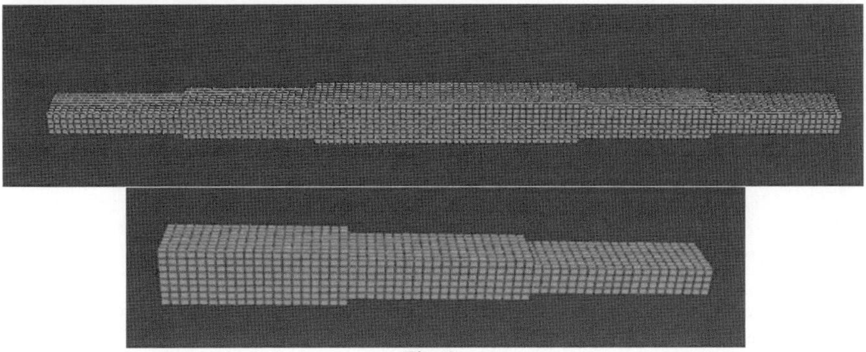

Fig. 4.

Table 1. In each case the deflection (m) span simply supported beam and the free end of the cantilever.

No.	Simply supported beam(m)	cantilever (m)
①	-0.000315919	-0.0063739
②	-0.000872122	-0.00174755
③	-0.00445428	-0.00894074

4. Conclusion

In this paper,it gives the midspan deflection calculation methods and formulas of variable section simply supported beam, obtaining the relationship between the simply supported beam and cantilever : $\omega = \frac{1}{2}\omega'$.

It can simply calculate the span deflection of tapered simply supported beam in bridge of simply supported beam, eliminating many tedious process.

References

1. Zhang dalun. Mechanics of Materials, volume One[M]. Tongji University Press, 1987.
2. Yang baoqing, Zhang liming, Wang xiaokun. The deflection formula of N segment variable cross-section cantilever beam[J]. CONSTRUCTION of ANHUI, 2012, 6(188).
3. Yuan xuezhong, Liu jiemin. The stiffness superposition of variable section cantilever[J]. 2011(2).

Congruence neural networks is used to diagnose fault

Fang Fang, Ji-Ren Xu, Huai-Hui Gao
Mechanical and electrical engineering college, University of Electronic Science and Technology,
Chengdu, Sichuan 610054, China
E-mail:jiren639@163.com
www.uestc.edu.cn

Replace power function with generalized congruence neural network and put forward a modified model and algorithm of generalized congruence neural network. Analyze structure, power function, algorithm of adjusting power and compare it with BP neural network. Comparing it with BP neural network in approaching sine function, it indicate the modified generalized congruence neural network has fast constringency speed and good stability which BP neural network has. The model and algorithm advance constringency speed and diagnosis precision.

Keywords: Congruence Neural Network; BP Neural Network; Rotary Machine; Fault Diagnosis.

1. Introduction

The traditional BP neural network adopts exponential Sigmoid incentive function, so its convergence speed is slow, and learning time is long [1]. Reference[1-3] puts forward a kind of generalized congruence neural network in 1998, and by using generalized congruence function instead of traditional neural network of incentive function, the purpose is to accelerate the convergence rate of the neural network. This paper calls it the first kind of generalized congruence neural network (abbreviated into GCNN1). But congruence function of GCNN1 is saw tooth function, and it has significant abrupt change, and once neural network node input into abrupt point, it can have obvious oscillations, because this kind of oscillation will happen periodically, eventually it leads to network misconvergence, and GCNN1 algorithm of back distribution error to adjust weight, may make variable range of weight unable to be controlled and may also cause network misconvergence. Reference[4] puts forward an modified generalized congruence neural network in 2001, and this paper called it the second kind of generalized congruence neural network (GCNN2). GCNN2 adopts an isosceles triangle function instead of saw tooth function, and it eliminates the abrupt point, and ensures the continuity of congruence change process. The back distribution error to adjust weight process takes modulus-operation, although it controls the variable range of weigh, but failed to change the defects of back distribution error fundamentally, and it also may lead to network oscillation and misconvergence.

It is necessary to research how to overcome the instability of GCNN1 and GCNN2. When the division generalized derivation, the principle of error back propagation, which refers to traditional BP algorithm[5,6] ,are introduced into generalized congruence neural network, and then we get the third kind of generalized congruence neural network (GCNN3)[7], which improves the generalized congruence neural network.

This paper will apply GCNN3 and BP neural network in fault diagnosis, and we can find that the approximation and steady performance of GCNN3 is equivalent to BPNN, and the convergence speed of the former is much faster than the latter.

2. Three Kinds of Generalized Congruence Neural Networks

2.1. *Neural network structure*

From reference [1-4], we can know the hidden nodes and output layer of three-layer GCNN1 and GCNN2 both have modulus, and it is necessary to determine node modulus of output layer which should is more than all desired output of neural network. The network structure is shown as figure 1(a) below. Only hidden nodes have modulus In three-layer GCNN3, and output layer node is linear summation device. When we cannot forecast all expected output of neural networks, linear summation avoids to consider modulus value of output node. The network structure is shown as figure 1(b) below. Only from the network structure, GCNN1 and GCNN2 is the same; and output layer of GCNN3 is different form GCNN1 and GCNN2, and the input and hidden layers are all same; and input and output layer of GCNN3 is same as BPNN, and their hidden layers are different, as shown in figure 1(c) below.

In figure 1,only draw a output node. Among them: a1,a2,...,al and o are input and output of neural network respectively, ω_{ji} and u_j are connection weigh between No.j node of hidden layers and No.i node of input layer, and output node respectively, and x_j and x_o are input value of No.j of hidden layer and output node respectively, and m_j and g_j are the modulus and incentive functions of No.j node of hidden layers of GCNN1 and GCNN2 the modulus and incentive functions; m0 and g0 are the modulus and incentive functions of output layer node of GCNN1 and GCNN2; L (.) is linear summation of output layer node of GCNN3 and BPNN; S (.) is incentive functions of BPNN hidden nodes;i=l,2,…,I;j=l,2,…,J.

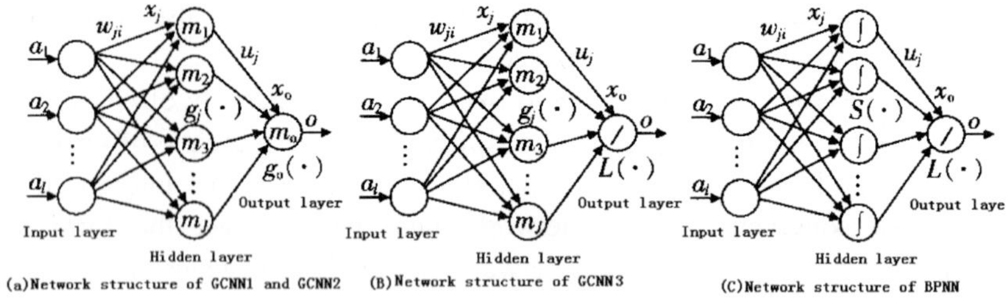

Fig. 1. The network structure of three kinds of GCNN and BPNN

2.2. *Incentive function*

Suppose m is modulus, and takes positive value generally, and X is input of neural network hidden nodes, and G is output of node. Call $n=[x/m]$ the value which is close to 0 after x divided by m, and the symbols of g and n are consistent with x. k≡n mod 2 is the result of integer n mode 2, namely k∈{- 1, 0, 1}. Operation of G mod is superior to addition and subtraction.

From reference [1-3], incentive function of hidden layer and the output layer of GCNN1 is shown in figure 2 (a) , and its expression is:

$$g \equiv x(G \bmod m) \tag{1}$$

From reference [4], incentive function of the hidden layers and the output layer of GCNN2 is shown in figure 2 (b), and its expression is:

$$g=\begin{cases} x(G \bmod m) & \text{N is even number} \\ m-x(G \bmod m) & \text{other} \end{cases} \tag{2}$$

Incentive function of output layer of GCNN3 is linear function, g = x, Hidden incentive function is shown in figure 2 (c), its expression is:

$$g \equiv \begin{cases} x(G \bmod m) & k=0 \\ m-x(G \bmod m) & k=1 \\ -m-x(G \bmod m) & k=-1 \end{cases} \tag{3}$$

Incentive function of BPNN output layer is linear function, g = x. Hidden incentive function is shown in figure 2 (c) shows, its expression is:

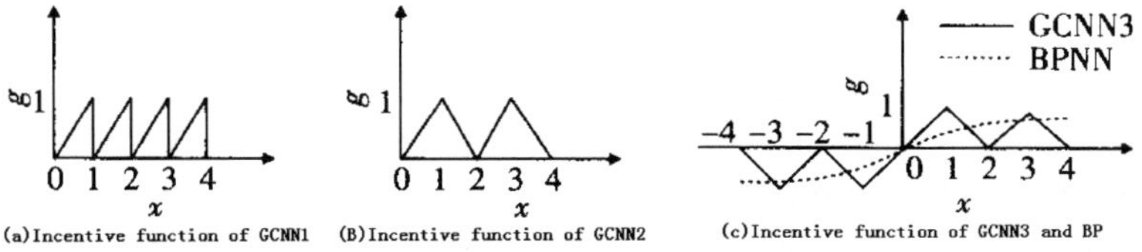

Fig. 2. The incentive function diagram of three kinds of GCNN and BPNN when m takes 1

$$S \equiv 2/(1+e^{-x})-1 \tag{4}$$

In dealing with a complex problem, part input or partial weight of neural network is negative often exists, and thus the input of nodes is negative inevitably. Formula (3) considers various situations of nodes input fully, and it is advantageous for GCNN3 to apply and promote.

2.3. *Algorithm of adjusting power*

GCNN3 adopts error derivation back propagation method which is similar to traditional BP algorithm. The excitation function of traditional BP neural network is continuous and differentiable. From figure 2 (c) we can find obviously, although incentive function of GCNN3 is continuous, not differentiable strictly in the sense of mathematics.

The derivative of formula (3) is :

$$g' = \begin{cases} 0 \\ 1 \\ -1 \end{cases} \tag{5}$$

Incentive function of GCNN3 output layer is linear function, and its derivative is 1 forevermore. Error back propagation can be expressed as follows simply:

A) The error of output layer d = y-o, and all g_j and g_j' is calculated according to type (3) and (5) respectively.

B) The correction value of ω_{ji}, $\Delta u_j = d\beta g_j$, $u_j(t+1) = u_j(t) + \Delta u_j(t)$.

C) The correction of ω_{ji}, $\Delta\omega_{ji} = d\alpha u_j g_j' \alpha_i$, $\omega_{ji}(t+1) = \omega_{ji}(t) + \Delta\omega_{ji}(t)$.

D) Repeat Adjust weigh value for every input data until the error square is less than the given number or come to the maximum value of study times.

If g and g' are replaced by S and its derivative S' of formula (4), and the algorithm is also the error derivation back propagation algorithm of BPNN.

3. Simulation Test [8]

To program GCNN3 and traditional BPNN algorithm to realize the approximation of sine function. Approximate function is $y = 0.2\sin(2\pi x/13)$, $x \in (0.01, 4)$. In the simulation experiments, make all parameters of GCNN3 and traditional BPNN to keep consistent as far as possible, and adjustment rate $\alpha = \beta = 0.05$, and number of nodes of input layer, hidden and output layer of the neural networks is 1,5 and 1 respectively, and Mould m_0 equals to 1.2, m_1 =0.03, m_2 =0.24, m_3 =0.45, m_4 =0.66, m_5 =0.87;weight between hidden and input layers and between hidden and output layer are ω_j and u_j (j = 1, 2, 5) respectively, The initial value of ω_j and u_j take random numbers in (- 0.5, 0.5). In GCNN3 and BPNN program, if all samples have learned 10 times, then stop, and every program run 20 times continuously respectively, and then choose the worst, medium and the best among them, and get approximation curve shown as Fig 3 respectively.

Table 1. The worst approximation weights of hidden layer of GCNN3 and BPNN

Algorithm	W1	W2	W3	W4	W5
GCNN3	-0.30	0.02	-0.25	-0.02	0.25
BPNN	0.31	0.34	-0.24	-0.07	-0.07

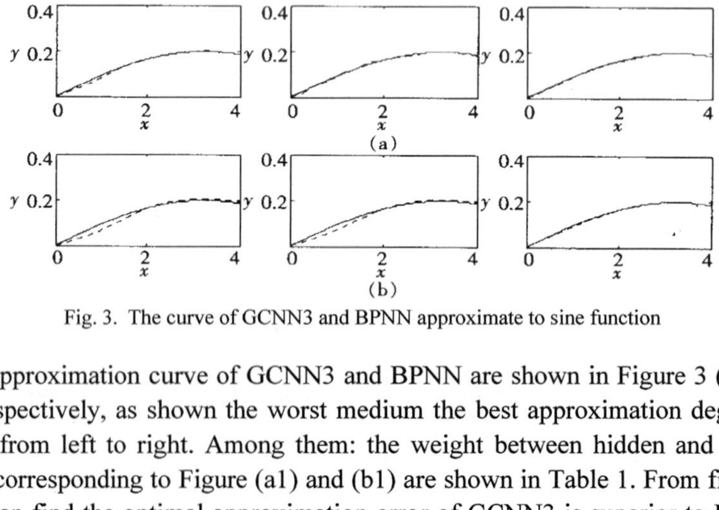

Fig. 3. The curve of GCNN3 and BPNN approximate to sine function

Approximation curve of GCNN3 and BPNN are shown in Figure 3 (a) and (b) respectively, as shown the worst medium the best approximation degree in order from left to right. Among them: the weight between hidden and output layer corresponding to Figure (a1) and (b1) are shown in Table 1. From figure 3 , we can find the optimal approximation error of GCNN3 is superior to BPNN. And GCNN3 is not sensitive to initialized weights and its stability is comparable with BPNN.

In convergence speed, running time of GCNN3 is faster than BPNN.

4. Experimental Results and Discussion [9]

The Standard patterns of common faults of rotating machinery are shown as Table 2. First of all, according to the empirical formula to determine the nodes of of hidden layer of BP neural network are in the scope of 9~27. Take the initial value as 25, and the network initial topology structure of fault diagnosis is 8-25-9 type, and nonlinear incentive function is Sigmoid function. The neural network learning algorithm which takes information entropy as cost function and error back propagation of GCNN3 learn the data of Table 2 respectively, and the calculation results are shown in Table 3. From table 3 we can find, learning speed of GCNN3 is obviously higher than BP neural network speed according to experience formula. Using BP neural network and GCNN3 to diagnose 3 kinds of fault samples on Table 4, from the diagnosis results of Table 5 and Table 6, we can find, the diagnostic accuracy of BP neural network is lower than GCNN3, and fuzziness larger relatively.

Table 2. Standard training samples

Fault mode	(0~0.39)f	(0.4~0.49)f	(0.5~0.99)f	1f	2f	3f	4f	≥5f	Output vector
Oil membrane oscillation	0.1	0.5	0.1	0.3	0.0	0.0	0.0	0.0	100000000
Rotation stall	0.2	0.1	0.2	0.3	0.2	0.1	0.1	0.0	010000000
Surge	0.7	0.0	0.0	0.3	0.0	0.0	0.0	0.0	001000000
Rubbing	0.1	0.1	0.1	0.2	0.1	0.1	0.1	0.1	000100000
Supporting loose	0.1	0.2	0.1	0.4	0.1	0.1	0.0	0.0	000010000
Axis crack	0.0	0.0	0.0	0.4	0.2	0.0	0.0	0.2	000001000
Imbalance	0.0	0.1	0.0	0.5	0.1	0.0	0.0	0.0	000000100
The oil film vortex move	0.1	0.4	0.1	0.3	0.1	0.0	0.0	0.0	000000010
Not middle	0.0	0.0	0.0	0.4	0.5	0.1	0.0	0.0	000000001

Note: f is work frequency, Hz

Table 3. Learning result comparison of GCNN3 and BP network

Network structure	BP network	GCNN3
accuracy	10-6	10-6
iteration	712	46
Training time	1′16″51	15″52

Table 4. Three kinds of fault mode

Fault mode	(0.0~0.39)f	(0.40~0.49)f	(0.50~0.99)f	1f	2f	3f	4f	≥5f
1	0.6	0.05	0.05	0.3	0.04	0.04	0.0	0.0
2	0.0	0.0	0.0	0.3	0.6	0.1	0.0	0.0
3	0.15	0.2	0.18	0.65	0.2	0.2	0.0	0.0

Note: f is work frequency,Hz

Table 5. Diagnosis result of BP neural network

Fault mode	Oil membrane oscillation	Rotation stall	Surge	Rubbing	Supporting loose	Axis crack	Imbalance	The oil film vortex move	Not middle
1	0.00000	0.209931	0.89616	0.00172	0.07036	0.00001	0.14609	0.01598	0.00007
2	0.00001	0.070496	0.00005	0.00062	0.00016	0.00335	0.00003	0.00001	0.87033
3	0.00384	0.002904	0.04663	0.00078	0.72625	0.00934	0.00241	0.498336	0.07942

Table 6. Diagnosis result of GCNN3

Fault mode	Oil membrane oscillation	Rotation stall	Surge	Rubbing	Supporting loose	Axis crack	Imbalance	The oil film vortex move	Not middle
1	0.00000	0.021469	0.97767	0.00046	0.03344	0.00002	0.00001	0.01439	0.00006
2	0.00001	0.09543	0.00001	0.00026	0.00028	0.00088	0.00005	0.00002	0.98343
3	0.00027	0.01138	0.01268	0.00008	0.99062	0.00748	0.47692	0.00361	0.00952

5. Conclusion

The traditional BP neural network adopts exponential Sigmoid incentive function, so slow convergence speed is low, and learning time is long. When the division generalized derivation, the principle of error back propagation, which refers to traditional BP algorithm, are introduced into generalized congruence neural network, and then we get the third kind of generalized congruence neural network (GCNN3).

Compared the generalized congruence neural network with traditional BP neural network to be approximate to sine function, it shows convergence speed of the latest improved generalized congruence neural network is fast, and stability is also the same as the traditional BP neural network. GCNN3 and BP neural network are used in fault diagnosis, it can be found that the approximation and steady performance of GCNN3 is equivalent to BPNN, and the convergence speed of the former is much faster than the latter. The model and the algorithm improves the convergence speed and diagnosis accuracy in the fault diagnosis of rotating machinery.

From the above example, we can find this method is feasible in fault diagnosis field of rotating machinery.

Acknowledgement

Jiren Xu and Huaihui Gao is from Electronic Engineering Institute of Hefei, 230037.

References

1. JIN Fan, “Architectures and algorithms of generalized congruence neural network,” Journal of Southwest Jiaotong University, 1998,6(2):119-125.
2. JIN Fan, “Study on principles and algorithms of generalized congruence neural networks,” // Proc of International Conference on Neural Networks and Brain.1998:441-444.
3. JIN Fan, “Basic principle and method of neural computational intelligence,” Chengdu: Press of southwest jiaotong university,2000:167-173.
4. HU Fei and JIN Fan, “Analysis on the Characteristics of Generalized Congruence Neural Networks with an Improved Algorithm,” Journal of Southwest Jiaotong University,2001,36(2):136-139.
5. SUN Xin, DING Xiang-qian and SHI Shuo, “Application of BP Neural Networks to Fault Diagnosis of Turbocharging System of Diesel Engine,”Application Research of Computers,2006,23(6):202-204.
6. ZHANG Limin, “Artificial neural network model and its application,” Shanghai: Fudan university press ,1993:32-37
7. CHEN Yong,WANG Guo-yin,JIN Fan,etal, “A novel generalized congruence neural networks,” // Proc of the 2nd International Symposium on Neural Networks.2005:455-460.
8. XUE Bin-dang, SUN Hui-guo and XUE Wen-fang etc, “Optimization Algorithm for Hidden Layer Nodes of Fault Diagnosis Neural Network,” Journal of Zhengzhou Univwesity of Technology, 1999,(1):36-38.
9. YAN Tian-yun, ZOU Shu-rong and YU Zhen-xia etc, “Study on comparison for GCNNs and BPNN,” Application Research of Computers,2008,(2):408-410.

Consolidation analysis of a composite foundation in double-layered soil

J. Ju*, X. B. Wang
Department of Construction Management, School of Management, Changshu Institute of Technology,
Changshu, 215500, China.
**E-mail: jujun_cit@163.com*

In this paper, a theoretical study is made of the consolidation of a composite foundation reinforced by piles in double-layered soil. A more reasonable initial condition for this kind of problem is derived from the equilibrium equation and the equal strain assumption. Then, the general analytical solution for excess pore pressure within the surrounding soil is obtained under the influence of several important factors. The general expressions for the whole average degree of consolidation of a composite foundation, both in terms of stress and deformation, are then proposed. With above theory, the special solutions for two particular cases, which consider constant loading and ramp loading, are given in detailed forms. Finally, a parametric study is made. The results show that the reduction in permeability and modulus of layer two, area replacement ratio and modulus of the pile reduce the consolidation rate of a composite foundation. However, the reduction in construction time accelerates the consolidation.

Keywords: One Dimensional Solution; Composite Foundation; Consolidation.

1. Introduction

The composite foundations reinforced by pile columns (e.g. deep mixing, concrete pile) which have great potential benefits not only to improve the bearing capacity and reduce total and differential settlements of superstructure, but also to accelerate the rate of consolidation of soft soils, have been used worldwide for highway, airport and so on. Consequently, studies on prediction of engineering properties of composite foundation still been attractive nowadays to both practitioners and researchers in Asia.

The consolidation theories of composite foundation are developed on the basis of that for sand-drained well foundations (Barron, 1948). Since Yoshikuni (1974) proposed the concept of 'stress concentration' for granular column reinforced foundations, this concept has been taken into account by most of the consolidation theories of composite foundation. In the past decades, various analytical solutions have been developed for this kind of problem (Xie et. al., 2001, Taiebat and Carter, 2001; Cai and Xu, 2002; Wang et al., 2002; Lorenzo and Bergado, 2003; Lei and Liu, 2008; Lu et al., 2010). These theories have made a great contribution to the design and assessment. However, almost all the previous consolidation theories assumed that the pile column is pervious which means that the water within the surrounding soil can flow into the pile. This assumption was not reasonable for cement-mixed columns as it is known that the permeability of cement-mixed columns is much lower than surrounding soil that they can be treated as impervious (Cai and Xu, 2002; Hou and Gong, 2000). Obviously, the piles are impervious if they are made by concrete. On the other hand, there seems to be no analytical solution for consolidation taking the layered characteristics of soil into consideration as due to mathematical difficulty. Thus, available expressions are generally limited to simple cases. To remedy the disadvantages in previous solutions, in this paper a more rigorous theoretical solution is developed to compute the degree of consolidation of composite foundations by considering the pile column of impervious and embedded in a double-layered soil, and the effect of time-dependent external loads. Then, the consolidation behaviour of pile foundation is investigated.

2. Governing Differential Equation

The unit cell of one-dimensional consolidation of composite foundation in double-layered soil is idealized as shown in Fig. 1. In order to obtain a simplified analytical solution, the following assumptions are made:

1) The equal vertical strain assumption is adopted, that is, the pile and the surrounding soil only deform vertically and have an equal strain at any depth
2) The pile is impervious
3) Lateral movement, soil arch effect, and disturbance of the soil induced by construction are not considered
4) The total average stress in a composite foundation is assumed to be constant.
5) At the moment of the load being applied, all the load is carried by the excess pore pressure in the surrounding soil. The initial excess pore pressure (u_{i0}) is equal to initial load (Q_0).
6) The effective stress within the soil is constant with depth
7) Other assumptions are the same as those used in Terzaghi's one dimensional consolidation theory

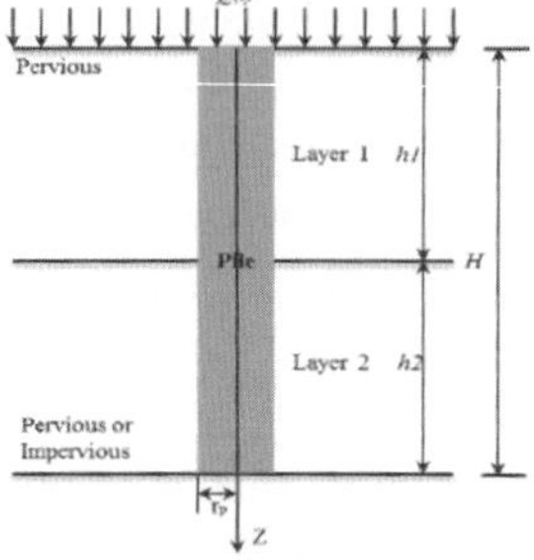

Fig. 1. Schematic diagram for the consolidation curveof composite foundation in double-layered soil.

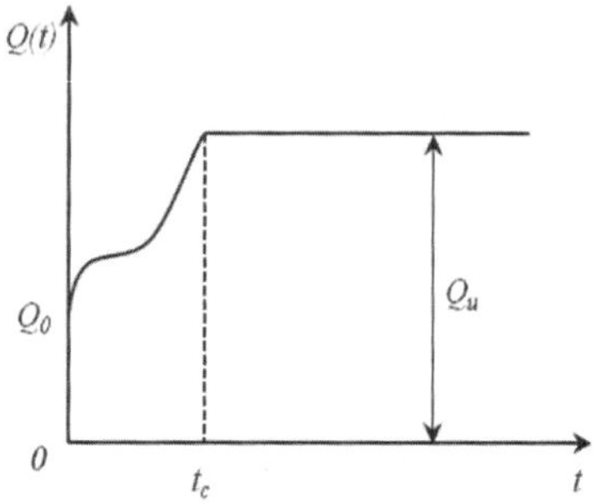

Fig. 2. Loading.

In Fig. 1, h_i is the thickness of the soil layer i (i=1, 2); $H=h_1+h_2$, the total thickness of the two layers. Define the original point of z coordinate on the ground surface. The thickness of the soil layer can be written as: $z_0=0$, $z_i=\sum_{j=1}^{i} h_i$, i=1, 2. $Q(t)$ is the uniformly distributed load applied on the surface, as shown in Fig. 2, in which, Q_0, Q_u are initial load and ultimate load, respectively. And t_c is the construction time.

Based on Terzaghi's one dimensional consolidation theory, the surrounding soil satisfies the following equation

$$-\frac{k_{vi}}{\gamma_w}\frac{\partial^2 u_i}{\partial z^2}=\frac{\partial \varepsilon_{vi}}{\partial t} \tag{1}$$

Where k_{vi} is vertical permeability of layer i (i=1, 2); u_i is the excess pore pressure within the layer i; γ_w is unit weight of water; ε_{vi} is volumetric strain of soil layer i.

At any time, both the pile column and the surrounding soil share the applied load in a pile foundation,

$$\mathrm{M}\sigma_{\mathrm{pi}}+(1-\mathrm{M})\sigma_{\mathrm{si}}=\mathrm{Q(t)} \tag{2}$$

Where σ_{pi} and σ_{si} are total stress within the pile and within the surrounding soil in layer i, respectively. And M is the area replacement ratio (the ratio of the pile column cross-sectional area to the unit cell area).

The equal strain assumption between the pile column and the surrounding soil yields the following equation:

$$\partial\varepsilon_{\mathrm{vi}}=\mathrm{m_{vsi}}\,\partial\sigma'_{\mathrm{si}}=\mathrm{m_{vp}}\,\partial\sigma_{\mathrm{pi}} \tag{3}$$

Where σ'_{si} is effect stress within the surrounding soil in the layer i ($\sigma'_{si}=\sigma_{si}-u_i$). m_{vsi} and m_{vp} are the coefficients of volumetric compressibility of soil and pile, $m_{vsi}=1/E_{si}$ and $m_{vp}=1/E_p$. E_{si} and E_p are elastic modulus of soil and pile.

Combining equation (2) and (3), σ'_{si} can be derived as

$$\sigma'_{si} = \frac{Q(t)-(1-M)u_i}{1+M(N_i-1)} \tag{4}$$

Where $N_i = m_{vsi}/m_{vp}$

The differential of equation (4) yield

$$\frac{\partial \sigma'_{si}}{\partial t} = \frac{1}{1+M(N_i-1)}\left[\frac{\partial Q(t)}{\partial t} - (1-M)\frac{\partial u_i}{\partial t}\right] \tag{5}$$

The rate of the soil volumetric strain change with time can be expressed as

$$\frac{\partial \varepsilon_{vsi}}{\partial t} = m_{vsi}\frac{\partial \sigma'_{si}}{\partial t} \tag{6}$$

Submitting equation (5) and (6) into equation (1), the follow relation can be obtained:

$$-\frac{k_{vi}}{\gamma_w}\frac{\partial^2 u_i}{\partial z^2} = \frac{m_{vsi}}{1+M(N_i-1)}\left[\frac{\partial Q(t)}{\partial t} - (1-M)\frac{\partial u_i}{\partial t}\right] \quad i=1, 2 \tag{7}$$

Equation (7) can be simplified as:

$$(1-M)\frac{\partial u_i}{\partial t} = c_{vi}\frac{\partial^2 u_i}{\partial z^2} + \frac{\partial Q(t)}{\partial t} \tag{8}$$

Where c_{vi} is the vertical consolidation coefficient of the composite foundation,

$$c_{vi} = [1+M(N_i-1)]c_{vsi} \tag{9}$$

where c_{vsi} is the vertical consolidation coefficient of layer i

Equation (8) is the one dimensional consolidation differential equation for this type of consolidation problem. In order to obtain the solution for equation (8), boundary conditions and initial conditions are necessary as follows:

$$z = 0:\ u_1 = 0 \tag{10}$$

$$z = h_1:\ u_1 = u_2, \tag{11a}$$

$$k_{v1}\frac{\partial u_1}{\partial z} = k_{v2}\frac{\partial u_2}{\partial z} \tag{11b}$$

$$z = H:\ \frac{\partial u_2}{\partial z} = 0 \tag{12}$$

$$t = 0:\ u_i = Q_0 \tag{13}$$

3. General Solution

Define:

$$a_i = \frac{k_{vi}}{k_{v1}},\ b_i = \frac{[1+M(N_1-1)]m_{vsi}}{[1+M(N_i-1)]m_{vs1}},\ \rho_i = \frac{h_i}{H},\ \mu_i = \sqrt{\frac{(1-M)b_i}{a_i}},\ c_i = \frac{h_i}{h_1}\ \text{i=1, 2} \tag{14}$$

Referring to Xie et al. (2002), the solution of the equation (8) may have the form of an infinite series as follow:

$$u_i = \sum_{m=1}^{\infty} g_{mi}(z)e^{-\beta_m t}[B_m + C_m T_m(t)] \quad \text{i=1, 2} \tag{15}$$

Where

$$g_{m1}(z) = \sin\left(u_1\lambda_m \frac{z}{H}\right) \tag{15a}$$

$$g_{m2}(z) = A_m \cos\left(u_2\lambda_m \frac{H-z}{H}\right) \tag{15b}$$

and β_m, B_m, C_m, A_m, λ_m, μ are unknown coefficients to be found out; $T_m(t)$ is a function of t, resulting from the item $Q(t)$ in equation (5). Obviously, equation (15) satisfies Equation (10) and (12). Using boundary condition represented by equation (11) and (13), the following relationship can be obtained

$$A_m = \frac{\sin(u_1\lambda_m\rho_1)}{\cos(u_2\lambda_m\rho_2)} \tag{16}$$

Where

$$\beta_m = \frac{\lambda_m^2 c_{v1}}{H^2} \tag{16a}$$

$$T_m(t) = \int_0^t e^{\beta_m \tau}\frac{\partial Q(\tau)}{\partial \tau}d\tau \tag{16b}$$

The positive eigen-values, λ_m, satisfy the following eigen-equation:

$$\sqrt{a_2 b_2}\,tg(u_1\lambda_m\rho_1)ctg(u_2\lambda_m\rho_2) = 1 \tag{17}$$

Furthermore, the following equations also can be obtained:

$$\sum_{m=1}^{\infty}(1-M)C_m g_{mi}(z) = 1 \quad (i = 1,\ 2) \tag{18}$$

$$\sum_{m=1}^{\infty} g_{mi}(z)B_m = Q_0 \tag{19}$$

Using the following orthogonal relation

$$\int_0^{h_1} m_{v1}\, g_{m1}(z)\cdot g_{n1}(z)dz + \int_{h_1}^{H} m_{v2}\, g_{m2}(z)\cdot g_{n2}(z)dz =$$

$$\begin{cases} 0 & m \neq n \\ \frac{H}{\mu_2\lambda_m}\sqrt{\frac{b_2}{a_2}} & m = n \end{cases} \tag{20}$$

Where

$$m_{vi} = [1 + M(N_i - 1)]m_{vsi} \tag{21}$$

C_m and B_m can be given as follows:

$$C_m = \frac{2\sqrt{a_2 b_2}}{a_2 u_2 \lambda_m(\rho_1 + b_2\rho_2 A_m^2)} \tag{22}$$

$$B_m = \frac{\sum_{m=1}^{\infty}(1-M)m_{vi}\int_{z_{i-1}}^{z_i} Q_0 g_{mi}(z)dz}{\sum_{m=1}^{\infty}(1-M)m_{vi}\int_{z_{i-1}}^{z_i} g_{mi}^2(z)dz} \tag{23}$$

Submitting all the coefficients into equation (15), the final results for u_1 and u_2 can be written as:

$$u_1 = \sum_{m=1}^{\infty} \sin\left(u_1\lambda_m \frac{z}{H}\right) e^{-\beta_m t}[B_m + C_m \int_0^t e^{\beta_m \tau} \frac{\partial Q(\tau)}{\partial \tau} d\tau] \tag{24}$$

$$u_2 = \sum_{m=1}^{\infty} \frac{\sin(u_1\lambda_m\rho_1)}{\cos(u_2\lambda_m\rho_2)} \cos(u_2\lambda_m \frac{H-z}{H}) e^{-\beta_m t}[B_m + C_m \int_0^t e^{\beta_m \tau} \frac{\partial Q(\tau)}{\partial \tau} d\tau] \tag{25}$$

4. Average Degree of Consolidation

After the solutions for excess pore pressure in the soil has been obtained, the expression for the degree of consolidation of a composite foundation can be proposed. Generally, the degree of consolidation can be defined either in terms of settlement or in term of effective stress. The former shows the rate of settlement while the later indicates the rate of the increase of effective pressure.

The total average degree of consolidation of the whole two-layered soil in terms of settlement, U_s, can be expressed as:

$$U_s = \frac{S_t}{S_\infty} = \frac{Q(t)}{Q_u} - \frac{\int_0^{h_1} u_1 dz + b_2 \int_{h_1}^{H} u_2 dz}{h_1 Q_u(1 + b_2 c_2)} \tag{26}$$

5. Solution for the Cases of Constant Loading

Based on the above theories, the solutions for two particular loading conditions: constant loading and ramp loading can be given in explicit forms and listed in Table 1.

Table 1. Solutions for the case of constant loading.

Excess pore water pressure and average degree of consolidation

$$u_1 = Q_u \sum_{m=1}^{\infty} C_m \sin(\mu_1 \lambda_m \frac{z}{H}) e^{-\lambda_m^2 T_v} \quad , \quad u_2 = Q_u \sum_{m=1}^{\infty} A_m C_m \cos(\mu_2 \lambda_m \frac{H-z}{H}) e^{-\lambda_m^2 T_v}$$

$$U_s = 1 - \sum_{m=1}^{\infty} \frac{C_m D_m}{1 + b_2 c_2} e^{-\lambda_m^2 T_v}$$

Coefficient A_m, C_m, D_m, E_m and eigen-equation

$$A_m = \frac{\sin(\mu_1 \rho_1 \lambda_m)}{\cos(\mu_2 \rho_2 \lambda_m)}, C_m = \frac{2\sqrt{a_2 b_2}}{\mu_2 a_2 \lambda_m (\rho_1 + b_2 \rho_2 A_m^2)}, D_m = \frac{\sqrt{a_2 b_2}}{\mu_2 \rho_1 a_2 \lambda_m}, E_m = \frac{b_2 - (b_2 - 1)\cos(\mu_1 \rho_1 \lambda_m)}{\mu_2 \rho_1 \lambda_m \sqrt{a_2 b_2}}$$

$$N_1 = \frac{m_{vs1}}{m_{vp}}, N_2 = \frac{m_{vs2}}{m_{vp}}, c_{v1} = [1 + M(N_1 - 1)] \frac{k_{v1}}{m_{vs1}}, a_2 = \frac{k_{v2}}{k_{v1}}, b_2 = \frac{[1+M(N_1-1)] m_{vs2}}{[1+M(N_2-1)] m_{vs1}}, \rho_1 = \frac{h_1}{H},$$

$$\rho_2 = \frac{h_2}{H}, \mu_1 = \sqrt{\frac{(1-M) b_1}{a_1}}$$

$$\mu_2 = \sqrt{\frac{(1-M) b_2}{a_2}}, T_v = \frac{c_{v1} t}{H^2}$$

6. Verification Against Existing Solution

Let N_1=N_2=1 and M=0, which means there is no pile in the foundation. Then the composite foundation in double-layered soil becomes the problem only considered the consolidation of the two layered-soil and the solutions listed in Table 1 should be reduced to those given by Xie (1994). The same results can be seen between two approaches as shown in Fig. 3.

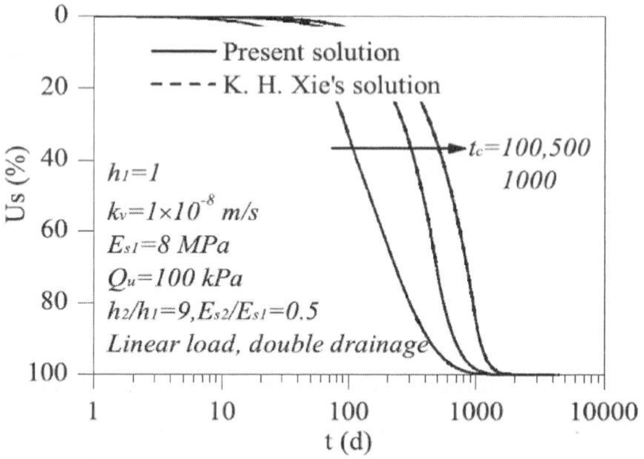

Fig. 3. Consolidation degree computed by Xie (1994) and present solution.

7. Parametric Study and Discussion

Based on the above solution, a computation program has been developed. The effects of several important influence parameters, such as k_{v2}/k_{v1}, E_{s2}/E_{s1}, E_p/E_{s1}, M, t_c on the consolidation behavior of a composite foundation are investigated in this section and a comparison of some solutions is made. Table 2 shows the basic parameters used in this analysis.

Table 2. Basic parameters used in the analysis.

Parameter	Value
The thickness of layer one, h_1 (m)	2
Modulus of layer one, E_{s1} (MPa)	5
Vertical permeability of layer one, k_{v1} (m/s)	1×10^{-9}
Modulus of pile, E_p (MPa)	100
Area replacement ratio, M	0.15
Constant load, Q_u (kPa)	150
h_2/h_1	5
E_{s2}/E_{s1}	0.5
k_{v2}/k_{v1}	5

Fig. 4 shows the influences of the permeability of second layer on consolidation behavior for the single-drainage situation, corresponding to k_{v2}/k_{v1}=10, 5, 1, 0.1, respectively. It can be seen that a reduction in the value of k_{v2}/k_{v1} can significantly reduce the rate of consolidation for the whole soil stratum as the excess pore pressure dissipation in the second layer is greatly reduced.

The value of E_{s2}/E_{s1} reflect the relative stiffness of the two layers. Fig. 5 shows the variation of this parameter just slightly effect the consolidation rate for the whole foundation. It can be obtained that the smaller the modulus of layer two, the more slowly the consolidation for the composite foundation.

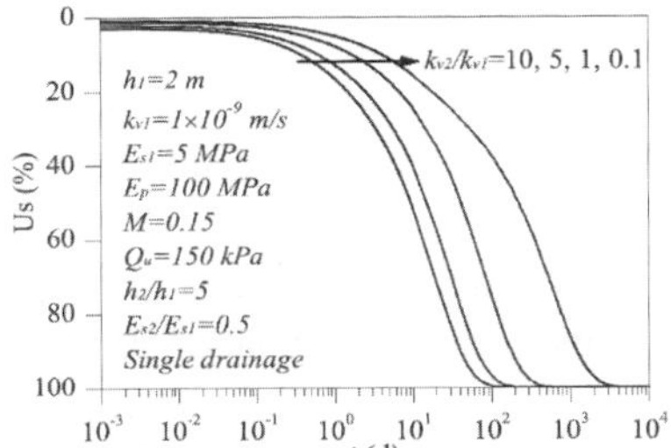

Fig. 4. Influence of vertical coefficient of permeability on the rate of consolidation.

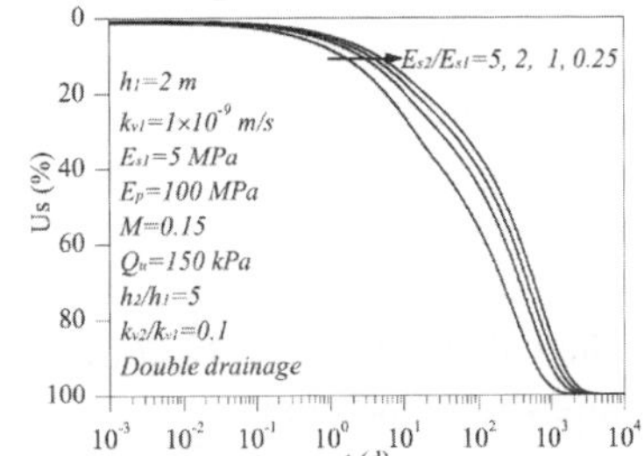

Fig. 5. Influence of the soil on the rate of consolidation.

As shown in Fig. 6, an increase in the value of pile modulus enhances the consolidation rate of a composite foundation. This is because less external load transfers into the surrounding soil when the modulus of the pile increases. Subsequently, the dissipation rate of excess pore pressure within the soil accelerated.

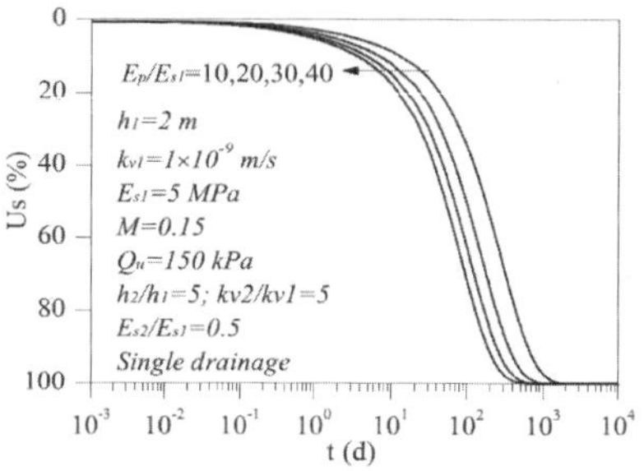

Fig. 6. Influence of modulus of pile replacement ratio on the rate of consolidation.

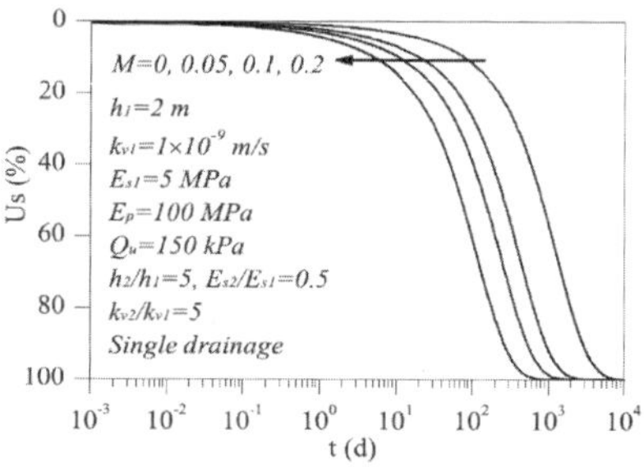

Fig. 7. Influence of area on the rate of consolidation.

The influence of area replacement ratio is plotted in Fig. 7, from which the following conclusions are drawn: the value of area replacement ratio (*M*) will significantly effect the consolidation rate of a composite foundation. It can be seen that the consolidation is greatly accelerated with an increase in the value of *M*. In the case of *M*=0, which means no pile in the foundation, the consolidation rate is at its minimum. While in the case of *M*=0.2, which means 20% of foundation soil is replaced by the pile, the consolidation rate is at its maximum.

8. Conclusion

In this article, a general solution was developed for computing the consolidation rate of a composite foundation reinforced with piles by taking into account double-layered soil and construction time effects. Comparison to the existing solutions verified that the present solution for this kind of problem is reasonable and can be reduced to Xie's solution. A parametric study shows that a reduction in the values of k_{v2}/k_{v1}, E_{s2}/E_{s1}, E_p/E_{s1} and M reduces the consolidation rate of a composite foundation.

References

1. Barron R.A., 1948. Consolidation of fine-grained soils by sand drain wells. Transacions of the American Society of Civil Engineer, 113:1718.
2. Cai Y.Q. and Xu C.J., 2002. Consolidation behaviour of cement- and lime-cement-mixed column foundations. Journal of Zhejiang University SCIENCE, 3(5): 507-512.
3. Hou Y.F. and Gong X.N., 2000. The permeability of cement-mixed material. Journal of Zhejiang University, 34(2): 189-193 (in Chinese).

4. Lei C. and Liu S.Y., 2008. Consolidation calculation of soft ground improved by T-shape deep mixing. Geocongress 2008: Geosustainability and Geohazard Mitigation. GSP 178, 620-627.
5. Lorenzo G.A. and Bergado D.T., 2003. New consolidation equation for soil-cement pile improved ground. Canadian Geotechnical Journal, 40(2): 265-275.
6. Lu M.M., Xie K.H. and Guo B., 2010. Consolidation theory for a composite foundation considering radial and vertical flows within the column and the variation of soil permeability within the disturbed soil zone. Canadian Geotechnical Journal, 47(2): 207-215.
7. Taiebat H.A. and Carter J.P., 2001. A semi-analytical finite element method for three-dimensional consolidation analysis. Computer and Geotechnics, 28(1): 55-78.
8. Terzaghi K., 1943. Theoretical soil mechanics. John Wiley and Sons, New York.
9. Wang R.C., Xie K.H. and Guan S.H., 2002. Analytical solutions for consolidation of composite ground with granular columns under time dependent loading. Journal of Zhejiang University SCIENCE, 36(1): 12-17.
10. Xie K.H., 1994. Theory of one dimensional consolidation of double-layered ground and its applications. Chinese journal of geotechnical engineering, 16(5): 26-38 (in Chinese).
11. Xie K.H., Gao P. and Xie X.Y., 2001. Consoliadtion theory for soft clays reinforced by cement of granular columns. Proceedings of the 10th international conference of the association for computer methods and advances in geomechanics, America, Tucson; 1263-1270.
12. Xie K.H., Xie X.Y. and Jiang W., 2002. A study on one-dimensional nonlinear consolidation of double-layered soil. Computer and Geotechnics, 29(2): 151-168.
13. Yoshikuni H. and Nakanodo H., 1974. Consolidation of soils by vertical drain wells with finite permeability. Soils and foundation, 14(2): 36-42.

The research on intelligent fusion optimization algorithm for hydraulic control system of jumbolter

Xiao-Yu Wang[1,2]

[1]*College of Engineering, Xi'an International University, 710077*
[2] *Electronic Engineering College, Xi'an Aeronautical University, Xi'an, 710077, China*
E-mail: lshtttt@126.com

This paper found the deficiency of classical Z-N method in optimization of PID parameters, proposed the intelligent optimization algorithm of PID parameters. The stimulation and experiment showed that the PID parameters getting from this algorithm are efficient, accurate and speedy. This paper used algorithm for hydraulic control system of jumbolter. The stimulation and experiment showed that when peak oil pressure of relief chamber is changed, optimum algorithm will find suitable and based on MAX power requirement to change displacement of relief surface, which lead to changing displacement of impact piston, at the same time, it will change the impact power and frequency of impactor, and realize impactor auto change working parameters based on the different situation under suitable parameters and MAX power working condition. The experiments show that it has a certain reference value for other control objects and control process.

Keywords: Intelligent Fusion Optimization Algorithm; Control Strategy; Variable Stroke Adjusting Mechanism; Performance Matching.

1. Introduction

The jumbolter is important tool for bolt supporting engineering construction, With the rapid development of geotechnical engineering in China, Airborne rotary hydraulic roofbolter be used in coal road way and Geotechnical anchorage engineering widely in recently. The most commonly used technique of tuning PID parameter is Z-N principle for hydraulic control system of Airborne rotary hydraulic roofbolter [1], Ziegler-Nichols technique is easy to implement but parameters need fine tuning. The optimization method based on genetic algorithm [2] and ant colony algorithm are also developed [3]. The optimization of PID parameters are improved over classical Z-N method, but the issues of premature convergence, stagnation, the global search ability is not high are still outstanding. The research for ant colony algorithm is mainly to improve the convergence speed of the algorithm through experiments and stagnation. For example, the PID parameter is be calculated by genetic algorithm, and then running the improved ant colony algorithm, by auto adjusting the route choice probability and pheromone updating rule, finally to search out the optimal value of the PID parameter by author in reference [4]. The premature convergence is avoided caused by local optimization of algorithm by defining the up and lower limits of information density on the path by using the MAX/MIN ant system in reference [5]. The PID parameter is be calculated by genetic algorithm, and then running the improved ant colony algorithm, by auto adjusting the route choice probability and pheromone updating rule, finally to search out the optimal value of the PID parameter by author in reference [6].

Reviewing the various situations, we proposed the tuning strategy of PID parameter based on intelligent fusion optimization algorithm in the paper and used for the research of control strategy hydraulic system of the jumbolter, After experiment and stimulation, the working parameter can be self-adapted in MAX power working condition by intelligent fusion optimization algorithm according the change of peak oil pressure of relief chamber in different working condition. The result is be proved that the control of strategy of airborne rotary hydraulic roofbolter hydraulic system is reasonable and feasible as intelligent fusion optimization algorithm is be used for tuning the PID parameter, the work efficiency is be improved and energy consumption id be reduced [7].

2. The Idea of Setting PID Parameters by Intelligent Fusion Optimization Algorithm

PID (Proportional-Integral-Differential Controller is currently the irreplaceable technology used in industrial process control and motion control owning to its advantage of simple algorithm, good robustness and convenient tuning. The most commonly used technique of tuning PID parameter is Z-N principle in practice. Ziegler-Nichols technique is easy to implement but parameters need fine tuning [1,6].

Reviewing the various situations, according the algorithm by author in reference4.the intelligent fusion optimization algorithm is be used for tuning the PID parameter, the initial value of PID parameter is be calculated by Z-N principle, the value is operation factor by genetic algorithm and then running the improved ant colony algorithm, by auto adjusting the route choice probability and pheromone updating rule, finally to search out the optimal value of the PID parameter [6].

(1) The step of tuning the PID parameter by intelligent fusion optimization algorithm

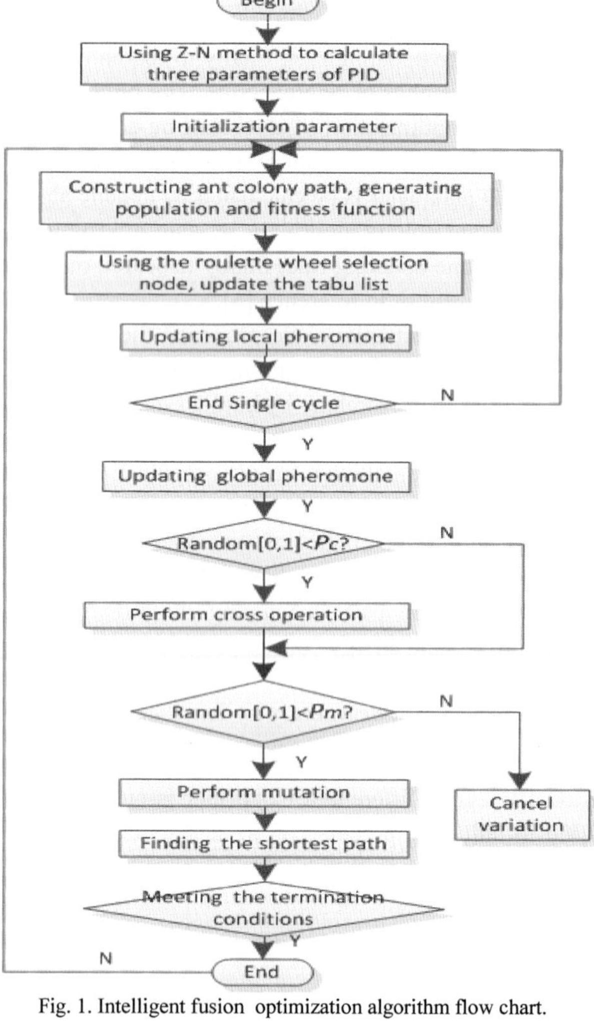

Fig. 1. Intelligent fusion optimization algorithm flow chart.

(2) The analysis of simulation

Reviewing the various situations, according the algorithm by author in reference4.the intelligent fusion optimization algorithm is be used for tuning the PID parameter, the initial value of PID parameter is be calculated by Z-N principle, the value is operation factor by genetic algorithm and then running the improved ant colony algorithm, by auto adjusting the route choice probability and pheromone updating rule, finally to search out the optimal value of the PID parameter [6].

A electro-hydraulic servo system of jumbler as an example G(s)=5268/s(s2+133s+110889), simulating the algorithm from transition function. In simulating test, setting system input as unit step signal. Setting $0 \leq \alpha < 5$, $0 \leq \beta < 5$, $0.1 \leq \rho < 0.99$, $1 \leq Q < 10000$,(α is pheromone inspiration factor, β is expect inspiration factor, ρ is pheromone evaporation factor and Q is constant for the number of ants left trial) ants number is 30, repeater times is 100. The scope of parameters of PID are : Kp is [0.00001 20], Ki, Kd is [0.00001 2].

For analysis performance of this algorithm on this paper, it is compared with N-Z, ACS, GA algorithm. Figure 2 is the unit step responds chart. Table 1 is the unit step responds table [6].

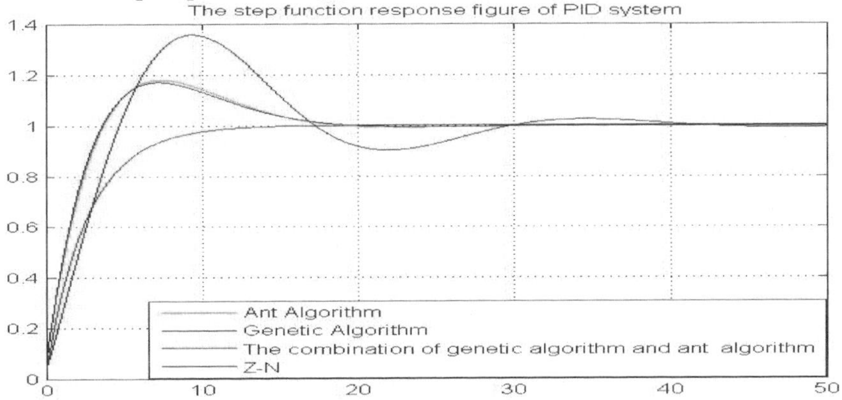

Fig. 2. Algorithm unit step responds chart.

Table 1. Comparison table of system unit step performance.

ALGORITHM	K_p	K_i	K_d	ts	ess	σ%
Z-N	4.58	1.6	0.8	37.35	8.2574e-8	35.9714
ACS	10.0	1.9944	1.6878	16.58	4.4409e-14	18.1963
GA	10.5228	1.9997	1.5159	16.72	3.3307e-14	17.2131
ACS-GA	8.0112	0.0071	0.6488	10.25	0.1676	0.0672

Reviewing table 1, comparing intelligent fusion optimization algorithm and other control algorithm, three PID parameter- KP, Ki and Kd are optimized, stability performance index of system static bias ess is most less, dynamic performance index of system control time ts and overshoot σ% are reduced dramatically. Therefore, the algorithm has best control performance and robustness [6].

The simulating results indicates the algorithm can search the optimal value in the scope repaid and efficiency, This result offer foundation to study the hydraulic jumbolter drive system optimum control strategy [8].

3. The Experiment Example

The key component of airborne rotary hydraulic roofbolter impact parts is impact piston, flow control valve, accumulator, and changeable travel control device [9,10]. The changeable travel control device diagram see Fig. 3. The optimal aim is the suitable system working oil pressure P_d and the pushing force F_t in MAX power of the jumbolter for the current working condition, mathematically modal see Eq. (1) and Eq. (2).

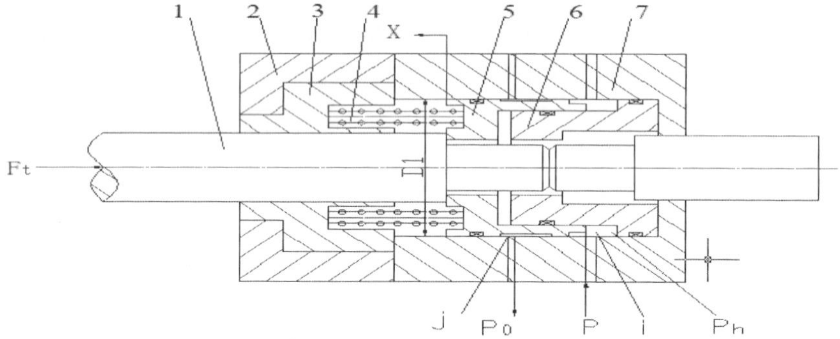

1-Drill Rod; 2-Organism; 3-Drill sleeve; 4-Regulating spring; 5-Buffer shell; 6-Buffer inner; 7-Intermediate

Fig. 3. Variable stroke adjustment mechanism model.

The working principle of variable stroke adjustment mechanism is: when peak oil pressure of relief chamber changed, working oil pressure P_d and the pushing force F_t will be controlled, ensure hydraulic impact in best distance, make power of jumbolter is in the best working condition [8,9].

(1) Mathematically modal of variable stroke adjustment mechanism

The mathematically modal can be set according the fig. 3 and dynamic balance of the moving body [8,9].

$$md^2x/dt^2=F_p - F_k - F_t - F_E - F_f \tag{1}$$

$$F_p = AP_h, F_k = k(x+x_0), F_t = P_t D_1^2 * \pi / 4 \tag{2}$$

In Eq. (1) and Eq. (2), m is the quality of buffer shell [kg]; x is the relief surface displacement of an impact device [m]; F_p is the Oil pressure, F_k is the Spring force, F_t is the pushing pressure; F_B is the viscosity resistance, F_f is the Seal friction [N]; P_h is the Buffer chamber pressure, P_t is the Hydraulic cylinder working pressure [Mpa]; A is the Section area of Buffer shell, [m^2]; K is the spring rate [N/m]; x_0 is the adjusting spring pre compression [m]; D_1 is the fit diameter of buffer cavity [m].

Simplified Eq. (1) and Eq. (2) (neglecting F_B and F_f). We can get Eq. (3) and Eq. (4) [8]:

$$x=0.0311014P_d-0.064P_t-0.002 \tag{3}$$

In Eq. (3), x is the relief surface displacement of an impact device

$$Sj = Sj' + x + x0 \tag{4}$$

In Eq. (4), S_j the impacting position distance of an impact device, Sj' is the return stroke of the impacting position [m].

(2) The objective function

Considering all kinds of characteristics of the working device, impact energy E and frequency f sleeplessly change in MAX power of the jumbolter satisfied for the current working condition. Take P_d and F_t as the value to be optimized, N maximum as the objective function. we can get Eq. (5) and Eq. (6). In Eq. (5), α is 0.29 [8,9].

$$N = E \bullet f = 10^6 \frac{(1-\alpha)^2 A_1}{k(1-2\alpha)} \bullet P_d S_j \bullet 10^3 \frac{\alpha}{1-\alpha}\sqrt{\frac{A_1(1-2\alpha)}{2k^3 m}} \bullet \sqrt{\frac{P_d}{S_j}} \tag{5}$$

$$\begin{aligned} P_d' &= P_d + k_1 \Delta P_h \\ P_t' &= P_t - k_2 \Delta P_h \end{aligned} \tag{6}$$

In Eq. (6), P_d' is the optimization of the system operating hydraulic limit value this time [Mpa]; P_t' is the optimization of the system pushing the hydraulic limit value this time, [Mpa]; K1 is 0.3 (Correction factor of working oil pressure), K2 is 0.1 (Correction factor of pushing oil pressure); ΔP_h is the deviation between this time and last time of peak oil pressure of relief chamber[Mpa].

The scope of parameters of airborne rotary hydraulic roofbolter are: Pt:5~10 [Mpa]; Pd:10~18 [Mpa]; N:20000 [W]; x:-0.020 [m]~+0.020 [m]; Sj0:0.03 [m]; x0:0.002 [m].

(3)Results of optimization algorithm

Setting the previous impactor impact parameters: ① P_d is 14 [Mpa], P_t is 7 [Mpa]; ② P_d is 15 [Mpa], P_t is 7.8 [Mpa]; ③ P_d is 17 [Mpa], P_t is 78.6 [Mpa]; when setting ΔP_h is 3 [Mpa]; -2 [Mpa]; -1 [Mpa]; 1 [Mpa]; 2 [Mpa]; 3 [Mpa].

Using intelligent fusion optimization algorithm, we can get initial operating parameters of an impact device and next impactor operating parameters optimization results are shown in table 2~3; table 4~5.

Table 2. Initial operating parameters of an impact device.

Working oil pressure P_d(Mpa)	Pushing oil pressure P_t(Mpa)	Impacting Energy E(J)	Impacting frequency f(Hz)	Impacting power N(kw)	Relief surface displacement x (m)	Impacting position distance S_j(m)
15	7.8	183	36.574	6.698	-0.0152	0.01684

Table 3. Next impactor operating parameters optimization results.

ΔP_h (Mpa)	-3	-2	-1	1	2	3
P_d'(Mpa)	14.206	14.917	14.950	15.281	15.522	15.854
P_t'(Mpa)	7.7206	7.805	7.920	7.832	7.884	8.012
E (J)	123.663	131.023	130.19	292.13	365.43	410.245
f(Hz)	42.119	42.945	43.217	29.493	26.786	25.811
N (kw)	5.208	5.625	8.847	8.614	9.786	10.589
x (m)	-0.02 (-0.0395)	-0.0199	-0.02 (-0.0274)	-0.0057	0.0005	0.00367
S_j (m)	0.0121	0.0121	0.0121	0.02634	0.0325	0.0357

Table 4. Initial operating parameters of an impact device.

Working oil pressure P_d(Mpa)	Pushing oil pressure P_t(Mpa)	Impacting Energy E(J)	Impacting frequency f(Hz)	Impacting power N(kw)	Relief surface displacement x (m)	Impacting position distance S_j(m)
17	8.6	435.89	26.857	11.71	0.00333	0.03533

Table 5. Initial operating parameters of an impact device.

ΔP_h (Mpa)	-3	-2	-1	1	2	3
P_d'(Mpa)	16.354	16.582	16.845	17.183	17.414	17.731
P_t'(Mpa)	8.362	8.473	8.525	8.661	8.793	8.890
E(J)	327.873	339.800	430.241	473.882	460.893	485.564
f(Hz)	29.783	29.667	26.780	26.031	26.733	26.392
N(kw)	9.765	10.816	11.521	12.3356	12.32	12.814
x(m)	-0.00437	-0.00376	0.0032	0.006	0.0045	0.006
S_j (m)	0.0276	0.0282	0.0352	0.038	0.0365	0.038

It can be concluded form chat2 to chat 5 that the impact energy E and frequency f can be controlled automatically by intelligent fusion optimization algorithm according to the hardness of working object. As the working object is soft or has been broken, drill rod rebound is more small, the peak oil pressure of relief chamber will reduce, then the displacement of impact piston also decreases, which lead to decrease impact energy and increase impact frequency by the controlling working oil pressure P_d and the pushing force P_t automatically. Conversely, As the working object is hard or has not been broken, drill rod rebound is more bigger, the peak oil pressure of relief chamber will increase, then the displacement of impact piston also increases, which lead to increase impact energy and decrease impact frequency by the controlling working oil pressure P_d and the pushing force P_t automatically[8].

4. Conclusions

Intelligent fusion optimization algorithm in the paper is be used for the tuning strategy of PID parameter, the theory is the initial value of PID parameter is be calculated by Z-N principle, the value is operation factor by genetic algorithm and then running the improved ant colony algorithm, by auto adjusting the route choice probability and pheromone updating rule, finally to search out the optimal value of the PID parameters. The simulating results indicates the algorithm can search the optimal value in the scope repaid and efficiency and the algorithm is be used for the control strategy research of airborne rotary hydraulic roofbolter. By engineering example, the intelligent fusion optimization algorithm can make working oil pressure P_d and the pushing force P_t best configuration according to peak oil pressure $\triangle P_h$ of relief chamber changing based on the different situation and MAX power principle. Impact energy E and frequency f sleeplessly change in MAX power of the jumbolter satisfied for the current working condition. The feasibility and correctness of intelligent fusion optimization algorithm is be verified and it can be referenced for other control objects and control processes[8].

Acknowledgment

This work was financially supported by Shaanxi Provincial Education Commission (Program No. 15JK 2129).

Reference

1. Wang yong-hua. Modern electrical control and the Application of PLC Technology[M]. Beijing-University Press, 2008.
2. Ho W K, Hang C C, Cao L S.Tuning of PID controllers based on gain and phase margin Specifications[J]. Automatica, 1995, 31(3): 497-502.
3. Wang P, Kwok D P. Optimal Design of PID Process Controllers Based on Genetic Algorithms [J]. Control Eng Prac, 1994, 2(4): 641-648.
4. Zhan Shi-chang, Wu jun. Parameters optimization design of PID controller based on ant colony algorithms[J]. Measurement & Control Technology, 2006, 23(01): 69~75.

5. Shu Tao, Liu Shaowei. Research on Parameters Adjustment of PID Controller Based on ACO-GA Hybrid Strategy[J]. Journal of Projectiles, Rockets, Missiles and Guidance, 2009, 29(01):73-76.
6. Wang Xiaoyu, Optimization Design Based on Self-adapted Ant Colony and genetic mix Algorithm for Parameters of PID Controller. International Journal of Hybrid Information Technology [J]. Vol. 8, No. 7 (2015), 411-422.
7. Wang Xiaoyu. The transformation of Hydraulic Roofbolter frequency Control system based on PLC and HMI. MACHINE TOOL HYDRAULICS[J], 2015, 43 (10): 172~174.
8. Wang Xiaoyu. The research control optimization strategy Based on mix algorithm For hydraulic drive control System of jumbolter[D]. Xi'an University of Architecture and Technology, 2014.
9. Li Hou-qing. The research on impact mechanism and hydraulic drive system for Heavy hydraulic rock drill[D]. Central South University, 2010.

Spatial order and affective stimulation-enlightenment of athenian acropolis on scenario expression of commemorative architectures

Fa-Yin Sun

School of City and architectural engineering, Zaozhuang University,
Zaozhuang, Shandong, P.R. China
E-mail: sunfayin@163.com
www.uzz.edu.cn

This paper proposes that the understanding and expression of commemorative quality of architectures shall be based on architectural scenario. Combining the architectural layout of Athenian Acropolis and its shocking influence on later generations, this paper conducts an in-depth analysis on the expression mode of commemorative architecture scenario from the perspectives of event, space and emotion.

Keywords: Commemorative; Scenario; Athenian Acropolis; Emotion.

1. Introduction

Athenian Acropolis, as the symbol of classic cultural concept in the world, has reached the highest level in the architectural complex of Holy Land of ancient Greece, in terms of hieron, column type and sculpture. Every visitor personally on the scene has been constantly impressed and sincerely praising such an everlasting magnificence, regardless of national boundaries or nationalities. Commemorative architecture is also a type of architecture that features great artistry and strong infection, the effect of which directly relates to the successful expression of commemorative architectural contents. Thus, it is of great realistic significance to study the way to improve the infection effect of commemorative architectures on visitors. The influence of Athenian Acropolis to later generational acts the best reference for commemorative architecture expression.

2. Connotation of Scenario

Architecture scenario, as the materialization of “specific scene in daily life”, indicates the specific scene in an architectural place which carries emotions. Therefore, architecture scenario indicates the architecture environment that carries specific emotions. Commemorative architecture scenarios can be divided into three levels: material level, event level and emotion level. Material level includes the materials that compose the scenario space and the scenario space composed by materials. It shall be visual, concrete and tangible. Event level: this event shall conform to the historical event or the fiction of event expressed by the scenario. Emotion level: it indicates the emotions that are placed on the scenario by the architect, that is contained in the scenario and that is generated in the heart of visitors in the commemorative architecture scenario. The success of a commemorative architecture depends on the consistency of the three aspects mentioned above, which is the deepest level to compose the architecture scenario and the major basis for the formation of scenario structure.

3. Spatial Level of Scenario Expression

Space, as one of the compositions of scenario, is the material carrier that composes the commemorative architecture scenario. From the perspective of function, a commemorative architecture acts as the space to exhibit mementos; its internal space plays as the spiritual space to cherish the memory of the history, as the container of materials and spirits. Visitors will cognize it perceptually when standing inside of it. The exhibition of a commemorative architecture is the space and the commemorative carrier. The definition of the scenario space boundary is the important issue for scenario construction, because space is boundless. Spatial boundary belongs to the division of visitors' psychological suggestions. From this aspect, the definition of spatial range can be divided as "strong" and "weak". When a space features greater sense of enclosure and smaller "open mouth" linking to other spaces, this definition can be defined as "strong", otherwise, "weak".

A commemorative space with "strong" definition can create a directly shocking effect to visitors. In a space with "strong" definition, there is no permeation between a physical image and its background space; when both images can reach an aesthetic balance, the commemorative quality becomes the unique target increasingly. Meanwhile, when two neighboring scenarios express distinctively different atmosphere for strong contrast, a space with "strong" definition can display greater advantage, making it easier to create a sudden emotional change of viewers during the process of visiting. A space with "weak" definition may make viewers confuse with several neighboring scenarios; but its advantage is as: the smooth transition and better continuity can make viewers grasp the visiting streamline more clearly and generate emotional tendency in advance. Moreover, the uncertainty of boundary can generate ambiguity; different viewers from different ranges of observation can lead to different understandings, creating different degrees of commemorative effects. The matching of strong and weak definition can dramatically strengthen the effect of sudden change.

A dominant method to organize the environment spaces is "spatial sequence". The design of scenario spatial sequence functions as the key to the success of scenario structure design. The target of commemorative scenario creation is to guide viewers' emotion and affection through the rendering of atmosphere, so that they could comprehend the theme of commemoration. Scenario processing can guide viewers step by step based on scenarios of various levels to put all of distracting thoughts away gradually and immerse themselves in the exhibition, to be shocked and by the scenarios with strong commemorative significance impressively. Athenian Acropolis is just in this nonlinear sequence.

Athenian Acropolis is composed by various independent hiera located on a mountaintop condescendingly, conveying the information that ancient Greeks were close to gods in the heaven and creating the majestic atmosphere of supreme national power. The passageway around the mountain is the overture; the long stairs act as the opening; the spectacular gate plays as the development; the victory hieron on the left is the visual change; Athena status as the opposite scenery of the gate is as the small climax; and Parthenon dominating the steps acts as the final climax. Walking along the axis of sacrifice, visitors will ascend step by step with the physical exhaustion and spiritual shock. When they arrive at Parthenon with their emotions that have been purified for several times, the awe-stricken and worshipful emotion will be naturally generated. When they approach with such an emotion, they will be impressed by the complicated yet magnificent Erechtheion, where the goddesses, wearing the elegant maxiskit fluttering in the air, are smiling silently, to demonstrate the national spirit of Greece and the aesthetic ideal. Such spiritual function streamline that runs through the entire scenario structure can guide the reaction process of viewers' mental mechanism to enter the realm along with the infection effect of spiritual function streamline.

4. Event Level of Scenario Expression

The event here doesn't only indicate men's activities that occur in the commemorative architecture, but also the commemorative events that are expressed under the theme of commemorative architectures and the resonation activities of viewers through participation. The introduction of events will make pure spaces complicated. In a commemorative architectural space, viewers can experience both the "shape" and the "meaning". Scenario is taken shape by one or more architectural space or above that includes events. The expression of an event shall be targeted at moving audiences. Then, how to process the events? Events shall be occurred in scenarios, and this leads to the creation of scenarios; events must absorb energy from conflict, this stimulates events and audiences' emotions. However, those events can't be just listed randomly, but in an organic way. How to make the choice, what is the sequence? To answer this question, the target of scenario shall be confirmed. What is the reason to combine those events? The reason shall be expressing the topic emotion. If the expression is unable to stimulate audiences' emotions, it will become self-complacence.

"Events" indicate changes, which can trigger the development of subsequent events as the "beginning" and the valuable "events". Changes can occur neither at the very beginning or subsequently, due to time maturation, when necessary foundation has been laid. Taking Athenian Acropolis as an instance, when visitors walking on the passageway around half of hilltop, they may feel like walking on the long stairways leading to the heaven with a devoted mode, to walk along this stairways and enter into the temple of gods, the gate is just such an "event". The salient wings are just like welcoming the visitors coming from all around to worship the gods. A wide porch is in the middle, next to colonnades on both sides, and the holy path leading to Acropolis.

Scenario events shall change meaningfully, instead of trivially. Events shall create meaningful changes, which are used to be expressed and experienced with certain value. For meaningfulness, changes shall be expressed and viewers shall make response to those changes. All of these can be measured by a value. Values are the souls of event handling and the common feature of humans' experience, which can be versatile and transformed from positive to negative mutually. Value in humans' experience can turn to the reverse side. Such a binary feature is event value. One who masters the event value will master the way of event processing and the charm of scenario processing. Thus, this requires designers to clarify the relationships among various independent scenarios in scenario processing of commemorative contents and designing each commemorative scenario.

5. Emotional Level of Scenario Expression

Emotion is the leavening culture of commemorative architectural works. Emotions will generate between viewers and architectures, along with the completion of audiences' transformation. The relationship between a viewer and the subject he/she reviews may undergo profound changes, when he/she concentrates his/her attention to the aesthetic object. Instead of an independently overall existence, the aesthetic object is injected with life and emotion. And the viewer is not the real person in daily life anymore, who has removed various considerations of practice and stake to just live in the object. This emotional change is accumulated through various links, which relates to all of links that will lead to the generation of emotional cognition, emotion and experience in humans' activities. It targets at generating the emotions of corresponding colors in the hearts of receivers and the degree of accumulated emotions and feelings.

Athenian Acropolis is located at a small hill, surrounded by the high enclosures, the annular pathway leading to the hilltop and the long stairways. The graduation of various scenarios has taken shape the sense of reverence and sense of holiness in a solemn atmosphere; the scenarios created by the great volume and the exquisite sculptures have strongly shocked the viewers psychologically. From overlook to the gradual approaching at the belt road at the bottom of the hill, from the long stone steps and gate to Winged Victory of Samothrace Hieron, from overlooking the bronze statue of Athena to Parthenon and Erechtheion Hieron, viewers' emotion of praise and worshiping with the changes of various scenarios are accumulated and rose to the climax of sincere praise and worship about Greek architecture and art. It is obvious that the entire scenario structures are run through by emotions, as the process of motional intention from generation, occurrence and development to conclusion. Emotions will definitely come to the end after several times of infiltration, contradiction, collision and development. During the process, emotions are adapted to the structural framework, the support and nourish it for its extension and expansion. The grandness and magnificence and the elegance and charm of the architectures of Holy land have run through the entire scenario.

In other words, the clue of the scenario structure of this memorial hall is a curve of the emotional development change, the qualitative change line and the line of environmental development change. The emotions that are conveyed by scenarios are formed and perceived with the features of scenarios and the distinct features of the intensity, balance and vitality of organization relation. If the emotion is suppressed, bullied or dramatically shocked for several times until being outburst due to no tolerance, then the quality, degree and times of bully will become the direct factors that lead to the emotional outbreak. Viewers will observe the scenario from the continued viewpoints along with their motion. Therefore, architects will arrange the unique scenario organization in the time stream to obtain the entirely emotional balance.

6. Conclusion

Quantity is related to the quality of scenario emotion. Generally speaking, "anger" can be accumulated to certain amount before its outbreak; and the quantity of "joy" shall not be too much in quantity. This means that negative emotion is in need of more quantity accumulation. In the creation of commemorative architecture scenarios, architects prefer to apply various series of scenario spaces, to generate the emotional impact force through neighboring spaces, material impacts, collision and contrast, to complete the corresponding relationship between quantity and emotion in the passage of time.

In conclusion, the scenario expression of commemorative architectures shall be based on three aspects: events, spaces and emotions to really move the viewers.

Conflict of Interest

The author confirms that this article content has no conflict of interest.

Acknowledgment

This work was financially supported by the Zaozhuang University excellent teaching team Foundation (20150005).

References

1. Han Gaofeng, Analysis on Feature Evolution and Development of Revolutionary Memorial Architectures in China, *J. Urban Architecture*, 7 (2008).
2. Qu Xiaoyu, Integration of History and Environment-Based on the Design of Menglianggu Battle Memorial, *J. World Architecture,* 5 (2008).
3. Xie Guangyun, Discussion on Urban Public Space in Athens in Classical Age , *J. Historical Research in Auhui*, 4 (2005).
4. Shi Jingkun, Scenario Expression of Commemorative Architectures, *J.Low Temperature Architecture Technology,* 4 (2007).

The loess of China kinds of microstructure and its seismic damage

Qing-Yi Zhang [2], Xiao-Peng Wu [4], Jin Deng [1,2,3,*]

[1]*Earthquake Administration of Gansu Province (Key Laboratory of Loess Earthquake Engineering, CEA)*
Lanzhou, 730000, China

[2]*Lanzhou Institute of Seismology, CEA, Lanzhou, 730000, China*

[3]*Geotechnical Disaster Prevention Engineering Technology Research Center of Gansu Province,*
Lanzhou, 730000, China.

[4]*Northwest Research Institute Co. Ltd. of China Railway Engineering Corporation,*
Lanzhou, 730000, China.

[*]*E-mail: dengjin@gssb.gov.cn*

Loess is formed through the loess plateau climate evolution and dust deposition in varied climate conditions. Influenced by monsoon, rainfall, temperature and dust, etc. by forming from different regional climate condition, the loess microstructure also have very big difference. By analysis the microstructure under different loess environment and its settlement size, preliminary to Chinese north-western loess microstructure types are divided into five categories: Near sand source fast sand-dropping forming loess microstructure (I), Strong sand-dropping and weak forming loess microstructure (II); Slow sand-dropping and cold weather forming loess microstructure (III); the warm and damp forming loess microstructure (IV); the strong warm and damp forming loess microstructure (V). Differences in particle size, particle size distribution, particle between the degree of consolidation, contact, etc. Microstructure types for the settlement had important effect on engineering properties.

Keywords: Loess; Microstructure; Climatic Environment; Seismic Subsidence.

1. Guidelines

The Qinghai-tibet plateau climate influence on the formation of loess, there are a lot of research results [1] [2] [3] [4]. Because of the different topography and rainfall distribution in loess plateau, the loess formation reflects by the regional characteristics of monsoon effect. Climate formation mechanism and the cause of the loess research, that is such as the literature [5] [6] [7] [8]. Loess strata changes can reflect the profile and the sand content in desert source area changes by distance, and can be used as alternative to climate change and desert border movements [9] [10] [11]. Before the more main focus of the regional distribution differences of loess is the particle size distribution and particle size composition of loess, can be divided into simple malan loess from the north to the south sand loess, loess and loess three belt [12]. Wang Yongyan [13] has carried on the detailed regional division to the grain size of loess. Combined with the feature of loess microstructure formation types and regional settlement of the size of the partition, can be used as an intuitive method of seismic hazard analysis[14] [15].

2. The Characteristic Microstructure Categories in Different Areas

Collection the aeolian deposits loess samples in the central and western of China, the typical loess is for loess plateau tableland, the sampling depth is about 5-6 meters. Due to consider the eolian loess deposition which the specific loess sample sand source, we try to choose loess tableland or flat near ground sampling, not included in the specific landform. For loess samples (kyky - 2800), scanning electron microscopy at high magnification test. Using dynamic triaxial test of loess samples at the same time get the seismic subsidence coefficient, the related literature[15].

Here are the main sedimentary loess in loess area of the microstructure pictures, divided the settlement coefficient index by microstructure classification according to the following:

2.1. *Near sand source fast sand-dropping and weakly pedogenical forming loess microstructure(I)*

Near sand source of rapid deposit loess microstructure I class structure, of main loess area is near the west of Inner Mongolia border by sand loess area. Near the Tengger desert in county is 130-180 km, such as Minxian, Haiyuan county, and the microstructure of loess is marked. Characteristics is of cold wet sand rapid dropping of weak loess microstructure. Particle accumulation is closely, weak cementation. Because near the source of sand and deposit for the northwest wind, sand particles is very large. As shown in figure 1the hickey sees 4 m depth loess samples of the SEM, sampling in Xi-ji in a landslide, loess samples of electron microscopy (SEM) image shows cementing is extremely insufficient, big almost no bonding force between particles, its seismic coefficient is of 4-6.1%.

Fig. 1. The Xi-Ji landslid loess, 4m

Figure.1 in the measurement of XIJi areas , the biggest grains size is 245-276μm. these areas are near sand source, deposition of sand and clear. Because of the loess particle size bigger, molecular attraction between loess particles is relatively small, almost no coupling between particles. for the potential landslide characteristics, and settlement, loess liquefaction hazard.

2.2. *Strong dust microstructure (II class)*

As shown in fig.2, Qing bai-shi loess microstructure. Part of lanzhou loess, such as Qingbaishi, GaoLAN, YuZhong, Jingyuan loess tableland of aeolian loess is belong to this type. Compared with the I class, the II class is far from the northwest to the main sand source. II class structure formed in the sand and strong down the dry and cold climate conditions.

Fig. 2. The Qing-bai-shi loess, 5m

II loess have elevated pores, pore is relatively close, the loess settlement coefficient of around 1.5-5.5% [20]. there is vibration deformation, landslide, collapse and liquefaction, such as mudflow danger.

2.3. *Slow droping dust's microstructure (III class)*

Yongdeng, Dingxi, part of lanzhou area of aeolian loess of silt particles loose weak cementation of loess structure (III class), is apart from the sand loess far from source than II class, inadequate cementation and porosity. As shown in figure 2 the Yong Deng's loess sand particle size is 58-70 microns. Because of the sand in the second category is far, the wind sand land grain structure of sand grain diameter smaller than the first one, the proportion of small, with dry climate, unable to close packing. Less rainfall and low temperature makes the weathering of loess is weak, grain formed between overhead of honeycomb, void ratio, dry density is low, the structure is an important characteristic of the thick loess deposits tubular.

Fig. 3. The Yong-deng loess, 4m

III types of loess porosity have a lot of overhead, therefore there is a strong shock of, is a strong earthquake in loess[21]. Such as permanent settlement of the loess seismic coefficient coefficient is of 7.99% [20], is affected by the earthquake damage is more serious areas. Such as figure 7 for 1995's 5.8 magnitude quake live forever in loess area of loess sample electron microscope images, the earthquake had a strong loess ground surface deformation and landslides.

2.4. *Little humid formed microstructure (IV)*

LiuPanShan west and east of Huajialing Jingning, Huining, west Gansu Jingyuan, Dongxiang, its' eolian loess, formed in relatively humid climate, is for weak strong eolian deposition in loess IV class microstructure. This microstructure forms in the slow sanding land and more humid environment. It has strong bonding strength between the particles than III class structure.

As shown in figure4 ,the Jing Ning's loess and sand particle size is of 35-106um. Around particles there are a lot of cementation of clay, or crumb of clay. At the time of the winter monsoon strength medium, the dust is slow, but the summer rainfall is more plentiful, especially nutrient-laden source of clay, higher than other regions.

Fig. 4. The Jing-ning loess, 5m(500 time)

Because only a small amount of overhead porosity makes the loess in these area is low settlement of loess. Settlement of coefficient of between 1.5 3.68% [20], has the potential risk of landslide and collapse.

2.5. *Strong temperature humidity of loess microstructure (V class)*

Huating and Xifeng district, large rainfall than LiuPanShan west, in the near east of LiuPanShan Xifeng loess's tableland area, those regions has humid climate. As shown in figure 5, for Xi-feng district in Gansu loess microstructure pictures. Maximum particle size about 40 μm, covered a large amount of clay, belongs to the strong bonding of overhead structure. Loess in this area due to the cementation is more sufficient, the Xi-feng loess settlement coefficient is only 0.9%, so these areas related loess, just create a slight vibration deformation.

Fig. 5. The Xi-feng loess, 5m(500)

3. Results and Discussion

(1) This paper analyzes the microcosmic structure of main eolian loess deposition area in Gansu province, loess depth of about 5-6 meters, according to the loess microstructure analysis shows that lists five types of microstructure type: nearly sand source rapid drop sand microstructure (I), strong dust the weak loess microstructure (II); Weak person slowly downhill dust into loess

microstructure (III); In the wet loess microstructure (IV); Strong temperature humidity of loess microstructure (V).

(2)Through the analysis the microstructure of the regional characteristics formation mechanism ,we can study the loess regional difference to the seismic deformation of strength, for the loess site construction engineering design, and puts forward corresponding protective measures. It has important reference meaning.

Acknowledgement

This work is financially supported by the 2013 Key Laboratory of Loess Earthquake Engineering laboratory open fund "Based on the microscopic mechanism of site response analysis" (KLLEE-12-001)), and the national natural science fund project (51578518).

References

1. H.Y. Lu, and X.WANG. The plateau monsoon variation during the past 130 kyr revealed by loess deposit at Northeast. Qinghai-Tibet (China)[J]. Global and Planetary Change, 2004 (41): 207-214.
2. W.E.Ruddima, and J.E. Kutzbach. Forcing of late Cenozoic northern hemisphere climate by plateau uplift in southern Asian and American west [J]. Journal of Geophysical Research,1989 (94): 18409-18427.
3. S.THOMAS,,A.J.IMON and Hua-yu LU. Sedimentation and diagenesis of Chinese loess implications for the preservation of continuous, high resolution climate records[J].Geology, 2006, 34 (10): 849-852.
4. J.J.Li, Wen Shixuan, Zhang Qing et al.Investigation on the uplift times, amplitude and styles of Tibet Plateau. Science in China (Series B),1979, (6): 608-616
5. H.TSOAR, and K.PYE. Dust transportan the question of desert loess formation[J]. Sedimentology,1987 (34): 139-153.
6. R.A Young, and Wiersma, J.L (1973) The role of rainfall impact in loess detachment and transport. Water Resoures 9 :1629–1635.
7. H.WANG, E.RANDALL. Hughes correlation of climate cycles in middle Mississippi valley loess and Greenland ice[J].Geology, 2003 (2): 179-182.
8. D.S. Liu, An Z.,Z. Bingquan. Accumulaiton and evolution of loess in China. In: Geochemistry Institute, Chinese Academy of Sciences ed. Preceding of the Third Quaternary Geology Conference. 1979. 1-31).
9. D.K.Rea, and S.Hovan. Grain size distribution and depositional processes of the mineral component of abyssal sediments lessons from the North Pacific[J]. Pa1eoceanograph, 1995 (10): 251-258.

10. D.H.SUN, , J. Bloemendal, D.K.REA. Grain-size distribution function of polymodal sediments in hydraulic and aeolian environments and nμmerical partitioning of the sedimentary components[J].Sedimentary Geology, 2002, 152(3-4): 263-277.
11. Y.Y. Wang. China loess the structure features and physical and mechanical properties of loess [M]. Beijing: science press, 1990.
12. J.J.. Zhao Sedimentary theory and the loess plateau environment evolution [M]. Beijing: science press, 2002: 56-70.
13. J.DENG,L.M. WANG, Z.Z. ZHANG, H. BING. Microstructure characteristics and forming environment of late Quaternary Period loess in the Loess Plateau of China[J]. Environmental Earth Sciences, Springer Berlin / Heidelberg, ISSN: 1866-6280 (Print), 1866-6299 (Online): 2009.04,21.
14. J. Deng, L.M. Wang, Z.Z. Zhang. Microstructure characteristics and seismic subsidence of loess. Chinese Journal of Geotechnical Engineering.2007, 27 (4): 542-548.
15. L.M. Wang, Loess dynamics, Beijing earthquake publishing house, 2005.

Intelligent architecture design and research based on smart city

Xue-Yong Wang[1,a], Zhen Peng[2,b], Xiao-Qian Yu[1,c], Jun Zhao[1,a], Lin Wang[3,a]
[1]College of Water Conservancy and Civil Engineering, Shandong Agricultural University,
Taian, Shandong, China
[2]Tai'an Architectural Design Institute Limited Liability Company,
Taian, Shandong, China
[3]Department of architectural engineering, Binzhou University,
Binzhou, Shandong, China
E-mail: [a]wxy199@163.com, [b]479095536 qq.com, [c]Xq1617@yeah.net

Constructions are main elements for forming the city, and the smart city is unable to run without intelligent architecture. In order to realize smart city, more reasonable architectural space, as one part of intelligent architecture, will be designed properly. This paper presents a depth investigation about smart city, to analyze architectural design, which based on the Internet of things and interactivity, and to research architectural space through reasonable layout, intelligent control and effective building energy conservation.

Keywords: Smart City; Intelligent Architecture; The Internet of Things and Interactivity; Architectural Space.

1. Smart City and Intelligent Architecture

Under the impact of global finance crisis in 2008, IBM came up with the Smart Earth, as one of intelligence project, which has been used for coping with global finance slowdown and recovering economy. For a focus of Earth's future development, city and the smart city is the foundation of a smarter planet[1].The smart city, supported by the next generation of information technology, managed to monitor and analyze different data, and seize them with both hands. Through that, a new city form, which could find changes of city living environment, improve overall control of urban resources, enhance coordination of every part and simplify operations on each aspect, will be established soon. A harmonious and win-win situation between people and city also could be achieved [2]. Managing city in a more intelligent way is a core of the smart city. That means, new generation of information technology, for example the Internet of things and cloud computing, should be used to improve urban running efficiency, to ensure sustainable development of cities, and to create a better city life.

Cities and buildings are inseparable, while at same time, intelligent buildings are significant for the smart city. In the 1980s, because of a rapid development of information technology, intelligent buildings appeared. In the construction of the smart city, building databases not only need to record machine account like building archives and drawings, but also contributing to storage building dynamic information during use. Even more important, dynamic management of the building, for example building's comfort, humanization and operating costs would be taken into consideration [3].

2. Intelligent Buildings Is Well-Combined in Technology and Architecture

Originated from the information age, intelligent buildings combine modern architecture with high technology. It is based on architecture. In addition, connecting intelligent computers, communications, information technology and architectural arts through system integration technique, so that managing information, optimizing architectures and serving users in information can be achieved by means of devices' automatic monitor. It provides a safe and comfortable life for people, which is environmentally friendly and convenient as well. At the same time, efficient living and working needs would be satisfied. Human wisdom assembles in intelligent buildings [4].

3. Architecture Design Based on the Internet of Things and Interconnection.

The Internet of things is a technology based on sensor network and RFID technology, and connect varies entities in physics. All kinds of entities in the Internet of things can communicate with each other over the network [5]. The smart city provides a depth interconnection of objectives, and defines supporting platform according to multi-users and multi-views. Meanwhile, it increases intelligence-handling logic and uses service technology to realize a multi-units functionality [6].

Great changes in working and living styles under the Internet of things have taken place, which make it possible for people can work and study anywhere, anytime, instead of limiting their activities in a single and fixed space. So that optimizing space environment can be achieved. There are interconnections between architectural space and construction equipment, people and objects can be friendly and intelligently communicating. Therefore, living and working environment will be safe, comfortable and livable. So, for the purpose of adapting to people's behavior in the context of information technology, forms of architectural space become more flexible and open. Owing to the rapid development of modern science technology, higher building quality, more innovative building modeling and more comfortable building space will be touched.

4. Intelligence in Building Design Strategies

4.1. *Intelligent layout of architectural space*

When organizing architectural space, some factors, such as building types, base weather conditions and building natures, need to give consideration. Optimizing spatial organizations according to construction types is the premise for intelligent building in the friendly-space design. For example, in the frigid North area in China, setting temperature butter space between outdoor environment and indoor rooms can reduce building energy consumption; in tropical climate, building natural ventilation is of paramount importance. Rooms should be penetrating. Well-designed natural ventilation buildings are comfortable and energy-saving. For another example, natural ventilation is a low-cost and sophisticated technical measures which adapt to climate. Compared with air-conditioning system, it reduce energy consumption while bring fresh air into rooms, and it's beneficial to people's physical and mental health [7]. Using natural ventilation and "stack effect [8]" properly can achieve passive energy-saving in the buildings. Based on above, in the building design, functional space needs to be organized smartly and optimizing building environments constantly. Smart and efficient architectural layouts contribute to reduce carbon emissions, improve working efficiency and use resources in more effective way. Buildings should be designed according to building types, base climate and occupants' working models. In addition, the function of architectural space should be thought when organizing building layout, to leave the room for architecture upgrading and reconstruction in building life cycles, and achieve intelligent architectural space design completely.

4.2. *Intelligent control of architectural space*

How to create a smart space environment? Only in the intelligent control system of space environment can integrate electric, multimedia and communication equipment into a user friendly interoperable solution; for it is not only intelligent, but also could manage space environment in a smarter way. The smart space combines automated control systems, computer networks with network communication technology. Smart control system gives owners or visitors a convenient way to manage and use architectural space. For example, by using mobile phones, Tablets, wireless remote control or voices recognition, users could control building space; operate scenes to enables linkage between multiple devices, which give users the greatest degree of efficiency, convenience, comfort and safety. Intelligent control of the space environment is mainly reflected in the following aspects.

4.2.1. *Intelligent security system of space environment*

Security technology is indispensable in the intellectual control system of space environment, and widely used in the community and building monitoring, burglar alarms and other fields. Main features of intelligent security system are theft-proof, fire protection and anti-gas leak. There are two working models, leaving alarms and home alarms. Additionally, in this system, some functions, such as scheduled deployment, scheduled removal, partitioned deployment and partitioned removal, are also included. When leaving alarm enabled, all the devices are working, host machine could make local alarm or mobile alarm under any cases, whether outdoors or indoors. When a fire or gas leaks occurring, the host machine will automatically contact the owner and alarm to the authorities, and the gas valve will be automatically closed [9]. Moreover, authentication is an important part of intelligent security control system, which is the necessary linkage between owners and visitors in mutual authentication. Then system will give different right to use architectures according to different identifies.

4.2.2. *Intelligent comfort degree of space environment*

The smart architectural space means a good relation between users, buildings and natures, which could create a comfortable environment for the people while protect the surrounding environment. The building will be in operation and maintenance phase after completed, including space management and environmental management, so the ecology of space environment is important for intelligent building. The smart architectural space will monitor and adjust temperature, humidity and air quality based on the season and daily weather variations, to achieve optimized comfortable building space. When organizing space layouts, designers should define the space environment clearly, and then, install comfort level sensors and controllers, realize intelligent comfort degree of space environment.

4.2.3. *Human-oriented design of space environment*

The intelligent space environment pays more attention to people's subjective experience, and the building facilities will give users a better experience. For example, lighting environment could interact intelligently instead of controlling it with fixed switches. When users getting up at midnight, the smart home enabling lighting systems suddenly. And the dazzling light is harmful to users. But smarter home is not only to make the switches perceive users' activities, but also adjust the brightness to 20% of normal according to objective cases. Traditional switches should also be more intelligent, but for a better experience, smarter improvement is necessary. The intelligent space environment embodies the concept of "people-oriented", no matter where users are, just an Internet-connected smart phone or Tablet, are beneficial to know home state and control automatic equipment freewheeling.

4.3. *Intelligent control of building energy consumption*

Building energy-saving means saving energy through heating, air conditioning, refrigeration and lighting, and it also includes other aspects, such as reducing the energy consumption used in adjusting indoor air and humidity, and changing the quality of indoor environments. Running the building equipment based on renewable energies, including solar, wind and geothermal energy, the intelligent energy-saving building technology integrates it into smart energy-saving material technology, to make a comprehensive unity. Indoor air conditioning could be centrally controlled by the smart system, which optimizing running time for air-conditioning and air-cooling system, thereby, the wastes of energy can be reduced. Energy management system will extract data of varies mechanical and electrical equipment thought the Internet of things and smart meters, which is helpful to understand the energy consumption of the buildings, even of each floor and room, to improve the building's energy-saving targets [10], to realize building energy saving at maximum, and to obtain a huge economic, social and ecological environment benefit.

5. Conclusion

The concept of smart city has been blended with the intelligent architecture design gradually while it is come up. Based on the development of intelligent buildings, the architect must make corresponding adjustments to assimilate new ideas, and connect architectures with the smart city, to translate the traditional linear design to integrate design. The city is inseparable from architecture, and the intelligent buildings are indispensable for the smart city. Architects need to link information technology with architecture design, and promote the development of smart city through intelligent buildings. The smart city has just started, and the intelligent architecture design owns an enormously potential. New building forms, which are highly-instrumented, highly-interconnected and highly-intelligent, will be achieved finally.

References

1. WU Xi-Bo, YANG Zai-Gao, *the Concept of Smart City and City Development*. Vol. 11 (Urban Studies, 2010), pp. 56-60.
2. WANG Jing-Yuan, *Survey of Data-Centric Smart City*. Vol. 51 (Journal of Computer Research and Development, 2014), pp. 239-259.
3. XIA Xiao-Bo, *Collection of Knowledge, Endowed with Wisdom-the Developing Thinking of Intelligent Building Which Adapt to the Smart City's Demand.* Vol. 167 (Smart City 2014), pp. 64-66.
4. Guangzhou Guang-Lian digital home industry Technology Research Institute, *Intelligent Buildings in the Smart City: Proper Combination of Science and Architecture*. Vol. s1 (Video Engineering, 2011), pp. 18-21.
5. XU Yang, WANG Xiao-Feng, *the Internet of Things Based Information Support System for Multi-Agent Decisions*. Vol. 25 (Journal of Software, 2014), pp. 2325-2345.
6. XU Jing, *Key Techniques of Smart City and Path to Realize*. Vol. 8 (Telecommunications Science, 2013), pp. 123-126.
7. WANG Zhan-You, *Research on Natural Ventilation Technique Used in Building*. Vol. 35 (Building Energy Efficiency, 2007), pp. 20-23.
8. ZHANG Zhong-Kuo, YAO Zhi-Fei, *Utilization of the Chimney Effect in Building Energy-saving*. Vol. 38 (Building Energy Efficiency, 2010), pp. 18-13.
9. ZHANG Xiao-Lin, ZHANG Bo-Han, *Analysis of Intelligent Home Solution.* Vol. 21 (Mobile Communication, 2014), pp. 38-41.
10. JIA Xiao-Ping, *Future of Architectures, Smart Development*. Vol. 1 (Investigation Design, 2015), pp. 36-39.

Meso-level 3d numerical study on single fiber pull-out test for interface heterogeniety

Y.F. Zhang*, X.R. Zeng

Faculty of Civil Engineering, Guangzhou University,
Guangzhou, 510006, China
**E-mail:zhangyafang2004@163.com*

A 3D single fiber pull-out model based on material meso-level and heterogeneity had been set up by using parallel computing approach with millions cells. AE (Acoustic Emission) evolution process, tensile stress nephograms and load-step-AE curves could, therefore, be obtained and discussed on models with various interface heterogeneities parameters. The results show that peak load and brittleness increase when the interface heterogeneity increases while toughness drops. In addition, high heterogeneity could weaken interface capacity for buffering tensile stress concentration when fiber is being pulled out.

Keywords: Interface; Heterogeneity; Acoustic Emission; Stress Nephogram.

1. Introduction

Technique of fiber reinforced concrete (FRC) is now widely adopted in civil engineering. Adding different kinds of fibers would enhance most mechanical properties of the concrete effectively [1]. The interface is formed when concrete is hydrating with fibers and it has completely different mechanical properties to concrete and fiber, thus it will cause a series of complicated physical and chemical reactions [2]. The interface bonding status is crucial to the macro performance of specimens. There are many approaches for characterizing the mechanical properties of interface however, single fiber pull-out test is the most common one, and many scholars got a lot of meaningful results by pulling the fibers out of the cement base [3-4]. Along with the rapid development of computer technique, simulating failure process numerically becomes a hot point in material science. For instance, Yang etc. [5] investigated shear stress distribution of interface by using FEM in 2003.

In this paper, meso-level single fiber pull-out numerical models with various interface heterogeneities had been studied by adopting RFPA^{3D} (realistic failure process analysis) and single fiber pull-out mechanical properties influenced by the change of interface heterogeneities had also been investigated.

2. Numerical Model

Normally, statistical method such like Weibull distribution function could be used to describe material heterogeneity[6]. The function is shown in Equation 1.

$$f(x) = \frac{m}{x_0}\left(\frac{x}{x_0}\right)^{m-1} \exp\left(-\frac{x}{x_0}\right)^m \qquad (1)$$

m is defined as heterogeneity index of material. Low m value stands for material tending to be heterogenous, while high m value stands for material tending to be homogenous. x_0 stands for the mean value of the mechanics properties like elastic modulus, strength and Poisson's ratio.

Furthermore, concrete is a typical quasi-brittle material. In this study, maximum tensile stress criterion and modified Mohr-Coulomb criterion have been adopted and the meso-level cells are abided by the elastic-brittle constitutive relationship.

The model is shown in Fig. 1. The numerical models are generated and divided elements by ANSYS first. Then the nodes information, load information were exported by macro command in ANSYS. Finally all the files information were imported to RFPA system.

Fig. 1. Numerical model.

Dimension of the concrete matrix is with 40 x 40 x 40 mm. The fiber has a diameter of 2 mm and its embedded depth is defined as 20mm. The interface thickness is 0.1 mm. The model, divided by SWEEP method in ANSYS, contains approximate 1.3 million cells. The vertical load has been applied by controlling the displacement of the top end of the fiber with 1.0^{-4} mm per step. Mechanical parameters are adopted with reference to Tang *etc.* [7]. Five different interface heterogeneity models (m=2, 10, 20, 50 and 100, respectively) have been calculated. Among these samples, the one with m=2 and another with m=100 are selected to represent for scenarios with low interface heterogeneity and high interface heterogeneity, respectively.

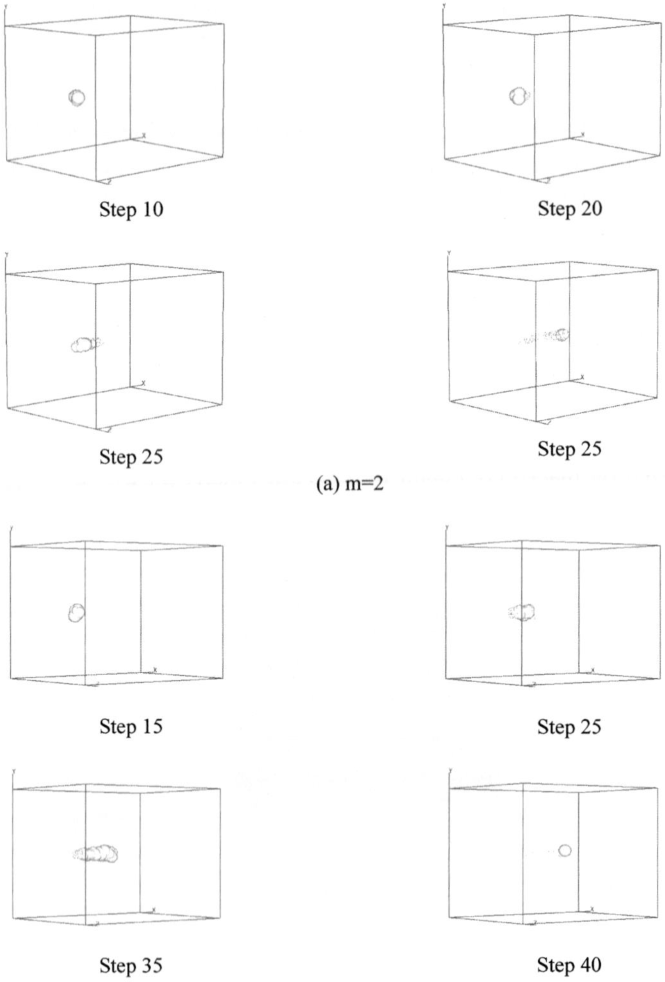

(b) m=100

Fig. 2. Pictures of AE evolution.

3. Results & Discussions

AE counts or AE energy release rate are closely related to material damage quantity and they are proportional to material damage [8]. The interior crack process and AE energy release condition could be shown by AE evolution pictures. Two different kinds of interface heterogeneity samples (*m*=2 & *m*=100) pictures of AE evolution are presented in Fig. 2. In these figures, bubbles are drawn at the position of AE, with their radius are set proportional to the energy released.

As can be shown in Fig. 2a, AE appears near the tip of fiber during the initial loading (Step 10). With broken weak cells on the interface near the tip of the fiber, the bubbles are then concentrated. This phenomenon can be interpreted as that more energy is released because the area near the tip of the fiber is stiffness mutation zone. With increasing load, some small bubbles appear in the inside interface which means that more interface cells of the matrix are damaged. The radius of bubbles is becoming smaller from the tip of the fiber to the inside matrix. The development of interface debonding trend is from the fiber tip to the interior matrix either. At Step 40, the bubbles in central interface becomes small, however, in the embedded end of the fiber, just a few larger bubbles appear. It implies that, most of the interface cells have been damaged. Then the fiber is pulled out from the matrix.

Fig. 2b shows that more interface energy releases near the tip of the fiber in Step 15 with larger bubbles. After interface debonding near the tip of the fiber, the interface inside the matrix releases more energy and the radius of bubbles become larger and larger from the fiber tip to the matrix. Larger bubbles are full of interface cells in Step 35. Plenty of interface cells are damaged and the interface energy releases fiercely. When reaches to Step 40, the AE disappears in the matrix, and the sample loses its bearing capacity.

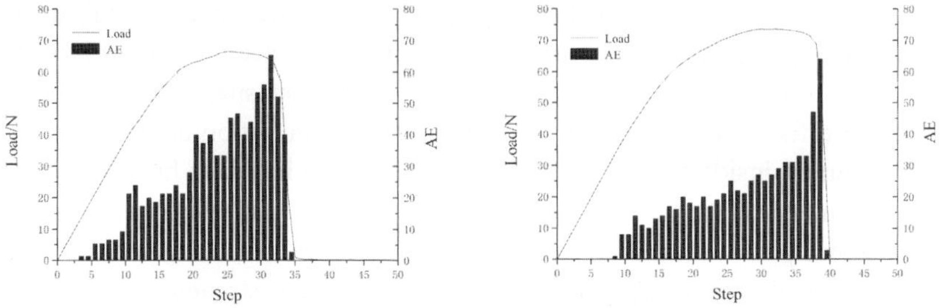

(a) m=2 (b) m=100

Fig. 3. Load-step-AE curves.

Load-step-AE curves of two samples are shown in Fig. 3. The influence of interface heterogeneity is significantly by comparing Fig. 3a and 3b. AE first appears at Step 4 of Sample with m=2 while in another sample, AE appears firstly at Step 9. In Fig. 3a, AE counts begin to rise gradually at elastic region of the curve. When the load reaches to the peak, The AE counts reache to the peak either. At this time, the load step is 32 and the peak AE counts are 49. While the curve turns into degradation region, obvious AE counts are founded because the interface cells are not completely damaged at degradation region. The sample does not lose its bearing capacity completely.

In Fig. 3b, AE count rises steadily but slowly beginning from Step 9. When the load reaches near to the peak point, AE counts rise greatly. The load step is 39 and AE counts are 64 at this step. After that, if read in conjunction with Fig. 2, the load curve turns into degradation region and AE counts disappear because plenty of interface cells are suddenly damaged that the sample presents obvious characteristic of brittle failure.

The change of interface heterogeneity has effects on the peak load. It could be seen that the peak load of the sample with m=2 is 66.5N and it is 73.5N for sample with m=100. Zhang *etc.* [9] pointed out that the change of material heterogeneity is related to the peak load, *i.e.* the peak load increases then the material heterogeneity is increasing. This conclusion is consistent with the result of the simulation in this paper.

In order to investigate tensile stress distribution of interior matrix, two tensile stress nephograms for samples with different interface heterogeneity are shown in Fig. 4. The area with large tensile stress value in the figures could be considered as the stress concentration zone. Comparing with Fig. 4a and 4b, it could be found that tensile stress concentration zone is larger in Fig. 4b, which is close to about 1/3 of the interior matrix area. Instead, the tensile stress concentration zone of Fig. 4a is small and exists only near the interface. This is because when interface heterogeneity is low, the discrete of interface mechanical parameters increases and the overall interface mechanics property is low as well. Cracks extend normally along with the low strength of the interface. Weak interface could efficiently buffer stress concentration and regulate stress redistribution inside the matrix. After increasing the interface heterogeneity, the overall interface mechanical property is enhanced and the overall interface mechanical parameters tend to the mean mechanical values given by Weibull distribution function. Interface has strong bonding strength, and it is not conductive to buffer stress concentration. Moreover, it is easy to damage the matrix due to the higher crack tip stress [2]. These results are consistent with the physical experiment conducted by Wu in 2000[2].

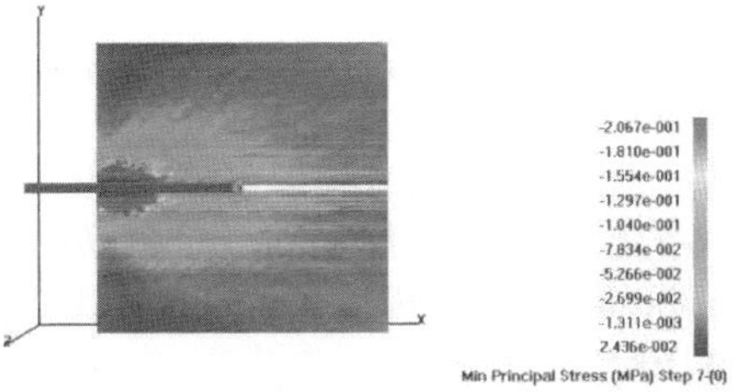

(a) m=2

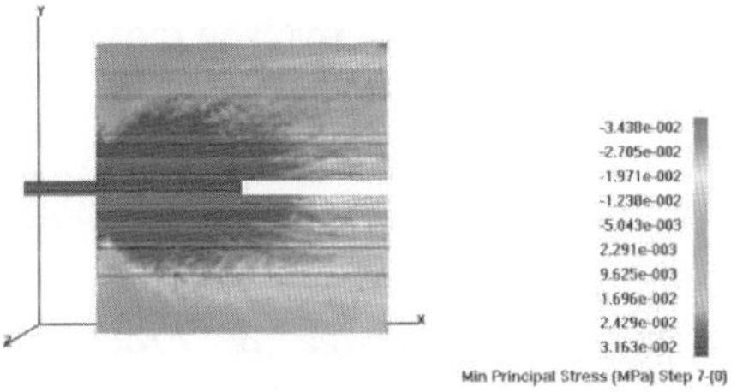

(b) m=100

Fig. 4. Tensile stress distribution inside the matrix.

4. Conclusion

3D FEM numerical simulation had been conducted and the influence of mechanics properties of single fiber pull-out tests had been investigated by changing interface heterogeneity. Conclusions can be drawn as follows.

a. With interface heterogeneity increased, the toughness of the samples drops but the brittleness increases.
b. Peak load increases when interface heterogeneity increases.
c. Interface heterogeneity has great impact on the distribution of tensile stress. With increasing interface heterogeneity, tensile stress concentration zone increases within matrix.

Acknowledgments

This study was supported by the National Nature Science Funding (Grant No. 51278136), Guangzhou science & technology innovation platform (Grant No. 15180005) and Guangzhou Education Bureau of YangCheng Scholar Funding (Grant No. 12A007G).

References

1. Wang C.K. *Structure of fiber reinforced concrete.* China Machine Press, Beijing (2004).
2. Wu R.J. *Composite materials.* Tianjin University Press, Tianjin (2000).
3. Shannag M J, Brincker R, Hansen W. *Pullout behavior of steel fibers from cement-based composites.* Cement and Concrete Research, 27(6), pp. 925-936 (1997).
4. Won Jong-Pil, Hong Byung-Tak, Lee Su-Jin, el al. *Bonding properties of amorphous micro-steel fibre-reinforced cementitious composites.* Composite Structures, (102), pp.101-109 (2013).
5. Yang Q S, Qin Q H, Peng X R. *Size effects in the fiber pullout test. Composite Structures,* 61(3), pp.193-198 (2003).
6. Weibull W. *A statistical theory of the strength of materials,* Stockholm (1939).
7. Tang C.A, Zhu W.C. *Damage & Fracture of Concrete-Numerical Simulation,* Science Press, Beijing (2003).
8. Jiao R.M, Tang C.A, Zhang G.M, et al. *Numerical test of influence of mesoscopic heterogeneity on macroscopic behavior of rock failure and seismic sequence types.* Chinese Journal of Geophysics, 46(5), pp. 659-666 (2003).
9. Zhang Y.F, Qi L, Tang C.A. *Numerical simulation in influence of heterogeneity on mechanical behavior of composites reinforced with fibers.* Journal of Wuhan University of Technology, 29(4), pp. 14-16 (2007).

Pile foundation test experimental program of Lanzhou new city science and technology innovation

Shuai-Hua Ye[1,2,a], Xiao-Nan Gong[1,b]

[1]*Research Center of Coastal and urban Geotechnical Engineering, Zhejiang University, Hangzhou 310058, China*

[2]*School of Civil Engineering, Lanzhou University of Technology, Lanzhou, 730050, China*

Email: [a]yeshuaihua@163.com, [b]xngong@hzcnc.com

This article expounded the pile testing about new city science and technology innovation in Lanzhou city (North), and proposed some testing schemes. Summarizes the predecessors' experience for pile foundation inspection, and formulate the pilot scheme about test pile bearing capacity of single pile vertical compressive test and loading plate test, engineering pile bearing capacity of single pile vertical compressive test and loading plate test, and equipped with corresponding figure.

Keywords: Pile Foundation Test; Static Load; CFG Pile; Loading Plate Test.

1. Introduction

With the rapid development of urban industrialization and commercialization in Lanzhou, the requirement of foundation engineering is becoming more strict. However, the development of pile foundation detection problem has not yet mature. Therefore, in this paper, beginning the CFG pile static load test for the scientific and technological innovation in Lanzhou north area, and loading plate test about vertical compressive ultimate bearing capacity of single pile test and of soil between piles of test pile in the selected area, design verification for the bearing capacity of composite foundation. And draw the corresponding test device.

2. Project Profile

According to the hydrogeologic and engineering geological prospecting institute of Gansu province, which provide science and technology innovation city of Lanzhou (north)the geotechnical engineering exploration report", proposed the whole site is surrounded by planning four roads, is in the shape of rectangular, north and south about 708m long and 338m wide from east to west. The two proposed sites including 2-20F, 50 buildings and supporting facilities. The proposed sites including 4 buildings of 8-11F and new energy and new material buildings in the North area; And the 13-15F building of energy conservation and environmental protection; A 15-20F building of the science and technology on the ground, with 1-2F the basement; Proposed biological technology, of which the ground 9-13F three biotech, the skirt building, dozens of underground parking lot to the whole basement.

3. Experimental Objective

According to "code for building pile foundation inspection technologies" (JGJ106-2003, J256-2003), and in accordance with the requirements, beginning the CFG pile static load test for the scientific and technological innovation in Lanzhou north area before construction, and loading plate test about vertical compressive ultimate bearing capacity of single pile test and of soil between piles of test pile in the selected area, design verification for the bearing capacity of composite foundation; At the end of the construction, detect the quality of engineering piles by the single pile vertical compressive static load test, determine the engineering characteristic values of pile vertical compressive bearing capacity of single pile, vertical compressive load between pile soil test to test the foundation bearing capacity of soil between piles; At the end of the construction, integrity testing of pile body by low strain.

4. Experimental Program

Test pile bearing capacity of single pile vertical compressive test and loading plate test.

Before construction, made 4 test piles according to the parameters of engineering pile in the selected area (2 is 22m, 2 is 18m), test of single pile vertical compressive ultimate bearing capacity with the method of static load, and choose three places for the soil between pile composite foundation loading plate test, and verifiy for the soil bearing capacity of composite foundation design. Test pile area propertied management center, the test pile and loading plate test layout as shown in Figure 1.

4.1. *Engineering pile bearing capacity of single pile vertical compressive test and loading plate test*

After the construction, adopt the method of static load test to detect the single pile vertical compressive bearing capacity characteristic value, single pile design characteristic values are 1700 kN and 1200 kN.

After construction, the number of tested pile foundation according to the provisions of code, the engineering pile of every single building, the number not less than 3, and the total number of pile foundation not less than 1% of the whole; The project construction monomer are 9, the number of engineering pile tested every single building is 3, a total of 27 root. The engineering pile of every single building, tested detailed location be established after completion of engineering pile construction according to the local position.

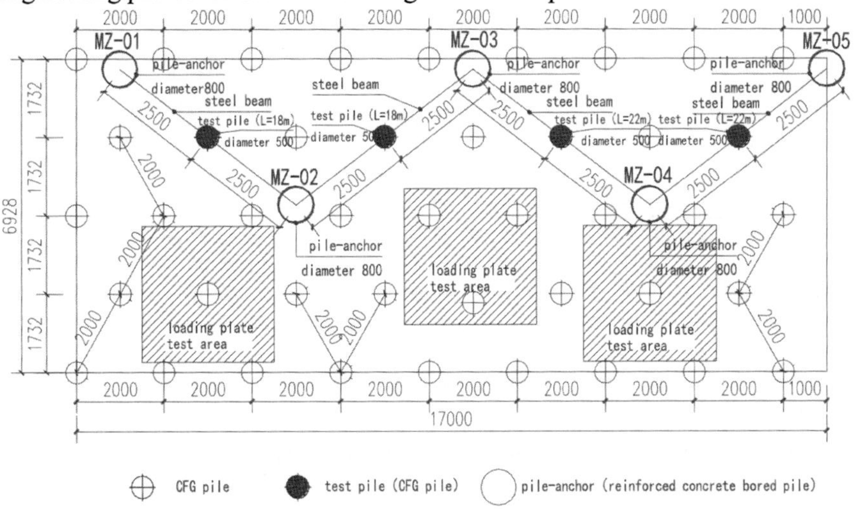

Note: the center of CFG pile spacing is 2 m

Fig. 1. The floor plan of test pile and loading plate test.

4.2. *Test apparatus and unloading system*

Bearing capacity of single pile vertical compressive test apparatus and unloading system. Bearing capacity of single pile vertical compressive test counterforce device can choose according to field condition of pile-anchor beam counterforce device, the pile-anchor adopt reinforced concrete filling pile, pile length 20m, diameter 800mm, a total of 5 root, numbered MZ-01, MZ-02, MZ-03, MZ-04, MZ-05. Load counterforce device of all the components checking computations on strength and deformation before the test, and checking computations on the pulling force of pile-anchor (the foundation soil, pile reinforcement, joint of pile), plane arrangement of the pile-anchor and test pile as shown in Figure 2, the position relationship of the pile-anchor reinforced and steel beam as shown in Figure 3, the main reinforcement of MZ-01, MZ-02 and MZ-03 using reinforced, the main reinforcement of MZ-04 and MZ-5 using reinforced, the reinforcement drawing of the pile-anchor as shown in Figure 4.

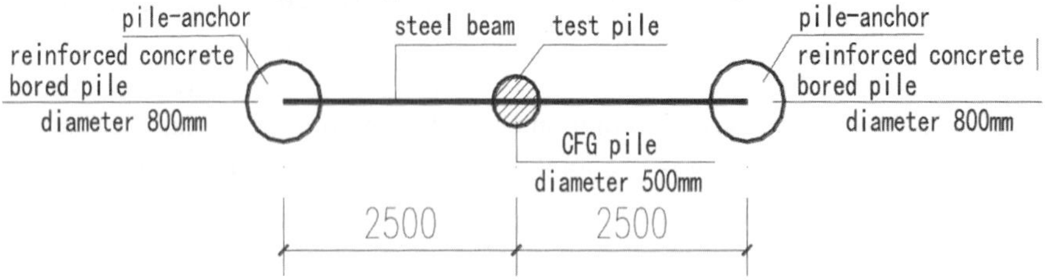

Fig. 2. The layout of pile-anchor and test piles.

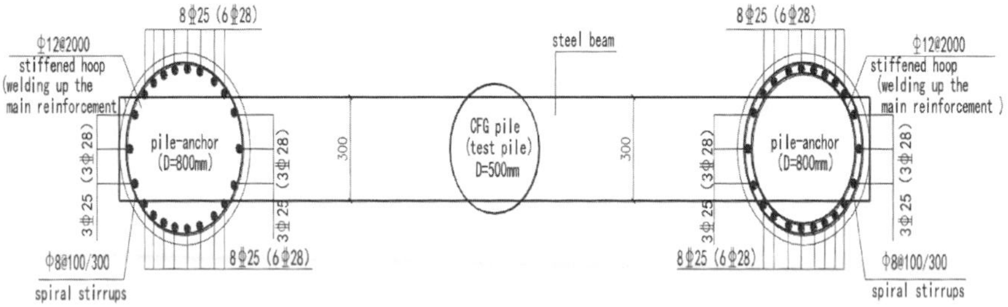

Fig. 3. The position relationship of pile-anchor reinforced and steel.

Test load using two hydraulic jacks parallel synchronous work, using the jack type, specification should be the same. The resultant force center of jack should be coincident with the pile axis. The settlement measurement adopt the displacement sensor.

5. Conclusions

(1) According to the test program, test pile vertical compressive ultimate bearing capacity of single pile test and loading plate test of soil between piles, program is reasonable.

(2) For low strain of pile integrity detection aspects still need to further improve.

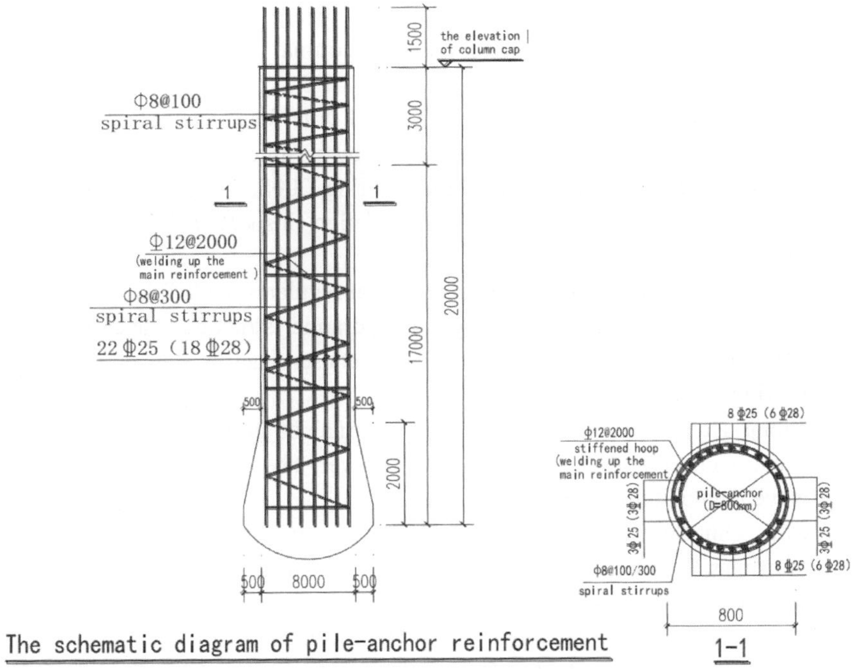

Fig. 4. pile-anchor reinforcement figure.

Acknowledgments

This work was financially supported by youth science and technology research fund projects in Gansu province "Study on the Characteristies of Pile Negative Friction in the Large Thickness Loess Area Considering the Collapsibility" (2014GS02363).

This work is supported by youth science and technology research fund projects in Gansu province "Study on the Characteristies of Pile Negative Friction in the Large Thickness Loess Area Considering the Collapsibility" (2014GS02363).

References

1. XIA Li-nong, LEI Ming, NIE Chong-jun. Field test of influences of load at pile top on negative skin friction behaviors [J]. Rock and Soil Mechanics, 2009, 30(3): 664-668.
2. LI Da-zhan, TENG Yan-jing, HE Yi-hua, et al. Vertical bearing behaviour of lager diameter belled pile in collapse loess [J]. Chinese Journal of Geotechnical Engineering, 1994, 16(2): 11-21.

3. WEI Jin, LI Zhe, HAO Mang-li, et al. Submerged static load test of pile foundation in collapsible loess area [J]. Journal of Chang'an University, 2011, 31(5): 63-67.
4. WANG Lan-min, SUN Jun-jie, HUANG Xue-feng, et al. Field tests on negative skin friction along piles caused by seismic settlement of loess [J]. Chinese Journal of Geotechnical Engineering, 2008, 30(3): 341-348.
5. Wu Mianba, Summary of pile foundation inspection[J]. Geotechnical foundation, 2001, 03: 2-8.
6. Ultrasonic testing concrete defect technical regulations, CECS21-90.
7. Luo Qixian. Ultrasonic testing of hydraulic structure concrete. High Efficiency Silicon Solar Cells (Water conservancy water power press, Beijing 1989).
8. Xiong Hongmei, Hao Xiaohong, in: Journal of North China Institute of Water Conservancy and Hydroelectric Power, The application of ultrasonic wave method to pile test, 66-69. (2002).
9. Chu Qingshun, in: Shanxi Architecture, on application of acoustic transmission method and reflected wave method in foundation pile checking, 104-105. (2010).
10. Zhou Depei, Wang Huanlong, Sun Hongwei. Micro-pile composite anti-sliding structure and its design theory[J]. Chinese Journal of Rock Mechanics and Engineering, 2009, 07: 1353-1362.

Static load test of a project CFG pile composite foundation

Shuai-Hua Ye[1,2,a], Xiao-Nan Gong[1,b]
[1]*Research Center of Coastal and urban Geotechnical Engineering, Zhejiang University, Hangzhou 310058, China*
[2]*School of Civil Engineering, Lanzhou University of Technology, Lanzhou, 730050, China*
Email: [a]yeshuaihua@163.com, [b]xngong@hzcnc.com

In this paper, carried on static load test of CFG pile composite foundation , according to the geological conditions of the region and determine the proper testing scheme, get CFG test piles' vertical compressive static load test load-settlement curve, and made further analysis on the curve. And solve the characteristic value of composite foundation bearing capacity .

Keywords: Pile Foundation Test; Static Load; CFG Pile; Characteristic Value.

1. Introduction

With the rapid development of urban industrialization and commercialization in Lanzhou,the requirement of foundation engineering is becoming more strict. It is particularly important to the est of rationality composite foundation. Therefore, in this paper, determined the vertical compressive bearing capacity of soil between piles and CFG single pile vertical ultimate bearing capacity, and determined the pile length of 18m and 22m single pile bearing capacity characteristic value.

2. Project Profile

According to the hydrogeologic and engineering geological prospecting institute of Gansu province, which provide science and technology innovation city of Lanzhou (north)the geotechnical engineering exploration report, proposed the whole site is surrounded by planning four roads, is in the shape of rectangular, north and south about 708m long and 338m wide from east to west. The two proposed sites including 2-20F, 50 buildings and supporting facilities. The proposed sites including 4 buildings of 8-11F and new energy and new material buildings in the North area; And the 13-15F building of energy conservation and environmental protection; A 15-20F building of the science and technology on the ground, with 1-2F the basement; Proposed biological technology, of which the ground 9-13F three biotech, the skirt building, dozens of underground parking lot to the whole basement.

3. Geological Data

According to the "geotechnical engineering exploration report", the survey maximum depth is 46.6m of the proposed site , according to drilling, the exploration depth of layer are mainly the quaternary unconsolidated sediments and mudstone, from top to bottom divided into seven layers, six interlayer, filled soil, powder soil, gravel, sand, clay, breccia, and breccia.

4. The Determination of the CFG Single Piles' Vertical Ultimate Bearing Capacity

1#-4#CFG test piles' vertical compressive static load test load-settlement are gradually applied, and the corresponding CFG test piles' vertical compressive static load test load-settlement curve as shown in Figure 1.

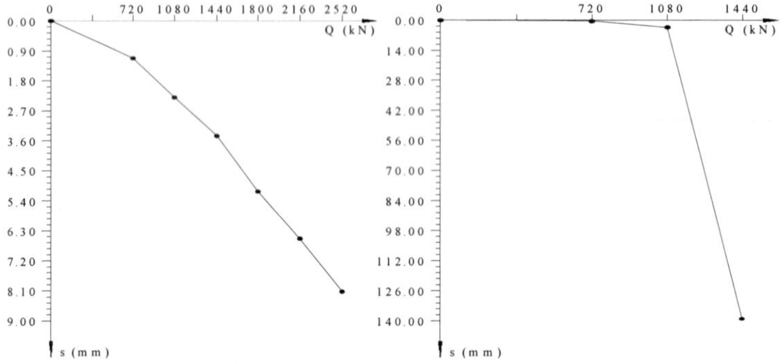

a) the Q-S curve of 1# pile

b) the Q-S curve of 2# pile

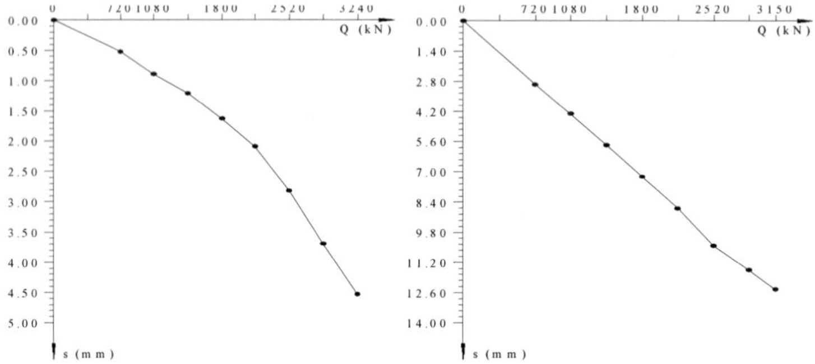

c) the Q-S curve of 3# pile

d) the Q-S curve of 4# pile

Fig. 1. 1#-4#CFG test piles' vertical compressive static load test load-settlement curve

According to the load-settlement and load-settlement curve shows that, in the natural state, 1#, 3#, 4# pile top vertical maximum stable loading quantities are 2520 kN, 3240 kN and 3150 kN, the corresponding settlements are 8.12mm, 4.53mm and 12.45mm, the Q~S curve is slowly, approximate straight line. Among them, 1# test pile(18m) load has reached the design load 2 times end loading condition, does not appear limit damage state, because of the CFG test pile's pile head appear broken phenomenon, lead to cannot continue to load, end load; 3#, 4# CFG test pile (22m) do not meet the design load 2 times end loading condition, but as a result of the CFG test pile's pile head appear broken phenomenon, lead to cannot continue to load; Therefore, 1#, 3#, 4# test piles' vertical ultimate bearing capacity were respectively measured values, and take the corresponding Q~S maximum load values: 2520 kN, 3240 kN and 3150 kN. In addition, 2#CFG test pile (18 m), in the process of loading premature failure state, the corresponding load is 1080 kN, the survey found that test pile has been destroyed in the process of the construction pile in, so 2# CFG test pile's test data is unreliable, cannot be used for subsequent.

5. The Determination of Soil Between Piles' vertical Compressive Bearing Capacity

1#-3# point's soil between piles vertical compressive static load test load-settlement summary are are gradually applied, and the corresponding soil between piles vertical compressive static load test load-settlement curve is shown in Figure 2.

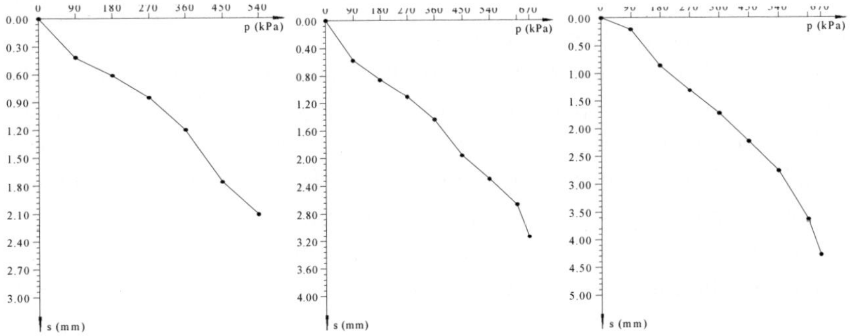

a) p-s curve of 1# b) p-s curve of 2# c) p-s curve of 3#

Fig. 2. 1#-3#points' soil between piles vertical compressive static load test load-settlement curve

As is shown in the design, CFG piles of this project have two kinds of pile types, therefore, in the experimental sites were selected three places, carried out the experimental of vertical compressive bearing capacity of soil between piles.

According to the load- settlement summary tables and load-settlement curve shows that, in the natural state, 1# point vertical maximum load is 540 kPa, 2#, 3# points' vertical maximum load are 670 kPa, corresponding settlements are 2.11mm, 3.14mm and 4.28mm, p-s curve is slowly, approximate straight line.

6. The Characteristic Value of Bearing Capacity Composite Foundation

By the "composite foundation technical specification" (GB/T50783-2012), the characteristics of composite foundation bearing capacity through integrated pile's vertical compressive load test and the soil between piles' vertical compressive load test, determined and combined with engineering practice experience, composite foundation bearing capacity characteristic value obtained according to the following formula:

$$f_{spk} = k_p \lambda_p m R_a / A_p + k_s \lambda_s (1-m) f_{sk} \tag{1}$$

$$f_{spk} = \beta_p m R_a / A_p + \beta_s (1-m) f_{sk} \tag{2}$$

$$\beta_p = k_p \lambda_p \tag{3}$$

$$\beta_s = k_s \lambda_s \tag{4}$$

$$m = d^2 / d_e^2 \tag{5}$$

In the formula: A_p - single pile's cross-sectional area (m^2); R_a - single pile's vertical compressive bearing capacity characteristic value (kN); f_{sk} - soil between piles' foundation bearing capacity characteristic value (kPa); m - composite foundation displacement rate; d - the pile's diameter (m); d_e - share of single pile foundation treatment area equivalent diameter(m); k_p - the correction coefficient of piles' actual vertical compressive bearing capacity of composite foundation, and have the relationship with construction technology, replacement ratio of composite foundation, the engineering properties of soil between piles, valued according to the regional experience ; k_s - the correction coefficient of soil between piles' composite foundation actual bearing capacity, and have the relationship with the engineering properties of soil between piles, pile construction technology,valued according to the regional experience ; λ_p - the developing coefficient of vertical compressive bearing capacity of piles, reflect when the damage of composite foundation, the developing level of pile's vertical compressive bearing capacity, valued according to the regional experience; λ_s -

the developing coefficient of soil between pile's foundation bearing capacity, reflect when the damage of composite foundation, the developing level of between piles' foundation bearing capacity, valued according to the regional experience and the engineering properties of soil between piles; β_p - the correction coefficient of piles' vertical compressive bearing capacity; β_s - the correction coefficient of soil between piles' bearing capacity.

7. Conclusions

According to the theoretical calculation results and field test, proposed science and technology innovation city of Lanzhou (north)composite foundation bearing capacity characteristic value as the following values:

(1) pile length 18m: single pile bearing capacity characteristic value is 1260 kN, bearing capacity of composite foundation characteristic value is 322.7 kPa.

(2) pile length 22m: single pile bearing capacity characteristic value is 1575 kN, bearing capacity of composite foundation characteristic value is 401.3 kPa.

Acknowledgments

This work was financially supported by youth science and technology research fund projects in Gansu province"Study on the Characteristies of Pile Negative Friction in the Large Thickness Loess Area Considering the Collapsibility"(2014GS02363).

References

1. XIA Li-nong, LEI Ming, NIE Chong-jun. Field test of influences of load at pile top on negative skin friction behaviors [J].Rock and Soil Mechanics, 2009,30(3): 664-668.
2. LI Da-zhan, TENG Yan-jing, HE Yi-hua, et al. Vertical bearing behaviour of lager diameter belled pile in collapse loess[J]. Chinese Journal of Geotechnical Engineering, 1994, 16(2): 11-21.
3. WEI Jin, LI Zhe, HAO Mang-li, et al. Submerged static load test of pile foundation in collapsible loess area[J].Journal of Chang'an University, 2011, 31(5): 63-67.
4. WANG Lan-min, SUN Jun-jie, HUANG Xue-feng, et al. Field tests on negative skin friction along piles caused by seismic settlement of loess[J]. Chinese Journal of Geotechnical Engineering, 2008, 30(3): 341-348.
5. Xiong Hongmei1 , Hao Xiaohong, in: Journal of North China Institute of Water Conservancy and Hydroelectric Power,The appl ication of ultrasonic wave method to pile test, 66-69.(2002).

6. Zhou Depei, Wang Huanlong, Sun Hongwei. Micro-pile composite anti-sliding structure and its design theory[J]. Chinese Journal of Rock Mechanics and Engineering, 2009, 07: 1353-1362.
7. Chu Qingshun, in: Shanxi Architecture, on application of acoustic transmission method and reflected wave method in foundation pile checking, 104-105.(2010).
8. Luo Qixian. Ultrasonic testing of hydraulic structure concrete. High Efficiency Silicon Solar Cells (Water conservancy water power press, Beijing 1989).
9. Ultrasonic testing concrete defect technical regulations, CECS21-90.
10. Wu Mianba, Summary of pile foundation inspection[J]. Geotechnical foundation, 2001, 03: 2-8.

Mechanical analysis for incremental launching construction of long-span continuous steel truss bridge

Hua Chai[1,a], Yu-Min Song[2,b]

[1]*Lawrence Technological University,*
21000 West Ten Mile Road Southfield, MI, USA.

[2]*College of Urban Railway Transportation, Shanghai University of Engineering Science,*
Shanghai, 201620, China

Email: [a]*582383070@qq.com,* [b]*song-yumin@163.com*

With the advantage of no-effect to shipping, shorter construction time and lower cost, the incremental launching construction method without supporting piers is widely used in the construction of steel and prestressed concrete bridges. At home and aboard, the cases of incremental launching construction that long-span bridges whose cantilever length exceeds 160m is rare, and the feasibility research on incremental launching construction method in long-span bridges is insufficient. With the example of a long-span bridge, whose main span is 180m, the feasibility of the incremental launching construction method without supporting piers is analysed by finite element method in this paper. The results show that, and the strength and stability of bridge will both meet requirements when the length of launching nose is 100m or 120m. So the method of incremental launching construction is feasible.

Keywords: Incremental launching Construction; Long-Span Bridge; Launching Nose; Strength; Stability.

1. Introduction

The incremental launching construction method without supporting piers is now widely employed in the construction of steel truss bridge and prestressed concrete bridge due to the advantages of such as low in cost, short in construction time and no-effect to shipping[1,2], and the construction technology has become more mature.

Millau Bridge has a cantilever of 171m and a span of 342m, the bridge was finally completed and operated successfully in 2004, which represented the advanced level of incremental launching construction[3].

In China, the incremental launching construction method was first used in Shanxi Dijiahe railway bridge (4×40m) in 1978. And by 2010, Zhengzhou Yellow River railway and highway shared bridge (120m+5×168m+120m) whose cantilever length attains 168m was completed[4], which means the application of the incremental launching technology is becoming ripe day by day. While, at present there are few cases of incremental launching construction that long-span bridges whose cantilever length exceed 160m, not to mention those exceed 170m.

A long-span steel truss bridge is taken as an example, in which the stress states of the bridge installed by incremental launching construction method is analysed by finite element method.

2. Project Description

The bridge adopts through type open deck continuous steel trussed girder, and the two main trusses are triangle trusses with vertical post. (See Fig. 1 and Fig. 2 for details)

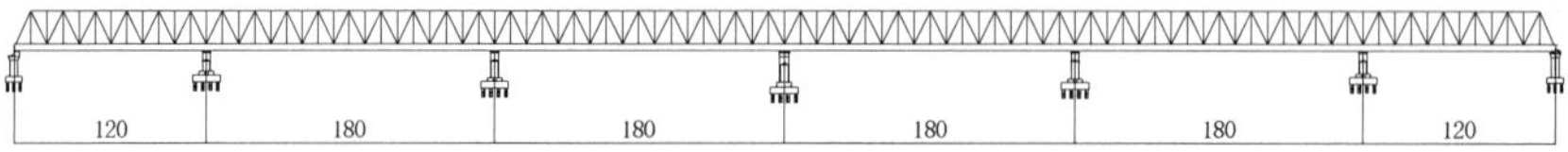

Fig. 1. Structure of main bridge (units:m).

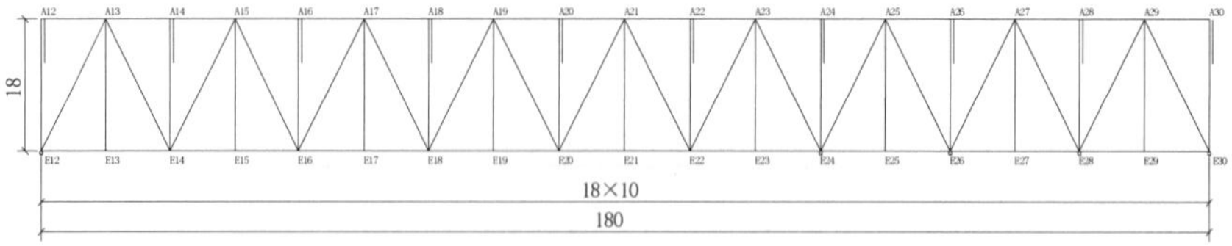

Fig. 2. General arrangement of the main truss (units:m).

3. Launching Nose Design

As the main temporary construction, the launching nose length is generally 0.5~0.7 times that of the cantilever[5] and is taken as 100m and 120m respectively in this paper. Steel truss beam with no-uniform depth are adopted in the bridge according to the connection between cantilevers and main trusses. The heights of steel truss beam in front and back of the cantilever are 9m and 18m, all the panel lengths are 10m. The length of the truss beam in front of the cantilever is 60m while the length of the truss beam back of the cantilever length are, in order, 20m, 40m and 60m. And both of the two main trusses are triangle trusses with vertical post. The truss space is 11.0m, the top bracing and the bottom bracing are installed on the of the main truss. Take 1/4 sectional dimension of the main truss members as the launching nose members' dimension. Fig. 3 illustrates the elevation arrangement of the launching nose(take the 120m-long launching nose as example).

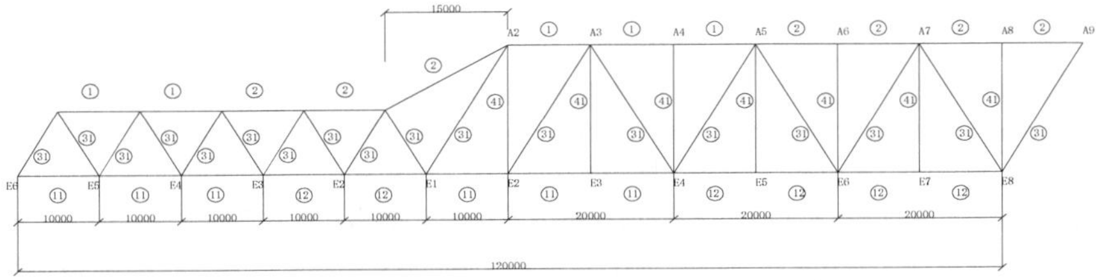

Fig. 3. Elevation arrangement of the launching nose(unit: m).

4. Constructional Mechanical Analysis

4.1. *The most unfavorable condition*

There are two critical stages during the incremental launching construction without supporting piers. One is the maximum cantilever stage, by this time the weight of the cantilever all act on the fixed end of the girder, which may result in overstressing of members. And the other is the stage when the launching nose end is pushed to the pier, during which large support reaction will act on the launching nose and lead to overstressing and even buckling of members. Under this condition, this thesis offers a feasibility analysis of the incremental launching construction based on the above two conditions.

4.2. *Calculating load*

In the calculation, three loads as follows are considered:

(1) Denote the unit weight of steel as 78kN/m3.

(2) Temporary load: 5kN/m according to the engineering experience.

(3) Wind load: According to relevant specifications[6], the wind pressure can be calculated as follows:

$$W=K_1 \cdot K_2 \cdot K_3 \cdot W_0=1.3\times 1.05\times 1.0\times 700=955.5(Pa) \tag{1}$$

4.3. *Material parameters*

Table 1. Main parameters of the steel.

E_g/MPa	E_v/MPa	υ
2.1×10^5	0.81×10^5	0.3

Table 2. Allowable stress of the steel.

steel	Q370	Q345
allowable stress	220	205

Two types of steel Q370q and Q345q are employed in the bridge. The main parameters such as elastic modulus E_g, shear modulus E_v and poisson ratio υ are shown in table 1 and their allowable stress[7] see table 2.

4.4. *Finite element model*

The calculation model is built by using truss elements to simulate the main stressed members and beam elements to simulate the other members. As is shown in Fig. 4, the finite element model includes 1145 nodes and 2829 elements (the number of truss elements and beam elements is 712 and 2117).

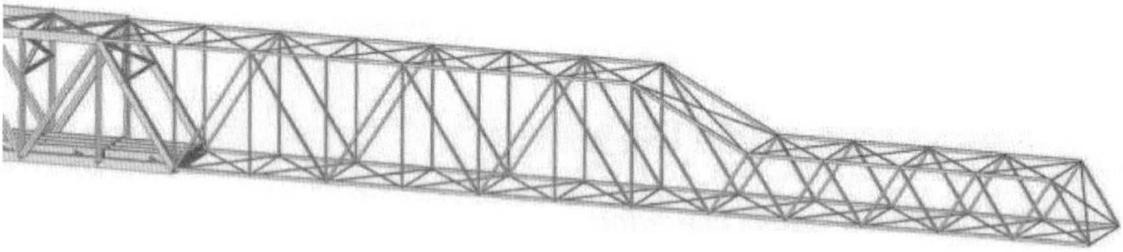

Fig. 4. Finite element model of the launching nose.

4.5. *Results of model calculation (L=100m)*

Under the condition of maximum cantilever stage (launching nose length L=100m), the calculation results under the combined action of self-weight, temporary load and wind load is shown in Fig. 5.

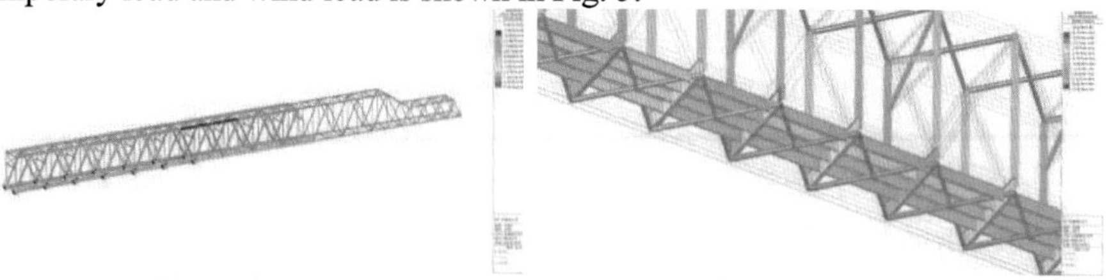

(a) truss elements (b) beam elements

Fig. 5. Stress results of maximum cantilever stage (L=100m).

As we can see in the Fig. 5:

(1) The maximum compressive stress of the truss elements is 218MPa and the maximum tensile stress is 189MPa, the structure is in safe condition.

(2) The maximum compressive stress of the beam elements is 252MPa and the maximum tensile stress is 318MPa, which is out of the allowed range. While the maximum stress was located in local members, so those members should be hardened to satisfy the strength requirement.

The first-order buckling mode and critical buckling coefficient under the condition of maximum cantilever are shown in Fig. 6.

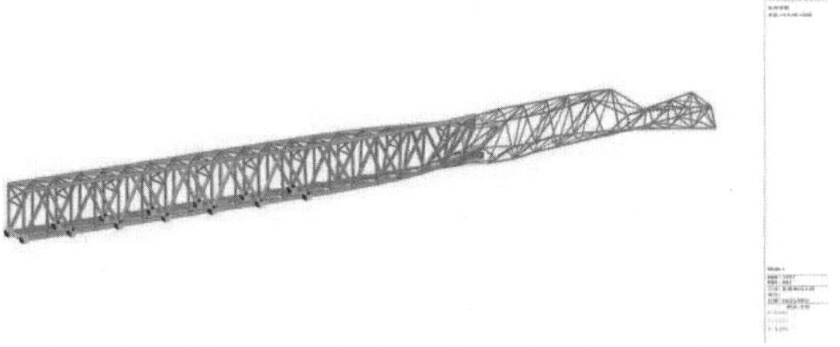

Fig. 6. The first-order buckling mode under the condition of maximum cantilever (L=100m).

From Fig. 6, under the maximum cantilever and varying self-weight condition, the critical buckling coefficient is 4.914 when the first-order bucking occurred on the cantilever. Hence, the good stability of the structure is proved.

At the stage when the launching nose end is pushed to the pier (launching nose length L=100m), calculation results under the combined action of self-weight, temporary load and wind load is shown in Fig. 7.

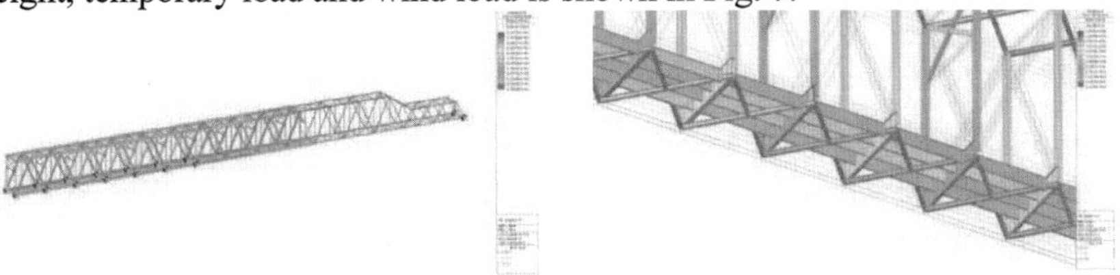

(a)truss elements (b) beam elements

Fig. 7. Stress state of the front end of the launching nose supporting (L=100m).

As we can see in the Fig. 7:

(1) The maximum compressive stress of the truss elements is 118MPa and the maximum tensile stress is 132MPa. So the structure is in safe condition.

(2) The maximum compressive stress of the beam elements is 121MPa and the maximum tensile stress is 149MPa. The structure is safe.

(3) When the front end of the launching nose is forced to move and supported on the pier, the stress of members decreases compared with the maximum cantilever stage.

The first-order buckling mode and critical buckling coefficient at the stage when the launching nose end is pushed to the pier are shown in Fig. 8.

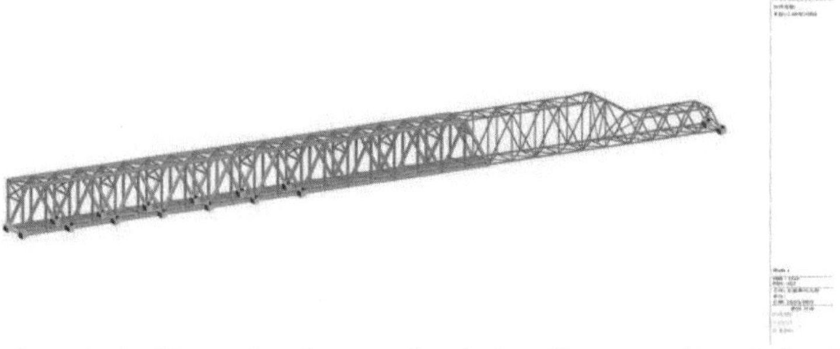

Fig. 8. The first-order buckling mode at the stage when the launching nose end is pushed to the pier (L=100m).

From Fig. 8, at the stage when the launching nose end is pushed to the pier and varying self-weight(while the temporary load and wind load are assumed to be constant value) condition, the critical buckling coefficient is 2.664 when the first-order bucking occurred, that is slightly larger than the specified value 2. Hence, the structure may be not so stable.

4.6. *Results of model calculation (L=120m)*

When the launching nose length is 120m, the main truss stress obtained is smaller than that when the launching nose length is 100m. Under the condition of maximum cantilever stage, calculation results under the combined action of self-weight, temporary load and wind load is shown in Fig. 9.

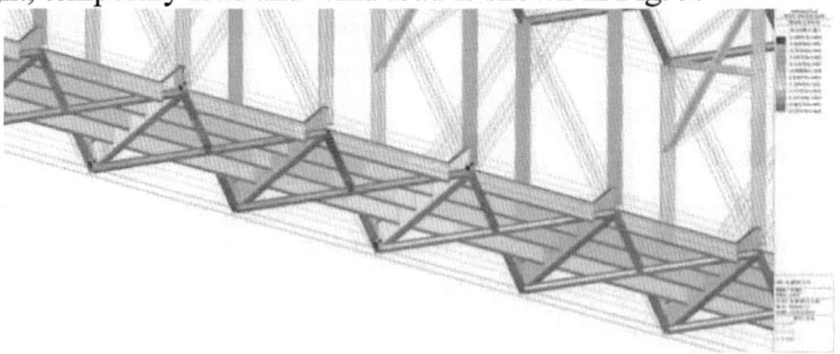

Fig. 9. Stress results of maximum cantilever stage (L=120m).

As we can see in the Fig. 9, when the launching nose length is 120m, The maximum compressive stress of the beam elements is 234MPa and the maximum tensile stress is 246MPa. Both are smaller than that when the launching nose length is 100m, so the structure is also in safe condition.

The first-order buckling mode and critical buckling coefficient under maximum cantilever and varying self-weight condition are shown in Fig. 10.

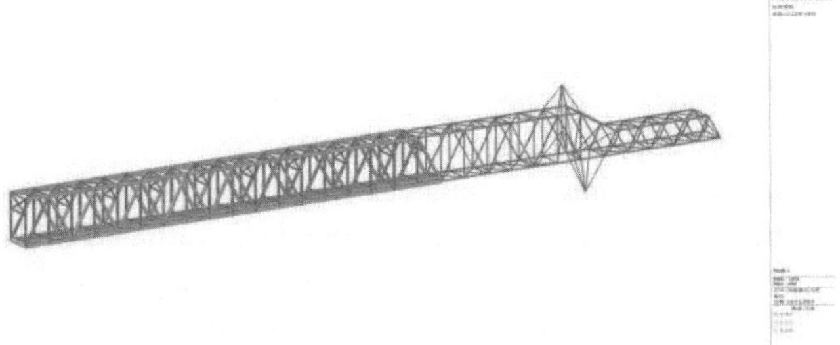

Fig. 10. The first-order buckling mode under the condition of maximum cantilever (L=120m).

Under the maximum cantilever and varying self-weight condition, the critical buckling coefficient is 2.232 when the first-order bucking occurred on the cantilever. The stability coefficient is smaller than that when the launching nose length is 100m. Although member buckling exists locally, and the stability decreases, the structure still stay safe.

When the launching nose end is pushed to the pier, the first-order buckling mode and critical buckling coefficient are shown in Fig. 11.

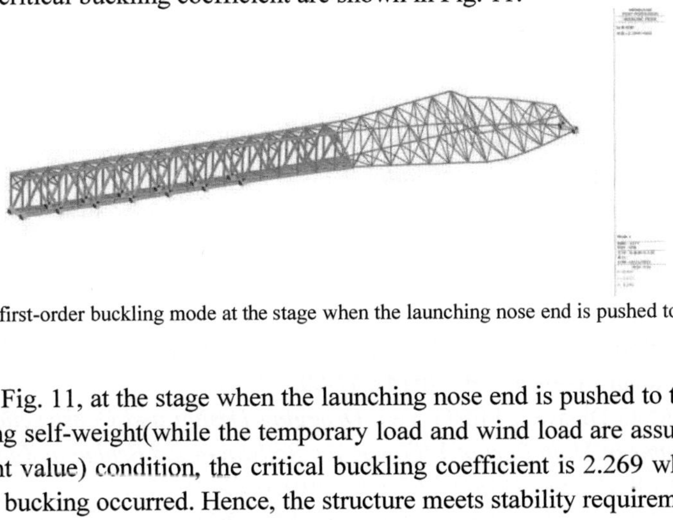

Fig. 11. The first-order buckling mode at the stage when the launching nose end is pushed to the pier (L=120m).

From Fig. 11, at the stage when the launching nose end is pushed to the pier and varying self-weight(while the temporary load and wind load are assumed to bc constant value) condition, the critical buckling coefficient is 2.269 when the first-order bucking occurred. Hence, the structure meets stability requirement.

5. Conclusion

Through the example of a long-span bridge, the strength and stability of the launching nose is studied when the length of launching nose length is 100m or 120m, come to the following conclusions:

(1) The strength and stability of bridge will both meet requirements when the length of launching nose length is 100m or 120m. Member buckling exists

locally, so those members should be hardened to satisfy the strength requirement. The launching nose should be further optimized.

(2) When the launching nose length is 100m, when the front end of the launching nose is supported on the pier, the stress of members decreases compared with the maximum cantilever stage. Therefore, in the design and construction of the launching nose, the site should be sufficient attention.

(3) Under the most unfavorable conditions, the results show that the structural strength and stability basically meet the requirements. So the pushing scheme is feasible, but needs further optimization.

(4) The stability analysis in this paper is based on the linear elastic supposition, then geometric nonlinearity and material nonlinearity should be considered in the further research.

Acknowledgments

The author would like to gratefully acknowledge the financial support from the National Nature Science Foundation of China (Grant No. 51478258) and the Youth Fund of Shanghai Education Committee (Grant No.A1-8500-14-0116), and thanks are given here.

References

1. LIU Guihong, *Design of Fully Welded Large Span Continuous Steel Truss Bridge over Railway*, in *Railway Construction Technology*, 2014, (1): 4-6.
2. JU Yunming. Constructi on Technologyof Large Span Steel Ttruss Bridge[J]. Construction & Design for Project, 2011, (7): 171-174.
3. XUE Xinguang. The Fast Construction Techniques for Cantilever-Erecting Continuous Steel Truss Beams[J]. Traffic Engineering and Technology for National Defence, 2012, 05(4): 47-50.
4. WANG Dianwei, FENG Yan, *Dragging Design and Construction of Steel Beams in Main Part of Yellow River Railway-Rood Bridge in Zhengzhou,* in *RAILWAY STANDARD DESIGN*, 2010, 05(5): 72- 76.
5. WANG Weifeng, LIN Junfeng, MA Wentian, *Ptimun Analysis of Launching Nose During Incremental Launching Construction of Bridge*, in *ENGINEERING MECHANICS*, 2007, 24(2): 132-138.
6. Ministry of Railways, TB10002.1-2005. *Fundamental code for design on railway bridge and culvert*. Beijing: China Railway Publishing House, 2005: 27-29.
7. Ministry of Railways, TB10002.2-99. *Code for design on steel structure of railway bridge*. Beijing: China Railway Publishing House, 2000: 7-9.

The study of reconstructed model and mechanical properties of porous sandstone

Qin Zhou*, Gang Wang, Hong Yan, Zhao-Yang Fan, Qing-Yun Wang, Yong-Ming Yang
School of mechanics and civil engineering, China University of Mining and Technology, Beijing,
State Key Laboratory of Coal resources and Safe Mining, China University of Mining and Technology, Beijing,
Beijing, 100083,China
**E-mail:hunanzhouqin@163.com*

Based on the statistic theory and image processing technology, a three-dimensional reconstructed model of pore structure for sandstone is established. On the basis of the model, the numerical simulation experiment is carried out to analyze the influence of pore structure parameters (pore size, shape and porosity) on the deformation, stress distribution and failure distribution of sandstone under the splitting load. Specific modeling process: firstly generating the spatial position of the pore with the statistical principle. Then using the finite element analysis software to establish the corresponding model and using the image processing technology in the image processing software to divide the model net. Importing the restructure 3D pore structure into the finite element analysis software and getting reconstruction of the pore model. The research results show that: the effect of void morphology on the damage of sandstone is obvious. With the change of the pore shape from the ellipsoid to the sphere, the fracture zone of the splitting disc is increased and more dispersed. With the pore shape is changed from sphere to ellipsoid, the maximum stress along the Y direction is firstly decreased and then increased. When the pore size increases, the failure mode becomes a single tensile failure from an induced-tension-shear failure, and the maximum stress along the Y direction is decreased. But the influence degree is weakened with the increase of the pore size of the model. Along with the increase of the porosity, the maximum stress along the Y direction is increased; therefore, the sandstone is easy to damage. But the degree of influence decreases with the increase of pore size. The results reveal the influence of the parameters of the pore structure on the mechanical properties of rock, which provide some reference for the study of the effect of pore structure on the rock physical mechanical properties.

Keywords: Rock Mechanics; Reconstructed Model; Numerical Simulation; Pore Structure; Mechanical Properties.

1. Introduction

Rock is a kind of natural porous material, and there are a large number of different sizes of the pores in the rocks, and the pores directly affect the physical, mechanical and chemical properties of rocks. Such as strength, elastic modulus, permeability, electrical conductivity, wave velocity, particle adsorption capacity and rock reservoir capacity, etc. Understanding and quantifying characterization of the influence of pore structure on the mechanical properties of rocks is very important for solving practical problems in petroleum, geology, mining, metallurgy, civil engineering and water conservancy projects [1-5].

In recent years, domestic and foreign scholars made effects on make the reconstructed model and developed considerable researches on the influence of internal rock pore on its macroscopic physical and mechanical properties. Zuo Jian-ping and other scholars used image analysis technique to analyse the phenomenon of the three points bending microscopic destruction test of the different depth basalt [6]. Yang Yongming and others adopted similar materials to simulate similar porous rocks, and analyzed the failure mechanism of rock and the changes of its pore structure [7]. Vernik and others studied the relationship between porosity with compressive strength and shear strength of sandstone, and presented the empirical relation model [8]. Clavaud J B and others analysed the effects of different rock pore geometry structure on permeability anisotropy, and found closed relationship in sandstone permeability anisotropy with sandstone inherent bedding [9]. Smith T M and others studied the properties of low porosity and low permeability sandstone, and got the conclusion that the influence of the pore geometry on the elastic modulus of sandstone is more than that of the porosity when the porosity is lower [10]. Palchik and others built a classical model to predict the coupling effect of porosity, elastic modulus and grain size on the uniaxial compressive strength of sandstone under high porosity soft sandstone [11]. M.S.Patserson and others found the trend of the transition of rock about brittleness and ductility with the confining pressure [12]. Ju Yang and others developed and verified the LBM method based on the CT image of pore structure of sandstone [13]. Li Xibing and others made split hopkinson pressure bar experimental study on the granite samples with circular and square holes, and obtained the characteristic parameters of the nuclear magnetic porosity and nuclear magnetic resonance images of granite samples [14]. Above researches revealed the macroscopic influence of pore structure on physical mechanics of rock in some extent. However, it is complex and different for pore structures of natural rocks distributed in different spans. At present, it lacks enough understanding toward the influence mechanism of pore on mechanical properties of rock, especially for the influence of pore shape and size on stress distribution, strength, deformation, and fracture of the rock.

In this paper, the statistic principle and image processing techniques are used to establish a three-dimensional reconstruction model of porous sandstone. On this basis, this paper makes a numerical simulation experiment for the splitting of porous sandstone and analyzes the influence law of pore size, shape, and porosity on internal stress field distribution, strength, deformation and fracture of sandstone, revealing the influence mechanism of pore structure on macroscopic mechanical property of rock. This paper provides a reference for the research on revealing the internal relationship between pore structure and mechanical properties of rock.

2. The Three-Dimensional Reconstructed Model of Porous Sandstone

In this paper, we utilize the statistic principle, the image processing and other methods to establish a three-dimensional pore model. On this basis, this paper makes a numeric simulation test for the separation of the pore stone. The three-dimensional pore model is reconstructed according to the following steps: according to distribution characteristics of natural sandstone pore position, it can be known that pore obeys to uniform distribution [15]. At first, the statistic principle is used to generate a random number sequence meeting uniform distribution as the space position of the pore. Secondly, the finite element analysis software is used to establish the preliminary corresponding model. Then, the model is divided into the grid model by the image processing software. Finally, the 3D restructured pore model is processed by the finite element analysis software. As is well known, natural rock is a kind of complicated heterogeneous pore material. Its macroscopic mechanical property, deformation and fracture are correlated with the number, size, shape, distribution and connectivity and other factors of the pore. Changing any factor may cause a change in the mechanical properties of the rock. To analyze the influence of one factor, therefore, this paper makes a research on the influence of pore shape, size and porosity on macroscopic mechanical property, deformation and fracture mechanism of rock by individually changing the pore shape, size and quantity in the model on the premise of maintaining basic mechanical properties basically unchanged. In total, two kinds of porosity and pore shape are designed. At the same time, four kinds of pore diameter are designed for the spherical pore, and 2 kinds of pore diameter for ellipsoidal pore. Specific parameters are shown as follows: the porosity includes 5% and 10%. The pore shape includes sphere and ellipsoid. The diameters of spherical pore are 1mm, 2mm, 3mm and 4mm respectively, and the sizes of the ellipsoidal pore are 1 mm*1 mm*2 mm and 1 mm*1 mm*3 mm respectively. As an example, Figure 1 gives the spherical pore model (diameter: 2mm; porosity: 5%,

10%) and the ellipsoidal pore model (size: 1 mm*1 mm*3 mm; porosity: 5%, 10%). Geometric dimensions of the model: 50mm (diameter) × 25mm (height). Figure 2 gives the relevant grid model divided.

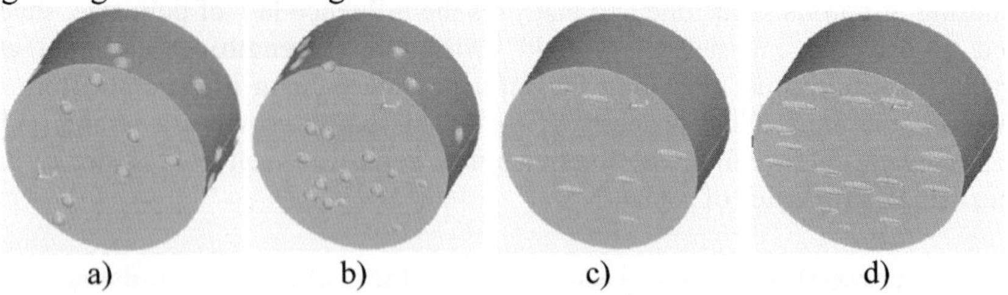

a) spherical pore (diameter:2mm; porosity:5%) b) spherical pore (diameter:2mm; porosity:10%) c) ellipsoidal pore (size:1mm*1mm*3mm porosity:5%) d) ellipsoidal pore (size:1mm*1mm*3mm; porosity: 10%)

Fig. 1. Reconstructed model of porous sandstone.

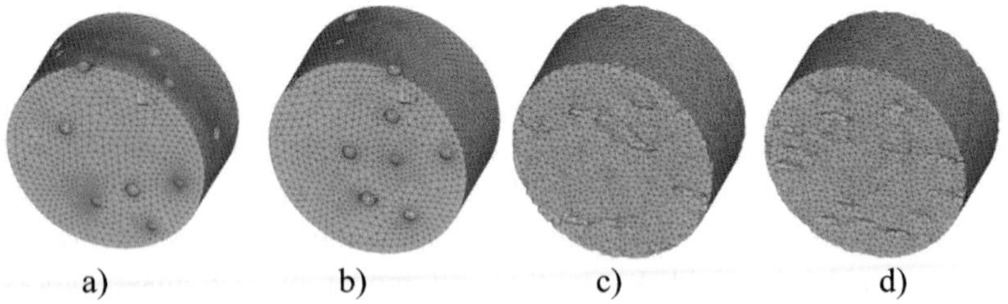

a) spherical pore (diameter: 2mm; porosity: 5%) b) spherical pore (diameter: 2mm; porosity: 10%) c) ellipsoidal pore (size: 1mm*1mm*3mm; porosity: 5%) d) ellipsoidal pore (size: 1mm*1mm*3mm; porosity: 10%)

Fig. 2. Grid model of porous sandstone.

3. The Numerical Simulation Experiment of the Reconstructed Model Under the Splitting Load

3.1. *The mechanical parameters of sandstone and boundary conduction*

This paper takes sandstone as the research object. According to the measured values of the mechanical parameters of sandstone, we determine the parameters of the model as following. The matrix is Mohr-Coulomb material model with a tensile strength of 1.17Mpa and the shear modulus of 7.0GPa, bulk modulus of 26.9GPa, sticky cohesion of 27.2MPa, and internal friction angle of 30 degrees. Using the split loading method, the uniformly distributed load is applied to the both side of the model. Figure 3 gives the schematic diagram of the load.

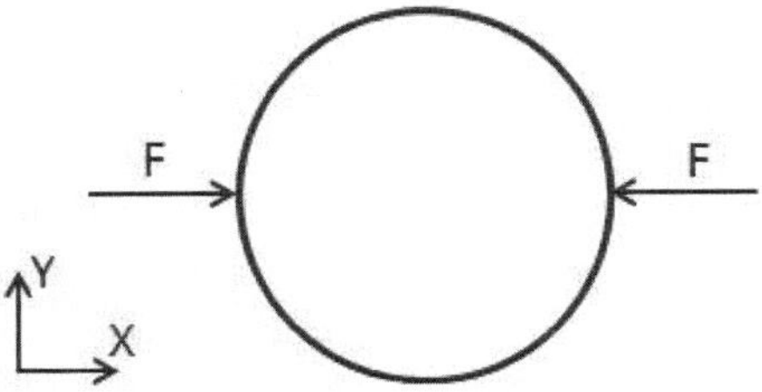

Fig. 3. Schematic diagram of the load.

In order to realize the numerical analysis of FLAC3D, we introduce the model divided into the grid in FLAC3D by self-made program. So we can make the numerical calculation of splitting loading experiment on the 8 spherical pore models and the 4 ellipsoidal pore models and give the results of the numerical calculation which are carried out for the basis of subsequent analysis.

3.2. *The influence of pore parameters on the damaged elements*

This paper analyzes the influences of the pore shape, size and porosity on the characteristics about plastic component distribution when the models are destroyed. Plastic failure distribution unit connected as shown in Figure 4.

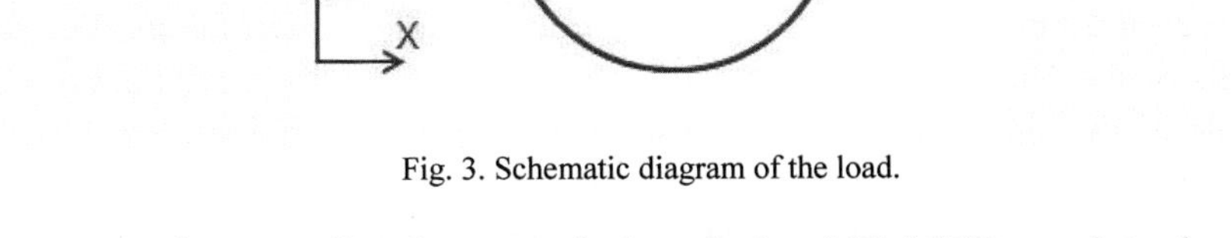

a) Porosity is 5%, diameter of 1, 2, 3 and 4mm respectively spherical pore

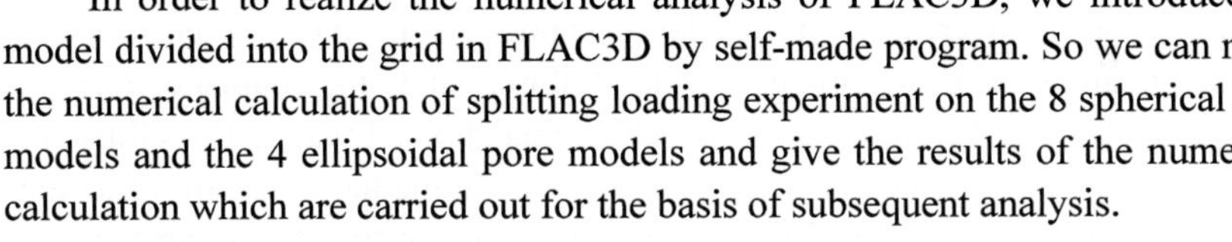

b) Porosity is 10%, diameter of 1, 2, 3 and 4mm respectively spherical pore

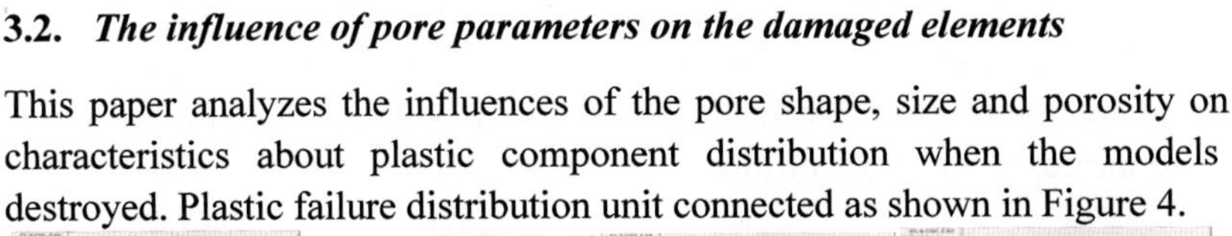

c) Porosity is 5%, size is 1 mm * 1 mm * 2 mm and 1 mm * 1 mm* 3 mm respectively ellipsoidal pore

d) Porosity is 10%, size is 1 mm * 1 mm * 2 mm and 1 mm * 1 mm * 3 mm respectively ellipsoidal pore

Fig. 4. The plastic failure elements distribution features.

The above results show that when the porosity is 5%, for the spherical pore models with the diameter of 1 mm, 2 mm and 3 mm, the failure mode is induced-tension-shear failure. When the porosity is 10 %, for the spherical pore models with the diameter of 1 mm, the failure mode is induced-tension-shear failure. When the porosity is 10 %, for the spherical pore models with the diameter of 2mm and 3 mm, the failure mode is single tension failure. For all models of the 4 mm diameter, the failure mode is single tensile failure. For all models of the ellipsoidal pore, the disc in the plastic transfixion area is very close to a straight line. For the models of the spherical pore, the pore numbers are decreased with the increase of pore diameter, and the degree of stress concentration is decreased. Thus, failure mode from induced-tension-shear failure changes into a single tensile failure, and the damage form is simplification. For the models of the ellipsoidal pore, because the ellipsoidal pore has long axis, the models are easy to damage along the long axis when they are loaded by external effect. Thus, the disc in plastic transfixion area is very close to a straight line. On the premise of keeping the same porosity, with the change of pore shape from ellipsoid to sphere, the number of the pore is increased and the place of distribution is relatively uniform, thus, the fracturing areas are increased and more diffuse.

3.3. *The influence of pore parameters on the maximum tensile stress*

Meanwhile, we also analyze the influences of the shape, size and porosity of the rock pore on the stress in vertical direction of the load when the pore models are damaged. Figure 5 shows the curve relationship between the maximum stress and the pore size for spherical pore models. Figure 6 shows the curve relationship between the maximum stress and the pore size for ellipsoidal pore models. Figure 7 shows the stress field distribution. At the same time, we extract the maximum tensile stress σ_y, we analyze the influence of the pore parameters on the maximum tensile stress. And we use the 1 mm diameter spherical pore as the special ellipsoidal pore that the long axis and short axis is also 1 mm.

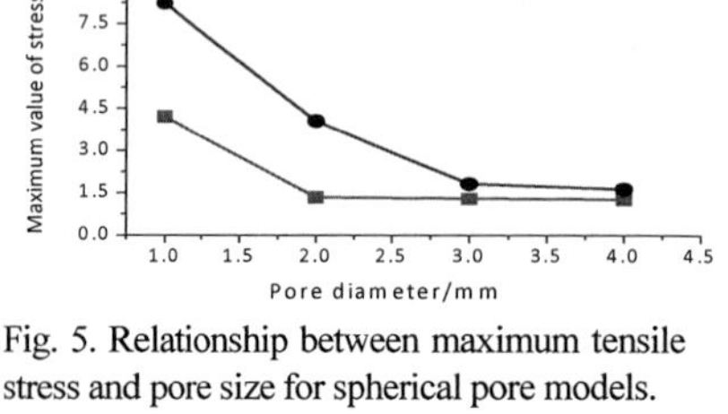

Fig. 5. Relationship between maximum tensile stress and pore size for spherical pore models.

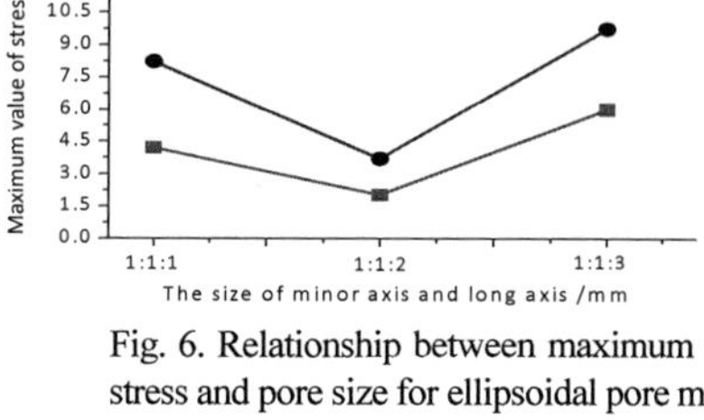

Fig. 6. Relationship between maximum tensile stress and pore size for ellipsoidal pore models.

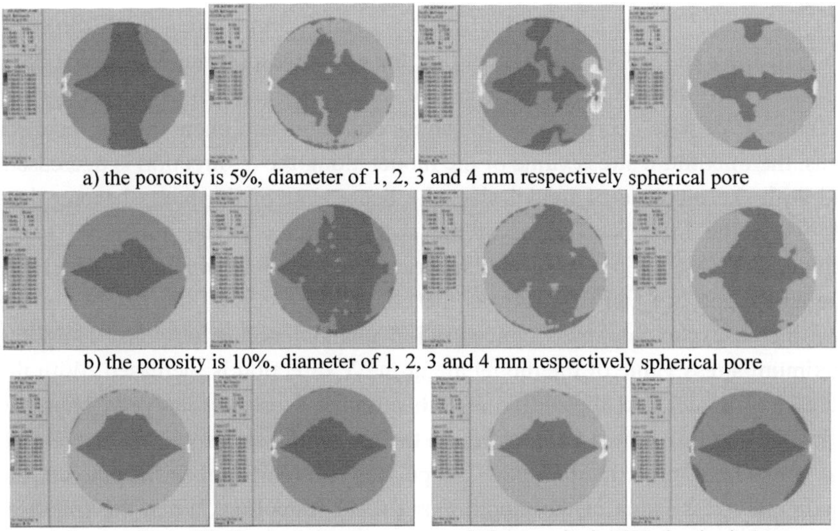

a) the porosity is 5%, diameter of 1, 2, 3 and 4 mm respectively spherical pore

b) the porosity is 10%, diameter of 1, 2, 3 and 4 mm respectively spherical pore

c) the porosity is 5%, the size is 1 mm * 1 mm * 2 mm and 1 mm * 1 mm * 3 mm respectively ellipsoidal pore

d) the porosity is 10%, the size is 1 mm * 1 mm * 2 mm and 1 mm * 1 mm * 3 mm respectively ellipsoidal pore

Fig. 7. The stress σy distribution of pore models.

According to the models of stress distribution, the center of all models is damaged by the tensile stress and the pore shape has a significant influence on the distribution of tensile stress. The ellipsoidal pore is easy to be destroyed along the direction of the long axis when it is damaged by the external force and the plastic transfixion area is very close to a straight line. Therefore, the tensile stress areas of the ellipsoidal pore model are mainly focused on the horizontal center line of two sides, there is no obvious extension. And the tensile stress areas of the spherical pore model have a different degree of expansion, even one part of the tensile stress areas are close to the upper and lower boundary region of the model.

According to the curve about the maximum stress along the Y direction, we can see: for the porosity is 5% and the diameter of 1 mm, 2 mm, 3 mm and 4 mm spherical pore models, the maximum stress values along the Y direction are 4.21 Mpa, 1.35 Mpa, 1.29 Mpa and 1.26 Mpa. For the porosity is 10% and the diameter of 1 mm, 2 mm, 3 mm and 4 mm spherical pore models, the maximum stress values along the Y direction are 8.23 Mpa, 4.05 Mpa, 1.82 Mpa and 1.63 Mpa. When the porosity is 5%, the maximum stress values of the ellipsoidal pores that long axis and short axis are 1 mm * 1 mm * 2 mm and 1 mm * 1 mm * 3 mm along the Y direction are 2.02 Mpa and 5.98 Mpa. When the porosity is 10%, the maximum stress values of the ellipsoidal pores that long axis and short axis are 1 mm * 1 mm * 2 mm and 1 mm * 1 mm * 3 mm along the Y direction are 3.70 Mpa and 9.73 Mpa. Comparing the spherical pore models which have the same porosity and shape, we can find that: as the pore diameter increasing, the number of pore decreases, so the stress concentration phenomenon reduced. Thus, the maximum stress value of the model along the Y direction is decreased, and with the increase of model size, the decrement of the pore become less relatively, the influence degree of the pore diameter to the maximum stress value is decreased. Comparing the spherical pore models which have the same size and shape, we can find that: as the porosity increasing, the number of the pore increases, so the stress concentration phenomenon is obvious. Thus, the maximum stress value of the model along the Y direction is increased. With the increase of model size, the additions of the number decrease relatively. Thus, the influence degree of the pore porosity on the maximum stress value is decreased. For ellipsoidal pore model, pore shape has a significant influence on the maximum stress value of the model along the Y direction. When the spherical pore gets into the ellipsoidal pore, the pore volume increases and the number of pore decreases. Thus, the maximum stress value of the model along the Y direction decreases. But as the irregular degree deepening, the ellipsoidal model is easy to be damaged along the long axis, the maximum stress value increases along the Y direction. Thus, we can conclude that the maximum stress value of the model along the Y direction decreases firstly then increases with the change of the pore shape from the sphere to the ellipsoid.

4. Conclusions

This paper uses the methods of statistic principle, image processing technology and correlation analysis software to reconstruct the porous rock model. Based on the three-dimensional model of the porous rock, we make the numerical simulation experiment on the rock of the deformation and failure under the condition of splitting. By changing the characteristic function of pore shape, pore size and porosity, we analyze the influence on the failure path, stress distribution and stress value. The results show that:

(1) The effect of the pore shape on the failure of the model is significant, which is related to the irregular degree of the model. Under the premise of keeping the porosity and the pore size unchanged, the failure region of the disk splitting is increased and more dispersed with the change of the pore shape from the ellipsoid to the sphere. In the same premise, the maximum stress along the Y direction is decreased firstly then increased with the change of the pore shape from the sphere to the ellipsoid.

(2) Under the premise of keeping the pore shape and the pore size unchanged, when the porosity increases, the maximum stress along the Y direction is increased, and the model is easy to destroy. However, with the increase of the size of the model, the influence of the porosity on the failure of the model is weakened.

(3) Under the premise that the porosity and the pore shape unchanged, when the diameter of the pore increases, the failure mode is changed from the induced-tension-shear failure to the simple tensile failure. The failure mode is simplified. And the maximum stress along the Y direction is decreased, and the failure is weakened. What's more, with the increase of the size of the model, the influence of pore size on the failure of the model is weakened.

Acknowledgments

The authors are grateful for the funding of the college students' scientific and technological innovation project of State Key Laboratory of Coal Resources and Safe Mining (SKLCRSM14CXJH06) and the college students' scientific research and innovation action plan project of Beijing (C201506014).

References

1. F. Yang, C. Q. Zhu, X. H. Wang, H. Z. Qu, F. X. Zhang and X. N. Niu. A capacity prediction model for the low porosity fractured reservoirs in the Kuqa foreland basin. *Petroleum Exploration and Development*, 2013, 4(30): 341-345.
2. Q. L. Hou, H. J. Li, J. J. Fan, Y. W. Ju, T. K. Wang, X. S. Li and Y. D. Wu. Structure and coalbed methane occurrence in tectonically deformed coals. *Science China Earth Sciences*, 2012, 55(11): 1755-1763.
3. X. Gan. Relation between the productivity and the degree of development of the fractures in low porosity and low permeability sandshale reservoirs. *Natural Gas Industry*, 2003, 23(5): 41-43.
4. K. X. Liu and X. Liu. Three-dimensional stress wave propagation in fluid-saturated porous media. *Acta Mechanic Sinica*, 2003, 35(4): 469-473.
5. R. H. Zhang, G. S. Yao and J. F. Shou. An integration porosity forecast model of deposition, diagenesis and structure. *Petroleum Exploration and Development*, 2011, 38(2): 145-151.
6. J. P. Zuo, N. B. Chai and H. W. Zhou. Investigation on failure behavior of basalt from different depths based on three-point bending meso-experiments. *Chinese Journal of Rock Mechanics and Engineering*, 2013, 32(4): 689-695.

7. Y. M. Yang, Y. Ju and H. J. Wang. Physical model and failure analysis of porous rock. *Chinese journal of geotechnical engineering*, 2010, 32(5): 736-744.
8. L. Verenik, M. Bruno and C. Bovberg. Empirical relations between compressive strength and porosity of siliclassic rocks. *Rock Mech Min Sic&Geomech Abstr*, 1993, 30(7): 677-680.
9. J. B. Clavaud, A. Maineult, M. Zamora, P. Rasolofosaon and C. Schlitter. Permeability anisotropy and its relations with porous medium structure. *Journal of Geophysical Research-Part B-Solid Earth*, 2008, 113(B1): 0148-0227.
10. T. M. Smith, C. M. Sayers and C. H. Sondergeld. Rock properties in low-porosity/low-permeability sandstones. *Society of Exploration Geophysicists*, 2009, 28(1): 48-59.
11. V. Palchik. Influence of porosity and elastic modulus on uniaxial compressive strength in soft brittle porous sandstones. *Rock Mech Rock Engng*, 1999, 32(40): 303-309.
12. M. S. Paterson. Experimental deformation and faulting in Wombeyan marble. *Geological Society of America Bulletin*, 1958, 69(4): 456-467.
13. Y. Ju, J. B. Wang, F. Gao and H. P. Xie. The Lattice Boltzmann simulations about CH4 microscopic seepage in porous rock under the deformation condition. *China Science Bulletin*, 2014, 59(22): 2127-2136.
14. X. B. Li, L. Weng, X. F. Xie and Q. H. Qu. Study on the degradation of hard rock with a pre-existing opening under static-dynamic loadings using nuclear magnetic resonance technique. *Chinese Journal of Rock Mechanics and Engineering*, 2015, 34(10): 1985-1993.
15. Y. Ju, Y. M. Yang, Z. D. Song and W. J. Xu. A statistical model for porous structure of rocks. *China Science*, 2008, 51(11): 2040-2058.

The study of the vitality of urban space in traditional communities -- -- a case study of chengdu yulin community

Yi Jian, Kun Zhang

College of Architecture and Environment, WangJiang Campus, SiChuan University, Chengdu, SiChuan, China
E-mail:iyam6@qq.com
www.scu.edu.cn

In the process of social transformation, as a victim of planning model and technology development, interpersonal communication has been put on an indirect position. Although contemporary community improved the quality of the living environment to some extent, due to various reasons, the vitality of the community cannot be maintained. But some traditional communities retains some of its vibrant urban space, satisfied the needs of current residents. In this paper, we took Chengdu Yulin community as a research object, analyzes its' historical background, residents, industrial facilities, planning mode, explore the key reason of their vibrant, hoping to provide a reference for today's city design.

Keywords: City Design; Open Block; Vitality; Yulin.

1. Introduction

"Residential Community", it is generally considered to be the social community which is dominated by residential construction, supplemented by businesses, culture and education, entertainment, public green land, and public facilities. It is limited to a specific region, contains a variety of social relationships and people who engaged in different social activities. Modern Chinese community has actually gone through many stages, from the self-contained "unit system", to the residential area planning under the condition of market economy.[1]

Fig. 1. Current Situation of Yu Lin

In the process of social transformation, as a victim of planning model and technology development, interpersonal communication has been put on an indirect position. The drop of residents' sense of belonging, neighborhood relationship becomes estranged, apathy and tense. Yu Lin, as the representative of our study carrier, built in 1980s, is located in the south of Chengdu, Si Chuan Province. Yulin has no clear boundaries, generally refers to the area surrounded by the First Ring Road, the Second Ring Road, Yongfeng Road, and the people's road south. Coincides with the transformation—from the "unit system" to the market system's "residential area planning", which makes it have both commonality of these two particular mode. Yulin preserves our ancestral cozy way of life, making it a rare reference sample of today's community design. (Fig.1)

2. Residential Community Evolution

Back to the beginning of the formation of Yulin Community, Chengdu expanded its' map through developing new industry. For facilitating worker's life, people built many ancillary dormitory. With the expansion of urban central areas, these originally industrial site were accepted into the project scope of urban construction, various types of residential communities were built around the old dormitory. Residents there also changed from single type to diversification. The change of neighboring groups and city development mode, became the direct stimulation of Yulin's develop.

3. Analyze the Vitality of Residential Community

3.1. *Diversity of spaces*

3.1.1. *The diversity of dwelling population*

Yulin community residents' heterogeneity is high. Material reveals that registered residents in Yulin are 12929 households, about 39645 persons, temporary population are about 16500 persons, floating population are beyond 5000 persons per day. Residents are engaged in diverse businesses, age bracket is distributed unbalanced while young people take up larger specific density. Natives composed of non-agricultural peasants, retired cadres, and the foreign population composed of factory workers, private businessman, and both intellectuals, artists and other special professional staff at the same time.[3] Yulin community has a relatively high population density, social strata represents itself complexity, making the levels of consumption structure unfolds multiple development trend.

Fig. 2. Traditional Business Maintained

3.1.2. *The diversity of live features*

Inside the Yulin community, old residential areas are divided into two parts according to the construction time – built in the 1980s and the 1990s. Among them, Yulin 2st Lane is considered to be a typical building of the 1980s'. As one of the oldest houses in Yulin community, Yulin 2st Lane now has become the representative of "dirty and messy" in the eyes of outsiders. Buildings there are primarily two to three layers. Despite the poor, dark and cramped living conditions, Yulin 2st Lane still becomes where the migrant workers chose to live for its advantageous geographical location, abundant resources, perfect infrastructure, and the low rent. People there broke the wall of the first floor to turn it to a shop and make a living. (Fig.2)

In contrast, the unit dormitory which represent the residential area built in the 1990s, single-family homes stands together to enclose separating courtyards. In this way, these courtyards can be more manageable. Inside these courtyards, there are various vegetation, also a small number of elaborated landscape architectures for the elderly to rest or talk, for children's play.

Yulin community commercial services system is made up of the business on the side of Yulin's avenues, roads, streets, lanes, forming a noisy open outward, inward private and peaceful atmosphere to the streets being the division 5. Yulin community's commercials are composed by the underlying type format (such as restaurants, supermarkets, clothing, and convenience stores) and leisure formats (including cafes, shops, cinemas, bars, can satisfy the residents' personalized demands). Because it's high spontaneous formability, their leisure, dining, entertainment and other service ratio is relatively reasonable, forming a virtuous complementarity among these formats.

3.2. *Form and scale*

3.2.1. *Suitable street network scale*

Yulin community is a hybrid open blocks, areas are divided by urban avenue, with an appropriate scale of road system, and the hierarchy of the whole community is perfect and clear. Represented by the People of South Road, which main function is traffic and the people flow over it is huge and with fast speed. The "avenue" scale is composed of 40m roadway and 12m pavement on each side, a large number of urban commercial space is distributed on both sides of the road. Take the Nijiaqiao Road as example, the "road" scale is composed of 14m roadway and 5m pavement on each side. The amount of space is moderate. Represented by Yulin Road East, the “street” scale is composed of 6m roadway and 3m sidewalks on each side. Medium-sized business spaces are densely distributed on both sides of the road. The form of streets is limited by buildings and the integrality of street public space is strong. As for the Yulin 1st lane’s "lane" system, which total width is 6m, is composed with a 4m roadway and 1.5m sidewalk on both sides of the road. And there exists some narrower laneway spaces which can only for people to go through (Fig.3). With simple spaces and pure function, it is a space which closed by the wall on at least one side, and has a few small or flow type of business spaces.

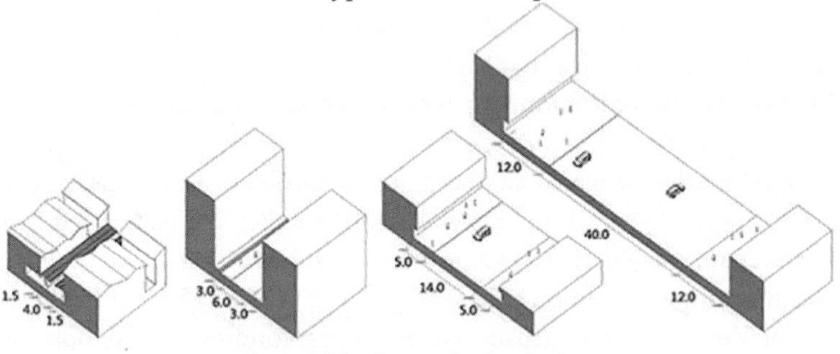

Fig. 3. Major Streets Section Scale

3.2.2. *Small-scale public space*

Venue, road, street, lane, these different levels of road scales, together form the YuLin community’s complete netty transport system. In the roadway spaces ,which are rarely exist nowadays , traditional small business such as book store, whetter’s store and stick-food store are dotted with, roadside stands, tranter were also able to be preserved (Fig.2). Also, it provide the safety supervision for this kind of long and narrow space.[7]

3.3. *Open and connection*

3.3.1. *Connectivity and liquidity*

With convenient and complementary transportation, Yulin community gains a strong accessibility. Meanwhile, the units inside the community are linked by streets, so that people can take walking as their way of daily commute. Different types of scattered virescence have strong interaction between each other, together constitute a large open courtyard. Both open for the residents of the city. Multiple sharing could be guaranteed (Fig.4).

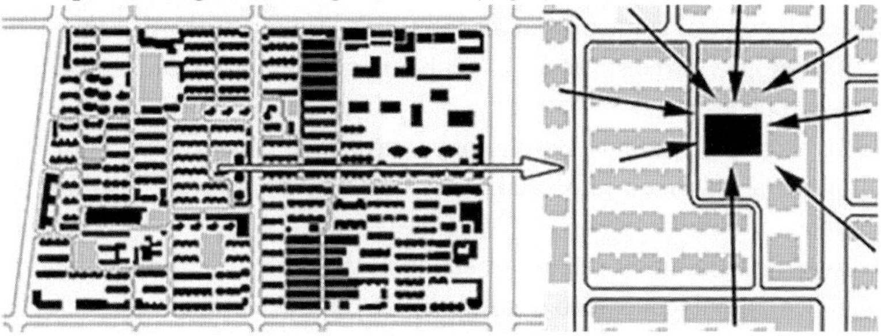

Fig. 4. Fitness and rest Spaces of Yulin

3.3.2. *The Establishment of Space Order*

Each living unit managed separately, form the whole Yulin community through Yulin's series and parallel roads system in a high degree of capillary way. Inward, there's a strong sense of settlement inside the residential area, residents have a strong sense of togetherness and ownership. Outward into the city streets, people still belong to their own community where both have a feeling of openness and regional identity. "Open Community - greening public spaces - living unit" refinement formed the "public - semi-public - private" living system which has different degree of openness and security living systems (Fig.5).

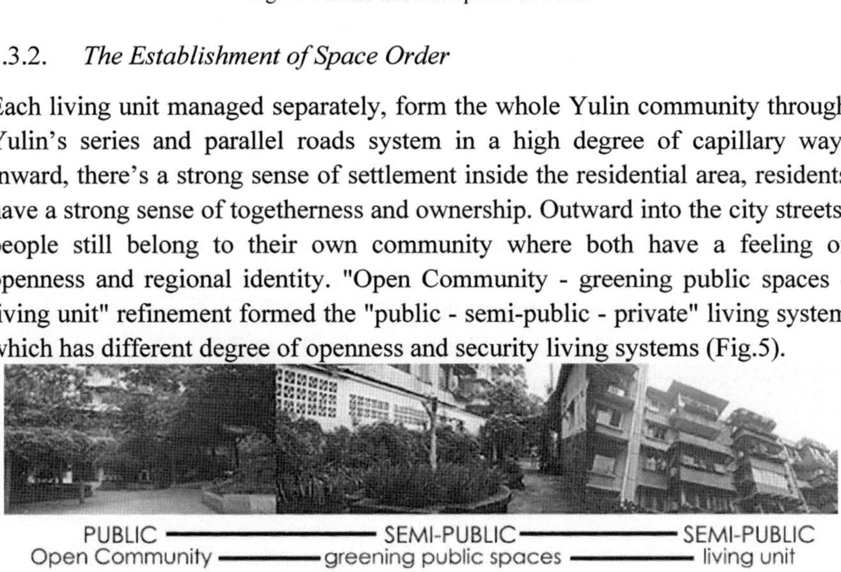

Fig. 5. Spatial Discrimination of Yulin Community

4. Conclusion

Yulin community has appropriate density of city function and comfortable living environment, it is compact arranged, intense developed, with a high level of pedestrian. Residents there have a strong sense of belonging and recognition, forming a strong attraction to people who live outside, establishing a virtuous cycle. In a nutshell, based on the research above, the key of Yulin community's vital atmosphere may come from the following respects:

(1)Residents' heterogeneity is high. Their various demanding directly leads to the community's diverse function, making it a self-sufficiency area not only in social and cultural contexts, but also in living and function aspects.

(2)Developed streets network, with an appropriate scale, good ratio of vehicles and pedestrian.

(3)Reasonably spatial morphology, strong openness and liquidity, big communities are divided into small ones in a step-down way, resulting a clear space order.

Nowadays, gradually relocation stares in the face of Yulin community. It's forming is a coincidence, also a corollary of the development of social system and culture. Hoping that the study of the vitality of Yulin would be an inspiration of the urban spaces design later.

References

1. Chen Xiaoxiao, Zhu Chuandi, Urban community research overview and outlooks in China [J]. Chongqing Social Sciences. 2007(1):108-115.
2. Hua Lanhong, Reconstruire la Chine, (SDX Joint Press, 2006).
3. Jan Gehl, Life between Buildings, (CA& B Press, 2002).
4. Ashihara Yoshinobu, External space design, (CA&B Press, 1985).
5. Li Yun, Zhang Kun, Research on identifiability of external space in residential areas, ShanXi Architecture, 2012, 07, pp.2-4.
6. http://gcontent.oeeee.com/0/80/08040837089cdf46/Blog/7a0/61c057.html.
7. Chen Liuqin. Research into the function of urban communities, Journal of Chongqing Technology and Business University (Social Science Edition), 2008, 10, pp. 57-64.

Mechanics properties of sand under triaxial test simulated by particle flow code

Jing-Yu Chen[1], Ying Hai[2]
[1]*School of Civil Engineering and Architecture, Jiaxing University*
Jiaxing, 314001, P.R. China
[2]*School of Mechanical and Electrical Engineering, Jiaxing University*
Jiaxing, 314001, P.R. China
E-mail: [a]jxxy_chen@126.com, [b]haiying@mail.zjxu.edu.cn

The triaxial tests of sand were simulated by PFC3D (particle flow code in three dimensions). The slip model is used to constitute samples of sand. The slip model provides no normal strength in tension and allowes slip to occur by limiting the shear force. Preliminary study on mechanical properties of the sand is done by PFC3D numerical simulation through endowing appropriate micro-properties. Good agreement between the results of numerical simulations and laboratory tests is shown. Macro properties of sand samples with various input parameters of particle elements are presented, and the results are valuable for developing sand constitutive models. It is shown that the microscopic parameters of sands affected markedly the macroscopic response of sands. The meaning of this research lies in the fact that PFC mode could break through the limits of real lab tests and make the microscopic research of sand possible.

Keywords: Sand; Mechanical Properties; Particle Flow Code; Slip Model; Microscopic parameter.

1. Introduction

Behaviors of sands and clays under loading, or mechanical propertities of soils, are a main research topic in the soil mechanics and geotechnical engineering. Complexity of meso-composition and meso-structures of soils result in complex mechanical propertities of soil. The mechanical propertities of soils should be studied in micro point of view, but it is too difficult to establish quantity relation between macro and micro levels.

Bardet and Proubet investigated the structure of persistent shear bands in granular materials by numerically simulating an idealised assembly of two-dimensional particles. The displacement, volumetric strain, void ratio, rotation of the particles, rotation of their neighbourhoods and contact orientation were examined within the shear band [1]. Zhou Jian et al simulated the stress-strain relations of sand and clay under plane strain state by two-dimention partical flow code (PFC2D) [2-3]. The particle flow code (PFC), which is a special discrete element method, gives an approach to study the macro mechanical response of granular materials by investigating its meso mechanics parameters [4-8].

In this paper, behaviors of sands under triaxial tests are simulated by a three-dimensional discrete element method, the stress-strain relations of sands under triaxial test stress state are modeled and studied with the three dimension partical flow code (PFC3D). Considering meso-structure of sand, the effects of meso mechanics parameters on macro behaviors of sand are studied, and the relationship between meso-structure and macro-response of sand are established. This resarch can let us to understand deeply the behaviors of sands and other granular materials under loading and help geotechnical engineer to do better design in geotechnical engineering.

2. PFC Simulation of Sand Behavior Under Triaxial Test

2.1. *Sand stress-strain relation simulation*

To simulate the stress - strain relation curve of sand, the PFC3D test sample is generated. A PFC test sample generation needs two steps: firstly, to establish a capacity space to constraint particles, this space includs a flexible frictionless cylinder on which confining pressure exert, and two rigid load plate to apply axial load; secondly, to generate a particle sample before a parameter algorithm is given. The three-dimensional particle test sample generation is shown in Fig. 1.

For saving computation time, a PFC3D test sample with the height of 4 mm and the diameter of 2mm is generated, the sample is composed of 2686 particles. In order to determine meso mechanics parameters of the PFC3D test sample, we use this sample and adjust these parameters to simulate the triaxial test stress-strain cuves of a medium density sand. Physical indexes of this medium density sand are: the average particle size d50= 0.18mm, uneven coefficient Cu = 2.8, particle specific gravity Gs = 2.63, the maximum void ratio e = 0.80, the smallest void ratio e = 0.40, the relative density of specimen Dr = 0.63, dry density = (1.7 + 0.01) g/cm3.

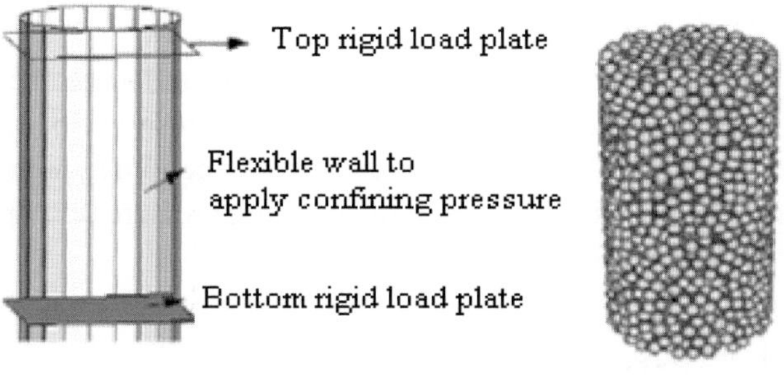

Fig. 1. Sketch map of triaxial test of PFC3D

Loading the principal stress, as the PFC3D test sample under the consolidation confining pressure 100, 300 and 500 kPa, respectively. The test sample keep balance in every step of loading until the end of test.Through trial and analysis of a large number of meso mechanics parameters, when the stress - strain relation of the PFC3D numerical test is approximate to the stress - strain relation of the medium density sand lab triaxial test, as shown in Fig.2, the parameters of the PFC3D test sample is eventually determined as shown in table 1.

Table 1. Sand parameters in PFC3D sample

Spicific density	Initial porosity n	Partical radius /mm	Partical diameter magnificatio n	Frictio n coeffi cient μ	Partical contact normal stiffness /N.m^{-1}	Partical contact stiffness ratio
2.63	0.36	0.075~0.1	1.66	1.2	2.530	1.0

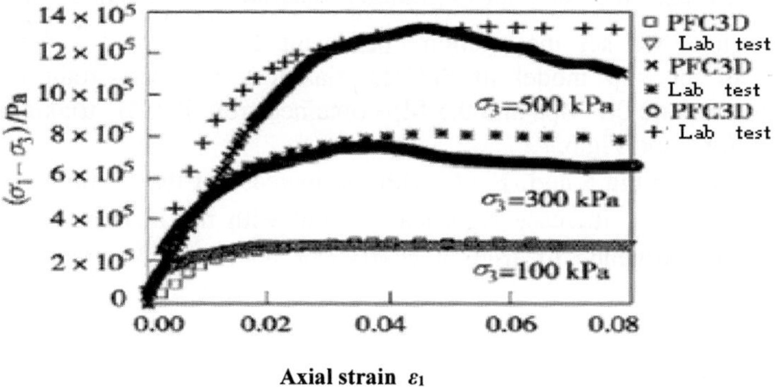

Axial strain ε_1

Fig. 2. Comparison between PFC3D model and lab test results

2.2. *The influence of porosity on the stress - strain relationship*

The porosity of sand is the first important factor to influence the behavior of sand under loading. So we select 4 test samples of porosity 0.35, 0.40, 0.45 and 0.50 to do PFC3D triaxial test. Stress-strain curve of confining pressure 0.1 Mpa are shown in Fig.3, stress-strain curves of confining pressure 0.5 Mpa are shown in Fig. 4.

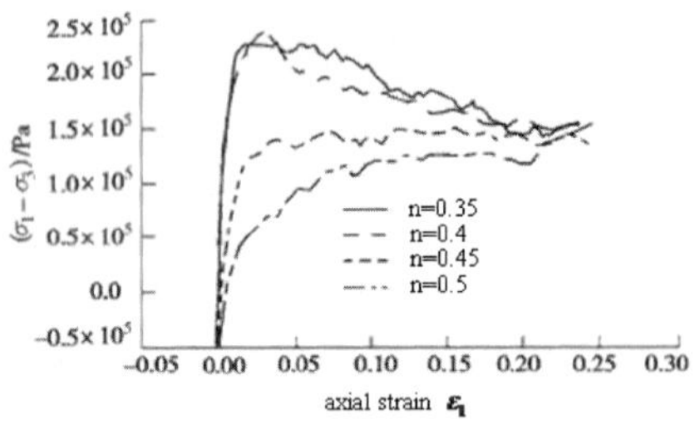

Fig. 3. Stress-strain curves with various porosities (σ_3 =0.1MPa)

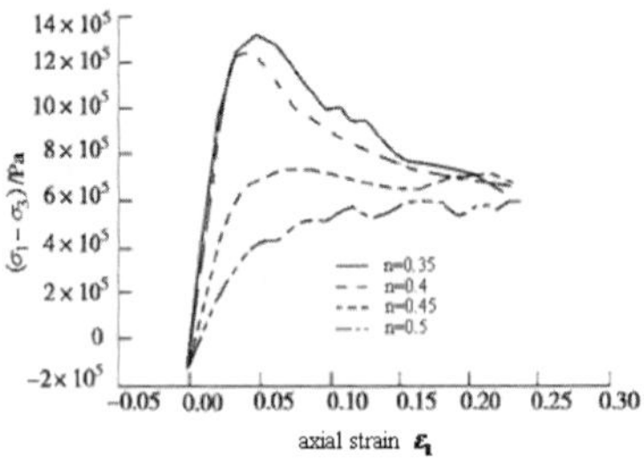

Fig. 4. Stress-strain curves with various porosities (σ_3 =0.5MPa)

As shown in Fig.3 and Fig.4, whether particles are loose or compact, with the increase of confining stress σ_3, the shear stress $(\sigma_1-\sigma_3)$ increases proportionaly, this is a common feature of sands and other granular materials.

2.3. *The influence of friction between particles on the stress - strain relation-ship*

The friction between particles of granular material is the another important factor to influence the behavior of granular material under loading. For to study this influence, we set the friction coefficient μ=1.1, 1.3, 1.5, 1.7 between particles in the slip model in PFC3D triaxial test. Stress-strain curves of confining pressure 0.1 Mpa and 0.5 Mpa obtained from PFC3D triaxial tests are shown in Fig.5 and Fig.6.

As shown in Fig.5 and Fig.6, with the increase of the friction coefficient, the peak strengths increase significantly, and with the increase of confining pressure, the influence of friction coefficient on the peak strength is more obvious.

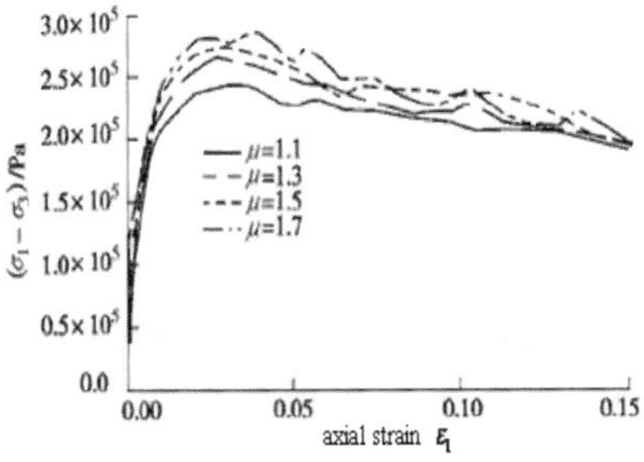

Fig. 5. Stress-strain curves with various Friction coefficients(σ_3 =0.1MPa)

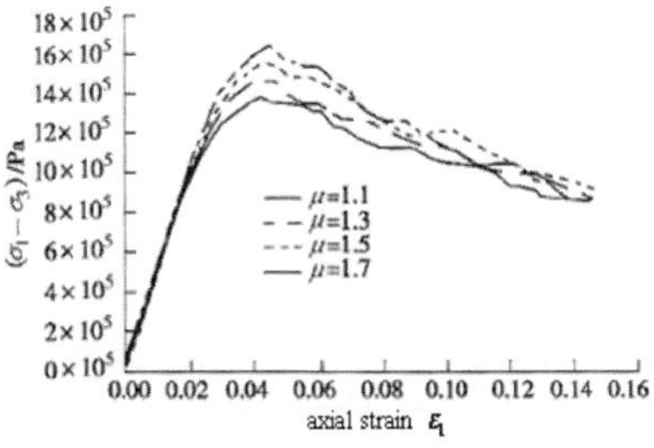

Fig. 6. Stress-strain curves with various friction coefficients(σ_3 =0.5MPa)

As shown in Fig.5 and Fig.6, with the increase of the friction coefficient, the peak strengths increase significantly, and with the increase of confining pressure, the influence of friction coefficient on the peak strength is more obvious.

3. Discussion and Conclusions

Based on the principles of three-dimensional particle flow (PDC3D), the PDC3D numerical simulation triaxial tests of sands carried out in this paper. These numerical test results similar to lab test are obtained, the effects of meso mechanics parameters on macro stree-strain relationships of sand are studied, the main conclusions of this study are as follows:

(1)With the decrease of porosity of a sand, the peak strength of a sand increases and its initial elastic modulus increases also.

(2)With the increase of contact friction coefficient between particles, the peak strength of a a sand increases, the initial elastic modulus also increases.

The DEM (Discrete Element Method) numerical simulation experiments can break through the equipment capacities and the condition limitations of indoor lab tests, and numerically simulate mechanical properties of sand from the meso point of view. The partical flow code(PFC), which is a special discrete element method and a powerful numerical simulation tool, can help researchers to explore influences of meso mechanical parameters on macro mechanical properties, and to estsblish the relationship between meso mechanical parameters and macro mechanical properties of sands.

Studies of other factors, such as the shape of the particles, the particle size distribution, etc., influencing the macro mechanical properties of sands, are further research work in the future.

Acknowlegement

This work is supported by ZHEJIANG PROVINCIAL NATURAL SCIENCE FOUNDATION (LY13A020005).

References

1. BARDET J P, PROUBET J. A numerical investigation of structure of persistent shear bands in granular media[J].*Geotechnique,* 1991, 41(4): 599–613.
2. ZHOU Jian, CHI Yu-wei, CHI Yong, XU Jian-ping. Simulation of biaxial test on sand by partical flow code[J]. *Chinese Journal of Geotechnical Engineering,* 2000, 22(6): 701–704. (inChinese).
3. ZHOU Jian, LIAO Xiong-hua, CHI Yong, XU Jian-ping. *Simulating plane strain test of soils by particle flow code[J].*2002, 30(9): 1044–1050. (in Chinese).
4. CUNDALL P A, STRACK O D L. A discrete numerical model for granular assemblies[J]. *Geotechnique,* 1979, 29(1): 47–65.
5. PULS M, PULSFORT M, WALZ B. Application of the PFC3D for determination of soil properties and the simulation of the excavation process in front of sheet pile wall constructions [C]//*Numerical Modeling in Micromechnics via Particle Methods,* London, 2004: 35–44.
6. PIERCE M E, CUNDALL P A, VAN HOUT G J, LORIG L. PFC3D modeling of caved rock under draw[J]. *Numerical Modelling in Micromechanics via Particle Methods,* 2003:211–217.
7. POTYOND D O, CUNDALL P A. A bonded-particle model for rock[J]. *Int J Rock Mech Mtn Sci,* 2004, 41(8): 1329−1364.
8. *PFC3D-Particle flow code in 3 dimensions*[M]. Minneapolis: Itasca, Inc, 2003.

Analysis and research on micromechanics model of FRP bridge panel

Kun-Ning Zhu[1,*], Chen Chen[2,†]
[1]*Jiangsu Province Communications Planning and Design Institute Limited Company, Nanjing, 210014, China*
[2]*Jiangsu Weixin Engineering Consultants Ltd., Nanjing, 210014, China*
*E-mail: *redkun@163.com, †cc0417@126.com*

Different micromechanics models are researched though the Finite element analysis of FRP bridge deck, and the result is compared with the load test of the deck. The results show that, the elastic modulus along the fiber direction of the deck calculated by the selected micromechanics models are the same. The holistic deformation and the local deformation computed by the finite element model, whose elastic modulus is computed by the rule of mixture, are very close to the load test values. The fiber reinforced cloth can't increase transverse modulus of the pultruded profiles. The rule of mixture model is the best model to simulate the pultruded FRP deck.

Keywords: Bridge Engineering; Bridge Panel; FRP; Finite Element; Micromechanics.

1. Summary

Micromechanics of FRP is used to analyse the relation between the macro properties and composition properties of layer, as well as the micro structure of the layer. It reveals why different combination of materials brings different macro properties. The micromechanics is the assistance of macromechanics. In micromechanics of FRP, the composite material is analysed as structural. The fiber and the matrix is considered as the basic element. And the mechanical properties of the FRP is studied though the geometry and arrangement of the fibers, the mechanical properties as well as the interactions between the fiber and matrix.

Material mechanics method, elastic mechanics and semi-empirical method both can be used to analyse the composite material. Different models and methods will lead to different results [1]. The commonly used micromechanics models are the rule of mixture model [2], the Chamis model [3-4] and Hill-Hashin-Christensen-Lo model [5-9]. This paper introduces the research of different micromechanics models. And the result is compared with the load test of the FRP deck [10]. The best model to simulate the pultruded FRP deck is confirmed. And this is also the foundation of the mechanical analysis of the pultruded FRP deck.

2. Layer Design of the FRP Deck

The FRP panel is made from E-glass fiber (1200Tex), fiber reinforced cloth(0°/-45°/45°/90°) and vinyl ester resins.

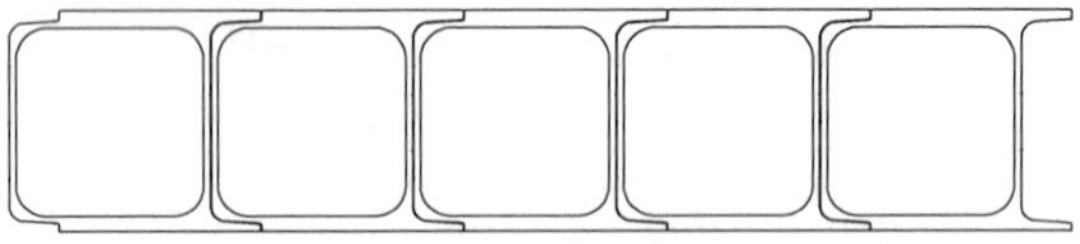

Fig. 1. Cross section of FRP bridge panel.

The design of the bridge panel includes layer design and cross-section design. The width of FRP profile is 200mm, and the hight is 218mm. The thickness of top plate is 17mm, and the bottom plate is 15mm. Both the top plate and bottom plate are made from 6 layers of E-glass fiber and 7 layers of fiber reinforced cloth. The thickness of each ply of E-glass fiber is 2.4mm, and the fiber reinforced cloth is 0.4mm. The web have a thickness of 7mm, and is made from 3 layers of E-glass fiber and 4 layers of fiber reinforced cloth. The thickness of each ply of E-glass fiber and fiber reinforced cloth is 1.8mm and 0.4mm. The fiber volume fraction is 0.55 in the E-glass yarn layer and 0.4 in the fiber reinforced cloth layer.

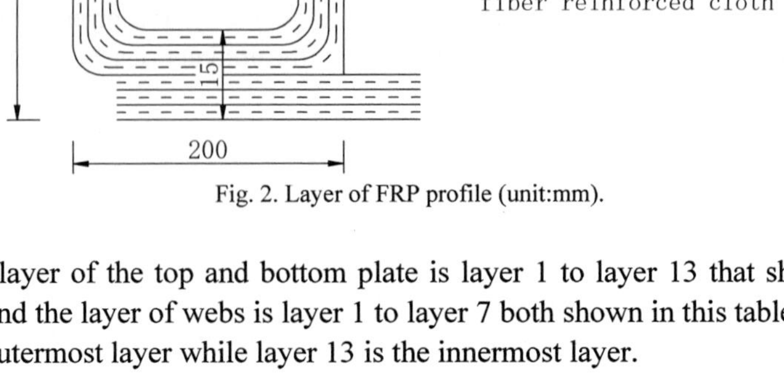

Fig. 2. Layer of FRP profile (unit:mm).

The layer of the top and bottom plate is layer 1 to layer 13 that shown in table 1, and the layer of webs is layer 1 to layer 7 both shown in this table. Layer 1 is the outermost layer while layer 13 is the innermost layer.

Table 1. Layer of FRP profile.

Layer	Material	Specification	Remark
1	fiber reinforced cloth	450g/m^2	fiber reinforced cloth(0/90 - 1/1)×2
2	E-glass fiber	1200Tex	
3	fiber reinforced cloth	450g/m^2	fiber reinforced cloth(-45/45 - 1/1)×2
4	E-glass fiber	1200Tex	
5	fiber reinforced cloth	450g/m^2	fiber reinforced cloth(-45/45 - 1/1)×2
6	E-glass fiber	1200Tex	
7	fiber reinforced cloth	450g/m^2	fiber reinforced cloth(0/90 - 1/1)×2
8	E-glass fiber	1200Tex	
9	fiber reinforced cloth	450g/m^2	fiber reinforced cloth(-45/45 - 1/1)×4
10	E-glass fiber	1200Tex	
11	fiber reinforced cloth	450g/m^2	fiber reinforced cloth(-45/45 - 1/1)×4
12	E-glass fiber	1200Tex	
13	fiber reinforced cloth	450g/m^2	fiber reinforced cloth(0/90 - 1/1)×2

3. Finite Element Model of FRP Bridge Panel

3.1. *Material parameters*

The FRP panel is made from E-glass fiber and vinyl ester resins. The material parameters of the fiber and resin are as follows [11-13]:

Table 2. Material parameters of fiber and resin.

Material	Elastic modulus /GPa	Shear modulus /GPa	Poisson's ratio	ρ/g/mm^3
E-glass fiber	74	29.5	0.255	2.55
vinyl ester resins	4.2	1.6	0.3	1.2

Table 3. Material parameters of single layer of FRP profile.

Micromechanics model	Layer	E_{11} /GPa	E_{22} /GPa	G_{12} /GPa	v_{12}
rulc of mixture model	E-glass fiber	42.59	8.78	3.33	0.275
	fiber reinforced cloth	19.57	19.57	2.57	0.098
Chamis model	E-glass fiber	42.59	13.98	5.36	0.275
	fiber reinforced cloth	21.41	21.41	3.98	0.138
Hill-Hashin-Christensen-Lo model	E-glass fiber	42.59	11.37	4.71	0.267
	fiber reinforced cloth	20.41	20.41	3.39	0.114

The material parameters of single layer of the FRP profile are computed by the rule of mixture model, the Chamis model and Hill-Hashin-Christensen-Lo model. Within the Chamis model, the material parameters of the fiber reinforced cloth is based on bidirectional proportion[14] of the fiber. The results of material parameters are listed in table 3.

Different micromechanics models effects the E_{22} and G_{12} of the E-glass fiber layer. The E_{11} of the yarn layer, which is computed by the selected micromechanics models, all equal to 42.59GPa, but E_{22}, G_{12} of the layer are different with each other. The E_{22} and G_{12} computed by the rule of mixture model are the smallest (E_{22}=8.78GPa, G_{12}=3.33GPa), and the value computed by the Chamis model are the biggest (E_{22}=13.98GPa, G_{12}=5.36GPa).

3.2. *FEA model of FRP panel*

The finite element model is built and analysis of the model is made. The layer of the FRP profile is simulated by orthotropic materials with the material parameters listed in table 3.

Fig. 3 shows the finite element model of the FRP bridge panel. Boundaries are added to the nodes that at a distance of 10cm to both ends of the panel to form simply-supported constraints. Uniformly distributed load is added at mid-span of the panel in an area of 0.2×0.6m (condition 1&2) and 0.1×0.1m (condition 3).

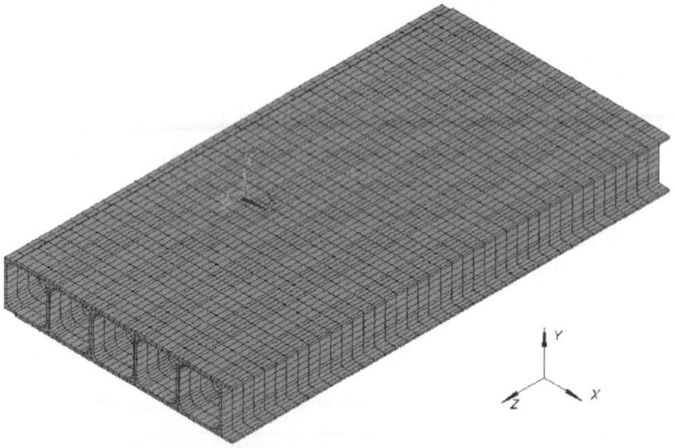

Fig. 3. Finite element model of FRP bridge panel.

The deformation of the deck, which is built with different micromechanics models, are studied. Both the holistic deformation of condition 1&2 and the local deformation of condition 1&3 are compared with the load test results.

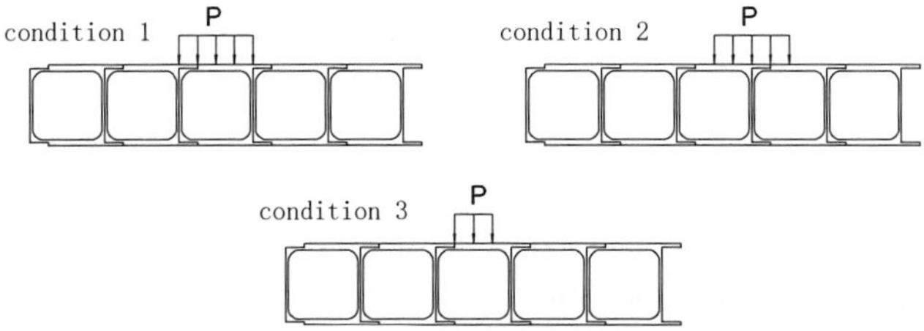

Fig. 4. Diagram of the load conditions.

4. Analysis of Deformation

4.1. *Holistic deformation*

Figures 5 and 6 show the holistic deformation of the FEA model of the deck and the load test results.

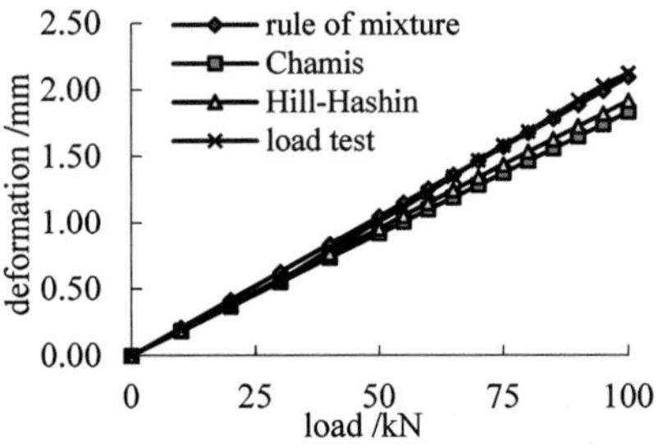

Fig. 5. Holistic deformation of condition 1.

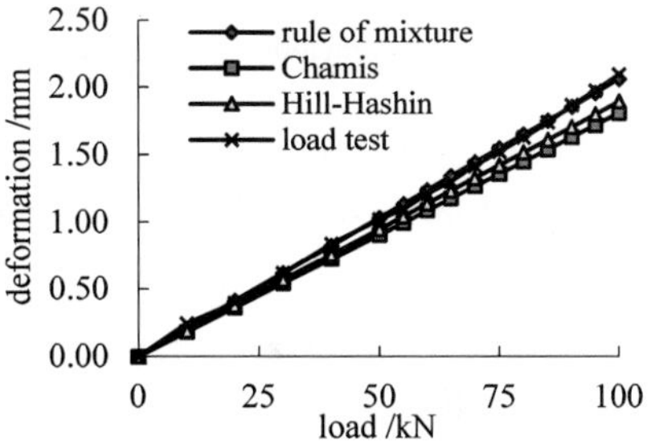

Fig. 6. Holistic deformation of condition 2.

Table 4. Comparison of holistic deformation.

	Load /kN	Deformation /mm	calculated value /mm		
			Rule of mixture	Chamis	Hill-Hashin
Condition 1	70	1.47	1.47	1.28	1.34
	100	2.13	2.09	1.83	1.92
Condition 2	70	1.42	1.44	1.27	1.33
	100	2.1	2.06	1.81	1.89

The holistic deformation of the FEA model of the deck, gain from the FEA model built with the rule of mixture model, is close to the load test value. In load condition 1, when the load is 100kN, the holistic deformation is 2.09mm (rule of mixture), is 1.8% smaller than the load test value(2.13mm), while the percentage based on the Chamis model and Hill-Hashin-Christensen-Lo model are -14.1% and -9.9%. In load condition 2, the holistic deformation gain from the FEA model are 2.06mm (rule of mixture), 1.81mm (Chamis) and 1.89mm

(Hill-Hashin-Christensen-Lo), compared with the load test value(2.1mm), the rule of mixture has the least error.

4.2. *Local deformation*

Figures 7 and 8 shows the local deformation (condition 1&3) of the FEA model of the deck and the load test results.

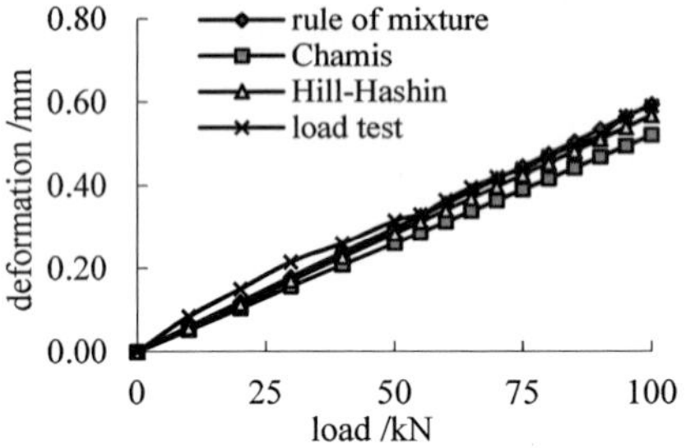

Fig. 7. Local deformation of condition 1.

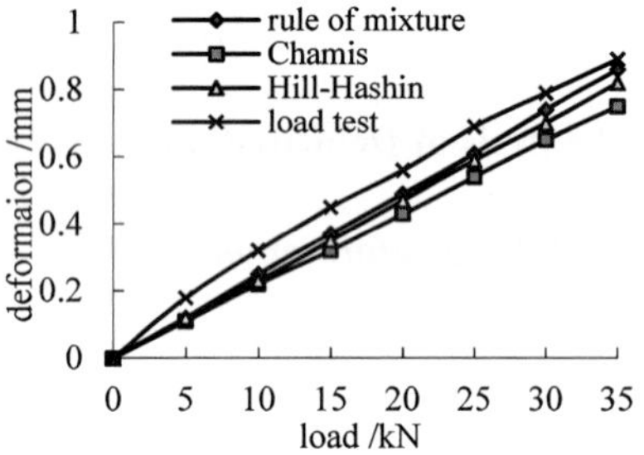

Fig. 8. Local deformation of condition 3.

The local deformation gained from the FEA model which is based on the rule of mixture model is the biggest and approximate the load test value.

Table 5. Comparison of local deformation.

	Load /kN	Deformation /mm	rule of mixture /mm	Chamis /mm	Hill-Hashin /mm
Condition 1	30	0.22	0.18	0.16	0.17
	50	0.32	0.3	0.26	0.28
	70	0.42	0.42	0.36	0.4
	80	0.47	0.48	0.42	0.45
	90	0.52	0.53	0.47	0.51
	100	0.59	0.59	0.52	0.57
Condition 3	5	0.18	0.12	0.11	0.12
	10	0.32	0.25	0.22	0.23
	15	0.45	0.37	0.32	0.35
	20	0.56	0.49	0.43	0.47
	25	0.69	0.61	0.54	0.59
	30	0.79	0.74	0.65	0.70
	35	0.89	0.86	0.75	0.82

The local deformation of the FEA model of the deck, which is built with the rule of mixture model, is close to the load test value. In load condition 1, when the load is 100kN, the local deformation gained from the FEA model which is based on the rule of mixture model (0.59mm) is equal to the load test value (0.59mm), while the percentage value contrast with load test is -11.9% (Chamis) and -3.4% (Hill-Hashin-Christensen-Lo). In load condition 3, when the load is 35kN, the local deformation gain from the FEA model are

0.86mm (rule of mixture), 0.75mm (Chamis) and 0.82mm (Hill-Hashin-Christensen-Lo), compare with the load test value(0.89mm), the rule of mixture has the least error.

The rule of mixture model can accurately reflect the material constants of pultruded FRP deck. This is mainly due to the pultrusion process, at the direction of layer, fiber reinforced cloth can't distribute flatly as design, but pleated. Fiber reinforced cloth can't increase transverse modulus value of the layer.

5. Conclusion

In this paper, the difference between the rule of mixture mode, Chamis model and Hill-Hashin-Christensen-Lo model is discussed, the deformation of FEA model is compared with load test value, and following conclusion is got:

1. The elastic modulus along the fiber direction of E-glass yarn layer computed by different micromechanics models are the same. The shear modulus and elastic modulus(vertical fiber direction in the layer) ,which are computed by the Chamis model has the maximum value, and the rule of mixture model reflect the minimum value.

2. The deformation of the FEA model of the deck, which is built with the rule of mixture model, is close to the load test value. The model can correct response the mechanical behavior of the deck.

3. The fiber reinforced cloth can't increase transverse modulus of the pultruded profiles.

4. The rule of mixture model can accurately reflect the material constants of pultruded FRP deck. The material parameters of pultruded FRP profile can be computed though the model, and this is also the foundation of macro analysis of FRP deck.

References

1. Zheng-Ming HUANG. *Introduction to micromechanics of composites.* (Science Press. 2004), pp. 17-22.
2. Tsai SW, Hahn HT. *Introduction to composite materials. Lancaster.* (Technomic Publishing Co., Inc., 1980).
3. Chamis CC. *Mechanics of composite materials: past, present and future.* Vol. 11 (J. Comp. Technol. Res. ASTM, 1989), pp.3-14.
4. Hopkins DA and Chamis CC. *A unique set of micromechanics equations for high temperature metal matrix composites.* (American Society for Testing and Materials, 1988), pp. 159-176.

5. Hill R. *Theory of mechanical properties of fiber-strengthened materials*. Vol. 12 (J. Mech. & Phys. Solids, 1964), pp. 199-212.
6. Hill R. *A self-consistent mechanics of composite materials*. J. Mech. & Phys. Vol. 13 (Solids, 1965), pp. 213-222.
7. Hashin Z, Rosen BW. *The elastic model of fiber-reinforced materials*[J]. Vol. 31 (J. Appl. Mech.,1964), pp. 223-232.
8. Hashin Z. *On elastic behavior of fiber reinforced materials of arbitrary transverse phase geometry*. Vol. 13 (J. Mech. & Phys. Solids, 1965), pp. 119-134.
9. Christensen RM, Lo KH. *Solutions for effective shear properties in three phase sphere and cylinder models*. Vol. 27 (J. Mech. Phys. Solids, 1979), pp. 315-330.
10. Kun-Ning ZHU, Shui Wan, Yu-Qing Liu. *Analysis and experimental study on statically loaded FRP bridge deck*. Vol. 27 (Engineering Mechanics, 2010), pp. 240-244.
11. Ding-Zhu WO, Ye-Shun LI, Xing-Ye WANG. *Encyclopedia of composites*. (Chemical Industry Press, 2000), pp. 11-105.
12. Yu-Long ZHANG. *Manufacturing technology of composite materials*. (China Machine Press, 2003), pp. 20-30.
13. Davalos J F, Salim H A, Qiao P, et al. *Analysis and design of pultruded FRP shapes under bending*. Vol. 27 (Composites Part B: Engineering, 1993), pp. 295-305.
14. Berthelot J-M. *Composite Materials: Mechanical Behavior and Structural Analysis*. (Springer-Verlag, 1999), pp. 267-391.

Comprehensive investigation of the application of the technology in complex foundation survey

Zhi-Gang Zhang[1,a], Wan-Shun Wang[2,b], Hong-Jun Li[3,c]
[1]*Institute of Scientific and Technical Information of China (ISTIC), Beijing, China*
[2]*China Institute of Water Resources and Hydropower Research (IWHR), Beijing, China*
[3]*Shanghai ChengKun construction engineering consulting co., LTD, Shanghai, China*
E-mail: [a]zzg@istic.ac.cn, [b]wangws_mail@126.com, [c]1636011@qq.com

The engineering survey is by one or more technical means to find out the planned area topography and geomorphology, strata distribution features, genetic types and engineering geological properties, provide the various physical and mechanical properties of soil, to analyze and evaluate the site engineering geological conditions, the stability and evaluate the suitability of the site. Based on a high-rise building foundation investigation as an example, the integrated use of soil, water, laboratory test (conventional physical properties test and indoor seepage test, mechanical test and the water quality analysis), static cone penetration test, standard penetration test, DMT test, vane shear test, water injection test, the confined water level observation, wave velocity test foundation survey and technical methods and means, such as the proposed site for a detailed comprehensive survey, to obtain the distribution of soil, foundation bearing capacity and site effect, and other important parameters.

Keywords: Integrated; Survey Technology; Complex Foundation; Application.

1. Overview

A commercial complex projects, the floor consists of three towers A1, A2, A3 and the highest 6 layer of business, including A1 the ground about 140 m high, A2 and A3 the ground about 100 m high, 4 floor underground. The largest single column load about 58000 kN, the maximum average basal pressure about 560 kN/m^2.Tower basement buried depth of about 23 ~ 24 m, podium and pure basement buried depth of about 22 m. The purpose of this survey is mainly a detailed check of this proposed site engineering geological and hydrogeological conditions, the control measures are put forward on the poor geology and special geology, provide sufficient geological basis for construction drawing design.

2. Exploration Hole Layout

(1) The main exploration hole according to the characteristics of the proposed object plane size, use the "grid" or " zigzag ", "plum blossom" glyph shape holes.Of high-rise office building (19 to 27 floors of the main building), the skirt building, underground garage foundation pit to control respectively.Exploration hole pitch control in no more than 35 m.(2) The proportion of borehole and static agent: according to the borehole and static probe hole ratio of 1:2, when exploration hole quantity is less, or design requirements, appropriately increase the proportion of drilling.(3) In the Ming city, the dark pond, hidden creek, the bad geological condition such as underground obstacles, should be integrated using small threaded hole drilling, exploration methods, such as encryption exploration hole, further find out the distribution range and character, etc. (4) To control the shallow crust soil layer (layer ②) level fluctuation and thickness changes, find out the proposed building, content within the scope of the Ming and the distribution of the hidden creek, this survey along the proposed structure of the basement line layout small threaded hole, hole spacing generally is 10 ~ 15 m, the hidden creek border control small threaded hole pitch for 2 ~ 3 m.(5) The largest engineering foundation pit depth of about 24 m, belongs to the level of the foundation pit, in order to provide foundation pit enclosure, excavation design and construction parameters, at the scene of the homogeneous foundation pit edge periphery decorate bore water injection test, vane shear test, the flat dilatometer test (DMT) bore holes, observation of confined water, wave velocity test, site test project.

3. Field Test (test) and The Indoor Experiment with Soil and Water

3.1. *Test in situ*

(1) The static cone penetration test

Static sounding can intuitively reflect the condition of the soil layer in a row, 10 cm per injection automatic test data, can be used to differentiate soil layer, find out the soil uniformity, especially have a full understanding of the situation of clay sand inclusion, more accurate evaluation of the strength and deformation behavior of the soil, accurate for pile foundation design parameters and feasibility analysis, etc., used for piling can also be used for evaluation of the foundation soil liquefaction characteristics.The survey and static ratio of two hole and drilling hole, two types of exploration and hole arrangement, static sounding holes were arranged in 42.

(2)The standard penetration test

Standard penetration test in sandy soil, powder soil, spacing generally is 2.0 m, for more than 20 m to saturated sand and powder soil liquefaction discrimination, penetration distance is 1.0 ~ 1.5 m, and sample disturbance particle analysis experiment was carried out.Proposed site area shallow if 20 m within the scope of existence into a layer of saturated sandy silt or sand soil, liquefaction discrimination.

(3) The flat dilatometer test (DMT)

Based on the results of flat dilatometer test (DMT) and according to the empirical formula can be obtained undrained shear strength of soil Cu, K_0 static lateral pressure coefficient, the deformation parameters of soil, horizontal to the related parameters, such as bedding coefficient, and can evaluate the classification and stress history of soil.This proposed site layout three flat dilatometer test (DMT) hole, hole deeply 60 m.

(4) The panel test

Field vane shear test is applicable to homogeneous saturated soft clay, used to determine the in-situ stress under the condition of undrained shear strength of soft cohesive soil Cu, estimate the sensitivity of soft cohesive soil St, provide the parameter of strength for the foundation pit design and construction of Cu.In survey layout three cross hole plate test, test distance of 1.0 m, hole deeply 40 m.

(5) Water injection test

In order to obtain required for foundation pit design of soil permeability index, decorate in investigation of three hole water injection test.Test depth according to 2.5 times the depth of foundation excavation is considered, each loop in the first test (2), (3), (4), (5, 7), a layer of soil (including each layer) water flooding experiment was carried out respectively, test section for a total of 30 segment.

(6) Confined water level observation

In order to more accurately evaluate confined water head pressure effect on the stability of foundation pit, require determination of confined water head work accordingly.In survey layout 3 confined water observation hole, 2 layer micro artesian water aquifer of (5), the first 7 layer, the first was 2 artesian water, confined water observation time of not less than 7 days in a row, for the design of the confined water head to provide more detailed information.

(7) Wave velocity test

In order to get the measurement of all kinds of dynamic parameters of foundation soil, the survey and hole arrangement and 3 a wave velocity test, using the puckering method, hole deeply 100 m.

(8)Small threaded hole

To find out hidden creek in the range of the proposed building and the distribution of underground obstacles, the survey around the foundation pit line layout small threaded hole, hole spacing generally is 10 ~ 15 m, hole depth is commonly 3 ~ 5 m.If hidden creek or underground obstacles in the process of exploration, add with holes in accordance with relevant regulations, the pitch of the border is not more than 3 m.The survey and decorate a small threaded hole total of 53.

(9) Water level measurement

Before the hole measuring water level at first, after the completion of 24 hours stable measuring water level, and at the end of the exploration and construction of all, unified survey was carried out on the water level of borehole stability, measurement error control within 5 cm.

3.2. *Indoor soil and water test*

According to the characteristics of the worksite properties and the test project is as follows:

(1) The conventional physical properties test

On the undisturbed soil samples are severe, density, water content, porosity, pore ratio test.The cohesive soil liquid limit, plastic limit test.The powder soil and sand soil points test, points test for particle size greater than 0.075 mm by sieving method of soil, size is less than 0.075 MMM soil by using the hydrometer method, percentage of particles.

(2) Indoor penetration test

Influence to foundation pit excavation depth within the scope of each soil layer provides the vertical and horizontal permeability coefficient (Kv, KH).

(3) The mechanical test

Mainly including consolidation test, triaxial consolidation undrained shear (CU) test, triaxial consolidation undrained shear (UU) solid fast, slow direct shear test and direct shear shear, unconfined compressive strength, static lateral pressure coefficient, the springback test project, and provide the corresponding mechanics index or parameters.

Consolidation test: provide e ~ p curve, the pressure is the first step on the soft clay is 25 kpa, other layers of 50 kpa;Imposed by the end of the level of pressure is greater than gravity pressure to add additional pressure, and the maximum pressure is greater than 400 kpa, slow method is used to test.Depth of the soil sample is in 0 ~ 25 m when the largest consolidation pressure is 600 kpa, 35 ~ 50 m largest consolidation pressure is 1000 kpa, 50 m to 65 m largest consolidation pressure is 1400 kpa, 65 m ~ 80 m largest consolidation pressure is 1800 kpa (to the first pet-name ruby layer upon layer surface), 80 m to 100 m (the first pet-name ruby layer) maximum pressure to 2000 kpa, 100 m (the first pet-name ruby layer) under the biggest pressure to 3200 kpa.

Triaxial consolidation undrained shear test (CU) : used to calculate the soil pressure of foundation pit, provide undrained consolidation state of total stress intensity index and effective strength index C, and Φ, value;At the same time can be made of Ko = 1 - sine phi, checking Ko.In the soil in the foundation pit influence scope.

Triaxial consolidation undrained shear (UU) test: to satisfy calculating stability of foundation pit excavation, soil not drainage consolidation is required to provide natural state strength index Cuu, Φ UU values, in (2), (3), (4), (5), End up layer.

Direct shear solid fast: provide solid quick direct shear test of each soil shear strength parameters C, Φ value (peak).

Direct shear slow shear: providing powder soil, sandy soil is slow shear test of shear strength parameters of the Cs, Φ s value (peak).

Unconfined compressive strength: provide the unconfined compressive strength of soft cohesive soil layer, the layer (2), (3), (4), (5).

Static lateral pressure coefficient, (2), (3), (4), (5, 7), a layer of Ko coefficient.Consolidation springback experiment: provides a layer of 5, 7), the modulus of resilience of Ee.

(4) Water quality analysis

For use on groundwater water samples for water quality analysis, Jane to corrosive evaluation of groundwater.

4. Survey Results

4.1. *The composition and characteristics of the foundation soil*

The investigation reveals the planned area of 130.05 m depth within the scope of the soil according to their age, cause, buried conditions and physical and mechanical performance factors such as comprehensive consideration can be divided into 10 layers, including the first 5, 7) layer, layer, the first pet-name ruby layer according to the differences in vestigated and engineering properties and can be subdivided into several layers.Each layer of the foundation soil based on the results of indoor soil test results and field exploration comprehensive name.

4.2. *Physical mechanics properties of foundation soil*

4.3.1. *The physical and mechanical property index of foundation soil*

Various physical and mechanical properties of soil parameters according to the field of drilling, in situ testing and laboratory soil test results is analyzed, layering, statistics, the sample of trade-offs, and considering the discrete data and experience of each index indoor soil test and standard penetration test adopts the arithmetic mean, at the same time, calculate the mean square deviation and coefficient of variation, the static sounding with small average value, the soil physical and mechanical properties parameters slightly.

4.3.2. *Bearing capacity of foundation soil*

The foundation soil bearing capacity design value fd in accordance with the relevant specification [1] [2] determined comprehensively and related experience;On foundation bearing capacity design value is used only for localized evaluation, the design should be based on the actual basis properties, size and buried depth is calculated, and considering the bottom layer of the strength and deformation;Characteristic value of subgrade bearing capacity value of fak (refer to the foundation soil bearing capacity design value of fd.Part of the foundation soil bearing capacity design value of soil (assuming D = 1.0 m, Dw = 0.5 m, strip foundation, B = 1.5 m).The soil foundation bearing capacity statistics are shown in table 4-1.

Table 4-1. Statistics of the bearing capacity of soil foundation

Layer sequence	Name of the soil	Compression index (average)		Intensity (average)		Than the penetration resistance (small average value)Ps（MPa）	The foundation bearing capacity Design valuefd （kPa）
		The compression coefficienta0.1～0.2（MPa-1）	The compression modulusEs0.1～0.2（MPa）	Quick direct shear (Solid)			
				cohesion (kPa)	Angle of internal frictionΦ(°)		
②	Further toner clay	0.49	4.18	18.0	18.5	0.61	75
③	Muddy gray silty clay	0.90	2.63	13.0	16.0	0.48	55
④	Gray silt clay	1.16	2.16	11.0	10.0	0.45	50
$⑤_{1-1}$	Gray clay	0.70	3.14	14.0	14.0	0.73	75
$⑤_{1-2}$	The grey silty clay	0.53	3.99	17.0	18.5	1.16	100

4.3.3. *DMT test results*

The survey carried out on each soil layer in the foundation pit influence scope of flat dilatometer test (DMT) (DMT), test the DMT - W1 type (DMT) instrument, test can be separately measured P0, P1 and P2, respectively is equal to the displacement of the diaphragm center 0.05 mm, 1.10 mm and return to the initial pressures of 0.05 mm), according to the specification [1] the related content can be estimated respectively unearthed class index I_D E_D with flat, horizontal stress index K_D, modulus, static lateral pressure coefficient K_0, undrained shear strength of C_U and lateral bed reaction coefficient Kh.

Soil type index: $I_D = \dfrac{P_1 - P_0}{P_0 - u_0}$

Horizontal stress index: $K_D = \dfrac{P_0 - u_0}{\sigma'_{V0}}$

Modulus with flat（kPa）: $E_D = 34.7\Delta P$

Static lateral pressure coefficient: $K_0 = 0.34\ K_D^n$

(Where n and status according to the soil sample, soil type index values as required)

The undrained shear strength (kPa) : $c_u =（-0.06 I_D^2 + 0.42 I_D + 0.19）\cdot \sigma'_{V0} \cdot（0.47 K_D）^{1.14}$

The compression modulus: $E_S = R_m \cdot E_D$ (The x value according to the index of soil type requirement)

Lateral bed reaction coefficient(kN/m3): $K_h = \Delta P / \Delta S$

The above formula:u0 For the hydrostatic pressure (kPa); σ'_{V0} For the effective overburden pressure (kPa);
$\Delta P = P1 - P0$（kPa）; $\Delta S = (0.00105) \cdot 2/3$.

4.3. *Groundwater situation*

4.3.1. *Diving*

Shallow water and confined water is closely related to the project construction, the main recharge sources of atmospheric precipitation and surface runoff, changes with the seasons, climate, rainfall, etc.The survey and the annual average can be according to 0.5 m high groundwater level depth gauge, the annual average low water depth at 1.5 m.

4.3.2. *Confined water*

The survey revealed the first 7 sandy silt layer, for the first I confined aquifer water, generally in cycles, water pressure on the head basically stable.Determined comprehensively through indoor soil test and field exploration, the original fitting is divided into the first 5. 2 powder soil layer of soil, due to the features of powder soil is not obvious, cross into (5) after 1-2 layer, not considering the confined water.

4.3.3. *The soil permeability coefficient*

This project deep foundation pit excavation, about 22 ~ 24 m, within the scope of influence of main soil for (2), (3), (4), (5) 1-1, 2, 3, 5) (5) 5. 1-4, 7) and 1 layer, indoor penetration test of the soil, the permeability coefficient of soil slightly.

Required for foundation pit excavation design for soil permeability coefficient, in indoor penetration test at the same time, the in ZS01 ~ 3 hole (2), (3), (4), (5) 1-1, 2, 3, 5) (5) 5. 1-4, 1 layer section 7) and precipitation head water injection test was carried out, soil water injection test results are shown in table 4-2.

Table 4-2. The soil water injection test result table

Layer sequence	Name of the soil	The permeability coefficientk(cm/s)		
		The minimum value	The maximum value	The average value
②	Further toner clay	1.35E-6	3.16E-6	2.26E-6
③	Muddy gray silty clay	1.56E-6	3.87E-5	1.49E-5
④	Gray silt clay	1.54E-6	3.21E-6	2.51E-6
$⑤_{1-1}$	Gray clay	1.23E-6	3.17E-6	2.13E-6
$⑤_{1-2}$	The grey silty clay	2.26E-6	2.61E-5	1.45E-5
$⑤_3$	The grey silty clay	1.60E-6	3.15E-6	2.37E-6
$⑤_4$	Sage green silty clay	9.62E-6	1.85E-5	1.41E-5
⑦	Gray green to gray sandy silt	7.72E-5	1.43E-4	1.23E-4
$⑧_1$	Gray clay	6.26E-6	7.06E-6	6.68E-6

4.3.4. *The groundwater temperature*

Buried depth of 4 m below the water temperature is stable, generally at 16 ~ 18°C

4.3.5. *Water and soil erosion evaluation*

Can determine according to the results of the groundwater quality inspection report: in III class environment, proposed site scope of groundwater is a slightly corrosive to concrete;To that of rebar in reinforced concrete in long-term immersion situation is corrosive, under dry-wet alternate with weak corrosive;Have weak corrosive to steel structure.

Because the proposed site groundwater buried shallow, near and no pollution sources distribution, so according to related specification [1], the foundation soil of steel corrosion in concrete and reinforced concrete evaluation and groundwater are the same.

4.4. *The site and foundation seismic effect*

4.4.1. *The site soil type and site category*

According to the proposed field wave velocity test results, the proposed site below ground vsi 20 m of each soil layer shear wave velocity is 122.0 ~ 167.2 m/s, the equivalent shear wave velocity of soil vse is 139.1 ~ 140.5 m/s.

According to the survey data show that ground cover layer thickness is greater than 130 MB.

According to the national standard [3], the ground under the 20 m soil vse equivalent shear wave velocity of 150 m/s or less, ground cover layer thickness is more than 80 m, can determine site soil types for soft ground soil, site category for IV site.

4.4.2. *The settlement of problems*

Proposed site foundation main stress distributed within the scope of a thick layer of high compressibility, low strength of silt soil, the settlement of soft soil is the main cause of earthquake damage.According to the national standard [4], when the bearing capacity characteristic value of 7 area more than 80 kpa, equivalent shear wave velocity is more than 90 m/s, don't consider the effect of soft soil settlement.

This vsi proposed site of shallow soft soil layer shear wave velocity is 122.0 ~ 167.2 m/s, more than 90 m/s settlement of discriminant criteria, and therefore under 7 degree seismic fortification intensity, don't consider the ground settlement of soft soil.

4.4.3. *The sandy soil liquefaction*

After the investigation, proposed site below 20 m depth range does not exist in layers of saturated sandy silt or sand soil, in seismic fortification intensity of 7 degrees when there is no possibility of liquefaction, therefore, determine the venue is not liquefied sites.

5. Conclusion

Through indoor and outdoor dozens of prospecting method, obtained the basic information of the proposed site, through the data can be uniformity of ground stability and suitability, foundation and foundation soil engineering characteristics.Analysis and evaluation on basis of proposed forms, including the choice of pile foundation bearing layer, pile type, pile foundation design parameters and the analysis of bearing capacity of single pile, pile foundation settlement estimation parameters, the feasibility analysis of pile, pile on the surrounding environment.Can analysis and evaluation on the stability of foundation pit, including foundation pit enclosure and construction, foundation pit tu chung, float sand, piping and foundation pit precipitation, etc.

References

1. China's local(shanghai) standard, Geotechnical engineering reconnaissance specification (DGJ 08-37-2012).
2. China's local(shanghai) standard, Foundation design specification(DGJ 08-11-2010).
3. CNS, Building seismic design code.(GB 50011-2010).
4. CNS, geotechnical engineering investigation specification(GB 50021-2001).

Numerical simulation research on the fracture behavior of basalt fiber reinforced reactive powder concrete

Yi-Hong Guo[a], Jian-Dong Jin[b], Yun-Long Lu[c]
College of Aerospace and Civil Engineering, Harbin Engineering University, Harbin, 145 NanTong Street, NanGang, China
E-mail: [a]gyhhrb@126.com, [b]1014932411@qq.com, [c]75489341@qq.com

Basalt fiber reinforced reactive powder concrete, as a new type of construction material, has a lot of outstanding performances. For example: high strength, good flexibility, good wear-resisting and impact resistance, it also shows good property in corrosion resistance. However, the study on basalt fiber reinforced reactive powder concrete is still in the initial stage, the numerical simulation research on the fracture property is basically in the blank stage. In this paper, the smeared cracking model was used in the numerical simulation. Cube compressive strength test, prism axial compression test, cylinder splitting tensile test and three-point bending beam fracture tests were done to get the relative parameters. The parameters obtained from the tests were used to develop the smeared cracking model by using the finite element analysis software ABAQUS. Based on the correct numerical model parameters, the ABAQUS finite element software was used to analysis the three-point bending beam with different span-depth ratio. The comparison of numerical results with experimental results indicates a good agreement.

Keywords: Basalt Fiber; Reactive Powder Concrete; Numerical Simulation; Fracture Energy; Smeared Crack Model.

1. Introduction

Normal concrete has some disadvantages such as large self-weight, low tensile strength, high brittleness and low toughness etc. Basalt fiber reinforced reactive powder concrete, as a new type of construction material, has a lot of outstanding performances. For example: high strength, good flexibility, good wear-resisting and impact resistance [1].However, the study on basalt fiber reinforced reactive powder concrete is still in the initial stage, and most of the studies are on the aspect of basic mechanical behavior, only few are on the aspect of fracture property, especially, the numerical simulation research on the fracture property is basically in the blank stage. In this paper, smeared cracking model was used to study the fracture property of BFRRPC, all the parameters needed were got by experiments.

2. Numerical Simulation of the Fracture Process of Basalt Fiber Reinforced Reactive Powder Concrete

2.1. *Constitutive model*

In this paper, smeared cracking model [2] was used as the numerical simulation model. The crack form of the smeared cracking model is to produce a large number of continuous and parallel micro cracks in a cell, while the cracks are not existed single or separated but gathered in a range. Parameters like the uniaxial compression and uniaxial tension softening curve values were needed to build the smeared cracking model in the numerical simulation software ABAQUS [3].In order to get these parameters, cube compressive strength test, cylinder splitting tensile test and rism axial compression test were done [4].

PO.42.5 cement, silica fume ,silica sand, super plasticizer were used to make the specimens.5 different kinds of basalt fiber volume additions (0%,0.5%,1.0%,1.5%,2.0%) were used. The test results were shown in Table.1.(The axial tensile strength of the specimens was calculated by axis tensile strength and splitting tensile strength conversion [5] Eq.(1).(k=0.921))

$$f_t = k \times f_{ts} \tag{1}$$

Table 1. The parameters of BFRRPC with different basalt fiber volume content

Name / Code	Elasticity Modulus E(GPa)	Poisson Ratio μ	Density ρ (kg·m^{-3})	Fracture Energy (N·m^{-1})	Uniaxial Tensile Strength f_t (MPa)	Uniaxial Compressive Strength f_c (MPa)	f_t / f_c
C0	21.95	0.281	2219.35	67.85	5.34	81.77	0.065
C0.5	20.49	0.291	2173.37	250.05	6.05	84.80	0.071
C1.0	21.61	0.303	2135.68	335.98	6.55	88.23	0.074
C1.5	21.81	0.291	2114.82	397.30	5.83	84.73	0.069
C2.0	22.90	0.283	2060.02	589.30	5.45	82.30	0.066

The results showed that with the increase of the BF volume content from 0% to 2.0%, the fracture energy was increased 868.5%.

In this paper,Yafeng Ma constitutive model of Uniaxial compression stress strain curve[6] was used. Eq. (2).

$$y = \begin{cases} ax + (5-4a)x^4 + (3a-4)x^5, (0 \le x \le 1) \\ \dfrac{x}{b(x-1)^2 + x}, \quad (x > 1) \end{cases} \tag{2}$$

By matching the results of test and theoretical model,the final values were got::a=1.15,b=7.The mathematical expression of the uniaxial compression curve was shown as Eq.(3)(Take C1.0 as example)

$$y=\begin{cases}1.15x+0.4x^4-0.55x^5,(0\le x\le 1)\\ \dfrac{x}{7(x-1)^2+x}, \quad (x>1)\end{cases} \tag{3}$$

Bilinear softening curve was used as the tension softening curve of the concrete fracture process zone. Its mathematical expression was shown as Eq.(4) [7].

$$\sigma=\begin{cases}f_t-(f_t-\sigma_s)\dfrac{\omega}{\omega_s}, & 0\le\omega\le\omega_s\\ \dfrac{\sigma_s(\omega_0-\omega)}{\omega_0-\omega_s}, & \omega_s<\omega<\omega_0\\ 0, & \omega_0<\omega\end{cases} \tag{4}$$

In this paper,bilinear softening curve suggested by Shiliang Xu[8] was used to figure out the parameters in Eq.(3). The bilinear softening curve formula suggested by Shiliang Xu was shown as Eq.(5)

$$\begin{cases} \omega_s = \dfrac{0.4\sqrt{\alpha_f G_f}}{f_t} \\ \sigma_s = \dfrac{\left(2-0.4\sqrt{\alpha_f}\right)f_t}{\alpha_f} \\ \omega_0 = \dfrac{\alpha_f G_f}{f_t} \\ \alpha_f = \lambda - \dfrac{d_{\max}^{0.9}}{8} \\ G_f = \dfrac{0.0204+0.0053\, d_{\max}^{0.95}/8}{\left(f_c/f_{c0}\right)^{0.7}} \\ \lambda = 10-\left(\dfrac{f_{ck}}{2f_{ck0}}\right)^{0.7} \end{cases} \tag{5}$$

Combining the test results with Eq.(3),the final mathematical expression of the uniaxial tension softening curve was fitted out as Eq.(6).(Take C1.0 as example)

$$\sigma = \begin{cases} 6.55-0.1046\omega, & (0 \le \omega \le 54.1) \\ 1.0508-0.00295\omega, & (54.1 < \omega \le 356.4) \end{cases} \tag{6}$$

2.2. *Numerical model establishing*

The geometrical shape of the specimen was shown as Fig.2.The length L=400mm,height D=100mm,the width t=100mm,the span S=320mm, the length of precast crack a_0=40mm, the width of crack is 5mm.The results were shown in Fig.4 and Table.2.

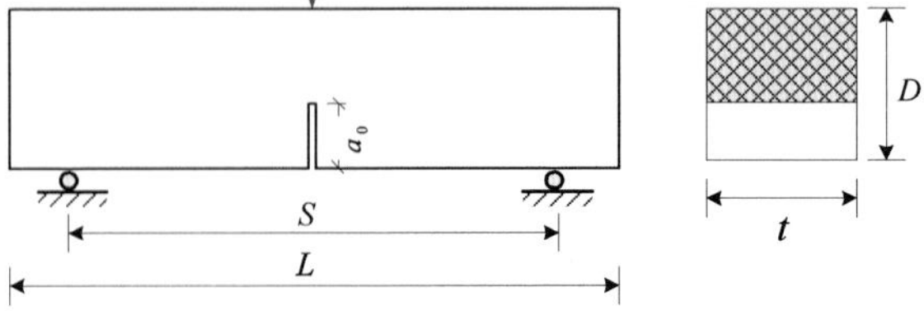

Fig. 2. Shape of Three-point bending beam

2.2.1. *Division of the model*

The division of the model was shown in Fig.1.The middle of the model was divided in fine grids while other part was coarse grids. The size of the middle units was 5mm×5mm×5mm.

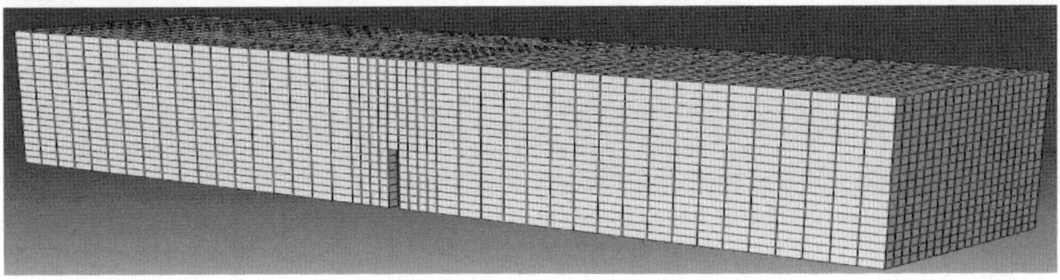

Fig. 1. Finite element mesh

2.2.2. *Determination of load and boundary conditions*

In this paper,Standard analysis module was used to analysis the interfaces separation of the material under Quasi-static situation[9,10].The normal behavior of concrete damage was set to linear,contact rigidity was 100.the contact type between both sides bearings and specimen was surface contact with surface(standard), The main surface smoothness was 0.2.During the simulation,displacement-control code was applied,two analysis steps were set in the analysis editing.In the first analysis step,the displacement was applied as 0.001mm while in the second analysis step was 0.9mm.the connecting format between beam and bear was hinge joint.

2.3. *Numerical simulating analysis and the comparison with test result*

Take C1.0 as an example,the fracture process was shown as Fig.3.

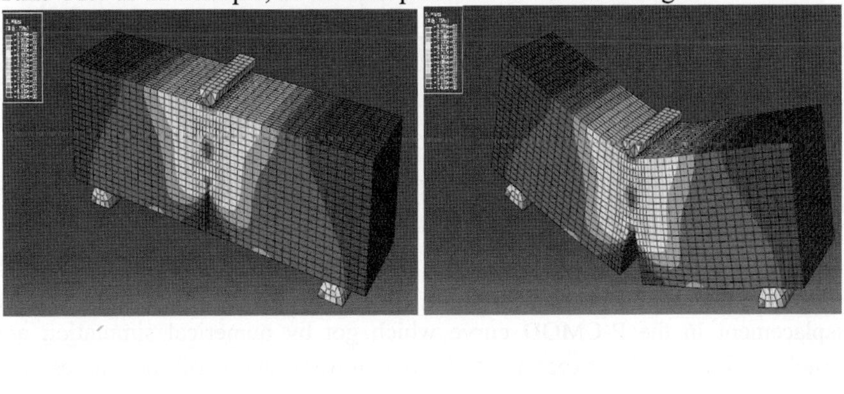

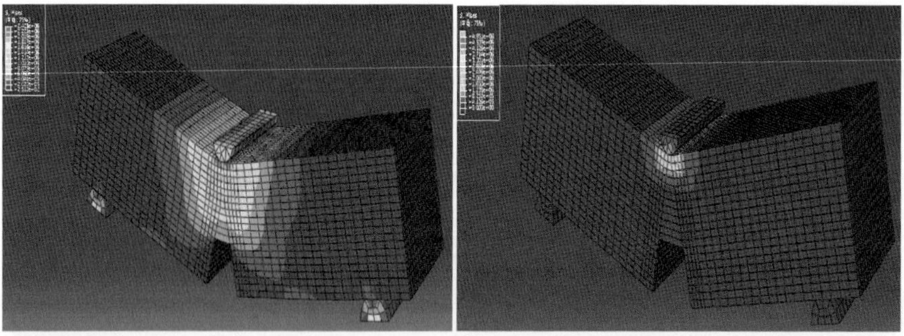

Fig. 3. Deformation during the fracture process of C1.0 and nephogram of major principle stress

Fig.3 told us that after the crack split, the pull stress on the fracture surface was not decreased to 0 at once, but decreased gradually from ultimate tensile strength to 0.With the increase of the crack width, the tensile strength of the crack surface was decreased. The tensile strength of the crack surface was totally lost when the crack width increased to a certain degree. With the displacement of the top of the model approach gradually to the maximum displacement in theory, the imposed load produced on the top was decreased gradually to 0,by this time the fracture process of the specimen was completely into the convergence phase.

The comparison between the Loading point load-Displacement curve(P-δ) and Load-Crack Mouth Opening Displacement curve(P-CMOD) got from experiment and numerical simulation was shown as Fig.4

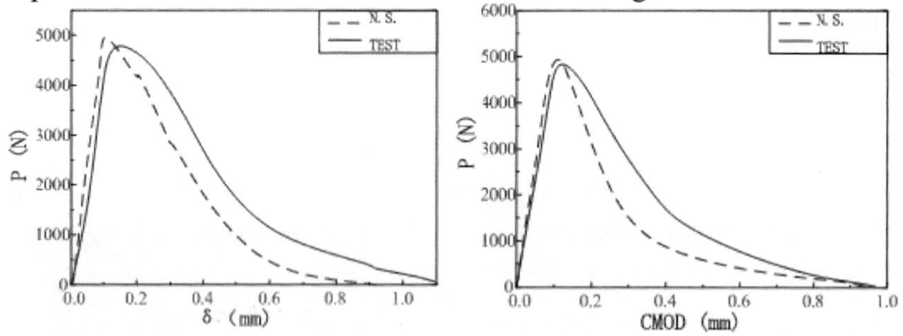

Fig. 4. Comparison between the (P-δ) curve and (P-CMOD)curve got from experiment and numerical simulation

From Fig.4 we can see that: Most time, the displacement of the loading point ,the maximum load in the P-δ curve and the crack mouth opening displacement in the P-CMOD curve which got by numerical simulation are closed to that got by experiment. In the upward phase of the curves, the numerical simulation results were little higher than that in experiment. It's mainly because the loading condition in experiment was uniform load, while in the numerical simulation study was displacement control of the loadin g point, which can release the stress in the fracture extension area of the BFRRPC. The decline phase of the curve of the experiment were fitted by using the Allometricl in Origin(a software)at the inflection point. It's because that the BFRRPC is still a brittle material to some degree, also other reasons such as the not enough rigidity of the test machine etc., so it's hard to get the whole decline curve of the concrete by experiment. That's why the result of the numerical simulation were obvious less than the result of experiment in the decline phase of the curves. The comparison showed that the smeared cracking model established by ABAQUS can simulate the fracture process of the BFRRPC well. Here are all the result comparisons between test and numerical simulation shown in Table.2.

Table 2. Comparison between test results and numerical simulation results

Code	P_{max}/KN		$CMOD_C/\mu m$		a_C/D		δ_{max}/mm	
	Test	N.S.	Test	N.S.	Test	N.S.	Test	N.S.
C0	3.343	3.491	491.9	369.4	0.588	0.674	0.361	0.324
C0.5	3.752	3.862	871.7	823.6	0.6785	0.762	0.897	0.742
C1.0	4.756	4.920	975.2	948.5	0.7243	0.814	1.12	0.918
C1.5	5.513	5.790	1438.7	1367.2	0.707	0.837	1.43	1.396
C2.0	6.379	6.426	1569.4	1489.6	0.718	0.862	1.58	1.474

2.4. *Numerical simulation analysis on the three-point bending beams with different span-depth ratio*

Based on the correct numerical model,4 different span-depth ratio were tested(3.2,4.2,5.2,6.2, labeled as C1-K3.2, C1-K4.2, C1-K5.2, C1-K6.2) All the specimens used the 1% basalt fiber volume addition, the sizes of all these specimens are shown in Fig1.

The Simulation analysis and calculation results of the three-point bending beam with different span-dcpth ratio were shown in Table.3.

Table 3. The Simulation analysis and calculation results of the three-point bending beam with different span-depth ratio

Code	a_0/D	P_{max}/KN	$CMOD_C/\mu m$	a_C/D	δ_{max}/mm
C1-K3.2	0.4	5.978	951.6	0.796	0.835
C1-K4.2	0.4	4.297	949.8	0.826	1.096
C1-K5.2	0.4	3.016	949.8	0.854	1.359
C1-K6.2	0.4	2.070	949.7	0.878	1.617

3. Conclusions

(1) item one, The numerical simulation analysis results matches well with the test results,the smeared cracking model applied in this paper reflected the fracture process of the three-point bending specimens real.

(2) item two, With the increase of the basalt fiber volume addition, the specimens showed the increasing ductility, the maximum load on loading point increased gradually as well. It means that the basalt fiber can strengthen the self-strength and toughness of the reactive powder concrete.

(3) item three, With the increase of the span-depth ratio: the maximum load on the loading point P_{max} were decreased, while the drop were decreased. The maximum crack mouth opening displacement $CMOD_C$ kept in 0.950mm around for all codes, which means the span-depth ratio has no influence on the failure mode of the specimen; The ratio between equivalent critical crack length and specimen's depth a_c/D were increased gradually; The maximum displacement on the loading point δ_{max} were increased significantly, basically linear increasing trend.

References

1. Sergey, Characters of basalt fiber and the application prospect in China.Glass fiber, 2005, (5): 44-48.
2. Rots J G. Smeared and discrete representations of localized fracture. International Journal of Fracture. 1991, 51: 45-59.
3. Jinchang Wang, Yekai Chen, The application of ABAQUS in civil engineering,Zhejiang University Press, 2006: 68-75.
4. Kaizhan Ke, study on the basic mechanical properties and application of fiber reinforced reactive powder concrete. master thesis, Fuzhou University, 2006.
5. Jian Yin, Shiqiong Zhou. Experimental study on the axial tensile strength and splitting tensile strength of high performance concrete. Journal of Changsha Railway University. 2001, 19(2): 25-29.
6. Yafeng Ma. Reactive powder concrete (RPC200)uniaxial compression constitutive system, master thesis, Beijing Jiao tong University, 2006.
7. Xu S L, Reinhardt H W. Determination of double-K criterion for crack propagationin quasi brettle fracture, Part II: Analytical evaluating and practical measuring methods for three point bending notched beam. International Journal of Fracture, 1999, 98: 151-177.
8. Xu S L. Determination of parameters in the bilinear, Reinhardt's and exponentially nonlinear softening curves and their physical meanings. Werkstoffe and Werkstoffpruefung in Bauwe sen, 1999: 410-424.

9. Wanquan Zhu, Qilin Zhao,Chunan Tang,Jiashou Zhuo, mechanical model and numerical simulation of the fracture process of concrete. Advances in Mechanics. 2002, 3(4): 579-598.
10. De Borst R, Nauta P. Non orthogonal cracks in a smeared finite element model[J]. Engineering Computations. 1985, 2(3): 35-46.

Analysis of negative skin friction resistance characteristic influence parameter of pile foundation under flooding condition

Shuai-Hua Ye[1,2,*], Chang-Liu Chen[3,†]

[1]*School of Civil Engineering, Lanzhou University of Technology, Lanzhou, 730050, China*

[2]*Research Center of Coastal and urban Geotechnical Engineering, Zhejiang University, Hangzhou, 310058, China*

[3]*China Mobile Group Gansu Co.,Ltd. Lanzhou, 730030, China*

Email: *yeshuaihua@163.com, †*13993149388@139.com*

There are many influence factors of negative skin friction resistance characteristic influence parameter of pile foundation, in this paper, through the finite element analysis, get the relationship curve of pile side friction's distribution and pile top load, the relationship curve of pile side friction distribution and the pile body's stiffness, and the relationship curve of pile side friction and flooding pressure, analysis feature characteristic of curve, get the corresponding conclusions.

Keywords: Pile Foundation; Negative Skin Friction; Influence Parameter; Finite Element Analysis.

1. Introduction

With the rapid development of urban industrialization and commercialization, the requirement of foundation engineering is becoming more strict. However, the development of negative skin friction resistance characteristic influence parameter of pile foundation has not yet mature. Therefore, in this paper, Established reasonable finite element model, it is concluded that pile top load, stiffness of pile and saturation pressure on the influence law of pile side friction distribution, and impact on the neutral position.

2. Analysis of Parameter Sensitivity

The play of pile lateral's resistance depends on the relative displacement of piles and soil. When the settlement of pile soil is greater than the displacement of pile body, pile lateral's resistance have been down, and in the same direction of the pile top load negative friction resistance; When the settlement of pile soil is less than the displacement of pile body, pile lateral's resistance will show up, could provide the upper load resistance.

3. The Influences of Pile Top Load

When the pile head have no load, pile lateral's soil subsidence is the main cause of pile soil interaction. Soil settlement of the pile body produce the negative skin friction, drag force caused by negative skin friction resistance making pile body compression deformation and settlement of pile end. And the emergence of pile body compression deformation and settlement of pile, making the relative displacement relationship of the pile and soil changed, and moving up of the neutral point of the pile. Positive and negative friction resistance offset that work together to achieve a new balance between pile and soil. Only when the settlement of pile and soil are stable, the pile of neutral point and the distribution of friction resistance along the pile shaft to stabilize.

In finite element analysis, pile top load respectively get 5000, 6000, 7000 KN, subject to flooding for 10 days, in order to study for the influence rule about change of pile top load and characteristics of negative skin friction resistance. The pile side resistance and pile top load curve as shown in Figure 1. Figure 1 reflect the distribution of different frictional resistance along the pile under pile top load. By the figure can be seen that, firstly, at the top of the pile is 7000 kN load of pile friction resistance is the largest. Secondly, with the increase of pile top load, the location of the neutral point obviously moved on. Thus, for the friction pile, the size of pile top load has a great influence for the characteristics of negative skin friction resistance. With the increase of pile top load, neutral point up, provide positive friction of the pile body below the neutral point part longer, the total value of friction resistance is increased. Analysis shows that, regardless of the pile top load determined the neutral position is the lowest. The existence of the pile top load can make the neutral point up, and with the increase of pile top load neutral point can significantly move up.

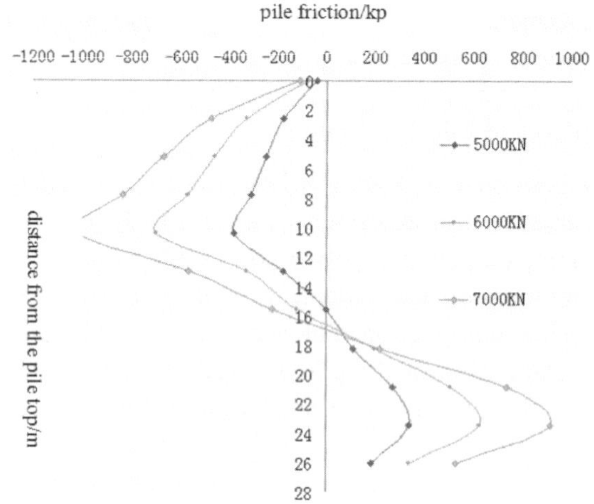

Fig. 1. The relationship of pile side friction's distribution and pile top load.

4. The Influences of Pile Rigidity

Displacement of the pile body is mainly made up of settlement of pile end and pile body compression deformation. According to Hooke's law, compressive components appear compression deformation and the size of compression deformation depends on the component materials' elastic modulus and the compressive stress on the component. When compressive stress certain, elastic modulus is the only factor affecting the component compression deformation. For the pile foundation of bear the negative friction, the compression modulus of pile body material can also influence the compression deformation of pile body, which affects the relative displacement between the pile and soil, influence the distribution of negative frictional resistance along the pile shaft. High compression modulus, pile body materials' compression is small, generally ignore this part of deformation, this is the same analysis of routine conditions. For CFG pile and cement-soil pile, pile body material compression modulus is low, the compression of pile is bigger, it will change the status of the pile and soil's relative displacement, shortening part of negative skin friction resistance, neutral point move up or even eliminate. And neutral point up or eliminate and make the effect caused by negative friction on pile under the pulling force to reduce or eliminate. This is a process of interaction influence each other. Therefore, the change of the compression modulus of pile body material had significant effect on the characteristics of negative skin friction resistance for pile foundation.

For negative skin friction resistance pile, amplitude of stiffness change is relatively small. In order to simplify the analysis, in only considering change of pile body elastic modulus for the impact of negative friction resistance characteristics of the pile.

In finite element analysis, the elastic modulus of pile material respectively get 6×10^5, 6×10^6, 6×10^7kPa and under condition by applying 6000 kN at pile top load, to study influence rule of the pile material elastic modulus' change for characteristics of negative skin friction resistance. Pile side friction and the stiffness of pile body curve as shown in Figure 2.

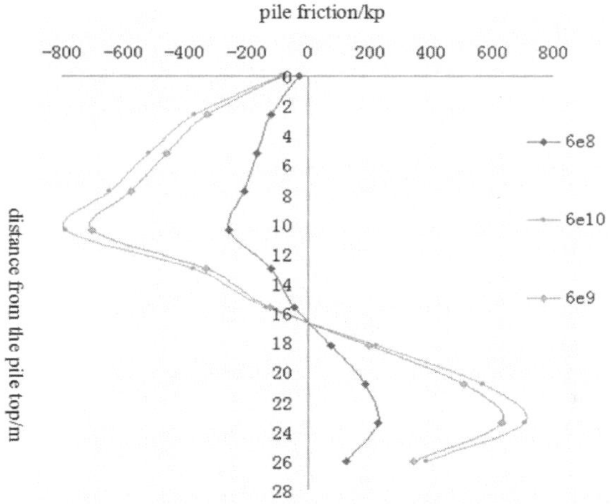

Fig. 2. The relationship of pile side friction distribution and the pile body's stiffness.

Figure 2 reflect the friction resistance's distribution along the pile shaft, which pile material of different elastic modulus under the condition of fixed load at the top of the pile. Can be seen from the diagram, firstly,when the loess elastic modulus is 6e10kPa, friction of pile is the largest. Secondly, low elastic modulus of pile neutral point position is shallow from the pile top, smaller values of the negative friction resistance, pile body parts' friction resistance value is smaller also. High elastic modulus of pile neutral position is deeper from pile top, the negative friction resistance value is bigger, pile body parts' friction value is bigger also. Thirdly, when the pile stiffness reaching a certain value, the develop level of positive and negative friction resistance is basically no obvious change.

5. The Influences of Flooding Stress

When the pile cap with load, the soil deformation is not the only cause of pile soil interaction. For collapsible loess regions, flooding pressure of the soil around the pile is also a cause of pile soil interaction. In this case, it is a practical value to study the development law about the characteristics of pile's negative frictional resistance with the soil around the pile under the condition of flooding pressure changes.

In finite element analysis, flooding stress of soil around pile respectively get 7000, 8000, 9000N and under the condition of applied 6000 kN load at the top of the pile and surrounding soil flooding for 10 days, to study the influence law of soil around pile in flooding pressure changes for pile negative friction properties. The relational curve of pile side friction and flooding pressure as shown in Figure 3.

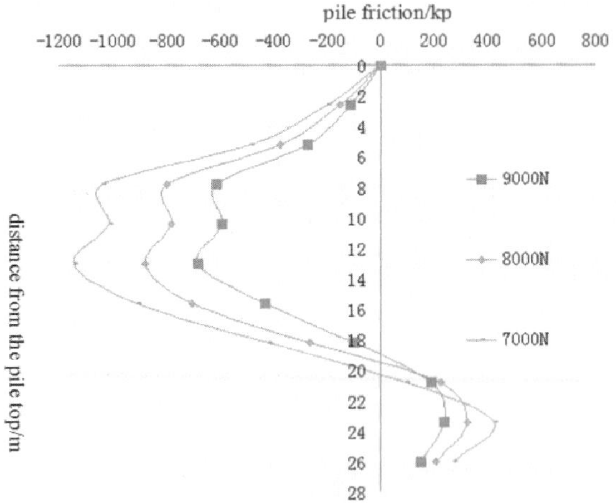

Fig. 3. The relationship of pile side friction and flooding pressure.

6. Conclusions

(1) For the friction pile, the size of pile top load has a great influence for the characteristics of negative skin friction resistance. With the increase of pile top load, neutral point move up, provide positive friction of the pile body below the neutral point part is longer, the total value of friction resistance is increased. Analysis shows that, regardless of the pile top load determined the neutral position is the lowest. The existence of the pile top load can make the neutral point up, and with the increase of pile top load neutral point can significantly move up.

(2) The greater flooding pressure, the neutral position is shallow from pile head, negative friction resistance value is small, pile body parts' friction value is small. The neutral point position of small flooding pressure is deeper from pile

top, the negative friction resistance value is bigger, pile body parts' friction value is bigger also.

Acknowledgments

This work was financially supported by youth science and technology research fund projects in Gansu province "Study on the Characteristies of Pile Negative Friction in the Large Thickness Loess Area Considering the Collapsibility" (2014GS02363).

References

1. XIA Li-nong, LEI Ming, NIE Chong-jun. Field test of influences of load at pile top on negative skin friction behaviors [J]. Rock and Soil Mechanics, 2009, 30(3): 664-668.
2. Lai Qionghua, Settlement of pile foundation practical calculation method[J]. Journal of rock mechanics and engineering, 2004, 06: 1015-1019.
3. WEI Jin, LI Zhe, HAO Mang-li, et al. Submerged static load test of pile foundation in collapsible loess area [J]. Journal of Chang'an University, 2011, 31(5): 63-67.
4. Ma Meng. Pile foundation dynamic stiffness factor analysis[J]. Rock and soil mechanics, 2015, 07: 1939-1944.
5. Xiong Hongmei1, Hao Xiaohong, in: Journal of North China Institute of Water Conservancy and Hydroelectric Power, The application of ultrasonic wave method to pile test, 66-69. (2002).
6. Ultrasonic testing concrete defect technical regulations, CECS21-90.
7. Luo Qixian. Ultrasonic testing of hydraulic structure concrete. High Efficiency Silicon Solar Cells (Water conservancy water power press, Beijing 1989).
8. Xiong Hongmei1, Hao Xiaohong, in: Journal of North China Institute of Water Conservancy and Hydroelectric Power, The application of ultrasonic wave method to pile test, 66-69. (2002).
9. Wu Mianba, Summary of pile foundation inspection[J]. Geotechnical foundation, 2001, 03: 2-8.
10. Zhou Depei, Wang Huanlong, Sun Hongwei. Micro-pile composite anti-sliding structure and its design theory[J] Chinese Journal of Rock Mechanics and Engineering, 2009, 07: 1353-1362.

Researches on cooperative mode of rural drinking water safety project — A case study of Dianjun district, Yichang

Yun-Pin Hu[1], Tian-Yun Zhong[1,*], Wei Chen[1], Jie Fu[1], De-Fu Zhang[2,†]
[1]*School of Civil Engineering and Architecture, Wuhan University of Technology, Wuhan, Hubei, China*
[2]*Huazhong University of Science and Technology, Wuhan, Hubei, China*
*E-mail: *zhongty110@qq.com, †zhangdefuzy@163.com*
www.whut.edu.cn

This paper aimed at putting forward a scientific management model to solve the rural drinking water safety problem. The comparative analysis method was applied to evaluate the performance of several management models according to the existing problems of high operation cost, difficult fee collection and backward operation management. The result was drawn that the cooperative model should be applied priority in the rural drinking safety project which can provide technical and management supports and help to explore a new way to achieve sustainable effectiveness.

Keywords: Rural Drinking Water Safety Project; Cooperative Model; Operation Organization; Sustainable Development.

1. Introduction

As farmer's production resources, rural drinking water is not only related to the practical farmer's quality of life and health level, but also plays an important role in promoting agricultural and rural economic development. Due to existing problems including lack of water and discharge of industrial and agricultural waste water in rural remote areas, governments at all levels begin to focus on rural drinking water safety. During the "Eleventh Five-Year Plan" period, the country totally invested 105.3 billion yuan in drinking water safety project related to 201 million rural people. During the "Twelfth Five-Year Plan" period, the total investment raised to 170 billion yuan to accelerate the pace of construction and solve rural drinking water safety issues comprehensively. However, China's rural drinking water projects are difficult to operate due to the characteristics of dispersion and considerable, some water supply project can't obtain the benefits and waste state funds. Improvement of management system and operation mechanism for achieving the sound operation of rural drinking water safety has become a major focus in rural drinking water safety project [1].

2. Basic Features of Rural Drinking Water Safety Project

In rural areas, rural residents live dispersedly and some local terrain are complex, resulting in a large number of small-scale rural water supply projects. Township or village water supply project can be implemented in the form of centralized water supply in some areas with better condition, the beneficial population can reach thousands; Decentralized village water supply project can only be implemented in some poor areas, water supply population is often only a few hundred or even dozens of people [2]. Water mainly refers to residential water consumption, water consumption in public buildings, fire water, roads and green water poured, Compared to urban areas, water consumption is very little in rural areas. Enterprise is also limited to a very few water rural water supply project. In some areas, the traditional concept of farmers' water does not completely change, reluctant to spend money on water. Except drinking tap water, some farmers have been using well water or river water as other domestic water. These factors lead size of rural drinking water security project to be small and can't form the scale advantages of centralized management and operation.

3. Situation and Issues of Rural Drinking Water Safety Project in Dianjun

Dianjun is urban dominated by traditional agriculture, located in south of the Yangtze, with a total area of 550km^2, under the jurisdiction of two townships, two towns and a sub-district, where infrastructure is very weak. With the complex area of geological type and the undulating ground, the most villages are dispersed widely, so the rural drinking water and water supply conditions are very simple, some villagers still directly drink river water, rain water and cellar water, which can't reach state standards of safe drinking water in respect of quality, supply guarantee rate and convenience.

With the project put into use, the focus of the construction work will be transferred to operation management. Due to the special nature of topographic and geologic conditions in Dianjun, the rural drinking water safety project were mostly single village water supply project and decentralized water supply project. To make the new water supply project benefit local people in long-term, the various operation and management mode of water supply projects were in-depth researched from Tucheng town, Lianpeng town, Aijia town, Qianbian town. The survey results shows that operation and management situation of the most water supply projects in Dianjun were not optimistic, some water supply projects operated in long-term loss, individual water supply project even couldn't run continuously because of the poor management after the construction, these problems stimulated drinking water problems of farmers again. There were existing operational management problems as follow:

3.1. *The cost price was high*

Average price of national rural drinking water safety project was 1.63 yuan/t, weight average cost was 2.3 yuan/t [3]. The residential water price standard was 1.44 yuan/t in Yichang. Jiangnan Water Company located in Dianjun charged 1.12 yuan/t (Table 1). According to "Rural drinking water safety project price calculation report", 77% of the drinking water safety project calculated the cost price is generally higher than the current price in Dianjun, the highest cost price of a water supply project was 4.5 yuan/t, higher than the standard of special industry in Yichang. It is clearly not accordance with farmers' spending capability. Therefore, the current water collection is difficulty and the compensation mechanisms are not yet formed, resulting in a shortage of engineering and maintenance requirements to achieve a good benefit.

Table 1. The price of main water plant and the drinking water safety project.

Enterprise	The last price (yuan/t)				File basis
	Resident	Administrative	Industry	Special Industry	
Yichang Water Plant	1.44	1.89	1.89	4.10	Environmental and Resource Regulation price[2012]103, Yichang regulated price[2012]96
Jiangnan Water Company	1.12	1.12	1.47	2.60	Yichang regulated price [2009]134, Yichang regulated price [2006]145
Rural drinking water safety project A	4.5	——	——	——	Rural drinking water safety project price calculation report in Dianjun
Rural drinking water safety project B	3.6	——	——	——	Rural drinking water safety project price calculation report in Dianjun

3.2. *The subject of management responsibility was missing*

At present, rural drinking water safety projects impose "top-down" management level in Dianjun. Forestry and water bureau coordinated the implementation of construction of rural drinking water safety project, the town water stations implemented specific tasks. Due to the strong role of the government to push construction, the construction of water supply projects owned quality and quantity. After the construction, the project ownership is generally transferred to the village collective who own rights to decisive specific water supply project management, management mechanism, engineering maintenance work. The phenomenon "one build, no one control" appeared, because of the low enthusiasm of the village collective. The water users lacked awareness of self-management and lost the right to know, decision-making, supervision due to can't participate in the basic operation and management of the project.

3.3. *The quality of water source was difficult to guarantee*

The most of rural drinking water safety projects were single centralized water supply project with no disinfection measures. With the development of agricultural production and breeding industry, a large number of chemical fertilizers and waste water got any emissions. Combined with a small amount of water and dispersed, water self-purification ability got poor. So, rural drinking water safety problem was particular in Dianjun. Survey results show that 54% of the residents are affected by the problem of water second pollution. Coupled with lack of a complete management system, good facilities, water quality test, the stability of the quality of water is affected seriously.

4. The Current Rural Drinking Water Safety Project Management Models

Rural drinking water safety project was funded and constructed by the government, so that the property of rural drinking water safety projects was owned by the government. In fact, the government couldn't manage all projects. Government should consider actual situation of the project for each type of water supply projects, water supply capacity, economic and technological conditions to select the appropriate mode of operation and management. "Peasant self-management" guidelines become the first choice. Currently, rural drinking water safety project management models as follows:

4.1. *Village management*

Rural drinking water supply projects were mainly single centralized water supply project in Dianjun. After the construction, ownership of project is generally transferred to the village collective to decide on the specific water supply project management, management mechanism, engineering maintenance and other work [4]. It's proved that the use of the village collective management model is simple and suitable for village with a good working basis and better economic conditions.

4.2. *Individual contract management*

Without changing the ownership of the project, the public tender is organized by a group of villagers to clear the rights, obligations and responsibilities in both sides. Water engineering contractor was responsible for the maintenance and collection. According to the collective total use of water, the contractor need pay a fee to the village for major repairs. Then the contractor achieved residual income. In that case, authority and responsibilities of investors and managers were delineated clearly, the fund of project management process can be effectively guaranteed.

4.3. *Water association management*

Water association is combined of the members of strong responsibility, prestige, management, having not-for-profit characteristics.[5] In recent years, water association was promoted to achieve the desired goals in many rural areas of our country. Water association is established as a village unit whose members are voted from the production group of the association. It can start the work of the collection, engineering maintenance, contradictions coordination to achieve a mutual assistance association, self-management and self-service, truly "farmers manage their projects by themselves."

5. Building Water Supply Cooperative Management Model of Rural Drinking Water Safety Project in Dianjun

At the "Eleventh Five-Year Plan" of the national rural drinking water safety project, the core of the policy was that water users participate in construction and management of the rural water project at the whole process. But at the "Twelfth Five-Year Plan", the core was engineering property rights. The government fully plays the market mechanism and further promotes the reform of the rural drinking water safety project management system. Since the past decade, the government has been the role of infrastructure in rural drinking water supply project and gotten good grades. In general, the village collective manages and maintains the water project but the result is not ideal. In accordance with principles “clear property rights, clear responsibilities, separating the government and enterprises”, corporate management is formatted as professional specialization management to standardize the relationship between property owners and business entities. Based on the main drinking water safety projects, building water supply cooperatives and clearing corporate entity are the current management trends.

5.1. *Feasibility analysis on the formation of the water supply cooperative*

5.1.1. *Homogeneity*

The homogeneity namely similarity of the farmers in the local range is the foundation of the farmers' cooperation. The higher is the homogeneity, the smaller is the difference between the farmers. Their goals will be the same in the same resource constraints. After cooperation, there will be a small difference between operational goals and single individual targets, the cost will be lower to eliminate difference. In other words, the stronger is homogeneity, the easier is collaboration. Effect of homogeneity is very complex and associated with local natural environment, customs, production and lifestyle. The service of centralized water supply is for a village. Villagers facing a "drink water" problem had high enthusiasm and great expectations, and have begun to operate. For example, a farmer who is named Fali Chen and from Tuchen town build the "Tianhe cellar" to solve the drinking water problems on his own.

5.1.2. *Social capital*

Social capital refers to the informal sharing values among members. The higher is the degree of trust between people, the greater is the possibility of cooperation between the inhabitants in this area.[6] In the current rural social capital, the role of civil elites and old party members is very special. They have a high prestige, broad economic vision and outstanding management ability among the mass, who hope to achieve common prosperity under their leadership. Under the leadership, village governance of public affairs is relatively democratic, cooperation is relatively easy to achieve among farmers. For example, Dahua garden cooperation was led by the economic genius, Shaohua Jiang, to implement "cooperative + base + farmers" business model and get the villagers richer.

5.1.3. *Legal guarantees*

Good system design is an important guarantee for the smooth function of cooperation. In July 1, 2007, the "Farmer Cooperatives Law of People's Republic of China" was enacted to make a clear institutional arrangement for the specific operation of cooperatives [7]. Each member has equal right to get service provided by cooperation. Therefore, the "Cooperatives Law" clearly make provision that members assembly own decision and voting elections, one person one vote, the investment of members isn't directly linked with members rights in cooperatives. Each member enjoys their basic right to vote, no one shall restrict and deprive. The provision differs from the corporate decision-making to protect the interests of vulnerable groups at the maximum possibility. Thereby weaker groups who are eager to cooperate are greatly inspired. According to the 2013 statistical yearbook of Dianjun, the new farmer cooperatives was up to 60, the total was 125.[8] It showed that momentum of development was strong, the market environment was good, the legal system guarantee was reliable, which was the practical foundation to establish cooperatives.

5.2. *The advantages of water supply cooperative*

As a legal entity, supply cooperative is independent from the government bureaucracy management system. It's better to construct flat institutional structures and implement bottom-up supply model (Figure 1).

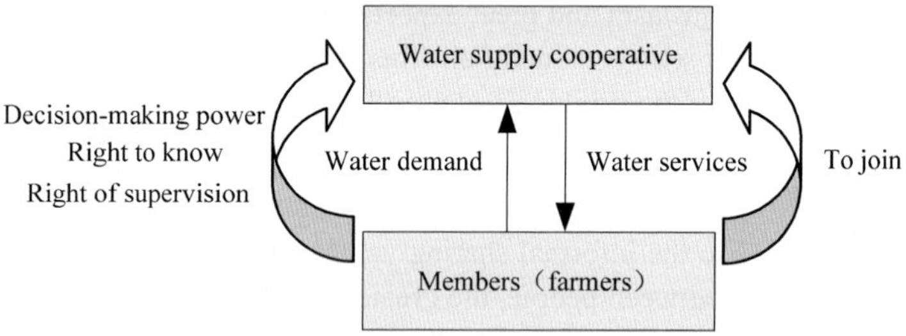

Fig. 1. Schematic of the water cooperative.

Firstly, the message can be feedback timely. Because of the only two levels of members and the cooperative, the supply of relevant information can be obtained quickly and timely feedback for supplying improper correction after the completion of water services. It's may be relatively quick and at lower cost. Secondly, management cost is low. The water cooperative is usually made up of several executive staffs, where there is a chairman of the Council and an acting supervisor. It's helpful for the staff to play their enthusiasm and creativity based on the actual local situation at low running cost. Thirdly, the water price is open to make. Under the cooperative system, the different water needs are often fully paid by the beneficiary, it's clear to be established in the form of constitution of the general cooperative assembly.

All in all, in the "bottom-up" supply model of the cooperative, the message can be feedback timely, management cost is lower, decisions of water price is more scientific and fairer. Compared to "top-down" government supply system, the "bottom-up" model is more effective.

5.3. *Establishing support mechanisms for water supply cooperative*

As a single village water supply cooperative, its strength is weak to operate and maintain the rural drinking safety project. For professional engineering equipment maintenance and water quality testing problems, cooperative as individuals is unable to seek market solutions at the lower cost. Therefore the government-led force is particularly important to build institutions to support cooperatives.

Focus on the daily work of a specific drinking water safety project, the cooperative is responsible for the operation and management, but the government provides engineering and technical support services. Traditional gold tower management model is transferred to the matrix management model. The idea is to transfer the government bureaucracy to collaborative project system. In the matrix structure of drinking water safety technical service work, the lengthways is cooperatives and the crosswise is technical support services group established by government (Figure 2). Safety engineering technology service center is a key member responsible for the docking work between cooperatives and technical services team, helpful for improving the efficiency and share resource technology.

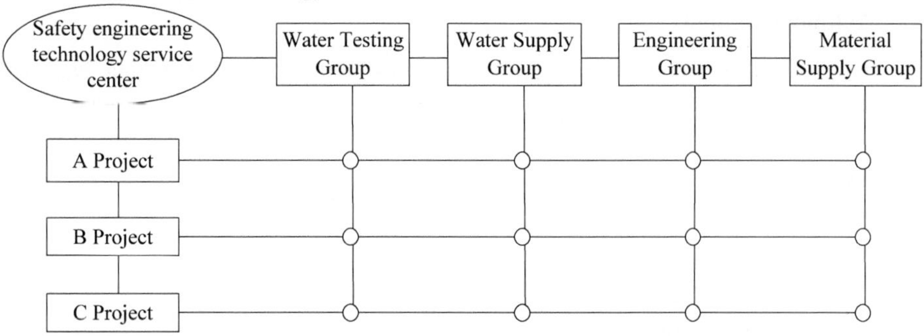

Fig. 2. Schematic of engineering and technical service center matrix.

Technical support services matrix management model can effectively apply to a temporary character of cooperatives engineering problems, but it does not mean that technical services center is a temporary organization. The rural drinking water engineering and maintenance work need to be done in the long time. When building a matrix organization, from short-term and long-term perspective, it not only provide temporary technical services, but also provide a regular cycle, which aimed at striving to build long-term stable technical services team.

6. Conclusion

The task of rural drinking water safety project construction is arduous. Dianjun government invested a lot of manpower, material and financial resources to solve rural safe drinking water supply problems. In order to play long-term benefits of rural drinking water safety project, it's important to explore and establish a scientific management model of rural drinking water safety project. In this paper, based on the status of rural drinking water projects in Dianjun, a single village cooperative with flat model was proposed to operate and manage projects with the assistance of matrix technical support service system. Development of cooperatives depends on government support, government needs to provide services in the capital, registration and technical training. Meanwhile government should do something to reduce administrative intervention. Cooperatives independently make decision and feedback problems in the "bottom-up" way, safety engineering and technology service center and cooperative system are the carrier of the government to provide support services. The two parts work together to ensure long-term operation of rural drinking water safety project.

Acknowledgment

This work is supported by grant S2013GMD100042 of the China National Science and Technology Assistance Program.

References

1. Qiuhua Dong, Explore the Long-term Management Mode to Ensure the Healthy Operation of the Project-Analysis on Typical Pattern of Rural Drinking Water Project in Zhejiang Province, Chinese Flood Control and Drought, 3 (2007).
2. Xuesong Li and Linxin Li, Game Analysis on Rural Drinking Water Safety Supply Mechanism, Ecological Economy, 5 (2011).

3. The "Twelfth Five-Year Plan" of Chinese Rural Drinking Water Safety Project (public draft), (China, 2012), http://www.doc88.com/p-776865903233.html.
4. Yanjiao Li, Research on Rural Drinking Water Safety Project Management Model in Zhangqiu, Master's thesis, Shandong University (Shandong, China, 2010).
5. National Development and Reform Commission, Affairs file of Ministry of Water Resources, Agricultural Water, 502 (2005).
6. Yuanhong Liao, Study on Internal Social Capital and the Performance of Farmer Cooperatives, Rural Economy, 7 (2011).
7. Ling He and Weili Zeng, New Mode of Rural Cooperatives Construction, Journal of Yibin University, 11 (2010).
8. Shuibing Cao, Yichang Yearbook, (Yichang, China, 2013) pp. 443–444.

Theoretical research on the effective length factor of columns in large space lattice structures

Peng Guo*, Bo-You Zhao, Qing Zhou, Fang-Jie Zhou, Xiao-Tong Zhang
China Aviation Planning and Design Institute (Group), Co., Ltd.,
12 Dewai Street, Xicheng District,
Beijing, 100120, China
**E-mail: guopeng123470@163.com*

Considering different boundary conditions of the column in large space lattice structures, theoretical formulas for the effective length factor has been derived by the equilibrium method. The finite element analysis method has also been applied to verify the theoretical solution. Seen from formulas, the effective length factor is mainly determined by the ratio of restrain stiffness and lateral stiffness of the column.

Keywords: Large Space Lattice Structure; Column; Effective Length Factor; Theoretical Derivation.

1. Introduction

Currently, many domestic and overseas researchers have done a great deal of work on the effective length factor of columns either in theories or by tests [1-3]. However, in terms of columns in large space lattice structures, no systematic results have been formed so far. It is obvious that the effective length factor is crucial for the design safety of columns in those structures. However, the factor can only be determined empirically now and further study is needed to prove the safety of the empirical method.

Typical large space lattice structures include hangars, terminals and so on. Columns of those structures are rigid connected with ground at the bottom and hinge connected with the roof at the top. The restrains between the roof and the columns are elastic. Considering different boundary conditions of columns in large span lattice structures, a theoretical derivation in terms of the effective length factor is presented in this paper. Relevant formulas and calculation tables are also available so as to provide reference in the design of hangar, terminals and other large space lattice structures.

2. Theoretic Derivation of Effective Length Factor for Large Space Lattice Structure Columns

To get a calculation model, the analyzed column should be isolated from the lattice structure and all the other parts be treated as elastic restraints applied on the top of the column. It can be presumed that the bottom of the column is rigid connected, the top of the column is elastic connected with five degrees of freedom and no axial deformation during the process. The calculation model is presented in Fig. 1.

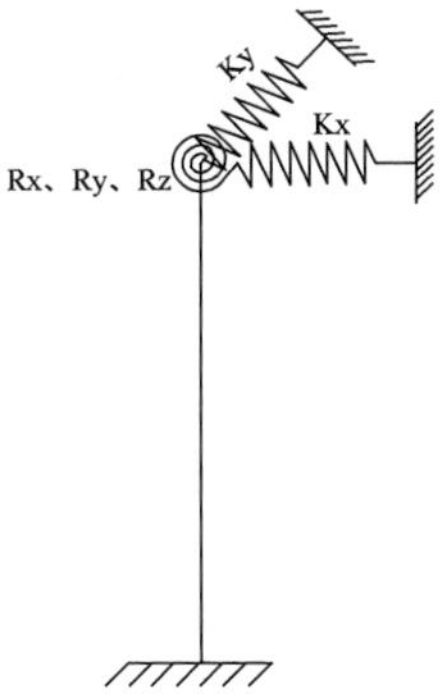

Fig. 1. Sketch of the calculation model.

For large space lattice structures, because of the hinge connection between the column and the roof, some bending and torsional restraints on the top can be ignored. So the calculation model can be further simplified as Fig. 2. In the simplified model, the top of the column is restrained by two lateral springs and one bending spring.

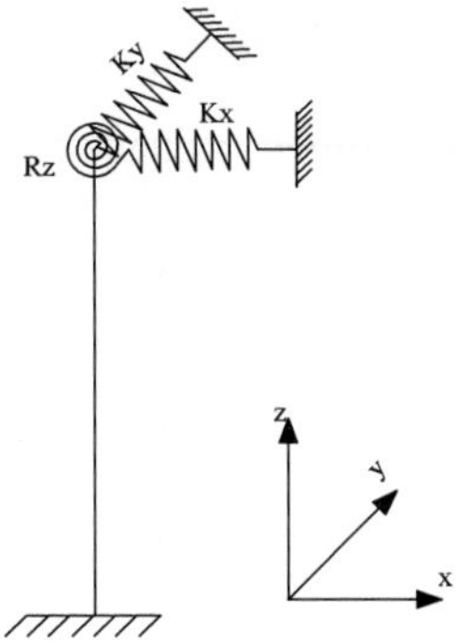

Fig. 2. Calculation model with lateral and torsional springs on top.

The equilibrium method is used for the theoretical derivation [4,5]. Fig. 3 shows the derivation model under the general condition.

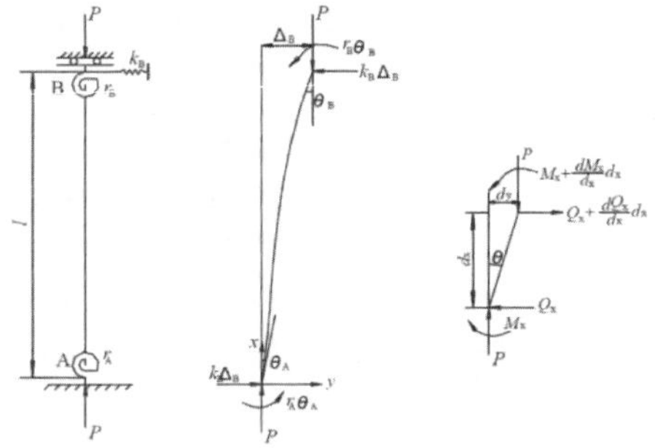

Fig. 3. Theoretical derivation model.

In Fig. 3, the axial compression member is restrained by springs at both ends. The spring constant of the bending spring at the top and the bottom are γ_B and γ_A respectively, while the constant of the lateral spring is k_B. The deformation of the column and forces applied on it are also showed in Fig. 3. In terms of the direction, it is stipulated that the clockwise rotation and right side displacement are taken as positive; bending moments and forces on the top are also positive when their directions are the same with the displacement. Otherwise are taken as negative. By selecting an isolated body, a force balance equation can be established:

$$Q_x - \left(Q_x + \frac{dQ_x}{dx} dx \right) = 0 . \tag{1}$$

Based on Eq. (1), we can get:

$$\frac{dQ_x}{dx} = 0 . \tag{2}$$

According to the bending moment balance equation, Eq. (3) can be derived:

$$M_x + Pdy + Q_x dx - \left(M_x + \frac{dM_x}{dx} dx \right) = 0 . \tag{3}$$

Based on Eq. (3), the following equation can be obtained:

$$Q_x = \frac{dM_x}{M_x} - P\frac{dy}{dx} . \tag{4}$$

From Eq. (2) and Eq. (4), it can be obtained:

$$\frac{d^2 M_x}{dx^2} - P\frac{d^2 y}{dx^2} = 0 . \tag{5}$$

Since,

$$M_x = -EI\frac{d^2y}{dx^2}, \tag{6}$$

Then,

$$EIy^{(4)} + Py'' = 0, \tag{7}$$

$$Q_x = -(EIy''' + Py'). \tag{8}$$

Eq. (8) is a four order differential equations which are applicable to axial compression members with any boundary conditions. Define:

$$k^2 = P/EI, \tag{9}$$

$$y = C_1 \sin kx + C_2 \cos kx + C_3 x + C_4, \tag{10}$$

$$Q_x = -PC_3. \tag{11}$$

As the boundary conditions defined, the column is fix at the bottom and translational restrained on the top. The Eq. (12) and Eq. (13) can be obtained:

$$y(0) = 0,\ y'(0) = 0, \tag{12}$$

$$y''(l) = 0,\ Q(l) = -PC_3 = -k_B y(l). \tag{13}$$

So,

$$k_B y(l) - PC_3 = 0. \tag{14}$$

Considering all boundary conditions, it can be obtained:

$$y(0) = C_2 + C_4 = 0, \tag{15}$$

$$y'(0) = C_1 k + C_3 = 0, \tag{16}$$

$$y''(l) = -C_1 k^2 \sin kl - C_2 k^2 \cos kl = 0, \tag{17}$$

$$k_B y(l) - PC_3 = k_B C_1 \sin kl + k_B C_2 \cos kl + C_3 (k_B l - P) = 0. \tag{18}$$

To guarantee the solution of above equations is non-zero value, it is required that the coefficient determinant equals to zero, that is:

$$\begin{vmatrix} 0 & 1 & 0 & 1 \\ k & 0 & 1 & 0 \\ k^2 \sin kl & k^2 \cos kl & 0 & 0 \\ k_B \sin kl & k_B \cos kl & k_B l - P & k_B \end{vmatrix} = 0 . \tag{19}$$

Eq. (19) can be simplified as:

$$kk_B l \cos kl - Pk \cos kl - k_B \sin kl = 0 . \tag{20}$$

Since,

$$k^2 = \frac{P}{EI} = \left(\frac{\pi}{\mu l}\right)^2 \Rightarrow \begin{matrix} k = \pi/\mu l \\ P = EI\left(\pi/\mu l\right)^2 \end{matrix} , \tag{21}$$

Eq. (20) can be rewritten as:

$$k_B \left(\frac{\pi}{\mu} \cos \frac{\pi}{\mu} - \sin \frac{\pi}{\mu} \right) = \left(\frac{\pi}{\mu l} \right)^3 EI \cos \frac{\pi}{\mu} . \tag{22}$$

Then, Eq. (23) can be derived:

$$\frac{k_B l^3}{EI} = \frac{\pi^3 \cos \frac{\pi}{\mu}}{\mu^3 \left(\frac{\pi}{\mu} \cos \frac{\pi}{\mu} - \sin \frac{\pi}{\mu} \right)} . \tag{23}$$

In Eq. (23), $k_B l^3 / EI$ is the ratio between the spring stiffness and the lateral stiffness of the column. This ratio is the determinant of the effective length factor μ.

The following is a brief discussion on Eq. (23):

When $k_B \Rightarrow 0$, it is apparently that:

$$\cos \frac{\pi}{\mu} = 0 \Rightarrow \frac{\pi}{\mu} = \frac{\pi}{\left(\frac{2}{2n+1} \right)} . \tag{24}$$

For the first order mode shape, due to its smallest critical load, the effective length factor is the largest. Therefore, when $n = 0$, $\mu = 2$. And this is equivalent to the cantilever condition in which the column is fixed at the bottom and free at the top.

When $k_B \Rightarrow \infty$, it can be obtained that:

$$\frac{\pi}{\mu}\cos\frac{\pi}{\mu}-\sin\frac{\pi}{\mu}=0\,. \tag{25}$$

Solving the Eq. (25) with the help of the MATLAB software, it can be obtained:

$$\pi/\mu=4.4934\Rightarrow\mu=0.699157\approx0.7\,. \tag{26}$$

This is equivalent to the condition in which the column is fixed at the bottom and hinged at the top.

Based on formulas, Fig. 4 shows the relationship between the effective length factor μ and the parameter $k_B l^3/EI$. In the graph, the vertical axis is μ and the horizontal axis is $k_B l^3/EI$. Since the coordinate value is dimensionless, the curve is applicable universally.

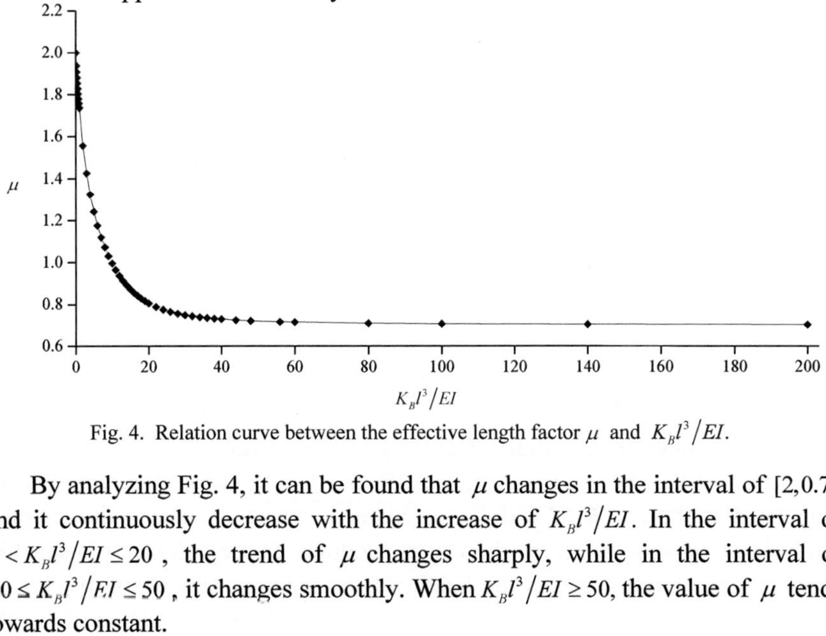

Fig. 4. Relation curve between the effective length factor μ and $K_B l^3/EI$.

By analyzing Fig. 4, it can be found that μ changes in the interval of $[2,0.7]$, and it continuously decrease with the increase of $K_B l^3/EI$. In the interval of $0<K_B l^3/EI\le 20$, the trend of μ changes sharply, while in the interval of $20\le K_B l^3/EI\le 50$, it changes smoothly. When $K_B l^3/EI\ge 50$, the value of μ tends towards constant.

The Eq. (23) can be written as cubic function as well:

$$(\pi/\mu)^3-\frac{k_B l^3}{EI}(\pi/\mu)+\frac{k_B l^3}{EI}\tan(\pi/\mu)=0\,. \tag{27}$$

For different values of $K_B l^3/EI$, the MATLAB software is used to generate function curves under the condition of $y=0$.

Where, $y=(\pi/\mu)^3-\frac{k_B l^3}{EI}(\pi/\mu)+\frac{k_B l^3}{EI}\tan(\pi/\mu)$.

The intersection point of curves is the function solution for the corresponding $K_B l^3/EI$. In Fig. 5, different line styles are applied to represent the function curve under the condition of different $K_B l^3/EI$ values. Because the function y contains the tangent function and $\tan(\pi/2)=\infty$, so only the interval of $[2,4.5]$ is considered. The horizontal axis represents the value of π/μ and the vertical axis represents the value of function y.

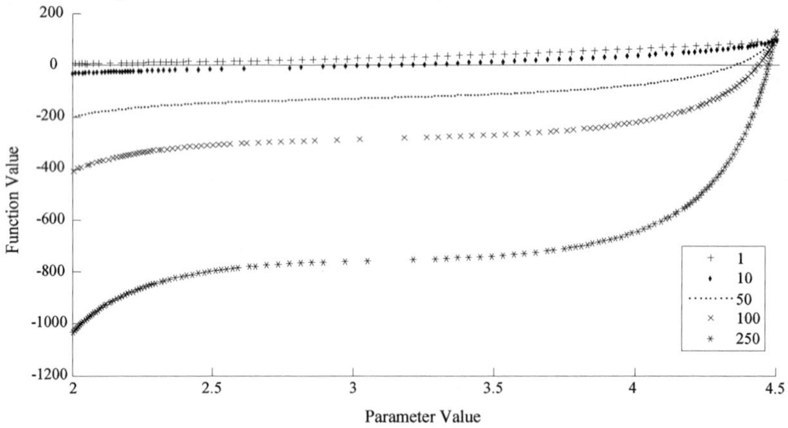

Fig. 5. Function curve with different $K_B l^3/EI$ values.

From Fig. 5 it can be seen that the function curves are similar in shape even the values of $K_B l^3/EI$ are different. When $K_B l^3/EI \geq 50$, the function solution π/μ is in the interval of $[4,4.5]$. Besides, with the increasing of the $K_B l^3/EI$, the π/μ is getting closer to 4.5. However, in any case the value of π/μ will not exceed 4.4934 (when $k_B \Rightarrow \infty$, π/μ=4.4934). In the condition of $K_B l^3/EI < 50$, the function solution lies within a relatively large interval.

Based on the above derivation, the effective length factor of columns can be calculated if the k_B, E, I and l are obtained, where k_B means the restrains stiffness of springs, E means the elastic modulus of the column, I means the inertia moment and l means the height of the column. Obviously, the key of the problem is k_B. Considering the difficulty in the direct calculation of k_B, the flexibility δ_B, the reciprocal of stiffness k_B, is introduced. So that the problem of solving k_B can turns into the calculation of δ_B.

In this paper, the finite element analysis method is adopted to calculate δ_B. As showing in Fig. 6, a column is selected and a unit force is applied on its top. With the help of finite element analysis software, structural responds under the action of unit force, which is shown in Fig. 7, can be gained. When the displacement at the top of the column is obtained, the k_B can be calculated then. (Fig. 6 and Fig. 7 are typical models of the hangar structure).

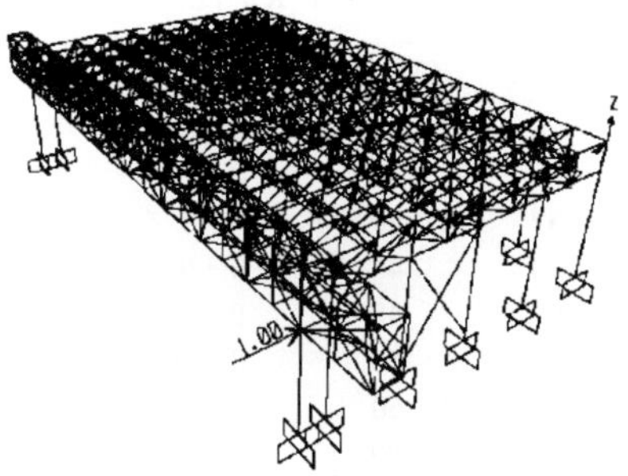

Fig. 6. Application of unit force.

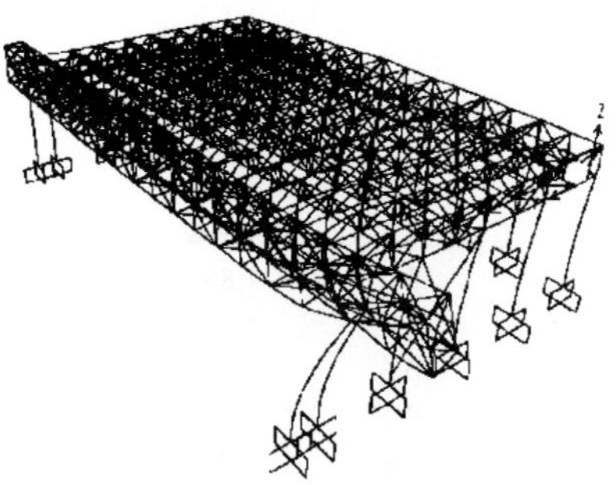

Fig. 7. Deformation caused by unit force.

Attention should be paid that the stiffness calculated by the above method is a combination of the elastic restraint stiffness and the lateral stiffness of the column. Fig. 8 shows such combination.

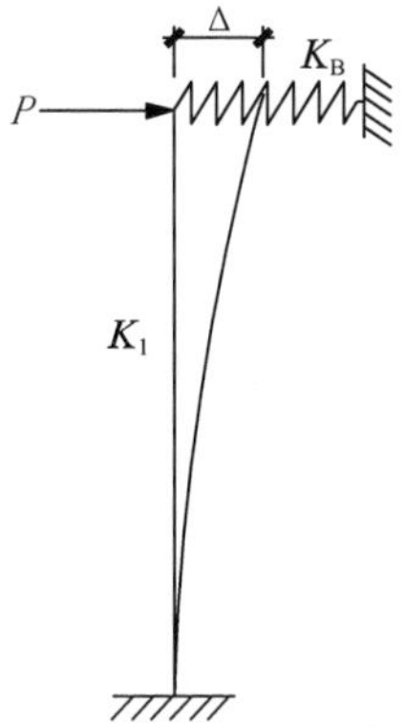

Fig. 8. Displacement at the top of the column.

Fig. 8 illustrates that the lateral displacement at the column top is always equal to the compression length of the spring. So the lateral stiffness of the column and the spring stiffness are in parallel and can be superimposed, that is:

$$(K_B + K_1)\Delta = 1 \Rightarrow K_B + K_1 = 1/\Delta . \tag{28}$$

So $K_B = 1/\Delta - K_1$.

Where, K_B is the stiffness of the spring, K_1 is the lateral stiffness of the column and Δ is the displacement produced by unit force.

On the basis of above, a calculation sheet is produced. When the elastic restraint stiffness, the column height, the material elastic modulus and the section inertia moment are known, the effective length factor μ can be obtained based on Table 1.

Table 1. Effective length factor μ corresponding with different $K_B l^3/EI$.

$K_B l^3/EI$	μ	$K_B l^3/EI$	μ	$K_B l^3/EI$	μ
0.1	2	10	0.996	36	0.736
0.2	1.937	11	0.965	38	0.733
0.3	1.908	12	0.937	40	0.73
0.4	1.881	13	0.913	44	0.726
0.5	1.854	14	0.892	48	0.722
0.6	1.828	15	0.873	56	0.718
0.7	1.804	16	0.856	60	0.716
0.8	1.781	17	0.841	80	0.711
0.9	1.758	18	0.828	100	0.708
1	1.736	19	0.817	140	0.705
2	1.557	20	0.806	200	0.703
3	1.426	22	0.789	250	0.702
4	1.324	24	0.776	300	0.702
5	1.243	26	0.765	350	0.701
6	1.176	28	0.757	400	0.701
7	1.12	30	0.75	500	0.701
8	1.072	32	0.744	1000	0.7
9	1.031	34	0.74	5000	0.699

3. Finite Element Verification on the Effective Length Factor of Large Space Lattice Structure Columns

Finite element analysis software which includes ANSYS and SAP2000 are used to verify the above theoretical formula.

The column is fixed at the bottom with all degrees of freedom restrained. The BEAM element is applied to simulate the column. On the top of the column, LINK 8 element is used to simulate the restraint between the roof and the column. It is important to make sure that the axis stiffness of LINK 8 element which is EA/l should equal to the spring stiffness which is k_B.

Table 2. Results comparison between the finite element analysis and the theoretical calculation.

Column section (mm)	Height (m)	K_B (kN/m)	μ (Theoretical)	μ (ANSYS)	μ (SAP2000)
300×300	9	10	1.8915	1.8915	1.8921
300×300	15	10	1.6098	1.6104	1.6105
300×300	15	200	0.7411	0.7411	0.7417
800×800	20	3000	0.7788	0.7788	0.7807
1000×1000	25	25000	0.7043	0.7041	0.7070

In order to compare the theoretical solution with the finite element result, some column sections and elastic stiffness are selected and listed in Table 2. The strength grade of concrete is C30. With the help of finite element software, the characteristic value of the first order bucking load of the column can be obtained. Then the effective length factor μ can be derived backwards from Euler's formula $P_{cr} = \pi^2 EI/(\mu l)^2$. Table 2 lists the theoretical solutions and the results of finite element analysis.

From Table 2, theoretical solutions are very close to the finite element results. The deviation is less than 1%.

4. Conclusion

Considering different boundary conditions of columns in large space lattice structures, a theoretical derivation of the effective length factor is conducted by the equilibrium method in this paper. Meanwhile, the finite element method is applied to verify the theoretical formula. Based on the comparison analysis in terms of the theoretical solution and the finite element analysis results, it is can be concluded that the theoretical formula is reliable and accurate.

In theoretical formulas, the parameter $K_B l^3/EI$ stands for the ratio between the restraint stiffness and the lateral stiffness of the column. And it determines the value of effective length factor μ. When the restraint stiffness is close to zero, the effective length factor is close to 2.0. While when it approaches to infinite, the effective factor tends to 0.7. In the interval of $0 < K_B l^3/EI \le 20$, the theoretical solution of μ changes rapidly, while when $20 \le K_B l^3/EI \le 50$, it changes slowly. When $K_B l^3/EI \ge 50$, it becomes stable.

References

1. Yuanqing Wang, Dayi Ding, Yongjiu Shi et al. Column Effective Length Factor And Girder Parametric Analysis for The Components of Xinqiao International Airport Terminal. Journal of Shenyang Jianzhu Univeristy. 2011, Vol. 27(3): 474-479.
2. Tiantian Lu, Xin Zhao, Jiemin Ding et al. Stability Analysis of The Shanghai Tower And Research on Effective Length of Super Column. Journal of Building Structures. 2011, Vol. 32(7): 8-14.
3. Jostein Hellesland and Reidar Bjorhovde. Improved Frame Stability Analysis With Effective Lengths. Journal of Structural Engineering, 1996, 1275-1283.
4. Ji Chen. Steel Structure Stability Theory And Design. Beijing: Science Press, 2006.
5. Shaofan Chen. Design Principle of Steel Structures. Beijing: Science Press, 2005.

Force analysis of finite element method for the aluminum alloy template of shear wall

Qiu Hao[1,2], Jin-Ping Zhuang[1,2], Xue-Feng Cai[1,2], San-Peng Zhang[1,2]
[1]Civil Engineering Department, Fujian University Of Technology,
[2] Fujian Provincial Key Laboratory of Advanced Technology and Informatization in Civil Engineering,
Fuzhou, Fujian, China
E-mail: qiuhao19910420@163.com
www.fjut.edu.cn.

Based on ABAQUS software, the pin connection between aluminum alloy templates simulated using MPC PIN element and spring element (Springs 2) ,and the deflection of aluminum alloy template simulated during the concrete placement process. Comparing with the results of simplified calculation, the analysis results of the finite element show that the finite element modeling of aluminum template with MPC PIN element and spring element (Springs 2) as well as the analysis results can be used as reference for the engineering application of aluminum alloy template.

Keywords: Aluminum Alloy Template; Springs Element; MPC Pin Element; Deflection; Stress.

1. Introduction

Aluminum Alloy is used as building templates, which still belongs to a kind of new building template system and has been used in engineering, but the research on its mechanical properties is relatively rare. The developed countries such as the United States and Canada and the emerging industrial countries such as Brazil, Malaysia, India have begun applying aluminum alloy template to construction. In recent years, China, Hongkong and Macao began to use Aluminum Alloy template system. The engineering projects Such as Wuhan Forte – Donghu international project, Wenchong project of Guangzhou Vanke (31 storey high-rise residential), Shenzhen Donghai International Center (apartment complex) and so on, applied Aluminum Alloy template to the construction. It has the advantages of high efficiency, good quality, good safety management, low cost, convenient transportation and environmental protection [1].

Significant differences between formwork system of aluminum alloy, ordinary plywood formwork and steel template are the pin connection between templates. the pin connection can transfer the shear force and bending moment. The internal force does not transfer between ordinary plywood formwork ,steel template, too. Due to the current calculation method of aluminum alloy template, its research has not yet been thoroughly. In consideration of the security concern ,the specification of calculation of aluminum alloy template ignores the function of pin connection. The aluminum alloy template is calculated according to the model of ordinary plywood templates [2][4][5]. To understand the force transfer mechanism of the aluminum alloy formwork system accurately and play the advantages of aluminum alloy template adequately, It is necessary to solve accurate analysis of the aluminum alloy template. Therefore, this article used the ABAQUS software to simulate the pin connection between aluminum alloy template of wall through the MPC pin element and springs element, in order to consider the action of the pin connection. the finite element results and simplified calculation results are compared, to provide reference for engineering application.

2. Simplified Calculation of Aluminum Alloy template

2.1. *The model of simplified calculation*

Take Aluminum Alloy template of shear wall as an example. the each side of the shear wall assembled by the size for 450mm x 2400mm of three blocks of aluminum alloy template. the thickness of shear wall is 200mm. Aluminum Alloy template is composed of panel, stiffener, transverse rib and rib. Connecting the dorsal bar, rebar and pin fixedly is the main assembly process of template. The space of dorsal bar is 800mm, its size is 80mm × 40mm × 2.5mm .the space of rebar is 900mm, its specification is M16 .they can fix the aluminum alloy template of shear wall, as shown in Figure 1, shown in Figure 2, shown in Figure 3.

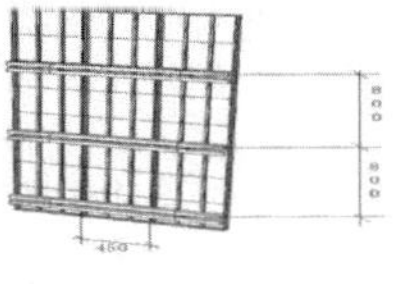

Fig. 1. Side view of assembly of Aluminum Alloy template

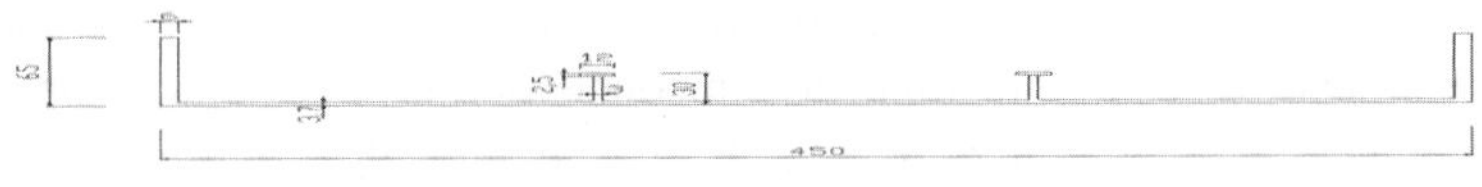

Fig. 2. Aluminum Alloy template profile

Fig. 3. Aluminum Alloy template

2.2. *Deflection and stress of Aluminum Alloy template*

The lateral pressure of Aluminum alloy template is 43kN/m^2 (including the deadweight of template) [10][7]. The method of deflection calculation takes the middle block of Aluminum alloy template of the shear wall into account ,whose span is 800mm.

2.2.1. *The position of the neutral axis and moment of inertia of Aluminum Alloy template[9]:*

$$y_c = \frac{2A_1y_1 + 2A_2y_2 + 2A_3y_3 + A_4y_4}{2A_1 + 2A_2 + 2A_3 + A_4} = \frac{1665 \times 1.85 + 30 \times 2 \times 28.75 + 2 \times 71.4 \times 15.6 + 2 \times 367.8 \times 32.5}{2 \times 30 + 2 \times 71.4 + 367.8 \times 2 + 1665} = 11.88\,mm. \quad (1)$$

$$I_z = \left(\frac{6 \times 61.3^3}{12} + 6 \times 61.3 \times 22.4^2\right) \times 2 + \left(\frac{12 \times 2.5^3}{12} + 12 \times 2.5 \times 16.87^2\right) \times 2 + \left(\frac{3 \times 23.8^3}{12} + 3 \times 23.8 \times 3.72^2\right) \times 2 + \left(\frac{450 \times 3.7^2}{12} + 450 \times 3.7 \times 10.03^2\right) = 794664.863\ \mathrm{mm}^4. \quad (2)$$

2.2.2. *Calculation of mid-span deformation of one-way slab[3]. the size of calculation of Aluminum alloy template is 450mm x 800mm.*

$$d = \frac{5ql^4}{384\,EI} = \frac{5 \times 18 \times 800^4}{384 \times 70000 \times 794664.863} = 1.72\,mm \quad \left[\frac{l}{400}\right] = 2\,mm. \quad (3)$$

2.2.3. *The calculation of cross panel side template stress[4].*

$$\sigma = \frac{My}{I_z} = \frac{1440000 \times 11.88}{794664.863} = 21.53\,Mpa. \quad (4)$$

3. The Finite Element Modeling of Aluminum Alloy Template

Based on ABAQUS software , the finite element model of shear wall aluminum alloy template is assembled. The ABAQUS software uses MPC pin element and springs element (Springs 2) to simulate pin connection between aluminum alloy wall templates. The modeling of panel , rib stiffener and edge rib is composed of solid element. The modeling of transverse rib is composed of the beam element

[3]. After aluminum alloy templates of shear wall assembled, the each side wall of the whole template edge is fixed end restrained, the contact part of the dorsal bar and template applies the hard contact method. The reinforced bar simulated by springs elements. Some parts of aluminum alloy template use solid element to model, whose mesh size is 30mm. The mesh size of transverse rib is 10mm, the mesh size of dorsal bar is 20mm. The lateral pressure is $0.043N/mm^2$. the following stiffness calculation of spring element to simulate the pin connection:

$$k = EA / l = E \times \pi r^2 / l = 206000 \times 3.14 \times 8.25^2 / 14 = 3144681.964 N / mm . \quad (5)$$

Type (5): E stand for the elastic modulus of pin; A stand for the cross-sectional area of pin; I stand for the length of pin.

The material properties of Aluminum Alloy template and dorsal bar refer the technical specification of Aluminum Alloy building template system in Fujian province (Draft for Comments) [8] and Domestic aluminum alloy material [6], as shown in Table 1, shown in Table 2.

Table 1. Physical properties of the Aluminum Alloy

Modulus of elasticity (N/mm2)	shear modulus (N/mm2)	linear expansion coefficient (1/°C)	Mass density (kg/m3)
70000	27000	23×10-6	2700

Table 2. Physical properties of steel

Modulus of elasticity (N/mm2)	shear modulus (N/mm2)	linear expansion coefficient (1/°C)	Mass density (kg/m3)
2.06×105	70000	12×10-6	7850

4. The finite element analysis and discussion of Aluminum Alloy template

4.1. *The analysis and discussion of Aluminum Alloy template deflection*

The ABAQUS software uses MPC pin element and springs element (Springs 2) to simulate pin connection between aluminum alloy wall templates. Displacement nephogram of two connecting elements is shown in Figure 4.

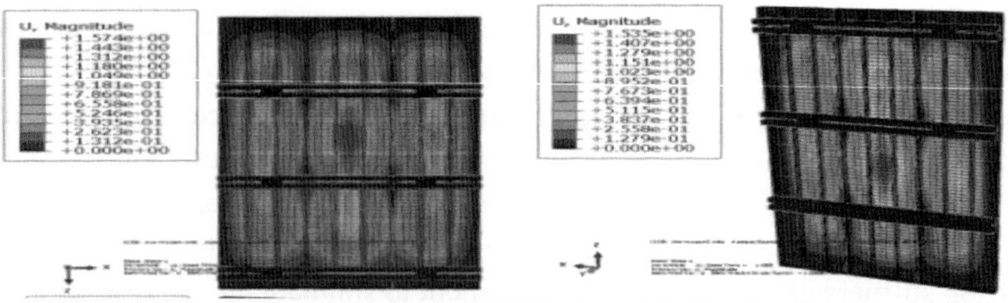

Fig. 4. Displacement nephogram of MPC PIN element (left) and spring element (right)

Through analyzing the distribution diagrams of displacement, the deflection of templates connected by MPC pin element is 1.57mm, the deflection of templates connected by Springs element (Springs 2) is 1.53mm. and the deflection of Aluminum Alloy template calculated by the theory of mechanics is 1.72mm. The analysis results of finite element and theoretical calculating value are in the allowed range of deflection.

The MPC pin is on behalf of hinge joint, when the hinge joint of templates rotate ,the contact surface of templates extrude each other to restrict Rotation range . This connection element is close to the force of the pin. Spring element (Springs 2) is semi rigid connection, six directions have spring stiffness. Spring element (Springs 2) and actual situation of stress of pin connection are also very close. So the results of finite element analysis of connection method of two kinds of elements has little difference.

4.2 *The analysis and discussion of Aluminum Alloy template stress*

The distribution diagrams of stress of two connecting elements is shown in Figure 5.

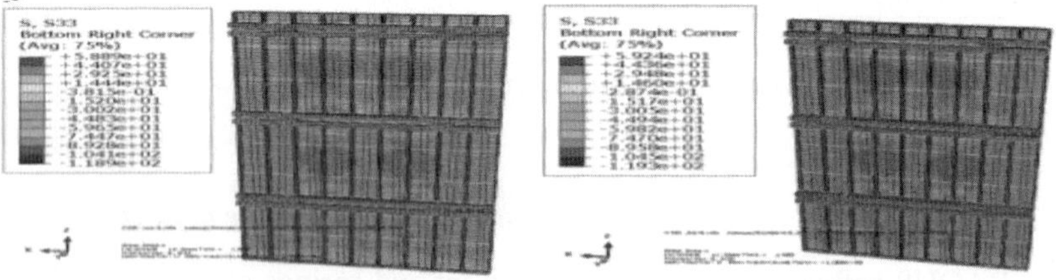

Fig. 5. stress nephogram of MPC PIN element (left) and spring element (right)

Through analyzing the distribution diagrams of stress , the stress of templates connected by MPC pin element is 16.10Mpa, the stress of templates connected by Springs element (Springs 2) is 15.10Mpa. and the stress of Aluminum Alloy template calculated by the theory of mechanics is 21.53Mpa. The analysis results of finite element and theoretical calculating value had some errors. The simplified calculation was based on hinged plates with two simply supported ends of unidirectional template of Aluminum Alloy , its result tends to be conservative . The analysis of finite element is in consideration of the mutual restraint of the templates, so that its deflection and strain is less than the calculation result of unidirectional template.

5. Conclusion

According to the comparison results of the Aluminum Alloy template theory calculation and finite element analysis, the finite element calculation of templates connected by spring element (Springs 2) and MPC pin element are in line with the stress characteristics of pin connection. The templates connected with two kinds of finite element can play a role in optimization about the analysis of the simplified calculation in the design involving the Aluminum Alloy template.

Acknowledgments

This work is supported by the National Natural Science Foundation of China(51478119), the general project of Natural Science Foundation of Fujian Province(2015J01182), the scientific research foundation of Fujian University of Technology. (GY-Z13119) and Industry-University Cooperation Key Project of Science and Technology of Fujian Province of China (2013H6003).

References

1. Jiaping Mi, Jun Yang. The Aluminum Alloy template and support - Canada Aluma system template [J], 2012,34 (4): 338-340.
2. Wu Shuigen, Ji Shaokai, Zheng Gang. Research on the development of aluminum frame wood formwork system applied to the high rise residential building shear wall. Building construction, 2013,35 (2): 134-136.
3. Li Wenfang. New Aluminum Alloy template test. Construction technology, 2013, Vol. forty-second, supplement: 279-280.
4. Dai Guiyang. Application of aluminium alloy formwork in building construction. Facility technology. 51-53.
5. Zhang Rui, Qiu Renbin, Xu Chao. Analysis of the advantages and disadvantages of aluminum alloy formwork. China building metal structure, phase seventh: 46-47.
6. Guo Xiaonong, Shen Zuyan, Li Yuanqi, Su Ci, Qiu Zhenge, Yao Nianliang. Study on constitutive relation and physical mechanical properties of

aluminum alloy materials for domestic structure [J]. Journal of building structures, 2007, 28 (6): 110~117.

7. Cai Xuefeng. Construction technology of civil engineering [M]. Beijing: Higher Education Press.2011-8.
8. the technical specification of Aluminum Alloy building template system in Fujian province (Draft for Comments) [S].2014.
9. Editing Group. Structural static analysis manual [M]. manual structural static analysis. Beijing: China Architecture & Building Press, 2001.
10. Jiao Yan. analysis of mechanical properties and wind vibration of Aluminum Alloy formwork system [D]: South China University of Technology.2015:15-16.

Finite element analysis of anti-bending of early age reinforced concrete beam based on bond-slip constitutive

Zun-Huang Yang[1,2,*], Jin-Ping Zhuang[1,2], Xue-Feng Cai[1,2], Zhan-Grong Zhang[1,2]
[1]Civil Engineering Department, Fujian University of Technology,
[2]Fujian Provincial Key Laboratory of Advanced Technology and Informatization in Civil Engineering,
Fujian Province, China
**E-mail: zun0168@fjut.edu.cn*
www.fjut.edu.cn.

Bond-slip constitutive of steel bar in the early age concrete is applied to the spring2 elements of ABAQUS to simulate the interaction between bar and concrete because of the existence of bond-slip phenomenon in the early age. Spring2 elements are set in the grids of concrete element and bar element which are coincident in the finite element model. What's more, bond-slip constitutive relation is used as the material of spring2 elements. The finite element models of 6 reinfored concrete beams are built to analyze the effect of age of concrete (3days, 5days, 7days and 28days) and ratio of reinforcement (0.43%, 0.84% and 1.39%) for mechanical properties of beam. It was found that cracking moment, yield moment and ultimate moment of concrete beams are gradually increasing with the extend of age of concrete. Three days ago, carrying capacity of early age concrete beam developed rapidly. And it tends to be stable five days later. Increasing the rate of reinforcement can dramatically improve the mechanical properties of the beams. When the rate of reinforcement reached to some degree, its influence will be down.

Keywords: Early Age Concrete Beam; Bonded-Slip of Steel Bar; Anti-Bending; ABAQUS.

1. Introduction

Ideology of raising the structure safety reliability has drawn more and more attentions in engineering circles about the life cycle. In the life cycle of structure, there are higher venture ratio of invalidation for early age concrete structure. It is important if early age concrete structure can meet the performance requirements of structural design for the whole structure in the life cycle. At present, mechanical properties about early age concrete material have been studied a lot at home and abroad [1-5]. However, there are less researches on the mechanical properties of early age concrete structural members [6-8]. Based on the experimental results, the numerical simulation of the early age steel reinforced concrete beam is carried out by using the nonlinear finite element analysis software ABAQUS [8]. The bond-slip, however, between steel bar and concrete is not taken into account in the process of the finite element simulation. Therefore, this paper considered the influence of the bond-slip between steel bar and concrete on the mechanical properties of early age reinforced concrete beam.

2. Finite Element Modeling

2.1. *The parameters of finite element model*

The influence of the age factor and the ratio of reinforcement on the mechanical properties of the early age reinforced concrete beam is considered in the ABAQUS simulation. 6 early age concrete beams were analyzed by finite element method, and the concrete parameters of the 6 reinforced concrete beams are shown in Table 1. The mechanical properties of reinforced concrete beams in the No.1~4 are mainly considered under different age. The tension area of beam is installed tension reinforcement with different size of cross section area to analyze the influence of different reinforcement ratio on properties of early age concrete beam. Section size, reinforcement form and loading device of the members are shown in Fig. 1.

Table 1. Table 1 Parameter list of early age reinforced concrete beam.

No	Specimen group	Concrete strength	Concrete age (days)	Tensile reinforcement diameter(mm)	Rate of reinforcement (%)
1	RCL1-3-14	C30	3	14	0.84
2	RCL2-5-14	C30	5	14	0.84
3	RCL3-7-14	C30	7	14	0.84
4	RCL4-28-14	C30	28	14	0.84
5	RCL5-3-10	C30	3	10	0.43
6	RCL6-3-18	C30	3	18	1.39

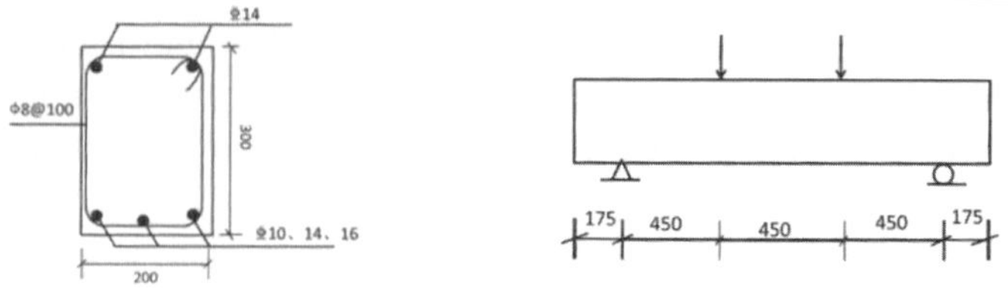

Fig. 1. Section size, reinforcement form and loading device of the member.

2.2. *The parameters of finite element model*

Concrete damage plasticity model of ABAQUS is used to simulate reinforced concrete beam in early age. Elastic modulus of early age concrete and uniaxial compressive constitutive model of concrete are refer to the data of paper [9], Poisson's ratio is 0.2. The GFI type is selected as tensile type of concrete, fracture energy can be used according to the literature [10]. The results of Wang Tiemeng's test [11] show that the axial tensile strength and compressive strength, within 3~28days after pouring, increase with the similar trend. With the increasing of age, the ratio of tensile strength to compressive strength increased at a very low rate between 0.095 and 0.087. The uniaxial tensile strength of concrete in the early age is based on the following formula:

$$f_t(t) = 0.087 f_{cu}(t) \tag{1}$$

In this formula, f_t is tensile strength of concrete; f_{cu} is compressive strength of concrete. The constitutive relation of stirrup and longitudinal reinforcement are set by ideal elastic-plastic model, yield strength is used according to the literature [9], Elastic modulus is 2.0×10^5 MPa, Poisson's ratio is 0.2.`

2.3. *Model building*

The concrete element is constructed with 8-node linear brick reduced integral unit (C3D8R), the steel bar is used a 2-node linear 3-D truss (T3D2). Steel cushion blocks are established at the loading points and the bearings. Stiffness of steel cushion blocks are infinite. The tie connection is adopted between the steel cushion block and the concrete. The embedded element is use to consider the interaction between compression reinforcement (or stirrup) and concrete. The spring2 element is used to simulate the interaction between the tensile reinforcement and concrete. Spring element is established at the three different directions in the coincident node of grid of steel bar and the concrete. Two springs which are perpendicular to the direction of longitudinal reinforcements are built with spring1 element, Their Stiffness are infinite. One spring along the direction of the longitudinal reinforcement is set as spring2. His bond-slip constitutive model and force-slip displacement are derived from the experimental data in the literature [12].

3. Result of Finite Element Analysis

3.1. *Cracking moment, yield moment and ultimate moment*

According to the results of analysis in ABAQUS. The moment whose slope has a mutation for the first time in the bending moment-deflection curve is defined as cracking moment M_{cr}. The moment whose corresponding's slope has a marked turnaround in the bending moment-deflection curve is defined as yield moment My, as per the *Standard for test method of concrete structures* (GB /T 50152-2012).δ_y is the deflection corresponded to the yield moment. Mu is the ultimate moment of concrete beams. M_c of No.1~4 is the ultimate moment of RCL4-28-14. Mc of No.5~6 is the ultimate moment of RCL1-3-14. Table 2 is the result of ABAQUS analysis.

According to the table 2, we can see that when the age is 3 days, ultimate bearing capacity of concrete beams has reached to the 84.65% of 28 days strength. Bearing capacity of early age beams develop rapidly in the young, and the strength had reached 92.03% of 28 days strength in the 5 days. The development of the ultimate bearing capacity of beams is not much after 5 days. So we can see that the cracking moment and the yield moment of the beams are gradually increased with the age of the beams in the early age. Cracking moment, yield moment and ultimate moment are also affected by the ratio of reinforcement. When the reinforcement ratio is increased from 0.43% to 0.84%, the ultimate bearing capacity of the beam is increased by 54%. The ratio of reinforcement is higher, and the value of bending moment is greater. According to the table2, we can see that when the reinforcement ratio is increased from 084 to 1.39, the ultimate bearing capacity of the beam is only increased by 5.07%. Hence, when the reinforcement ratio reaches to some degree, the influence of reinforcement ratio on the mechanical properties of the beam will be reduced.

Table 2. Crack moment, yield moment, ultimate moment and yield displacement.

No.	Specimen group	M_{cr} (kN·m)	M_y (kN·m)	*Mu* (kN·m)	δ_y (mm)	M_u/M_c
1	RCL1-3-14	14.60	44.01	45.34	5.06	84.65%
2	RCL2-5-14	17.13	48.33	49.29	4.21	92.03%
3	RCL3-7-14	18.75	49.22	51.33	3.76	95.84%
4	RCL4-28-14	22.88	51.64	53.56	3.10	100.00%
5	RCL5-3-10	12.48	28.95	29.37	3.28	64.78%
6	RCL6-3-18	16.69	45.45	47.64	3.91	105.07%

3.2. *Comparisons of bending moment-deflection curve*

Fig. 2(a) is a result of beams in the different age after finite element analysis. Fig. 2(b) is a result of beams with different reinforcement ratio after finite element analysis. According to the Figure 1, we can see that the bending moment-deflection curve of early age concrete beams have three stages which is the same as the mature concrete beam. From the initial stage of loading to the failure of the specimen, the early age concrete beams experience the elastic stage, elastic-plastic stage and the failure stage. From the results of Fig. 2(a) can be seen that with the growth of age, the ultimate bearing moment of concrete beam is also gradually increasing. And with the increase of age, the flexural rigidity of reinforced concrete beam is also gradually increasing, the slope of the bending deflection curve is gradually decreasing. It can be seen from the Fig. 2(b) that the reinforcement ratio has great influence on the ultimate bending moment and the stiffness of the reinforcement beam. With the increase of the ratio of reinforcement, the section stiffness of the beam is gradually increasing. However, when the reinforcement ratio reaches to some degree, the influence of reinforcement ratio on the mechanical properties of the beam will be reduced.

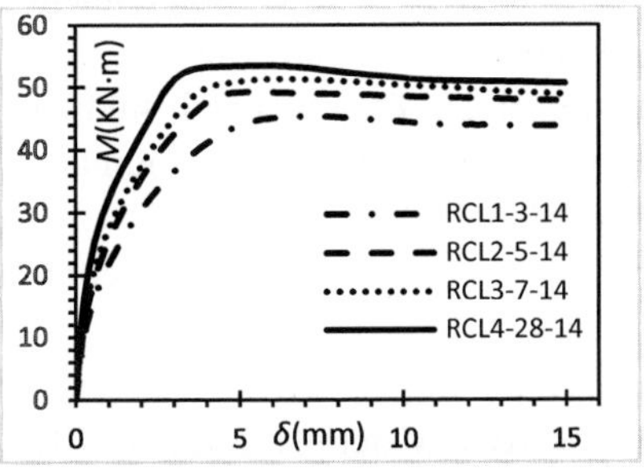

(a) Results of curves under different ages

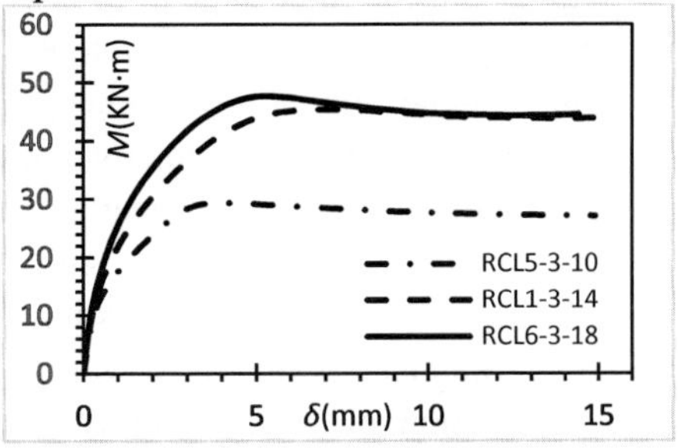

(b) Results of different reinforcement ratios

Fig. 2. Comparison of bending moment-deflection curve at mid-span.

4. Conclusions

(1) Bond-slip constitutive relationship in early age between concrete and reinforcing bar is introduced in spring2. The simulation results are consistent with the general mechanical property of reinforced concrete beams subjected to bending, so we can conclude that simulation results have a certain credibility.

(2) The cracking moment, the yield moment and the ultimate bending moment of the reinforced concrete are significantly improved in the first three days, and they tend to be gentle after 5 days.

(3) Increasing the reinforcement ratio can observably improve the mechanical properties of early age concrete beams, but the help is not very obvious after rate of reinforcement reach a certain degree.

Acknowledgment

This work is supported by the National Natural Science Foundation of China (51478119), the General Project of Natural Science Foundation of Fujian Province (2015J01182), the Scientific Research Foundation of Fujian University of Technology (GY-Z13119) and Industry-University Cooperation Key Project of Science and Technology of Fujian Province of China (2013H6003).

References

1. Francis A, Oluokun, Edwin G, Burdette, et al. Early age concrete strength prediction by maturity-another look[J], ACI Material Journal, 1990, 87(6): 565-572.
2. Jin-Keun Kim, Sang Hun Han, Young Chul Song. Effect of temperature and aging on the mechanical properties of concrete: part I: experimental results[J]. Cement and Concrete Research, 2002, 32(7): 1087-1094.
3. Jin Nanguo, Jin Xianyu, Zheng Yanguo, et al. Experimental study on fracture property and microstructure of early concrete [J]. Journal of Zhejiang University: Engineering Sciences, 2005, 39(9): 1374-1377.
4. Jin Xianyu, Tian Ye, Jin Nanguo. Early age property and cracking control of concrete[J]. Journal of Building Structures, 2010, 31(6): 204-212.
5. Sam Slatnick, Kyle A R, Kevin J F, et al. Evaluation of autogenous deformation of concrete at early ages[J]. ACI Materials Journal, 2011, 108(1): 21-28.
6. Hu Xiaopeng, Niu Ditao, Zhang Yongli. Experimental research on bond performance of early-age flash concrete [J]. Journal of Building Structures, 2013, 34(4): 152-157.
7. Ying Yimiao, Yang Junjie, Chen Xiong, et al. Cracks and internal damage of early-age concrete columns under aftershock[J]. Building Structure, 2011, 41(3): 72-74.
8. Yuan Quan, Yang Zhengkun, Cai Jie, et al. Experimental study and FEA on mechanical behavior of H-steel reinforced concrete beams at early-age [J]. Building Structure, 2014, 35(3): 193-200.
9. Tian Mingge, The research on the performance of reinforced concrete frame structure during construction [D]. Hunan University, 2008.

10. Genikomsou A S, Polak M A. Finite element analysis of punching shear of concrete slabs using damaged plasticity model in ABAQUS [J]. Engineering Structures, 2015, 98:38-48.
11. Wang Tiemeng. Control of cracking in engineering structure [M]. Beijing: China Architecture & Building Press, 1997:1-22.
12. Song X, Wu Y, Gu X, et al. Bond behavior of reinforcing steel bars in early age concrete [J]. Construction & Building Materials, 2015, 94:209-217.

Study on ordinary detonating tube detonators applied by hole blasting technique

Rui Guo

College of Civil and Environmental Engineering, University of Science and Technology, Beijing, Haidian Dist. Beijing China
E-mail: guorui4860@126.com
www.ustb.edu.cn

Based on the complexity belt rock pillars stripping engineering of Capital Yun Ore Company, the thesis puts forward the ways of ordinary detonating tube detonators applied by hole blasting technique. Through the field practice, it proves that ordinary detonating tube detonators applied by hole blasting technique reduces the rate of chunks of rock and the harmful effect of blasting vibration, and also improves the effect of the broken rock and shovel loading rate.

Keywords: Rock Pillars Stripping; By Hole Blasting; Shovel Loading Rate.

1. Introduction

Traditional hole blasting technique is using high strength, high precision millisecond detonating tube detonators, and outside-hole with low grade delay millisecond detonator connection, while in hole using high-end delay detonator detonating, implementing cannon hole one by one initiation blasting [1]. In reducing the single ring of explosive charge, it also greatly reduces the harm of blasting seismic wave. As the price is higher, high precision millisecond detonating tube detonator limit the technology promotion. This article uses the common millisecond delay detonating tube detonators to try to change the traditional hole by hole blasting tandem connection, make hole blasting realized.

According to the body occurrence characteristics and mining design of the deployment of Capital Yun Ore Company, the extension of underground mining will form two wings on □, □ mining ore belt from the empty area to the border. The north open-pit in south slope is bounded, while the south boundary is huge isolated rock pillars on small Miao Gou fault zone. Rock pillars on the top has width of 320 m, 200 m wide, and northern depth of 280 m, the southern tip is more than 300 m deep. Looked as “U” shaped, from north to south slope, big rock pillars around are empty area, fault isolation, after cutting, originally deep mining of ground pressure is unable to transfer to the surrounding rock mass, so the total weight will directly effect the deep ore rock mass, the area is large and thick ore body occurrence nuclear units, is relying on the future mining of the main ore body. At an average depth of 300 m, rock pillars gravity causes the pressure at the bottom at least above 800 ~ 900 t/m^2, as the two wings to the middle or unilateral withdrawal in advance, no transfer of ground pressure will decrease of compression of body as a result of retreat mining area, the deep mining of mine pressure doubled, produce roadway, catastrophic collapse. Collapse disasters such as rock burst, high pressure, deep main ore body recovery will be extremely difficult, even a large number of ore body will not be able to recycle. Therefore, in the underground mining subsidence area take open-pit blasting stripping rock pillars.

2. Principle Hole Blasting Technique

The premise of hole blasting technique is that the first blasting hole delay time must be greater than the total time the ground explosive detonator delay needed. In every hole by hole blasting process, the detonation is independent, each hole by clicking on the design initiation delay time, initiating first shot creates more free surface for later shot, reduce the resist line, explosive detonation explosion shock wave and detonation gas fully broken rock, raise the utilization ratio of explosive energy and reduce the blasting vibration, so as to achieve the purpose of improving crushing effect [2].

2.1. *Blasting mechanism of rock fragmentation*

Currently, internal and external research on the theory of differential blasting are mainly contained four aspects of the hypothesis.

2.1.1. *The new free surface hypothesis*

The hypothesis, first explosive package has caused some degree of damage in rock mass, around the blasting area formed a certain width of crack and additional free surface, which have created favorable conditions for subsequent cartridge blasting the rock fragmentation. Due to the explosive package produced new free surface caused the change of the subsequent charge the minimum resistance line, changed the subsequent charge blasting direction vertical original free surface (no longer). Therefore, to some extent, hole by hole blasting can reduce the width of the thrown distance and the detonation of the rock pile, and can make full use of the kinetic energy of the rock, in the process of movement caused collide with each other, which improves the quality of the broken well.

2.1.2. *The stress wave interference hypothesis*

Theoretically, if the adjacent charge blasting at the same time, due to the interaction of stress wave, in the middle of the two cartridge will form a certain area of rock mass stress free or stress zone. Thus prone to large. But if make the adjacent two charge interval initiation time, ohlendorf explosive package that provoked a compression wave in rock mass, and from the free surface reflection condensed wave, detonate subsequent charge, which not only can eliminate stress free zone, and also can increase the area of tensile stress, so as to improve the degree of broken pieces.

2.1.3. *The residual stress hypothesis*

The hypothesis starts from the beginning of the rock subjected to explosion, until its burst, and further the process of crushing, to analyze the differential blasting mechanism. Which main contents include: first, explosive package to produce the explosive stress wave in rock mass of dynamic stress field and produce a series of cracks formation; Second, the effect of rock mass under high temperature and high pressure of detonation gas, the cracks further expand, but with the expansion of the detonation gas, pressure decrease; Finally, after explosive package should be prior to explosive package has not yet disappeared before initiating the static stress field, use of explosive charge in the first of the formation of residual stress in the rock mass to improve the quality of fractured rock. The hypothesis and free surface hypothesis is complemented each other, combined two kinds of hypothesis can explain the characteristics of. differential blasting in rock fragmentation mechanism.

2.1.4. *The seismic wave interference hypothesis*

Differential blasting, the adjacent hole (or two pairs of hole, two groups of blast hole) have a short interval between making waves interfere with each other, thereby reducing the blasting seismic effect. If successively primer time interval is equal to half of the seismic wave cycle, the seismic effect is weakened. Observation of a large number of engineering practice proved that many segment of short-delay blasting is an effective way to reduce the blasting seismic effect [3].

2.2. *Open blasting design*

2.2.1. *The steps parameter and perforation equipment*

According to the features of the Capital Yun Ore Company, column stripping treatment planning and preliminary design, the outdoor work bench height H = 12 m, work bench slope Angle alpha α = 65 ~ 70 °. With aperture Φ = 150 mm high wind pressure open-air DTH drilling rig drilling and perforation.

2.2.2. *The design of blasting parameters*

(1) The Specific Charge

Open-pit stripping rock mass is mainly for the plagiarize gneiss, rock hardness coefficient is 8 ~ 10, an average of 9. Requirements after blasting rock breaking the large degree of line length is 500 mm, large rate is 4%. According to the requirement of the rock hardness and blasting fragmentation, library reference B.H. figure Nie Zoff rock blasting scale and rock blasting fragmentation scale, determine the specific charge of ore for q = (0.52 ~ 0.68 kg/m^3), design for q = 0.6 kg/m^3 [4].

(2) Hole Linear Meter Charge

Hole linear meter charge calculated according to the type:

$$q_1 = \frac{1}{4}\pi\rho d^2 \tag{1}$$

Type: for π , apply for π = 3.14;

ρ - hole charge density, anfo explosives off for rho = 0.86 g/cm^3;

The 1.15 g/cm^3 emulsion explosive,

d - hole diameter; 150 mm.

By computing: Φ 150 mm diameter hole loading emulsion explosive, linear meter charge of 20.0 kg/m.

Φ 150 mm diameter hole filling anfo explosives, linear meter charge is 15.0 kg/m.

(3) The Length of Hole Filling

Rate of blasting chunk is to reduce the number of hole plug section, ensure blasting without \"punch\", improve the utilization rate of explosive energy of blast hole plugging length l_2 = (20 ~ 30)d, d - hole diameter; Blasting operation process, based on actual water level in holes, hole plugging with 3.8 ~ 4.2 m advisable, anhydrous hole jam 4 m, there is water hole plug 4.5 m is advisable.

(4) The Drill Hole Depth

Actual drilling depth to add gun hole pays deep bench height; Cannon hole pays deep generally take $\Delta h = (8 \sim 12)d$, according to Capital Yun ore blasting by lithology of Capital Yun, take 10 times the aperture drilling ultra deep, normal step height H = 12 m, calculated:

Cannon hole pays deep$\Delta h = 10d = 1.5$ m;

Hole depth $L = H + \Delta h = 13.5$ m.

(5) Chassis Resistance Line

According to the aperture value, Chassis resistance line is commonly (25 ~ 35) times of hole diameter, considering the surrounding rock as the application of job requirements and rig in front of the broken blocks of perforation safe distance, take 30 times the aperture, namely:

$W = 30d = 4.5$ m;

(6) Charge weight per blast hole is

Under the single charge according to the calculation:

$$Q = q_1(L - l_2) \tag{2}$$

Computing: water hole, the block length is 3.8 ~ 4.2 m, loading explosives, puckering control Q = 140 ~ 145 kg;

Hole with water, plug length from 4.0 to 4.5 m, loading emulsion explosive, puckering control Q = 189 ~ 200 kg;

Blasting area finally could clip made with big hole, puckering charge can be appropriately, with increased 5 ~ 10% advisable.

(7) Hole spacing and row spacing

In order to improve rock breaking effect, big density coefficient using the small resistance line wide aperture fabric hole, hole according to the triangle layout between each row, hole density coefficient control between 2 ~ 3, the pitch of the press type calculation:

$$a = mW_1 \tag{3}$$

Type: - hole density coefficient.

Hole way using triangle hole (plum blossom) cloth, is calculated:

$$b = \sqrt{\frac{s}{m}} \tag{4}$$

Type: s -hole area.

Open stope blast hole layout: pitch = 3 m, = 7 ~ 8 m, front row triangle hole. Main production blasting main technical parameters see table below:

Table 1. Summary table of open-pit mining owners blasting technology parameters.

Types of explosives	units	anfo explosives	emulsion explosive
Hole diameter	mm	150	
Bench height	m	12	
Kong pays deep	m	1.5	
Specific charge	kg/m^3	0.6	
Linear meter charge	kg/m	15.0	21.0
Holepacking density	g/cm^3	0.86	1.15
Chassis resistance line	m	4.5	
Hole spacing	m	7~8	
Hole row spacing	m	3	
Hole filling length	m	3.8~4.2	4.0~4.5
Single-hole area	m2	21~24	
Charging length	m	9.3~9.7	9.0~9.5
Single-hole controll	kg	140~145	189~200

3. The Network Connection

When using normal plastic detonating tube detonators, a blasting control of around 10 t, blasting 1 ~ 2 area every day, blasting area according to the 5 s hole, guarantee the degree of loose blasting heap, is advantageous for the shovel. Different detonators by hole blasting network diagram is as follows:

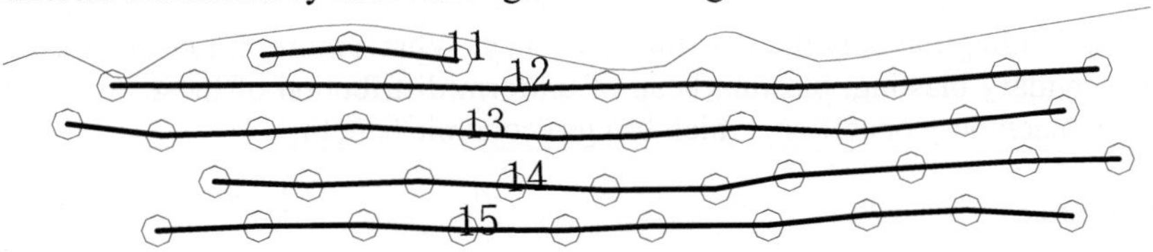

Fig. 1. In every row hole blasting cap don't arrangement.

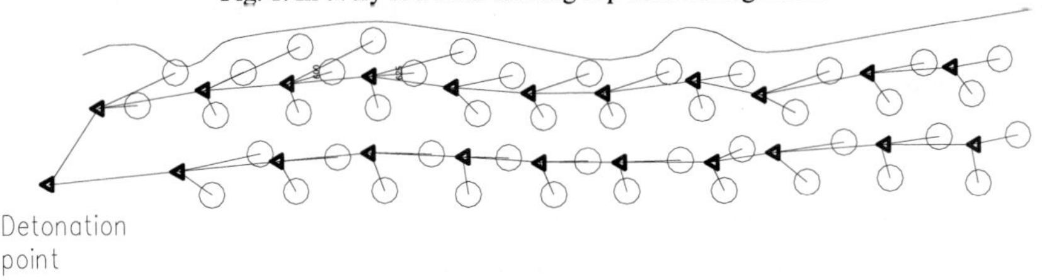

Fig. 2. Each row hole blasting connection arrangement.

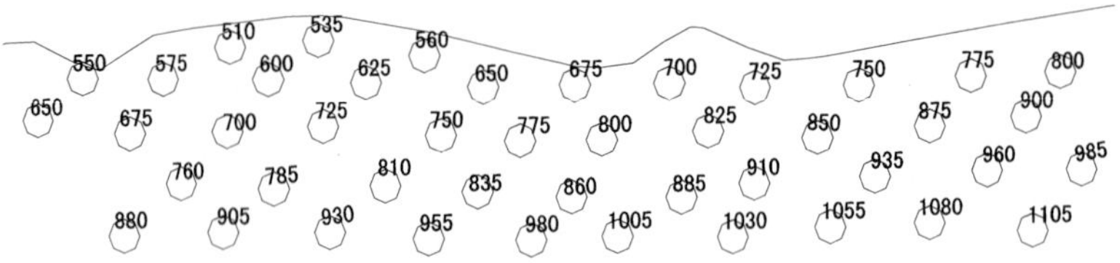

Fig. 3. The water-hole blasting moment arrangement.

3.1. *Analysis of blasting effect*

After blasting, blasting heap surface uplift 3 meters, back row fracture connectivity, form 2 meters cave groove, the overall forward 20 meters, side 5 meters or so, after about 3 meters, rock breaking is uniform, large rate is low. Loose blasting zone has no form of blasting slungshot, free surface in the steps of the new form, the less life phenomenon.

Fig. 4. blasting effect.

4. Conclusion

Through using common millisecond detonating tube detonators by hole blasting, and compared to traditional tandem high-precision millisecond detonating tube, the thesis find that rock crushing effect is better, which cause smaller blasting vibration damage, and reduces the blasting engineering cost, improves the shovel loading rate and the blasting efficiency. What's more, it reduces the cost of secondary blasting, blasting heap of uniform distribution of loose, and make easily machinery operation, which brings economic benefit.

References

1. Ding Linmin,Wang Maoling, Chen Yingfeng, Ordinary Millisecond Detonating Tube Mining Technologies to Realize by Hole Blasting Technique. [J] 201409 14 (5) : 83-84.
2. Dian-zhong liu. Blasting Practical Handbook [M]. Beijing: Metallurgical Industry Press, 2003:310-318.
3. Hole Blasting Technique Applied Basic Research [D]. Liaoning: Liaoning University of Engineering Science, 2010:5-7.
4. Xu-guang Wang. The Blasting Design and Construction [M]. Beijing: Metallurgical Industry Press, 2011.

Finite element analysis of mechanical behavior of the fabricated lattice template support

Zhang-Rong Zhang[1,2,*], Ji-Zhong Zhou[1,2], Jin-Ping Zhuang[1,2], Xue-Feng Cai[1,2]
[1]*Civil Engineering Department, Fujian University Of Technology,*
[2]*Fujian Provincial Key Laboratory of Advanced Technology and Informatization in Civil Engineering,*
Fuzhou, Fujian, China
**E-mail: zzr2014@fjut.edu.cn*
www.fjut.edu.cn.

This paper used the finite element analysis software ABAQUS, to research the elastic stability's performance and ultimate bearing capacity of fabricated lattice template support under axial compression. The finite analysis model used the eigenvalue analysis to obtain the first step mode. Then, considering the semi-rigid connection which used spring element 2 and the initial structure imperfection to establish the nonlinear static analysis model. Meanwhile, considering the two main factors of reinforcement bar and cross-braced to make the parameter analysis. The analysis model shows that the failure parts were the nodes which were mainly occurred in the middle parts of the component. On the basis of the analysis results, the settings of cross-braced can enhance the overall stability and longitudinal rigidity.

Keywords: Lattice Template Support; Finite Element Analysis; Semi-Rigid Node.

1. Introduction

In the paper, the fabricated lattice template support has many advantages such as simple structure, easy installation and less steel material. Adjustable supports can adjust the height of components which can adapt to complex conditions. At the same time, such fabricated support system is good in integral security, because lattice column' proportion is unfold, which can reduce the slender ratio and can increases the moment of inertia. Meantime, the columns which surround the component are also mainly under axial compression when the load eccentricity ratio increased. It can effectively reduce the collapse accident. So, using the appropriate template support system is meaningful to reduce the casualty accidents caused by the scaffold collapse.

2. The Establishment of Finite Element Analysis

According to the size, material properties and boundary conditions of the component, we can analyze it by using finite element analysis software. The component is shown in Fig. 1.

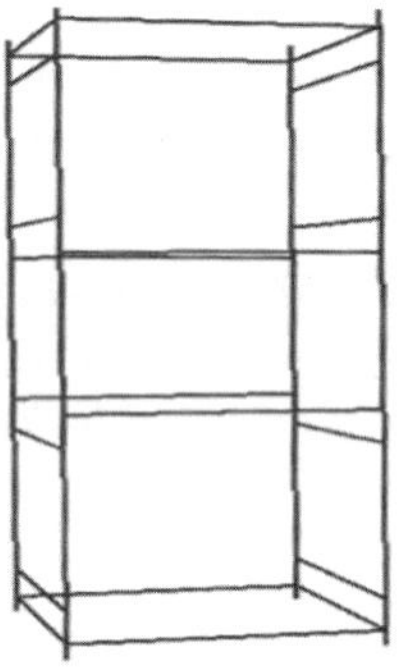

Fig. 1. Lattice framework.

2.1. *The size, parameters and properties of the material*

(1) Properties of the material: the vertical bar's latitude is 1.2*m*, its cross range is 1m, the total height is 3.9*m*.

(2) Parameter values of the material: the vertical bar's diameter is 48mm, its length is 1m, and the bar's thickness is 3*mm*; the cross bar's diameter is 43mm, its length is 1m, and the bar's thickness is 2.5*mm*.

(3) The selection of material properties: In the elastic analysis, the elastic modulus is 206000*MPa*, Poisson's ratio is 0.3, and the tensile strength is 235*MPa*. In elastic-plastic dynamic analysis, using the second-order elastic-plastic model because the steel bar is low-carbon mile steel. As Fig. 2 is shown, the dotted line represents the actual stress-strain curve; the solid line represents the simplified stress-strain curve [1]. In the Fig. 2, f_p represents proportional limit, f_y represents yield limit, f_u represents ultimate tensile strength.

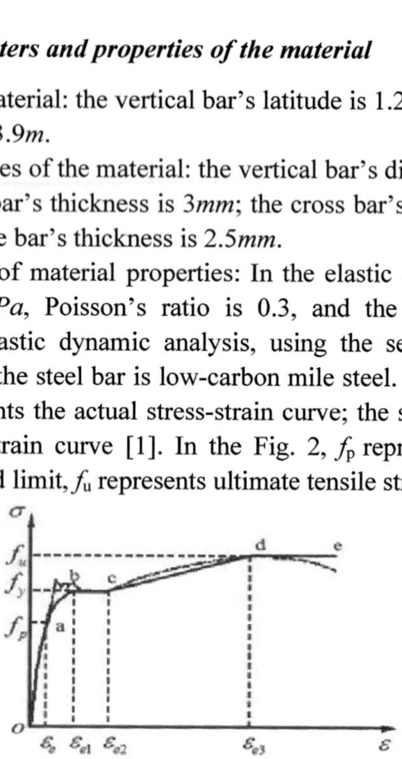

Fig. 2. Constitutive model of the material.

2.2. *Unit type and boundary conditions*

(1) Unit type: Vertical bar and cross bar were both using two nodes beam element B31 which has translational stiffness and rotational stiffness each in x, y, z direction. It is based on Timoshenko beam theory which can simulate quadrangle, circle, annular and so on, can be successfully apply in the nonlinear finite element analysis of large deflection [2]. The nodes between the upper pole and lower pole used six spring elements to simulate semi-rigid connection. The springs along the direction of displacement represented slippage, and the springs along the direction of rotation represented torsion [3].

(2) Boundary conditions: Constraining the displacement of the bottom component in x, y, z. Restraining the displacement of the top in x, z and applying the load in the negative direction of y.

2.3. *Method and result of finite element analysis*

(1) Method: In general, unstable failure will occur before compression members reach the ultimate bearing capacity. What's more, unstable failure is one of the most important issues that should be considered in the process of analysis and design of steel structures. The main factors for the unstable failure of the axial compression members are initial imperfection (initial bending, initial eccentricity, end restraint and so on). The initial imperfection leads to the decrease of stability bearing capacity of compression members. Aiming at this problem, eigenvalue buckling analysis will be used to gain the initial deformation of the first order buckling mode. It is applied to the nonlinear static analysis to obtain the critical load.

(2) Consequence: The first five orders buckling mode of buckling analysis is shown in Fig. 3.

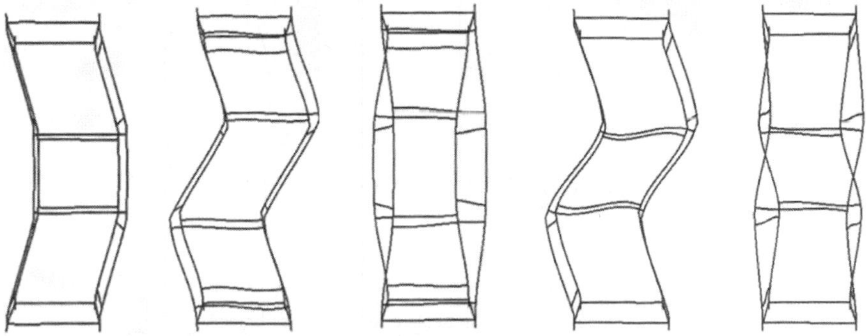

(1) first order (2)second order (3)third order (4)fourth order (5)fifth order

Fig. 3. Buckling mode of members.

The results of nonlinear static analysis were shown in Fig.4. What we can see in this figure is that the global deformation is similar to the first order buckling mode. The first flexural failure occurred at two endpoints of middle frame. The main reason for this failure is that initial imperfection leads to the larger bending moment in the middle part of the upright stanchion. Critical loads of component are 145.6*kN* and don't reach the yield strength of compression members. The maximum lateral deformation is 36*mm*.

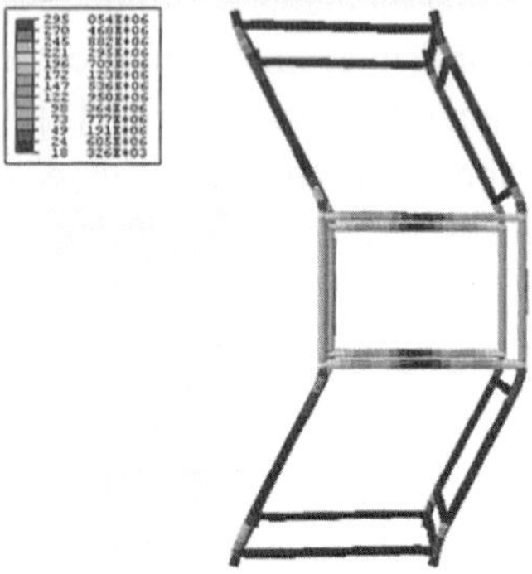

Fig. 4. Mises stress of model.

3. Parameter Analysis

Cross-braced, strengthened frames, initial imperfection and connection stiffness are influencing factors for stability bearing capacity of members. Strengthened frames and cross-braced are considered as the two main influencing factors in this paper. Comparison between this two main influencing factors are shown in the Fig. 5.

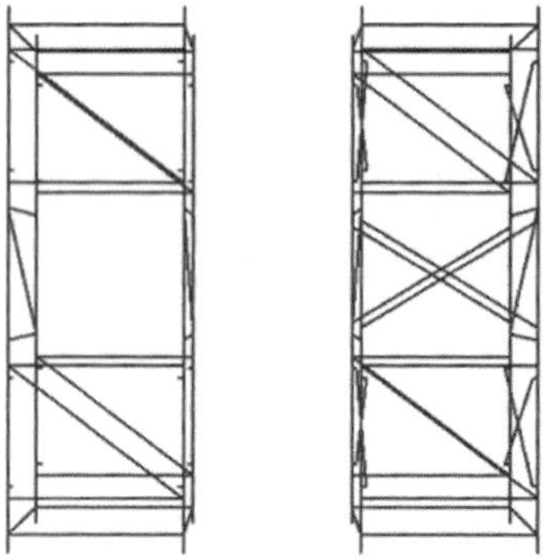

Fig. 5. Comparison of slant rod and cross-braced.

Capacity analysis of the bearing capacity of the components, with inclined rod and with inclined rod and cross support, are carried out by using the finite element analysis. The result is shown in the Table 1.

Table 1. Stable bearing capacities of components with different configurations.

Structure form	Eigenvalue buckling $P_{cr}(kN)$	Ultimate bearing capacity $P_u(kN)$	P_{cr}/P_u
No slant rod and cross-braced	155.72	145.68	1.07
With slant rod	141.52	133.43	1.06
With slant rod and cross-braced	183.32	172.13	1.06

From table 1 we can see that the slant rod does not improve bearing capacities of components. But showed a tendency of decreasing. The bearing capacities were down 9.1% in the table 1. The cross-braced not only can enhance the longitudinal stiffness and overall stability[4], but also can improve the bearing capacity of the member. Its bearing capacity increase by 17.7%. In addition, The stiffness of connection node between the cross-braced and component will have an impact on the bearing capacity.

4. Conclusions

(1) Although P_{cr} (the result of eigenvalue buckling analysis) is higher than P_u (the result of nonlinear static analysis), the values of P_{cr}/P_u almost keep unchanging. The comparative analysis of different parameters of features shows that the result of buckling analysis and the result of nonlinear static analysis is consistent.

(2) Enhancing the stiffness of the local area would decrease the bearing capacity and lead to safety hazards.

(3) The stiffness of the cross support could largely impact on the bearing capacity of the component.

Acknowledgment

This work is supported the National Natural Science Foundation of China (51478119), the Foundation of Fujian Provincial Department of Housing and Urban-Rural Development (2014-k-01) and the Cooperative Foundation of Fujian University of Technology (GY-H-14010).

References

1. Wang Wenda, Han Linhai, *Nonlinear finite element analysis on mechanical performance of concrete filled steel tubular frame structure*. Journal of Building Structures, 2008 (6):75-83.
2. Liao Yanbo, *Experiment and research on concrete filled steel tubular laced columns under axial compression*. Tsinghua University, 2009.
3. Cai Xuefeng, *Research and application on formwork supports system under super-high, long-span and heavily load condition*. China Architecture & Building Press, 2012.
4. Liu Xiyue, *Stable bearing capacity of gabled steel tube scaffolding structure*. Central South University, 2010. 1947 (2003).

Analysis on foundation pit stability under heavy rainfall by numerical simulation

Feng-Zhan Cui

School of Mechanics and Civil Engineering, China University of Mining and Technology (Beijing),

Beijing, China

E-mail: 2316064905@qq.com

www.cumtb.edu.cn

In view of big impact and huge losses on foundation pit caused by complicated conditions, this paper analyzed the stability of foundation pit under the condition of heavy rainfall by numerical simulation software FLAC3D. Through simulation, we gained the displacement curves when slope reinforced by 3 m, 5.8 m, 11 m, 3 m + 5.8 m, 3 m + 11 m, 5.8 m + 11 m, 3 m + 5.8 m + 11 m depth bolting. The results showed that, when slope was reinforced by 5.8 m depth bolting, the displacement was almost minimum. Considering cost and ease of construction, reinforced by bolting in 5.8 m depth was determined as the economic and reasonable way to control slope deformation.

Keywords: Foundation Pit Stability; Numerical Simulation; Reinforced Supporting; Bolting; Slope Deformation.

1. Introduction

As the improvement of urbanization in our country, efficient use of underground space was the most important ways to solve three major crisises of urban population, resources and environments, and to implemente sustainable development. At present, various uses of underground space, which were involved in deep foundation pit engineering, have been developed and utilized during city construction in the world. Therefore, study on foundation pit stability has become an important content of urbanization [1-2]. Foundation pit accidents not only caused casualties, construction period delay, investment increasing, but also generated adverse impact on urban construction and enterprise image. There were many factors induced to foundation pit accidents [3-4]. Uncertainty was one cause of foundation pit accidents. There were two rainstorms in Beijing on July 8 and July 21, 2012. These two rainstorms induced serious impact on foundation pit and caused many disasters, such as slope sliding and rupture. Through field monitoring [5-7] and numerical simulation [8-10], scholars have researched on this problem and got a lot of results. Based on this, relying on one foundation pit in Beijing, this paper analyzed the effects of two heavy rainfall disasters on foundation pit instability, and probed control measures further.

2. Modeling Schemes

(1) Analysis on displacement of slopes without supporting, which was the basis for comparing different supporting ways.

(2) Analysis on displacement of slopes under designed supporting ways (Slope protected pile, with 800 mm in diameter, 1.6 m in spacing and 21.0 m in length. Three rows of anchors: the first row was 4.67 m underground, the second row was 8.87 m underground and the third row was 12.87 m underground.).

(3) Analysis on displacement of slopes with weakened strata mechanical parameters (This was used to simulate the effect of rainstorm) under improved supporting ways, which were increased anchors at different depth than designed supporting ways).

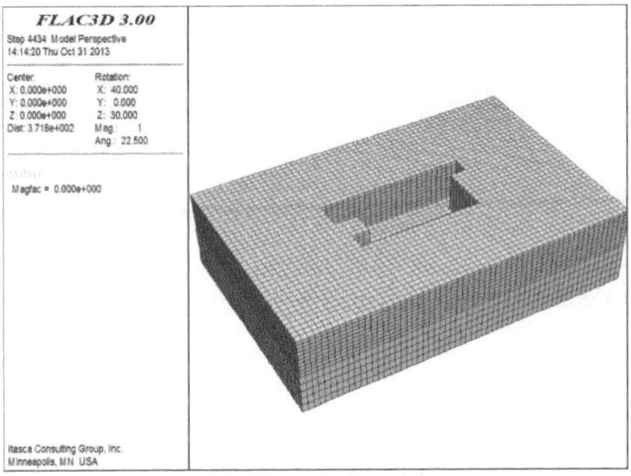

(a) Numerical model

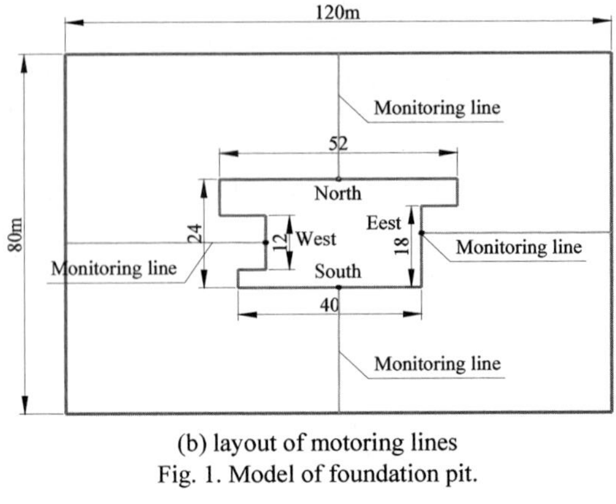

(b) layout of motoring lines

Fig. 1. Model of foundation pit.

3. Numerical Model

According to simulation schemes, combined with the shape and size of foundation pit, the model was established and shown in Fig. 1(a). For the center of foundation pit as the centre of model, the length along east and west direction was 120 m and the width along north and south direction was 80 m. The height was 30 m and the foundation pit was 15 m in deep. Soil was simulated with Mohr-Coulomb failure criterion. Protecting pile was simulated by pile command. Anchor was simulated by cable command. To monitor the displacement of slopes, monitoring lines were set up in the model, which were shown in Fig. 1(b).

4. Results Analysis

4.1. *Analysis on slope displacement under no and designed supporting ways*

Under the action of rainstorms, in no and designed supporting states, displacement curves of north and south slopes along depth direction were shown in Fig. 2.

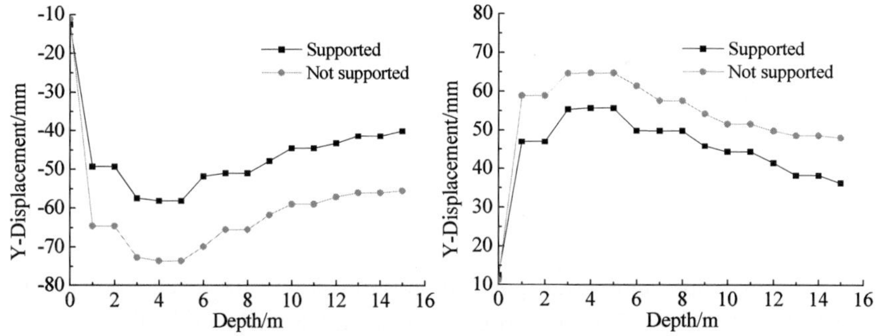

(a) Displacement in north slope (b) Displacement in south slope

Fig. 2. Displacement curves in no and designed supporting states.

Displacement curves of two slopes were almost the same. Along the depth direction of foundation pit, slope displacement increased firstly, and then decreased. Within the scope of 4 to 5 m in depth, slope displacements reached peak. In north slope, when no supporting, the peak displacement reached 73.57 mm. And when in designed supporting states, the displacement decreased by 21.04% and the peak reached 58.09 mm. In south slope, the displacement was slightly smaller than north slope. When no supporting, the peak displacement reached 64.74 mm. And when in designed supporting states, the displacement decreased by 14.03% and the peak reached 55.66 mm.

4.2. *Analysis on slope displacement under different reinforced supporting ways*

Slopes of foundation pit moved after rainstorm. To control deformation and keep normal use of foundation pit, bolting was used to reinforce slopes. In order to determine reasonable reinforcing positions, places of 3 m, 5.8 m, 11 m were chosen to add anchors. Based on this, three models were set up. Through simulation, displacement of slopes was gained and analyzed as follows.

In designed supporting state and reinforcing scheme (Based on designed supporting way, 3 m, 5.8 m, 11 m in depth were reinforced by anchors) states, displacement curves of north and south slopes along depth direction were shown in Fig. 3.

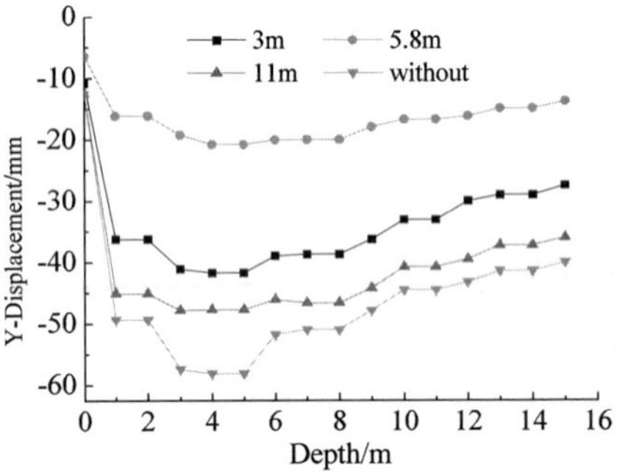

(a) Displacement in north slope

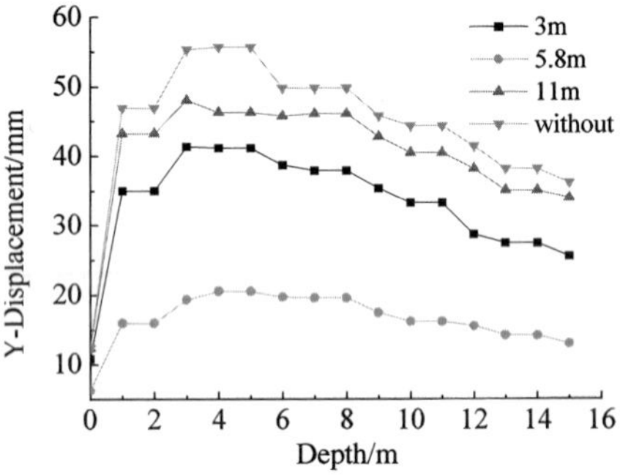

(b) Displacement in south slope

Fig. 3. Displacement curves in reinforced supporting states.

In north slope, under designed supporting ways, the peak displacement reached 58.09 mm. And when reinforced by 3 m, 5.8 m, 11 m depth bolting, the peak displacement decreased to 41.71 mm, 20.8 mm and 47.68 mm respectively. So we can get that bolting in 5.8 m depth showed the maximum effective in slope reinforcing. In this state, the displacement decreased to 35.8% of designed supporting state. In south slope, when reinforced by 3 m, 5.8 m and 11 m depth bolting, the peak displacement decreased to 41.15 mm, 20.55 mm, 46.3 mm respcetively. Consistently, bolting in 5.8 m depth showed best effective in slope reinforcing.

4.3. *Analysis on slope displacement under further reinforced supporting ways*

In order to probe better reinforced approaches, bolting model in different depth of 3 m + 5.8 m, 3 m + 11 m, 5.8 m + 11 m, 3 m + 5.8 m + 11 m were set up. Through simulation, displacement curves of north and south slopes along depth direction were shown in Fig. 4. In north slope, when reinforced by 5.8 m depth bolting, the peak displacement reached to 20.8 mm. When reinforced by 3 m + 5.8 m, 3 m + 11 m, 5.8 m + 11 m, 3 m + 5.8 m + 11 m depth bolting, the peak displacement decreased to 19.57 mm, 21.73 mm, 21.73 mm, 18.15 mm respectively. In general, in these five states, the displacement was much close to each other. In south slope, displacement in those four further reinforced states reached to 19.32 mm, 21.42 mm, 18.69 mm, 17.88 mm respectively. So we can get that based on 5.8 m depth bolting, further reinforced states exhibited little effect on slope deformation control. Considering cost and ease of construction, reinforced by bolting in 5.8 m depth was the economic and reasonable way to control slope deformation.

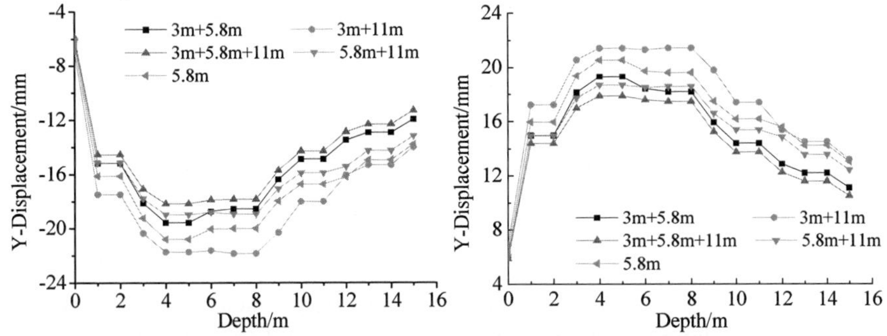

(a) Displacement in north slope (b) Displacement in south slope

Fig. 4. Displacement curves in further reinforced supporting states.

5. Conclusions

Stability of foundation pit under rainstorm was analyzed through numerical simulation. Main conclusions were as follows:

(1) Slope displacement under no supporting and designed supporting ways was got and demonstrated the main role of pile and anchor in keeping slope stability.

(2) Slope displacement curves when reinforced by 3 m, 5.8 m, 11 m, 3 m + 5.8 m, 3 m + 11 m, 5.8 m +11 m, 3 m + 5.8 m + 11 m depth bolting were gained. Based on these, it was determined that when slope were reinforced by 5.8 m depth bolting, the displacement was almost minimum. Considering cost and ease

of construction, reinforced by bolting in 5.8 m depth was the economic and reasonable way to control slope deformation.

References

1. M. Jing, Research on the optimization design and engineering application of detail retaining structure for deep excavation, Chongqing University (2012).
2. W. F. Zheng, Research on mechanical performance of prestressed anchor flexible supporting system in deep foundation pit, Dalian University of Technology (2007).
3. Z. G. Li and X. J. Wu, Discussion on calculation method for environmental protection around deep pit under different protection levels, Chinese. *J. of Rock Mech. and Eng.*, 27: 1074-1080 (2008).
4. C. L. Ding, S. H. Zhou and S. D. Zhang, Centrifuge model experimental study on influence factors of deformation stability for confined water foundation pit, *J. of Tongji Univ. (Natural Science)*, 33: 1586-1591 (2005).
5. F. Y. Liang, F. Chu, Z. Song and Y. S. Li, Centrifugal model test research on deformation behaviors of deep foundation pit adjacent to metro stations, *Rock and Soil Mech.*, 33: 657-664 (2012).
6. N. Wang and M. Huang, Deformation Monitoring Artificial neural network model of deep foundation pit considering the excavation effect, *J. of Shanghai Jiaotong Univ.*, 43: 990-993 (2009).
7. Z. W. Wang, G. B. Liu, X. D. Wang and J. M. Zai, Study on the deformation of a deep metro excavation under complex environment, *Chin. J. of Geo. Eng.*, 28: 1263-1266 (2006).
8. L. Shen, Y. N. Lu and J. Y. Yue, Deformation characteristics of oversized deep foundation pit: comparison of numerical simulation with the instrumentation results, *Chin. J. of Underg. Space and Eng.*, 1: 538-542 (2005).
9. Y. H. Fu, X. D. Wang and J. M. Zai, FEM analysis on tmie-effects of deep excavations, *J. of Nanjing Univ. of Tech.*, 27: 32-36 (2005).
10. S. L. Sun, S. M. Wu and H. J. Pei, Orthogonal design study for numerical modeling of deformation of multipropped deep foundation pit, *Rock and Soil Mech.*, 26: 1771-1774 (2005).

Energy saving reconstruction of existing residential buildings in hot summer and warm winter zones-experience learnt from home and abroad

Jian-Xun Deng[1,a,*], Zhi-Zhou Shen[2,b]

[1, 2]School of Architecture & Civil Engineering, Xiamen University, Xiamen, 361005, Fujian, China

Email: [a]djxun21cn@sina.com, [b]uniquevector@163.com

Combined with cases of energy-saving existing building from home and abroad, this passage discusses about the reconstruction condition of existing buildings and makes a comparison by analyzing the existing buildings in terms of energy-saving technologies and relevant policies, which points out the existing energy-saving problems of existing buildings in Hot Summer and Warm Winter Zones in our country. Also, it sums up the experience of energy-saving reconstruction and finally proposes energy-saving measures.

Keywords: Existing Buildings; Hot Summer and Warm Winter Zones; Energy Saving; Measures.

1. Introduction

Sustainable economic development puts forward new requirements for our energy-saving buildings. The energy plays a more and more important role for economic development, but the growth rate of China's energy production lags behind the growth rate of GDP, which constraining economic development. Because of a lack of energy, our country needs to spend a lot of foreign exchange to buy coal, oil, natural gas and other high-quality energy. The level of economic development becomes increasingly high. The improvement of the quality of people's life has put the issue of the lack of energy to the fore front with the problem of supply and demand becomes more and more prominent. For the healthy development of economy sustaining and continuing, the approval of the construction of high-energy buildings should be strictly controlled, promoting green energy efficient building is the future trend.

2. The Current Status Quo of the Existing Building Energy Saving

At present, China's existing buildings mostly focus on the economy while ignoring the residential area in harmony with the surrounding environment issues, and residential building design plagiarism is pretty serious, During the same period the originality of the internal and external design of the building is not enough, not according to the relationship between environmental characteristics and architectural building with the external geographical environment, which considers little from the energy point of view to guide the design, resulting in building energy consumption and environmental resource sustainability are not in harmony with each other, excessive energy consumption resulted in not only social energy waste and environmental pollution, but also unnecessary financial burden to the residential. Therefore, energy-saving should be planned as a primary aspect. What the plan to save means fully taking into account the relationship of the characteristics between architecture and the external environment, and the main guiding principle of planning is based on energy-saving, fully using natural resources to achieve the overall design to create good conditions.

China has made some achievements in existing building energy saving. Country is vigorously promoting energy saving of existing residential buildings. Particularly heat metering and energy saving in the northern heating areas which requires large amounts of coal resources in existing residential buildings. Energy saving in the hot summer and cold winter and hot summer and warm winter area mainly focus on windows and doors, shading, natural ventilation and other aspects. At present, China actively explores the mode of energy-saving of the subtropical region and make experiments in some areas. Nevertheless, China's existing buildings still confront a serious situation. On the one hand, the quality of people's life is in urgent need to improve the quality of construction services; on the other hand, low levels of building energy efficiency mandatory standards, which are currently being pursued is only equivalent to the early 1990s level of Germany, and energy consumption indicators are two times than in Germany. In addition to the subtropical region lack of fresh air and hot water supply systems, the shading, ventilation and other passive energy-saving measures have not been effectively applied, which not only increases energy consumption, but not also improves interior comfort index.

3. Analysis of Building Energy Efficiency at Home and Abroad

3.1. *Cheng Kung University green magic school [1]*

Xiamen and Taiwan are both belong to the subtropical region, and their geographical features are largely in the same way, the energy-saving building experience in Taiwan has a highly referred value for the subtropical region. We now regard Cheng Kung University in Taiwan as an example to discuss about energy-saving experience in Taiwan.

3.1.1. *Kiln kitchen ventilation system*

Chong Hua Hall in schools is an ancient kiln kitchen ventilation system, which replaces the air conditioning system in the winter, the air quality in the room can take full advantage of natural ventilation.

Specifically: Firstly, set the row vents behind the podium to bring in the fresh air, and then set three chimneys which shaped like fireplace in the back wall of the guest, thereby forming a low-to-high air flow field that effectively promotes air circulation. In the south chimney, the designer design a transparent glass window to improve the upward flow rate of air, the black paint aluminum window can be used to absorb heat sent out by the solar radiation, thus it forms "stove kiln" combustion laminar wind field without power, it can reach a comfortable ventilated environment in the winter.

Using CFD software to simulate the Chong Hua kiln hall kitchen ventilation system, during the period from November to March in the winter, the wind speed inside the room is around the range of 0.1 ~ 0.6m/s, which people feel more comfortable, and the times of ventilation are up to 5-8 which is in a certain condition of the most health level. When the weather gets hot, heat power and wind speed will enhance indoor ventilation, which is in line with health requirements.

Using the DOE program to simulate energy consumption of air conditioning in winter, the analysis confirmed that stove kiln ventilation system can be executed when the outdoor temperature is below 28°C, annual air-conditioning power consumption can be reduced by 27%. Control equipments cost a total of 20 million RMB, and the period of payback is about three years, most of the existing air-conditioning systems are far from energy-saving systems.

3.1.2. *Roof energy-saving technologies*

The construction of roof overhangs is deep in Cheng Kung University, which can make a good impact on external shading effect, the sun does not directly shine into the interior, which helps to reduce the use of air conditioning. The roof is stepped, which grows some plants and flowers that including mainly sedum drought-tolerant plants in Taiwan. So from the point of efficient view, it costs a little on maintenance. And the soil is composed of granular sludge in reservoir, because of porous materials, this approach will not only reduce the structure load, but also the water-saving effect is obvious. Planting flowers on the roof can absorb solar radiation as well as reduce the heat of roof, thus effectively reducing the energy consumption of buildings.

3.1.3. *Air conditioning system energy-saving technologies*

The indoor air conditioner and ceiling fan are both used in the system. When the outdoor temperature is above 31°C, the air conditioning is enabled, when it is around the range of 27°C ~ 31°C, ceiling fans are used for ventilation, when it is below 27°C, you can open the window for ventilation, and then air conditioning energy consumption is greatly reduced. In sub-tropical climate, human physical sensory is comfortable in 32°C ventilated environment. After questionnaire study investigated by the designer, the three systems model can be extended and learned from.

3.1.4. *Other energy-saving technologies*

Solar photovoltaic movable leaf-technology applications, using solar energy collection panels for the building to provide lighting and other electrical equipment.

3.2. *"Bugginger Str" in Freiburg Germany, existing residential energy-saving demonstrated project*

Germany has been ahead of other countries in existing building energy saving, there is a certain sufficient experience during 30 years history of energy saving, because energy scarcity in Germany, the government attaches great importance on energy conservation in existing buildings, there are two points we can learn from, namely, the first is the provision of preferential policies and incentives in the transformation, the second is innovative research in advanced technology of building energy efficiency [2].

Bugginger Str is a public rental housing which is nearly 50-year history and developed by the Freiburg City Construction Company, and the structure is a common concrete frame and shear wall structure. In 2009, the company decided to increase occupancy rates to improve housing living environment in accordance with market demand survey. First, they change the structure of housing construction. Based on the original room type, they utilize exterior corridors to add a new room, referring to technical standards for passive house renovation. After the transformation, the use of an area increase nearly 1800m^2, the number of households increase by 40, theoretically, the building's heating energy consumption reduce by nearly 50kwh / m^2.a. It costs €1,600 per square meter, and the cost of pure energy saving account for 37.5%.

3.2.1. *Considering the state of energy-saving building, the following aspects of the case are worth learning:*

(1)Funding:

The federal government subsidizes one million euros to fund the transformation of existing house, which shows that the German government attaches great importance on the rational use of energy, they provide a great support for the construction of energy conservation, improve the enthusiasm for involving in the project with each other in energy efficiency of the building project. In the process of energy conservation in existing buildings, the buildings are equipped with solar equipment, fresh air heat recovery equipment and heat pump equipment. In eight years, the Government's annual subsidy accounts for 2%, or about 16% of the total investment cost. Because of the great financial support of the German government, the German energy saving of existing buildings develop very fast.

The mechanisms of Policy support for energy-saving existing buildings are not very sound. Lack of related government's incentives, on one hand, resulting residents pay little attention on energy conservation in existing buildings, on another hand, real estate companies are also likely to overlook the environment sustainable development, the whole life cycle of the building did not receive due attention and energy saving investment is still at the primary level.

(2)Technical standards:

Germany is very strict in energy conservation of existing buildings, requiring a separate energy efficiency standards. In Bugginger Str project, with the related technical support, it make considerable reduction on energy consumption for heating, reducing from 65kwh/m^2.a to 15kwh/m^2.a, and now most of the area in China is 65% of the heating energy consumption of building standards. With respect to German standards, roofing, exterior walls and windows of the heat transfer coefficient is twice as high as the standard of German, in the development of mandatory building standards, the level of standards in energy-saving needs to be raised.

(3)Energy-saving technologies and measures:

Aiming to significantly improve the thermal insulation effect, in the process of transformation of Bugginger Str, the thermal insulation materials include mineral wool panel, aerated concrete blocks and AEROGEL aerosol (coefficient of thermal conductivity is 0.013W/m·k). The method of construction is using aerated concrete block to make up exterior wall with masonry, and exterior surface of the building selects the material of mineral wool panel with 200mm thick, the upper side is made with alkali-resistant mesh cloth, combining painting with thin plaster for construction. Because AEROGEL is more expensive than mineral wool panel by 10 times, it is generally used in the details of construction sites. The gravity of wall material is 150kg/m^2 aerated concrete parapet and gypsum slurry coating, which not only improve the level of anti-sound, but also improve the façade of tightness. For effectively using solar energy, the company has been installed water heaters for each household, it is also used with geothermal and added with capped room floor, achieving the effects of economic and environmental protection. Finally, balcony preform is a feature of the transformation, which uses the method of connecting the bridge, it can effectively increase the effect of insulation, and solve the high cost of insulation measures and the problem of complex construction.

3.2.2. *The measures of energy-saving with existing residential building:*

(1)Architectural planning and design:

Subtropical region mainly located in coastal areas, in summer and autumn ,the direction of the wind is southeast or southwest , the direction of general living building are designed to the north and south orientation, so that it not only helps to increase the indoor circulation of air, but also improves the quality of living experience. Because of ventilation, for such building, the summer sunlight has a little impact on it, which is conducive to the southern area of subtropical district like Xiamen and Zhangzhou that mainly considering refrigeration in summer .At the same time, in the north area of subtropical district ,solar radiation can reduce heating consumption in replace of the need for air conditioning in winter . Therefore, in the subtropical region, the building should be designed towards south and north direction.

(2)Energy-saving measures for the facade of the building:

It accounts for 20%-25% in the overall building energy consumption, to some extent, and the thermal performance depends on energy consumption of the external walls. In the subtropical region, it not only need to add measures of the building's exterior wall insulation, but also improve the efficiency of the wall insulation. The external walls need to improve the coefficient of heat conductivity and the index of thermal inertia, increase the total attenuation of temperature harmonic and reduce the average temperature of the inner surface the of external walls, thereby reducing energy consumption of air conditioning in summer. [3]The energy consumption of German building accounts for 75% of total energy consumption, electricity consumption accounts for 25%, which is similar to use more energy of air conditioning in subtropical region in southern China in summer, which need better insulated measures .Their aim is to achieve thermal insulated walls and ensure air tightness of doors and windows. Regarding to this, Bugginger Str use good insulation of aerated concrete materials, and plaster coating layer is made to achieve airtight for windows and doors. Of course, more and more new materials are worthy to be developed and utilized at present.

For subtropical region, heavy materials facades are used as generally measures for the outer insulation , exterior and interior insulated measures is not as effective as the insulation of external walls which belong to the air-conditioned area. For walls made by lightweight materials, especially using air conditioning to cool the building in summer, it is suitable to use the wall of heat-resistant lightweight material and heat-resistant insulated wall. For the south of subtropical area, when the wall uses the material of aerated concrete that has self-insulated effects, it cannot make an effective impact on energy consumption.

It is worth mentioning that Dryvit company from US, since it creates thermal insulation system, it has completed a large number of existing buildings renovation project. Their invention that is Exterior Insulation and Finish System (EIFS) has been applied to B&B Theater that lies in the hometown of Mark Twain .In the process of design and construction, the Dryvit company use exterior wall system of EPS that makes the building "fundamentally changed", and the process of construction is simple. It combines the energy-saving and beautiful appearance of the building together, which won the praise from all sectors of society, and Chinese enterprises should learn from it [4].

(3)Energy-saving measures for roof:

The technology of greenery insulation is particularly effective for south of the subtropical region which is exposed in the long-term sunshine. In some region of Taiwan, retrofitting of existing buildings is realized by planting green plants on the roof to achieve ecological insulation effect, reducing structural radiation intensity from the sun, while green plants absorb solar radiation and have transpiration, and reduces the amount of heat conduction from ambient radiation of the sun, which will help to improve the quality of exterior thermal environment of the building, thereby reducing the urban heat island effect.

Energy-saving for roof in subtropical region also includes the impounded roof ,wet roof, sun roof with reflective coating, sunshading roof and heat insulation roof with thermal insulation. Due to the impact of self-weight of the impounded roof , the building will bear a great load, generally do not recommend existing buildings to use, and setting the automatic water supply for the reservoir means produce a certain power consumption. Compared to the impounded roof, ceramic wet roof has a better operability, the construction request for little and insulation performance is good, it is possible to reduce the temperature of the inner surface of the roof about 3°C. Because of strong solar radiation in the subtropical region, the roof is coated with a reflective coating and built with sunshading measures, reducing the building cooling load, especially top-floor room. Added with insulation on the roof, it can make the conductivity of thermal and coefficient of heat storage lowered, reducing heat absorption and dissipation, and it can better insulation property.

(4)Energy-saving technology for the outer window:

Comparing windows and other peripheral structure of the building, the pressuretightness is poor and the transmittance of solar radiation is high which is the weakest part in building insulation. For hot summer and warm winter region in our country, the ventilation should be mainly considered. It is different from German climate, and the requirements are also different. In the process of energy-saving of Bugginger St building ,outer windows use glass with 3 layers that the coefficient of heat transfer is 1.0 W/m·k. In terms of doors and windows, the German pay more attention to pressuretightness and heat insulation, although this does not match with our southern climate requirements, but the process of hollow glass or multilayer is worth learning. The coefficient of heat transfer of double window is much smaller than a single layer of aluminum alloy windows, which can achieve the purpose of thermal insulation and sound insulation. At the same time, effective measures of sunshading for windows should be taken, and it should increase the area of the wall and reduce the area ratio of window to wall, in order to improve the overall heat insulation of external structure.

The study finds that Low-E glass meets the need of insulation in subtropical region very well. Windows of sunshading Low-E glass reduce more energy consumption than ordinary windows. The effect of ordinary energy-saving windows of three layers are better than two-way ventlight and air louver windows, and the effect of energy-saving of two-way ventlight is better than air louver windows. Comparisons between brushing transparent glass with transparent insulation coating, ordinary hollow glass, monolithic Low-E glass and hollow Low-E glass, referring building energy efficiency of the contribution rate, found that the effect of brushing transparent glass with transparent insulation coating is the best, reconstruction cost is minimized. To reduce energy consumption, the proximate matter of window frame should choose plastic steel and insulated aluminum alloy, and the structure of window frame should choose multi-cavity windows, and the type of windows should choose fixed window and casement window [5].

(5)Renewable energy sources:

Energy efficiency of the building should not only save from the start, but also try to explore new green renewable energy, such as solar energy, tidal energy and wind energy, etc. At present, the use of solar technology is more mature. The region of hot summer and warm winter in our country has short summer and long winter, and has long time and strong sunshine, which enjoys abundant solar energy resources. In the process of energy saving in existing buildings, if set up solar panels on the walls and roof, achieving solar power can not only save electricity energy, but also save non-renewable resources needed for thermal power generation, which reduce the amount of using coal ,as well as reduce emissions of carbon dioxide in the environment, thereby improve the greenhouse statement. Currently, the solar water heater is the most fundamental and economical equipment, it has become an emerging industry, and solar water heaters have been commercialized. At present, in many cities, considering the application of solar technology has become the consensus in the architectural design. For private homes, the way which combines the construction with solar collectors is generally used with two-way circulation system, and the two-way circulation system has a separate indoor storage tanks with heat exchangers and auxiliary heating systems. Solar water heaters can meet 70% demand of the hot water. Such system demands high requirements for collectors to bear capacity, only a small number of users use non-pressure-cycle systems. Currently collectors are generally divided into flat-plate collectors, glass evacuated tube collectors and heat pipe evacuated tube collectors, heat pipe evacuated tube collectors has the highest conversion efficiency, as well as it is the most expensive. Mainly heat pipe evacuated tube collectors consist of heat pipe, heat plate, glass, metal caps and getter .The efficient heat transferring components use the antifreeze copper pipe of water heat .It will not frost damage at an ambient temperature of -20°C ~ -30°C.The "sun strips" of albronze is used as the absorber plate, the surface sputters with aluminum-nitrogen-oxygen-selective absorbing coating, and the pipe uses high-strength borosilicate glass, which has compression resistance. Users of solar water heater at home not only invest in collector, but also invest in the water tank. Therefore some companies produce thermal storage tube, which can serve as both a heat absorber and a water tank.

4. Conclusion

Energy saving in developed countries is mostly done on the basis of the existing buildings, and they do not dismantle old buildings to build a new one. The current existing buildings in China have a large scale and face with the problem of high energy consumption, which should become the focus of building energy efficiency. At the same time, our country, after the reform of housing property rights, is confronted with a prominent issue that the disperse rights of existing residential buildings hamper implement energy-saving measures. On the basis of in summing up domestic and international experience, we should encourage governments, institutions, individuals and any other diversified investment approaches to engage in the energy-saving renovation of existing buildings, so as to effectively solve the problems of funding, and we look forward to rational and efficient energy-saving measures to achieve renovations in area of hot summer and warm winter.

Acknowledgement

Foundation: Xiamen Construction and Technology Projects (Item Code: XJK2013-1-3).

Jian-Xun Deng, male, Title: assistant professor, Research Field: engineering management, green building management, property development and investment decision.

Zhi-Zhou Shen, male, Title: master student, Research Field: civil engineering construction and management.

Reference

1. Linxian De, 65% energy-saving green building at a level of diamond,which is Green Magic School Cheng Kung University in Taiwan [J]. New building, 2010,(2): 77-81.
2. Tian Lingjiang, Housing Construction and Building Energy Efficiency Enlightenment in Switzerland and Germany, [J], Housing Industry, 2011 (Z1): 80-84.
3. Hu Daming, Ye Qing, Bo Zengwen, Liu Junyue, Luo Gang.Based on Economic of Energy of Thermal Insulation in Shenzhen [C]. National building energy-saving technology and design conferences. National energy-saving technology of buldings and design conference in Fujian, 2005: 135-139.
4. Mu Haoming, Foreign Cases of Energy Saving in Existing Building- A Project of Wall Insulation Renovation of American Theater.
5. Chen Elan, The appliance of Low-E glass in the subtropical region, [D], Guangzhou: Guangdong Industrial University, (2012).

Remodeling and reflection historic district — Taking Qianmen Street as an example

Tie-Fu Chen*, Zhi Wang†

Tongji University College of Architecture and Urban Planning, Shanghai, 200092,
AN-DESIGN Architects, Historical Architecture Reservation Design and Research Dept.,
Beijing, 100097
Institute of Agricultural Scientism Information, Beijing Agriculture and Forestry Academy of Science,
Beijing, 100097
*E-mail: *14516522@qq.com, †angelwang1009@163.com*

Qianmen area is no longer in those years of prosperity spectacular, there is a historical reason, a more important reason is its commercial positioning fuzzy, overall positioning error caused. This paper studies the Qianmen Street commercial development context, to extract the elements of Qianmen Street economy and cultural prosperity. We would like to see through the nature of the phenomenon and find Qianmen Street "soul", which stimulate the revival of the potential for the revival of the historic district of cultural, economic revival and morphological remodeling, providing development ideas and implementation strategies.

Keywords: Historic District; Qianmen Street; Business Evolution; Remodeling.

1. Introduction

Qianmen Street was founded 570 years ago, it is the only way forroyal ritual, hunting, patrol at Ming and Qing Dynasties, known as "Heaven Street". Sedimentary nearly 600 years, making Beijing Qianmen Street became the architectural culture, business culture, cultural hall, opera culture, folk culture precipitate profound characteristic historic district, is one of Beijing landmark. Today, however, Qianmen Street no longer have the same spectacular, so many people confused by its development. History Commercial Streets become ordinary street shops, tasteless attractions. By combing historical context of Qianmen Street, the author were compared for business forms and characteristics in different periods, to extract the Qianmen Street's economic and cultural boom operating mode. We would like to see through the appearance essence, Claim this lost Qianmen Street "soul" unlcash the potential of Qianmen

Street revival, provide ideas and implementation strategies for the development of the historic district of remodeling.

2. Overview of Qianmen Street

Qianmen Street is the most famous commercial street in Qianmen area. It is a traditional Beijing Commercial Street. Located in Beijing's central axis, starts at the northern end of Qianmen Door, reach the southern end of the bridge crossing. Jiajing twenty-nine years (1550), before establishing outside the city, this street was the Royal Road which Ming Royal to the Temple of Heaven, after outside the city was established the street was the main north-south street. It was called as Zhengyang Street by the republic at the Ming and Qing Dynasty, commonly known as Qianmen Street. Since the Republic of China after the liberation, Beijing Qianmen Street has been one of the three most important commercial Old Town streets. It’s officially named Qianmen Street in 1965.

Qianmen area is a commercial circle that the center is Qianmen Street, which includes both sides of the fish mouth, Dashanlan Street, Jewelry Street and other traditional city street bazaar. Many old and famous brands located here, it’s one of the urban areas of Beijing's most historical and cultural charm.

3. Causes of Qianmen Street Boom

Qianmen Street in history to be Beijing's most bustling commercial street, the reasons are mainly following:

(1) Special location

It was the only way for royal ritual, hunting, patrol at Ming and Qing Dynasties, known as "Heaven Street". Qianmen Street was also the only way for Beijing businessman. After the Ming Dynasty moved the capital to Beijing, Beijing's commercial center has moved to Qianmen Street. In addition, the expansion outside the city after the Ming Dynasty, the government building houses to attract resident stay in here, and also to attract businessmen operating. Wealthy businessmen in Zhejiang, Suzhou province and other places moved to here, brought prosperity to the region many passengers and goods, and promote the rapid development of all aspects of the Qianmen Area.

(2) Old and famous shop or time-honored brand gathered

Historically, Qianmen Street is a gathering place of time-honored brands, these Old and famous are: Founded in the Qing Tongzhi-three years, as the first Chinese eat named "Quanjude"; founded in Kangxi Dynasty eight years, the country's most famous traditional Chinese medicine hall called "Tong ren-tang Pharmacy"; one of Beijing's oldest opera house, built in 1796, "Guangde Opera

House Park," and so on, over twenty Old and famous shops, including the food, clothing, and other life forms.

4. The Outstanding Problems Facing the Qianmen Street

2008 years ago, the street door reopened. Now, years later, it has experienced a brilliant century, but was decline of Qianmen Street, still did not revive the glory. Its ability to achieve revival, still faces the following outstanding issues:

(1) Lack of cultural accumulation neither fish nor fowl.

Qianmen Street attractive is that the old Beijing's history and cultural accumulation however, reopened Qianmen Street, there are many international brands have entered the marketplace Qianmen Street's inherent cultural and historical accumulation has been greatly hurt. And also, the local business climate has been heavy damaged to. There are all kinds of incompatible and inconsistent between Qianmen Street and international brands. The original consumer groups are rapidly lost.

(2) Loss of cultural traditions, a vague position.

Many folklorists believe that the transformation of Qianmen Street is demolished and sacrifice of alleys as the price, lost their original culture label. As the positioning fuzzy, poor management, this ancient street has been reduced to a "dull" attraction.

5. Ideas and Suggestions for Qianmen Street Revival and Remodeling Development

(1) The Collating of Qianmen Street's value, reasonable positioning its remodeling function.

Business location is a critical decision-making, this is about the historic district remodeling and development success. Qianmen area as commercial district has a rich tradition of historical value, cultural value, artistic value and commercial value, resource value and social value. Must fully understand the basis of these values, in order to objectively and reasonably given the overall positioning of the front door area.

I believe that the Qianmen Street's revival and remodeling can be positioned as follows: to protect and display the "Civil Culture" as the core, focused on "Beijing old" culture. traditional old, characteristic industry, folk customs as such Qianmen Street both pay their last respects, tourism, recreation, education, dining, shopping and other functions, and ultimately become a commercial, leisure, entertainment, experience is a comprehensive history of one of the commercial cultural areas.

(2) Enhance the cultural connotation of Qianmen Street, making it a tourist destination for the through connection with "Forbidden City, Tiananmen Square and the front door."

Qianmen Street retains the full traditional pattern of streets, many streets, alleys still follow the traditional name of the Ming and Qing Dynasties, become a traditional memory of the city. Courtyards there are an unique urban fabric of traditional Beijing city, are very rich in geographical characteristics.

Qianmen Street as the geographic coordinates of the capital, occupys an excellent central location. Its wide streets scale reinforces the axis of the charm of the ancient city.

Focus on refining the cultural connotation Qianmen Street, depth excavation its special historical position, take advantage of its rich cultural heritage and other physical resources, and actively guide, training people's attention on Beijing folk customs and traditional style. Thereby forming a travel chain throughout "Forbidden City, Tiananmen Square and Qianmen Street", reproduce the old spectacular and reputation of Qianmen Street.

(3) Based on tradition, reasonably and effectively development of Qianmen Street intangible cultural heritage.

Qianmen Street once the "Civil Culture" and "marketplace atmosphere", with rich human touch. This human touch and rich meaning of life that makes Qianmen Street reflect a spirit of place, so that the street is full of multi-color, multi-exotic atmosphere of public life, but also makes people produce their sense of identity and sense of belonging.

Remodeling Qianmen Street, the main way to make it re-energize is to "the continuation of the old city social life, heritage spirit of place." to form "Beijing school culture" typical and representative, it is necessary to organize the collection of information by mining its form, manner and other activities, and then re-creation, then, demonstrated.

Development of experiential tourism. By arrangement visitors to participate in commercial or production activities or the development of private goods and other hotel visitors or residents to exchange with the owner, to understand old Beijing customs, traditions and crafts. This makes many of the traditional crafts, folk customs are preserved not only in the exhibition, handed, carry forward, inheritance, but also enhance tourism in the process of entertainment, fun. The effect of making the same strain theme, bring tourists "an unique" cultural tourism experience.

Acknowledgments

This is where one acknowledges funding bodies etc. Note that section numbers are not required for Acknowledgments, Appendix and References.

References

1. http://www.bj.xinhuanet.com/bjpd sdwII1 /2009—10/15/content—17956624.htm.
2. Jia Miao, Zhang Bin. Qianmen Area Historical Origin[J]. Architectural Creation, 2007(12):56-60.
3. Zhou Nan. Qianmen Street: commercial renaissance Rye[J]. Chinese market, 2009(43): 16-17.
4. Zhou Chun-yan. Old Beijing, New Beijing --Urban Consumption nature of the business landmark refracted[J]. Sales and Marketing Manager, 2013(08): 32-39.
5. Selection paved rebirth road[J]. Investment Beijing, 2008(01): 71-74.
6. http://news.xinhuanet.c0m/shuhua/2013-10 /15/c 125540405.htm.

Simulation of a discrete pontoon floating bridge induced by moving loads

Jun Zhang, Chuan Liang, Ming-Kang Gou, Lin Li, Ying Zhang
Wuxi First Scientific Research Institute,
Wuxi, 214035, China
E-mail: zj_163mail@163.com

The discrete pontoon floating bridge is assumed as a simply supported beam with several SDOF mass-spring systems of pontoons to study the dynamic responses of the bridge subjected by moving loads. The equation of motion in matrix form for the discrete pontoon bridge subjected by moving loads is formulated with Hamiltons's principle and the assumed mode method. Simulink models are introduced to solve the above equation. The results shows that it is easy to take the present approach in simulation of behaviors of a discrete pontoon floating bridge under both one moving load and multi-moving loads, which provides an effective technical means for people to solve the design and calculation of this type of bridge.

Keywords: Discrete Pontoon Floating Bridge; Moving Load; Dynamic Response.

1. Introduction

A discrete pontoon floating bridge is one type of floating bridge on a series of constraints of floating pontoons shown in Fig. 1. When vehicles traverse this type of floating bridge, not only vehicles and the bridge beam interact with each other, but also pontoons, water and the bridge beam work together. Behaviors of this type of floating bridge are complex.

For many years, studies on such structural systems have been of primary interest to many engineers, designers, and researchers. Li Guohao [1] (1992) and Ladislav Fryba [2] (1972) studied dynamic responses of a simply supported beam bridge induced by moving loads with analytic method. Lee [3] (1995) studied dynamic response of a multi-span beam on one side point constraints subjected to a moving load. Seif and Inoue [4] (1998) studied the behaviors with time-domain method. Zhang et. al.[5] (2008) studied this type bridge with analytic models.

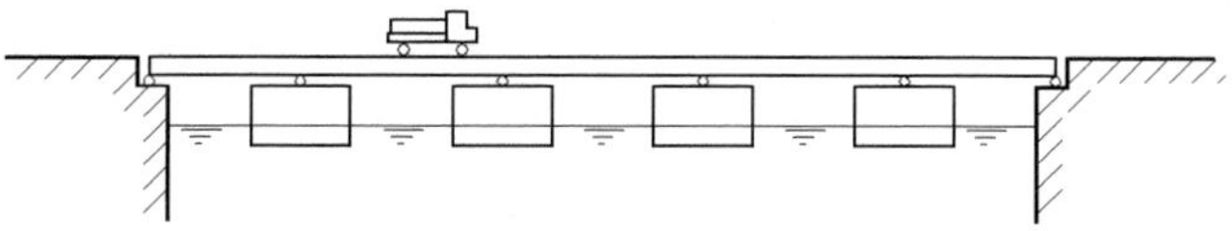

Fig. 1. The discrete pontoon floating bridge.

In the present study, the bridge beam can be looked on as a simply supported beam on a series of constraints as mass-spring systems based on the assumption of rigid pontoons subjected by fluid loads composed of effects of added mass, damping coefficient and water buoyancy. The equation of motion in matrix form for the discrete pontoon bridge subjected by moving loads is formulated with Hamiltons's principle and the assumed mode method. Simulink models are introduced to solve the above equation.

2. Formulation of the Model

There is a discrete floating bridge of length L with N rectangular pontoons. A load of F_v moves on the bridge at constant speed of v from left to right (shown in Fig. 2). The assumptions made in the following formulation are that transverse deflections are small so that the dynamic behavior of the bridge is governed by the Euler beam theory.

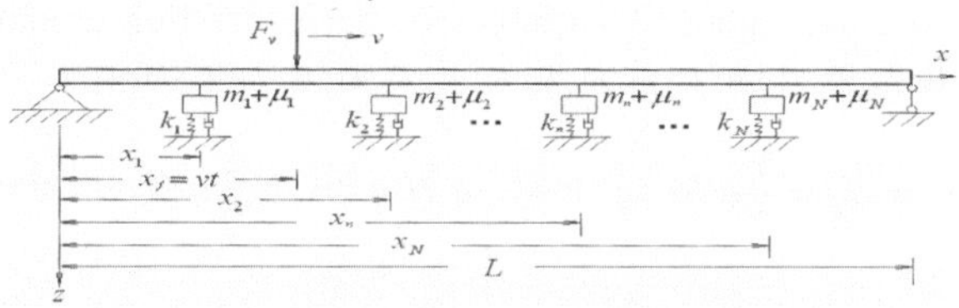

Fig. 2. Definition sketch of a floating bridge.

The deflection of the bridge under the moving load is u, which enables energy due to the kinetic energy, the strain energy, the potential energy due to the pontoon supports, and the work done by the moving load F_v to be expressed as:

$$T = \frac{1}{2}\int_0^L m_0(x)\left[\frac{\partial u(x,t)}{\partial t}\right]^2 dx \tag{1}$$

$$T_n = \frac{1}{2}(m_n + \mu_n)\left[\frac{\partial u(x_n,t)}{\partial t}\right]^2 \tag{2}$$

$$V = \frac{1}{2}\int_0^L EI(x)\left[\frac{\partial^2 u(x,t)}{\partial x^2}\right]^2 dx \tag{3}$$

$$V_n = \frac{1}{2}k_n u^2(x_n,t) \tag{4}$$

$$W = F_v u(x_f(t),t) \tag{5}$$

where x_f is the location of the moving load in *x*-direction, x_n ($n = 1,2,\cdots,N$) are the locations of pontoons, m_0 is the mass of the bridge beam per unit length, m_n and μ_n are natural mass and added mass of the n-th pontoon, EI is the flexural rigidity of the cross section of the bridge beam, k_n is the characteristic coefficient of buoyancy spring for the n-th pontoon.

Using the assumed mode method , the quantity $u(x,t)$ can be expressed as

$$u(x,t) = \sum_{j=1}^{J} q_j(t)\psi_j(x) \tag{6}$$

where $\{\psi_j(x), j = 1,2,\cdots,J\}$ are spatial function that satisfy the boundary conditions at the two ends of the bridge. In the present study, $\psi_j(x)$ are assumed as the modal function of a uniform beam simply supported at both ends. $\{q_j(t), j = 1,2,\cdots,J\}$ are general variables of corresponding assumed modes of the bridge. Equations (1) to (5) can be expressed in matrix form:

$$T = \frac{1}{2}\dot{\boldsymbol{q}}\,\boldsymbol{M}\,\dot{\boldsymbol{q}}\,, T_n = \frac{1}{2}\dot{\boldsymbol{q}}\,\boldsymbol{M}_n\,\dot{\boldsymbol{q}}\,, V = \frac{1}{2}\boldsymbol{q}\,\boldsymbol{H}\,\boldsymbol{q}\,, V_n = \frac{1}{2}\boldsymbol{q}\,\boldsymbol{\Theta}_n\,\boldsymbol{q}\,, W = F_v\,\boldsymbol{\psi}^T\boldsymbol{q} \tag{7}$$

where, $\mathbf{q} = \{q_1, q_2, \cdots q_J\}^T$, $\boldsymbol{\psi}$ $\quad {}_1(x_f),\ {}_2(x_f),\ ,\ {}_J(x_f)\ ^T$,

$(\boldsymbol{M})_{ij}$ $\quad {}_0^L m_0(x)\ {}_i(x)\ {}_j(x)dx$, $(\boldsymbol{M}_n)_{ij}$ $\quad (m_n \quad {}_n)\ {}_i(x_n)\ {}_j(x_n)$,

$(\boldsymbol{H})_{ij} = \int_0^L EI(x)\psi''_i(x)\psi''_j(x)dx\,, (\boldsymbol{\Theta}_n)_{ij} = k_n\psi_i(x_n)\psi_j(x_n)$.

The Lagrangian of the bridge can be written as

$$L = T + \sum_{i=1}^{N} T_i - V - \sum_{i=1}^{N} V_i + W \tag{8}$$

The Euler-Lagrange equation is

$$\frac{d}{dt}\left(\frac{\partial L}{\partial \dot{\boldsymbol{q}}}\right)-\frac{\partial L}{\partial \boldsymbol{q}}=0 \tag{9}$$

Substituting Eq.(7)into Eq.(8)and Eq.(9), the resulting equation is:

$$(\boldsymbol{M}+\sum_{n=1}^{N}\boldsymbol{M}_n)\ddot{\boldsymbol{q}}=F_v\psi-\boldsymbol{H}\boldsymbol{q}-\sum_{n=1}^{N}\boldsymbol{\Theta}_n\boldsymbol{q} \tag{10}$$

3. Simulink Model of Solution

Eq. (10) can numerically be solved by Simulink model in Matlab software in the present study. Simulink has become the most widely used software package in academia and industry for modeling and simulating dynamic systems [6]. For example, Xiao and Shen [7] (2005) simulated a simply supported bridge under a moving mass, Zhang et al [8] (2010) simulated the behavior of a ribbon pontoon bridge. Eq. (10) can be regarded as the simulink model with mathematical formulation of the system. The main model with $J=5$ is shown in Fig.3. M-file of "DFBridge.m" is a function program to solve the quantity of $\ddot{\boldsymbol{q}}$ in Eq.(10) with inputs of $\boldsymbol{q}$ and $F_v\psi$.

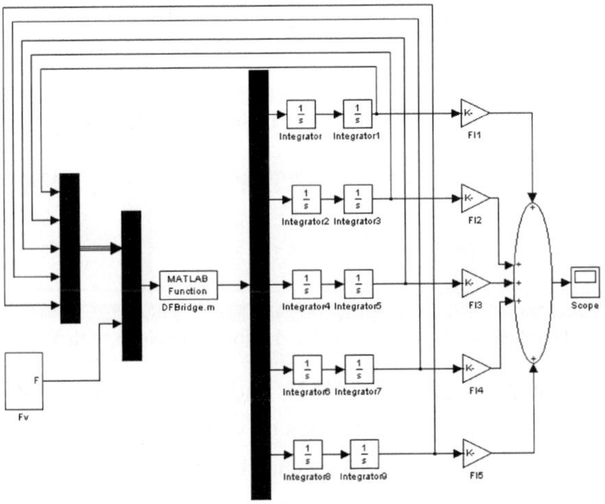

Fig. 3. Model of simulink.

4. Examples

The discrete-pontoon floating bridge (shown in Fig. 4) consists of three cuboid pontoons placed 50m apart: the total bridge span is 200m. The line density

m_0 of the bridge beam is 1725kg/m, the Young's modulus E is 206GPa, and the principal second moment of area of the cross section is 0.8m^4. All pontoon are similar with mass of 10000kg, length of 10m, breadth of 5m and draft of 2.5m. The added mass can be set to 1.5 times the value of natural mass of the relative pontoon.

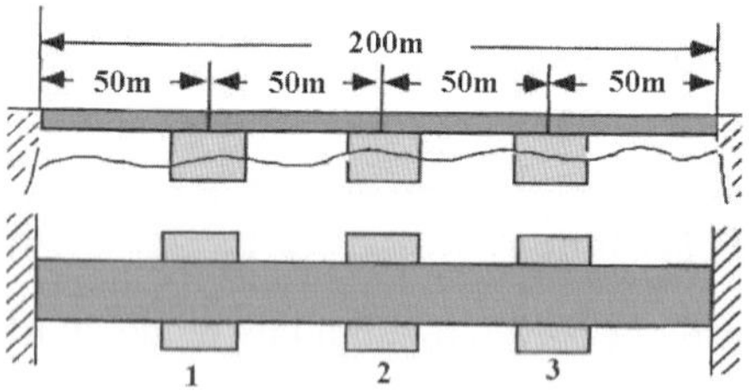

Fig. 4. Model of simulink.

When a moving load of 200kN travels the length of the bridge with speed of 5m/s and 10m/s, the responses of vertical motions of all the pontoons are presented in Fig. 5 and Fig. 6, in which the value of abscissa indicates the time(s), and the value of ordinate indicates the displacement response (m).

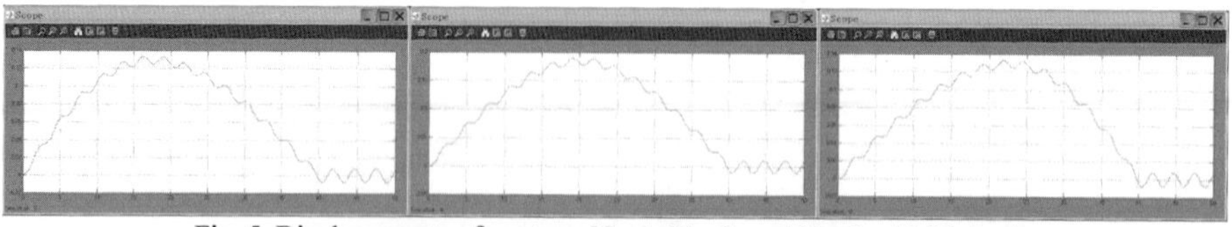

Fig. 5. Displacements of pontoon No. 1, No. 2, and No. 3 with V=5m/s.

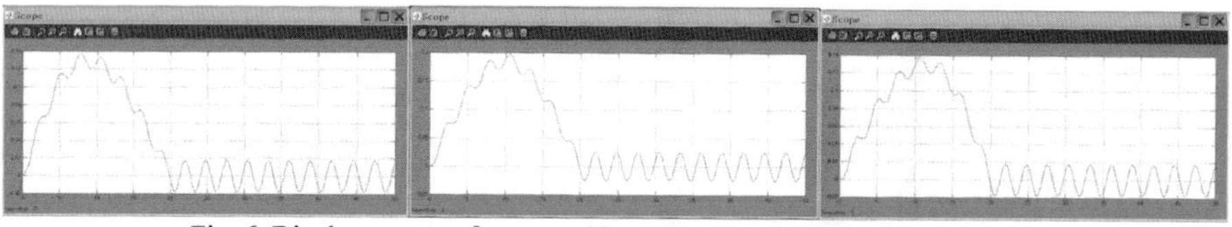

Fig. 6. Displacements of pontoon No. 1, No. 2 and No. 3 with V=10m/s.

When two moving loads of 200kN and 300kN travel the length of the bridge with speed of 10m/s, the responses of vertical motions of all the pontoons are presented in Fig. 7 and Fig. 8 in two case of different spatial intervals.

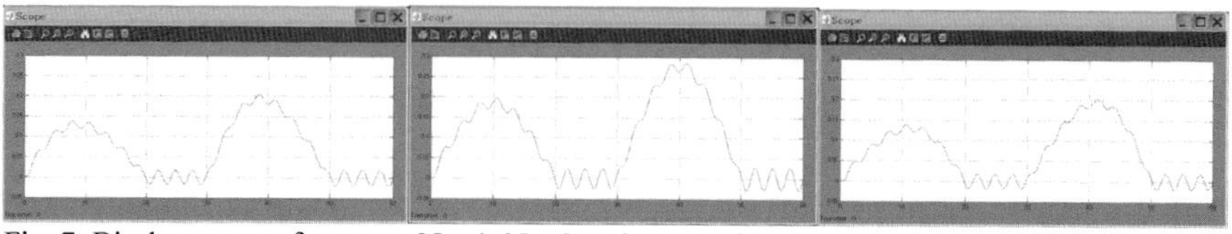

Fig. 7. Displacements of pontoon No. 1, No. 2 and No. 3 with V=10m/s under two loads in case of 300m interval.

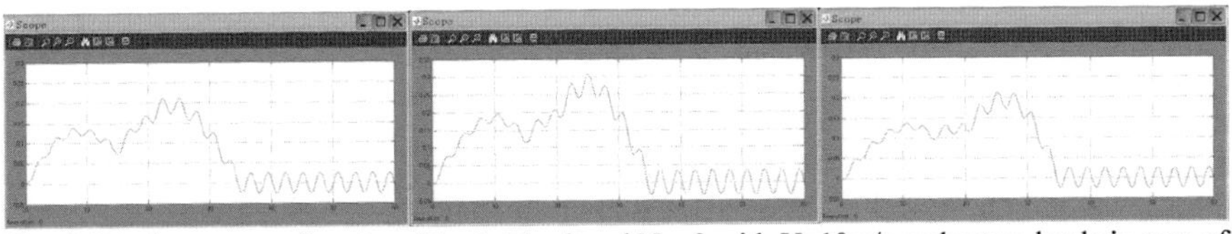

Fig. 8. Displacements of pontoon No. 1, No. 2 and No. 3 with V=10m/s under two loads in case of 150m interval.

5. Conclusion

The problems of dynamic responses of a discrete pontoon floating bridge subjected by moving loads are studied with Simulink models. Results indicate that it is easy to take the present approach in simulation of behaviors of a discrete pontoon floating bridge under both one moving load and multi-moving loads.

References

1. Li Guo hao. *Stability and Vibration of Bridge Structures*. Beijing: China Railway Publishing House, (1992).
2. Ladislav Fryba. *Vibration of solids and structures under moving loads*. Noordhoff International, Groningen, (1972)
3. Lee H P. Dynamic response of a multi-span beam on one-side point constraints subject to a moving load. *Computers & Structures*, Vol. 55(4), (1995), pp: 615~623.
4. Seif M S,Inoue Y.Dynamic analysis of floating bridges. *Marine Structures*, Vol.11, (1998), pp:29~46.
5. Zhang Jun, Miao Guo-ping, Liu Jian-xun, Sun wen-jun. Analytical models of floating bridges subjected by moving loads for different water depth. *Journal of Hydrodynamics, Ser. B*, Vol. 20(5), (2008), pp: 537-546.
6. John F. Carnder. *Simulation of Machines Using MATLAB and Simulink*. Wadsworth Group, 2001.
7. Xiao Xinbao, Shen Huoming. Dynamic simulation of bridge subjected to moving load. *Journal of Vibration and Shock*, Vol. 24(1), (2005), pp: 121-123 (in Chinese).
8. Jun Zhang, Guoping Miao, Ziwen Zhou, et al. Simulink based simulation of the behavior of a ribbon pontoon bridge. *Journal of Marine Science*, Vol. 9, (2010), pp: 328-333.

Study on irregular spherical latticed roof of silo

Xian-Cheng Wang[1, a], Meng-Yang Zhang[2], Fang Yuan[2], Peng Su[2], Xing-Pei Liang[2]
[1]*Henan University of Urban Construction,*
Ping Ding Shan, 467000,He Nan, China
[2]*School of Civil Engineering and Architecture, Henan University of Technology,*
Zhengzhou, 450001,Henan,China
E-mail: [a]wangxc1964@126.com, 13838093086@163.com

The paper which is based on combination of polyhedron theory studys the fundamental laws and the configuration characteristics of irregular triangular spherical shell. Then it realized the design and visual modeling of irregular spherical latticed silo roof on the computer. The silo roof is a kind of space assembly structure. Its construction is convenient so that it has short construction period and low cost. It also has attractive appearance. Above all it has broad application prospects.

Keywords: Irregular Net Shell; Spherical; Silo Roof; Polyhedral; Primitive.

1. Introduction

With the increase of silo in its diameter and height, the silo roof will face enormous difficulties during the construction. For example, the concrete silo roof need to set up full steel tubular scaffold so it has high construction cost and long construction period. Irregular spherical latticed silo roof can solve these problems. It has convenient assembly. It just needs to manufacture several bars with different lengths in the factory, then assemble and hoist them on site. Therefore, tubular scaffold can be omitted and the construction cost will be reduced.

After analyzing the current situation and development of irregular spherical latticed silo roof, SU [1] pointed out the following content. Irregular latticed shell can be beautiful architectural image and compose large space with small and single bar, and they have rational stress. The structure which can be long span also has the characteristics of beam structure and thin-shell structure. When irregular latticed shell applied in the configurations of silo roof, it not only saves materials, shortens the construction period, also can be recycled again. It has very important significance for the development of the silo structure in our country.

The paper also studies the configuration principle and method of irregular spherical latticed silo roof. It also sets up a model, and then gets the spherical silo roof. All above provide new thought and method for the design and construction of silo roof.

2. The Configuration Principle of Irregular Spherical Shell

According to the combination of polyhedron theory [2], the arbitrary polyhedron shell has 12 pentagons which distribute equally on the sphere. The lines made up by the centroid of three adjacent pentagons constitute the primitive. The number of triangles in the primitive determines the surfaces of the polyhedron shell.

As shown in Figure 1(a), 12 pentagons distribute regularly on irregular spherical surface. Between the centroid of the adjacent pentagon and the center of sphere is invariable, as in Figure 1(d), $\angle aob = \angle aoc = \angle boc$. Here it stipulates that the distance between the centroids is *L*, the radius of the corresponding sphere is *R*, as in Figure 1(a), *AB*=*L*, *OA*=*R* and *D* is the middle point of *GH*. It is necessary to get the circular *OAB* and *OGH* in order to make the relationship more clearly, as shown in Figure 1(b) and (c).

$$OD = OE + DE = \frac{L}{2} + \sqrt{BD^2 - BE^2} = \frac{L}{2} + \sqrt{L^2 - R^2} \tag{1}$$

$$OD = \sqrt{R^2 - \frac{L^2}{4}} \tag{2}$$

$$L = \frac{\sqrt{10\left(5 - \sqrt{5}\right)}}{5} R \tag{3}$$

$$\angle aob = \angle aoc = \angle boc = \arccos\frac{\sqrt{5}}{5} \tag{4}$$

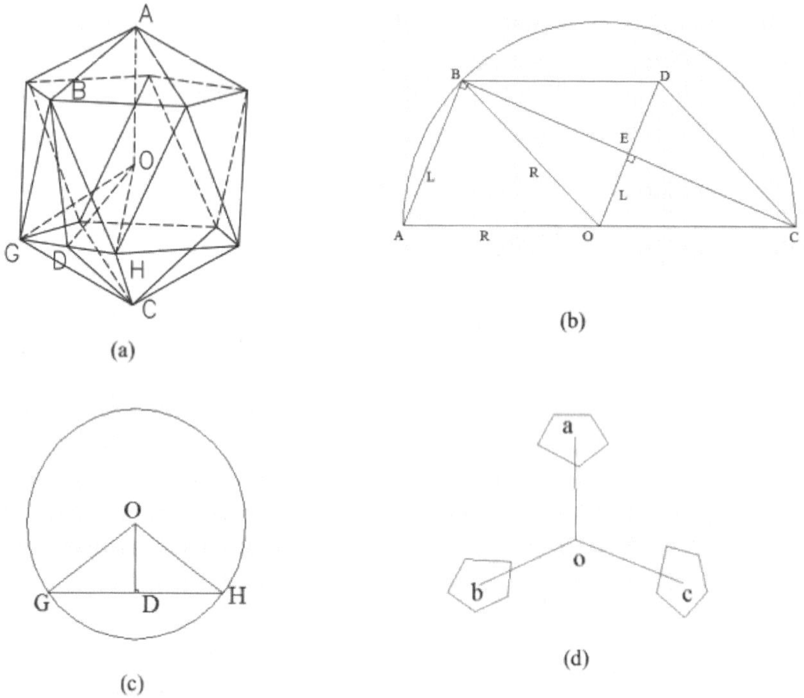

Fig. 1. The figure of pentagon internal angle and length.

The position of the primitive center must be clear in primitive topology relations and the calculation of node position. Its coordinate can be achieved through the geometric mathematical method. In the process of drawing primitives, it mainly adopts column coordinate system and the local coordinate system to make full use of the geometric features such as symmetry and antisymmetry. After building primitive, the spherical shell model can be built through adopting the two coordinate systems to achieve rotation, mirroring, replication.

3. The Construction Method of Primitive

If the number of triangles in the primitive can be known in advance, the primitive will have many different types of triangles. The length of edge in primitive is determined by the shape of the triangle. As shown in Figure 2, the two primitives belong to the polyhedron with 240 surfaces. They all contain 9 triangles so they have the same topological relation. In Figure 2(a), the types of triangle length are less so the primitive has more advantages. When the triangular polyhedron shell has more than 180 surfaces, it will get reasonable layout of point, line, and edge in the primitive if the edge bc, de, fa are halved. Therefore, the edge bc, de, fa need to be divided equally.

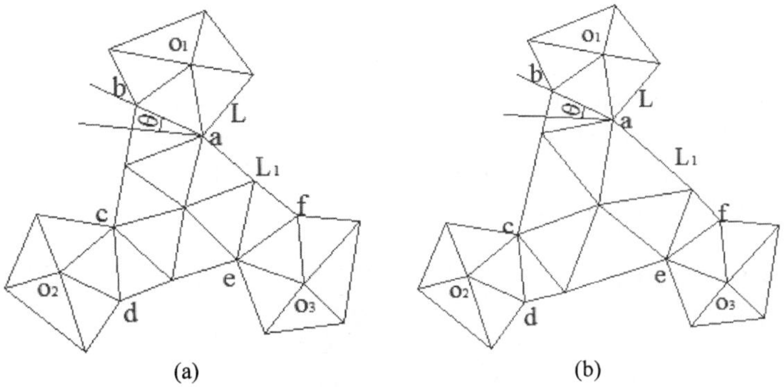

Fig. 2. Primitive schematic diagram.

In the picture 2, θ is the angle that the pentagon goes around its centroid and it is relation to the shape of internal triangle. So the length of pentagon edge is in connection with θ. As to the irregular triangular shell, the length L

$$L = f(\alpha, \theta, M) \tag{5}$$

In the formula, R is the spherical radius, α is the angle which is formed by the two radiuses that the centroid of the adjacent pentagon and center of a sphere lie in. θ is the angle that the pentagon go around its centroid. M is the specific value between the pentagonal edge length L and the triangular edge length L_1 connecting two pentagons.

Through the topological relationship of structure and calculation [3, 4], the length of triangle and the pentagon in the primitive can be determined.

As the picture 2 shown, the primitive is hexagon abcdef and the length of its edges can be known through all the above ways. It is important to get the nodes orientation in the hexagon except to determine the orientation of the nodes on the edge bc, de, fa. Concrete ways are as follows: The first step is topology planning and it need to confirm the quantity and the rough position of the nodes. The second step is to determine the specific location of nodes and it can be realized by two methods. One is calculation, it get the coordinate of a certain node by combining the coordinate of four spheres. Another is drawing by the computer to obtain the wanted node through confirming the known three points and the relationship between the three points.

4. The Construction of Spherical Shell and the Design of the Silo Roof with Truncation

After determining the length of pentagon and the nodes in primitive, the main work on manufacturing primitive is almost completed. The paper takes the function of the program ANSYS to get the specific program about establishing irregular triangular shell and the model of the silo roof with truncation. The fig. 3 and fig. 4 can be designed according to the construction principle of the model. The fig. 3 is about primitives, integral model of the polyhedron with 540 surfaces and the fig. 4 is about the spherical silo roof with truncation.

Fig. 3. Primitives and overall configuration diagram of the polyhedron with 540 surfaces.

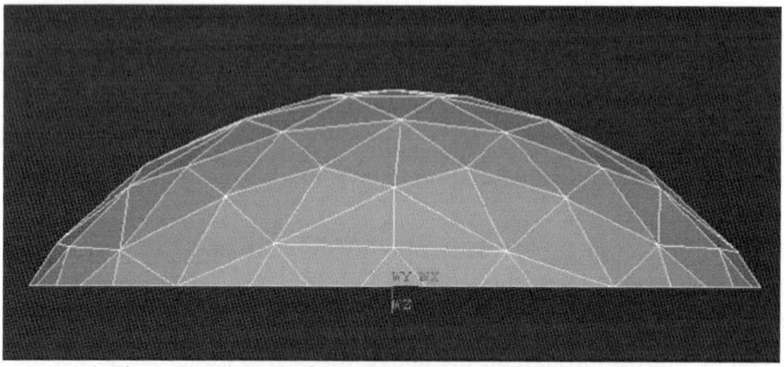

Fig. 4. The diagram of the spherical silo roof with truncation .

5. Conclusion

Irregular spherical latticed silo roof is easy to install, and it has beautiful appearance and short construction period. It also overcomes many shortcomings of the traditional silo roof and has wide application prospect.

Acknowledgments

This work is supported by key scientific and technological project of Henan Provenience (132102210450).

Work partially supported by grant (51578216) of the National Science Foundation of China.

References

1. C. Shu. Geometric Configuration Research of Imitating Nanometer Combinational Polyhedrons Net Shell. Luo Yang: Henan University of Science and Technology, 2007.
2. H. Li. The Studies of Static Behaviors of Irregular Lattice Shell. Zheng Zhou: Henan University of Technology, 2011.
3. P. Su. Irregular Triangular Polyhedron Space Frame Roof Structure Type Positions and Mechanical Analysis. Zheng Zhou: Henan University of Technology, 2014.
4. P. Su. Research and Development of Irregular Latticed Warehouse Roof. Zheng Zhou Construction, 2014(9):79-79.

A novel method for free response signal with measured acceleration

Wei-Zhen Chen, Yong-Qing Qian, Tao Hu
School of Electrical & Electronic Engineering, Wuhan Polytechnic University, Wuhan 430023, China
E-mail: qyqwhu@gmail.com

In this paper, the novel method for free response signal of large-scale system is proposed. On the basis of random decrement method on the basic idea, two key parameters are discussed. One is the length of the system's output signal N, the other is the reference value of the amplitude A. in order to obtain more precise free response signal of system, using the result of eigenvalues analysis to modify the primary free response signal. Processing the measured acceleration from real system, this novel method can get approximate free response signal.

Keywords: Free Response Signal; Random Decrement Method; ARMA Method.

1. Introduction

With the development of the times, the problem of large-scale structure's health condition is paid more and more attention by many engineers and researchers. As we all known, system's free response signal can imply system's characteristics. That means, if free response signal of system is gotten, the modal parameters of system are identified. Problem is change [1-4].

Currently, many methods are used in this field. Such as frequency domain method, time domain method, time-frequency domain method, etc. Random decrement method is a vibration analysis method developed in the 1970s. It can not depend on the excitation signal, only the output signal can get free response signal of system approximately. The basic idea of random decrement law raised in free decay signal of the structure vibration signal from the random vibration response of the structure. In many time-domain identification methods are in need of free response signal structure. These signals can be used to obtain transient excitation method can also be extracted from the environment excitation signal. If the environment is seen as random excitation, then the method of extracting free response signal from the random response signal is called the random decrement method. With the increase demand for real-time online, time domain method has been widely researched and applied, including random decrement method, Ibrahim time domain, least square complex index method, ARMA time series analysis and eigenvalues realization algorithm, etc.

In this paper, the novel method for free response signal of large-scale system is proposed. On the basis of random decrement method on the basic idea, two key parameters are discussed. One is the length of the system's output signal N, the other is the reference value of the amplitude A. in order to obtain more precise free response signal of system, using the result of eigenvalues analysis to modify the primary free response signal. Processing the measured acceleration from real system, this novel method can get approximate free response signal.

2. Random Decrement Method

For the structure of a linear system, at any excitation of forced vibration response of a measuring point signal can be expressed as

$$y(t) = y(0)D(t) + \dot{y}(0)V(t) + \int_0^t h(t-\tau)f(\tau)d\tau \tag{1}$$

Here, $D(t)$ is the free response signal of the system, which the initial displacement is 1 and the initial velocity is 0. $V(t)$ is the free response signal, which the initial displacement is 0 and the initial velocity is 1. $y(0)$ is the vibration initial displacement and $\dot{y}(0)$ is the initial velocity. $h(t)$ is the system unit impulse response function, $f(t)$ is excitation input of system.

Select an appropriate constant A to the interception of a measured random vibration signal $y(t)$, the intersection of a series of different available time $t_i(i=1,2,\cdots,N)$, the response time t_i from the beginning can be seen as a linear superposition of three parts, the free response by the time the initial displacement, the free response by the initial velocity and the force response by the start of random excitation. which i

$$y(t-t_i) = y(t_i)D(t-t_i) + \dot{y}(t_i)V(t-t_i) + \int_{t_i}^t h(t-\tau)f(\tau)d\tau \tag{2}$$

Since the excitation $f(t)$ is stationary, the beginning of time does not affect its random nature, then moved $y(t-t_i)$ to coordinate a series of time starting origin, obtained a series of sub-sampling function corresponding stochastic process $x_i(t)(i=1,2,\cdots,N)$, namely

$$x_i(t) = AD(t) + \dot{y}(t_i)V(t) + \int_0^t h(t-\tau)f(\tau)d\tau \tag{3}$$

Statistics average of $x_i(t)(i=1,2,\cdots,N)$

$$x(t)=\frac{1}{N}\sum_{i=1}^{N}x_i(t)$$
$$\approx E[AD(t)+\dot{y}(t_i)V(t)+\int_0^t h(t-\tau)f(\tau)d\tau]$$
$$\approx AD(t)+E[\dot{y}(t_i)]V(t)+\int_0^t h(t-\tau)E[f(\tau)]d\tau \quad (4)$$

Since excitation $f(t)$ is pure stationary random vibration and means is 0, and the system response $y(0)$ and $\dot{y}(0)$ both stationary random vibration, which means is 0. That means $E[\dot{y}(t_i)]=0$, $E[f(\tau)]=0$. Then, $x(t)\approx AD(t)$.

Thereby obtaining an initial displacement A and the free response signal with zero of initial velocity.

3. A Novel Method for Free Response Signal

From the above mentioned, the core of random decrement method is to segment collected signal at first, then using superposition to erase uncertain factors and strengthen determine factors. But there are two key elements in the implementation process, one is how choose the appropriate segment length, the other is about selecting the amplitude A.

According to the basic idea of random decrement method, segmented according to the length of the decay time of the impulse response of the system may be made the results more accurate. In engineering, we tend to think $3\sim5$ is the end of the transition process. This indicator with inertia about the size of a particular system.

A single degree of freedom vibration differential equation as an example:

$m\ddot{x}+c\dot{x}+kx=f(t)$ Then, $\lambda_{1,2}=-\zeta\omega_0\pm j\omega_0\sqrt{1-\zeta^2}$, $x=Ae^{-\sigma t}\sin(\omega_d t+\varphi)$

Thus, $$\begin{cases} A=\sqrt{x_0^2+(\frac{\dot{x}_0+\sigma x_0}{\omega_d})^2} \\ \varphi=\arctan\frac{x_0\omega_d}{\dot{x}_0+\sigma x_0} \end{cases}$$

Here's the speed with related $\tau=\frac{1}{\sigma}=\frac{1}{\zeta\omega_0}=\frac{2m}{c}$ attenuation. Therefore, according to the specific object of study, according to this criterion to select the length of time segment, which is the length of the sub-sample signals. If consider the sampling frequency f_s, length of the sub-sample signal selectable N $\frac{3}{_0}$ $\frac{5}{_0}$, wherein the damping ratio ζ and natural frequency ω_0 can reference design value.

For selected amplitude A, according to the free vibration response to the resulting

$$\delta_j(\tau)=\frac{1}{N}\sum_{i}^{N}x_j(t_i+\tau),(j=1,2,\cdots,N) \tag{5}$$

when $\tau=0$, $\delta_j(0)=\frac{1}{N}\sum_{i}^{N}x_j(t_i)=A,(j=1,2,\cdots,N)$, then $A=\frac{1}{N}\sum_{i}^{N}x(t_i)$.

In fact, using this method cannot get perfect free response signal. If using ARMA with analysis the eigenvalues results to modify the signal, better free response signal would be gotten [5].

4. Measured Acceleration Data Analysis

One of observation overpass in Fig. 1. Acceleration vibration signal can be collected by sampling points [6]. After signal processing by removing zero drift, vibration signal can be gotten in Fig. 2.

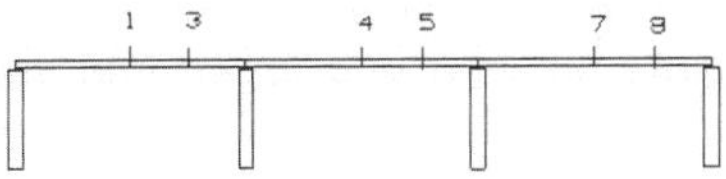

Fig. 1. Observation points distribution map.

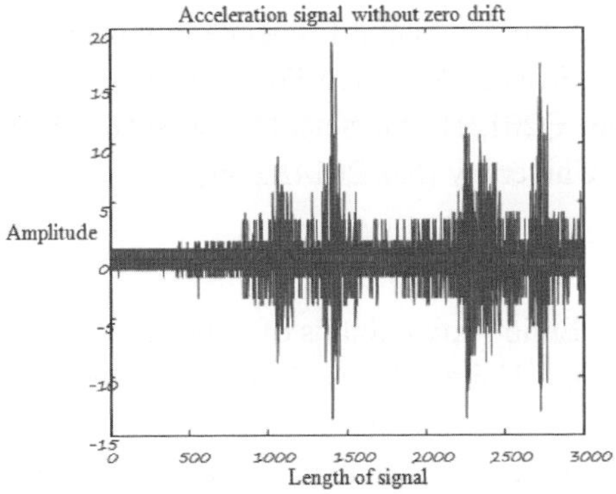

Fig. 2. Acceleration signal without zero drift.

Using random decrement method to process the signal, free response signal can be gotten. From the result in Fig. 3, which is on the left. After using the novel method mentioned before, the result is better, which is on the right in Fig. 3.

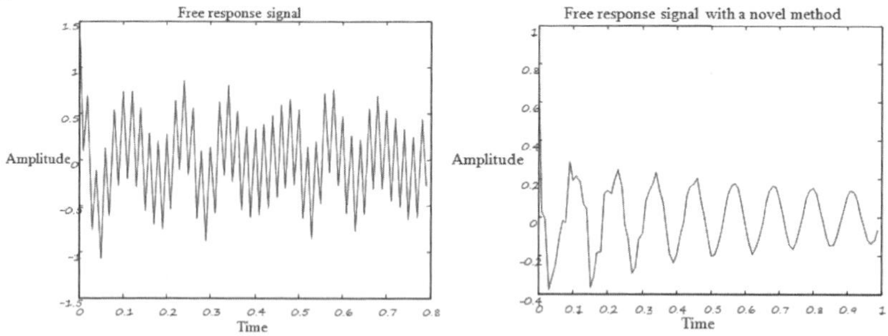

Fig. 3. Free response signal with different method.

Processing the measured acceleration from real system, this novel method can get approximate free response signal.

5. Summary

The free response signal is one of important factor in identification of large-scale structure health condition. In this paper, using analysis the eigenvalues results by ARMA method, based on random decrement method to form a novel method. From the result of processing the acceleration signal, which obtained from real system. The better free response signal with this novel method can be gotten. It will help us to solve the problems in this field.

Acknowledgments

This work supported by the natural science foundation of Hubei Province (No. 2014CFC1131), Science Research Foundation of Education Department of Hubei Province (No. Q20121811). Scientific Research Starting Foundation of Wuhan Polytechnic University (No. 2014RZ39).

References

1. Darryll Pines, A Emin Aktan. Status of structural health monitoring of long-span bridges in the United States. Proc.Struct.EngngMater. 4, 2002, 372-380.
2. Peeters B, Maeck J, De Roeck G. Stochastic system identification for operational modal analysis: A review. Journal of Dynamic Systems, Measurement, and Control 2001; 123(4): 659–667.

3. Andrew Smyth, Mei-Liang WU, Multi-rate Kalman filtering for the data fusion of displacement and acceleration response measurements in dynamic system monitoring. Mechanical Systems and Signal Processing, 2007, 21, pp. 706-723.
4. Wei-Zhen Chen, Bing-Wen Wang, Xiao-Aa Hu. Vibration-based structural damage identification using grey forecasting model. Applied Mechanics and Materials, 2012, 128-129:1114-1117.
5. Wei-Zhen Chen. A novel method to estimate ARMA time-varying parameter using grey forecasting model. Advanced Materials Research, 2012, 532-533, pp. 876-880.
6. Wei-Zhen Chen, Jing-Lu Hu, Long Zhou. Fast integral algorithm for bridge displacement using measured acceleration by wireless sensors. WIT Transaction on Engineering Sciences. 2014, 87, pp. 477-487.

The method to absorb vibration of uniform mass beam in the action of simple harmonic loads

Y. J. Chen
Department of Civil Engineering, University of Foshan,
18 Jiang Wan Road, Foshan 528000, Guangdong, China
Email: chenyj1310@163.com
www.fosu.edu.cn

One of the main work in structure design is to absorb vibration of the structure. This paper put up a mass-spring system linked at the midspan of the beam in the action of simple harmonic loads. It is shown by the paper's study that when letting its stiffness k and concentacted mass m2 meet the needs of the condition k=m2θ2, the vibration would not be took place in the beam. The propose ideas could get reference for the engineering structure design.

Keywords: Method; Vibration; Uniform Mass Beam; Simple Harmonic Load.

1. Introduction

Vibration is one of the common situation in engineering structure. The adverse effects to structures may be resulted when the structure is under the vibration, for instance, to damage the structure, to influence the machining accuracy, to make people uncomfortable and so on. Therefore to decrease the vibration and to eliminate the vibration are very important in engineering design. Some researchershave studied some methods. Their concept of structural vibration control is to absorb vibration energy of the structure by introducing auxiliary devices. Various types of structural vibration control theories and devices have been recently developed. Cacciola, P. et al.[1] dealed with the vibration control of existing structures forced by earthquake induced ground motion. It is proposed for the first time to exploit the structure-soil-structure mechanism to develop a device, hosted in the soil but detached from the structure, able to absorb part of the seismic energy so to reduce the vibration of neighbourhood structures. The design of the device is herein addressed to protect monopile structures from earthquake induced ground motion. Xie, X. D. et al. [2] developed an optimal design of a piezoelectric coupled cantilever structure attached by a proof mass subjected to harmonic motions to achieve efficient energy harvesting for applications in high-rise buildings. Energy harvesting is realized from the electromechanical coupling effect by the piezoelectric patch mounted on the cantilever. Zhang, X.C. et al. [3] proposed a new kind of combined lead extrusion magnetorheological (CLEMR) damper. A mathematical model employing the current, excitation amplitude, and frequency as inputs is proposed and verified by comparing experimental and numerical results. Chen, Y.Q. et al. [4] put a fluid viscous dampers (FVDs),which are highly effective in reducing the response of both the main structure and the secondary system. Both the structural response and economic analysis show that using FVD to absorb seismic energy not only satisfies the Chinese seismic design code for a "rare" earthquake, but is also the most economical way to improve seismic performance both for one-time direct investment and long term maintenance. Cheung, Y.L. et al.[5] studied one drawback of the traditional dynamic vibration absorber which can only absorb the vibration of a continuous structure such as a beam at the point to which it is attached. Jalili, N. eta. [6] studied the use of multiple delayed resonators (DR) in suppressing tonal vibration of multi-degree-of-freedom mechanical structures. The DR is a recently introduced, actively tunable vibration absorber which converts a conventional passive absorber into a marginally stable resonator. This structure absorbs all the vibratory energy at its point of attachment. Aida,T. [7] researched a new shell-type dynamic vibration absorber presented for suppressing several modes of vibration of the shallow shell (main shell) under harmonic load. It consists of a shallow shell (the dynamic absorbing shell), under the same boundary condition and with the same shape as those of the main shell, with connecting springs and dampers in the vertical direction between the main and dynamic absorbing shells. Ducarne, J. et al. [8] dealed with the reduction of structural vibrations by means of synchronized switch damping techniques on piezoelectric elements. Piezoelectric patches are attached to the vibrating structure and connected to an electrical circuit that includes a switch. The latter allows to continuously switch the piezoelectric elements from an open-circuit state to a specific electric impedance, synchronously with the mechanical oscillations. Liu, W.Q. et al. [9] designed the seismic reduction by using viscous dampers to dissipate the kinetic energy of the structure both longitudinally and transversely for Hangzhou Bay Bridge. Using the vibration reduction system and aiming at the reasonable optimal goal, the purpose to reduce the seismic responses in south and north navigable spans has been achieved. Keltie, R.F. et al. [10] has been developed to determine locations of masses required to reduce the vibration level at an arbitrary location on a structure based on optimization techniques. The method is demonstrated using a simply supported beam driven by a harmonic point force.

This paper studies a simple method to eliminate the vibration of structure by means of D.Alembert´s Principle.

2. Principle and Design of Absorb Vibration

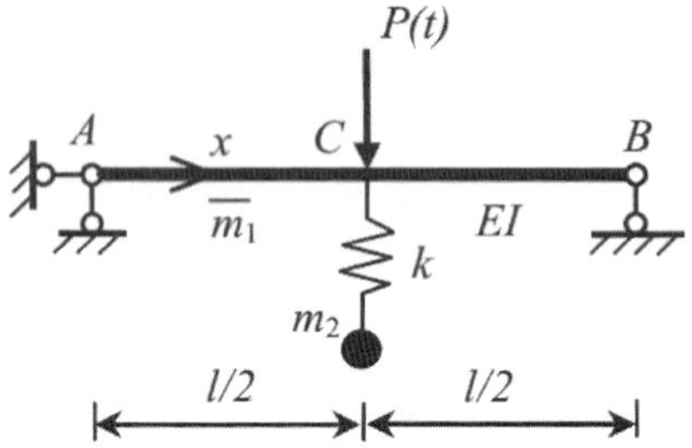

Fig. 1. Mode of simple supproted beam with mass-spring system.

Fig. 1 is a simple supproted beam, where $\overline{m}_1$ is the mass at unit length of the beam and is a constant, l is the span length, EI is the moment stiffness, A is the origin of coordinates x. In order to eliminate the vibration of the beam, a mass-spring system is linked at the midspan of the beam. We discuss below how to design the mass-spring system.

The dynamic displacement of concentrated mass m_2 is $y_2(t)$. The dynamic displacement of the simple supported beam is $y(x,t)=y_1(t)f(x)$, where $y_1(t)$ is the dynamic displacement of the midspan of the beam, $f(x)$ is the function of vibration mode. According to D.Alembert´s Principle, the static caculation method could be used to compute the structure by adding to the distribution inertia forces $I_1(x) = -\overline{m}_1 \dfrac{d^2 y_1}{dt^2} f(x)$ at the uniform mass beam and the concentrated inertia forces $I_2 = -m_2 \dfrac{d^2 y_2}{dt^2}$ at the mass point m_2. It could be got by the deformation coordination condition of all system and the equilibrium condition about the mass point m_2:

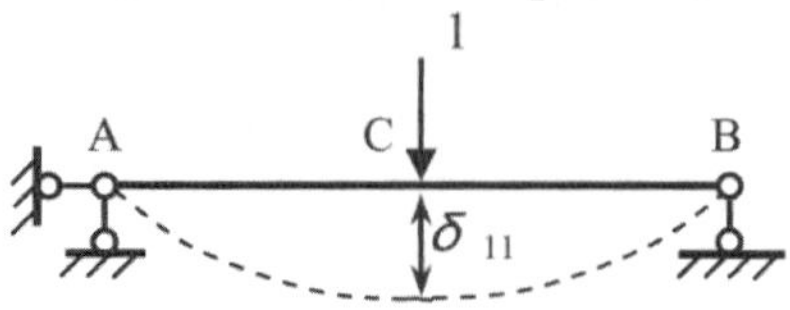

Fig. 2. Schematic diagram of δ_{11}.

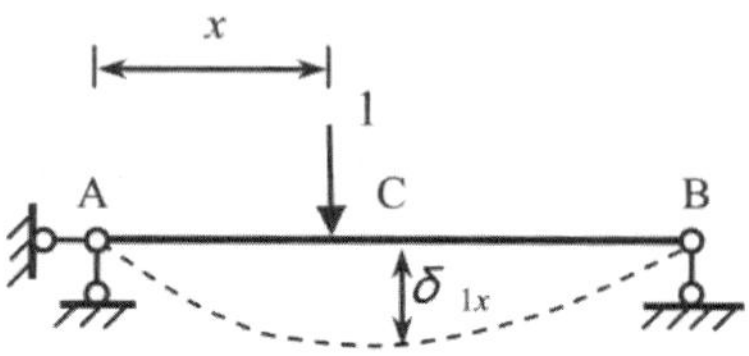

Fig. 3. Schematic diagram of δ_{1x}.

$$\begin{cases} y_1 = [P(t) + I_2]\delta_{11} + \int_0^l I_1 \delta_{1x} dx \\ I_2 + k(y_1 - y_2) = 0 \end{cases} \tag{1}$$

where δ_{1x} is the deflection at the point C produced by the unit vertical force acted at the point from the origin of coordinates for x, δ_{11} is the the flexibility coefficient at at the midspan of the beam.

According the deflection of the beam in the action of the unit load at the midspan of the beam, the deflection δ_{1x} could be assumpted as follow

$$\delta_{1x}(x) = \begin{cases} \dfrac{x}{4EI}\left(\dfrac{l^2}{4} - \dfrac{x^2}{3}\right) & 0 \le x \le \dfrac{l}{2} \\ \dfrac{(l-x)}{4EI}\left(\dfrac{l^2}{4} - \dfrac{(l-x)^2}{3}\right) & \dfrac{l}{2} < x \le l \end{cases} \tag{2}$$

Then

$$\int_0^l I_1 \delta_{1x} dx = -0.00252 \frac{l^4 \overline{m}_1}{EI} \frac{d^2 y_1}{dt^2} \tag{3}$$

The flexibility coefficient δ_{11} could be got by means of the diagram-multiplication method

$$\delta_{11} = \frac{l^4}{48EI} \tag{4}$$

It could obtained by taking the formula (3),(4) and inertia forces to the formula (1)

$$\begin{cases} y_1 + \dfrac{0.00252 l^4 \overline{m}_1}{EI} \dfrac{d^2 y_1}{dt^2} + \dfrac{l^4 m_2}{48EI} \dfrac{d^2 y_2}{dt^2} = \dfrac{l^4}{48EI} P(t) \\ m_2 \dfrac{d^2 y_2}{dt^2} - k y_1 + k y_2 = 0 \end{cases} \tag{5}$$

This is the motion differential equation of the structure system.

If the load is the simple harmonic load

$$P(t) = P \sin \theta t \tag{6}$$

where P is the maxium load, θ is the frequence of load, the special solution of equation (5) is

$$\begin{cases} y_1 = A_1 \sin \theta t \\ y_2 = A_2 \sin \theta t \end{cases} \tag{7}$$

where A_1 and A_2 are the amplitude of the middle of the beam and mass point m_2 respectively.

To bring the formula (6) and (7) to the equation (5) could give

$$\begin{cases} A_1 \left(1 - \frac{0.00252 l^4}{EI} \bar{m}_1 \theta^2 \right) - \frac{l^4}{48EI} m_2 \theta^2 A_2 = \frac{Pl^4}{48EI} \\ \left(k - m_2 \theta^2 \right) A_2 - k A_1 = 0 \end{cases} \tag{8}$$

The equation (8) is the algebraic equation group about A_1 and A_2. Its solution is

$$\begin{cases} A_1 = \dfrac{\frac{Pl^4}{48EI}\left(1 - \frac{m_2\theta^2}{k}\right)}{\left(1 - \frac{m_2\theta^2}{k}\right)\left(1 - \frac{0.00252 l^4}{EI}\bar{m}_1\theta^2\right) - \frac{l^4}{48EI} m_2 \theta^2} \\ A_2 = \dfrac{\frac{Pl^4}{48EI}}{\left(1 - \frac{m_2\theta^2}{k}\right)\left(1 - \frac{0.00252 l^4}{EI}\bar{m}_1\theta^2\right) - \frac{l^4}{48EI} m_2 \theta^2} \end{cases} \tag{9}$$

It is obvious that when $k=m_2\theta^2$, $A_1=0$. That is to say $y(x)=A_1 f(x)\sin\theta t=0$. It is shown that the simple supported beam will not take place vibration. Negative sign in the formula $A_2 = -\frac{P}{m_2\theta^2}$ expresses the movement direction of the mass m_2 to be in contrast to the load $P(t)$.

According to above results the inertia force of the mass m_2 is $I_2 = I_2^0 \sin \theta t$, where $I_2^0 = m_2 \theta^2 A_2$ is amplitude the of I_2. So $I_2^0 + P = 0$. It is shown that since the mass-spring system could completely absorb the energy of the dynamic load $P(t)$, the beam would keep still.

Above conclusion is to the first vibration mode of the beam structure. If the second mode of the beam structure might be eliminated, the mass-spring system could be linked at the one fourth span of the beam where the displacement is the maxmum in the second mode of the beam. The mass-spring system for the second mode of the beam is designed according to the same idea above.

3. Conclusion

This paper investigates the principle of a dynamic vibration absorber as a means to reduce the vibration of a simple supported uniform mass beam caused by the simple harmonic load at midspan of the beam. It is shown that the efficient of the mass-spring system is the best when the stiffness k and the concentacted mass m_2 meet the condition $k=m_2\theta^2$.

References

1. Cacciola P., Espinosa, M.G., Tombari, A., Vibration control of piled-structures through structure-soil-structure-interaction, *Soil Dynamics and Earthquake Eingieering*, 77 (2015) 47-57.
2. Xie, X. D., Wu, N., Yuen, K. V., Wang, Q. Energy harvesting from high-rise buildings by a piezoelectric coupled cantilever with a proof mass, *Int. J. Engineering Science*, 72 (2013) 98-106.
3. Zhang, X.C.; Xu, Z.D. Testing and modeling of a CLEMR damper and its application in structural vibration reduction, *Nonlinear Dynamics*, 70 (2012) 1575-1588.
4. Chen, Y.Q., Cao, T.Z., Ma, L.Z., Luo, C.Y. Structural vibration passive control and economic analysis of a high-rise building in Beijing, *Earthquake Engineering and Engineering Vibration*, 8 (2009) 561-568.
5. Cheung, Y.L., Wong, W.O. Isolation of bending vibration in a beam structure with a translational vibration absorber and a rotational vibration absorber, *J. Vibration and Control*, 14 (2008) 1231-1246.
6. Jalili, N., Olgac, N. Multiple Delayed Resonator vibration absorbers for multi-degree-of-freedom mechanical structures, *J. Sound and Vibration*, 223 (1998) 567-585.
7. Aida, T., Aso, T., Nakamoto, K., Kawazoe, K. Vibration control of shallow shell structures using a shell-type dynamic vibration absorber, *J. Sound and Vibration*, 218 (1998) 245-267.
8. Ducarne, J., Thomas, O., Deu, J.F. Structural Vibration Reduction by Switch Shunting of Piezoelectric Elements: Modeling and Optimization, *J. Intellgent Material System and Structures*, 21 (2010) 797-816.
9. Liu, W.Q., Xu, X.L., Wang, R.G., Wang, Z.J., Wu, X.L. Vibration reduction design of the Hangzhou Bay cable-stayed bridges, *Structural Engineering and Mechanics*, 24 (2006) 339-354.
10. Keltie, R.F., Cheng, C.C. Vibration reduction of a mass-loaded beam, *J. Sound and Vibration*, 187 (1995) 213-228.

On rural residential standardization design

Kun Lan[1,*], Yang Chen[1,†], Ze-Hao Zhang[2,‡]
[1]*School of Settlements and Civil Engineering, Xi'an Jiaotong University, Shaanxi, China*
[2]*School of Public policy and Administration, Xi'an Jiaotong University, Shaanxi, China*
*E-mail: *lankunxsd@foxmail.com, †chenyangxjtu@foxmail.com, ‡zehaozhang@foxmail.com*

It is well known that China urbanites' main assets are their house properties which rural properties are homestead instead. However, due to the root cause of the large difference from urban to rural land value is the standardized form of house properties. This article focuses on the rural house property standardization's implication, characteristics and significant influence that helps us to analysis the main problems, damages and reasons of the process of rural residential standardization. Meanwhile, it also puts forward some suggestions to promote the standardization of rural standard residential. In order to enhance the new rural residential functions and values through standardization construction, it could avoid the low-level constructions and save manual labor, lands, and environment resources. Provide a value-added way for the farmers and rural prosperities go out of poorness and bring a new type of modernized agricultural and urbanization to rural areas.

Keywords: Rural Housing Standardized; Rural Residential; Real Estate Value-Added Design; Construction.

1. Introduction

To promote the standardization design of rural residence, especially to accelerate the process of urbanization for the new countryside which could bring standardization planning, design and function for rural residential that close to city standard. The construction of new urbanization and new rural areas, bring the transformation and development of low cost and high value-added for the farmers to make millions of farmers get benefits from the development process of the new urbanization. Thus, the real estate industry, which cause the main difference between cities and rural area can provide an equal opportunity of land property management for the formers who living off the land. Therefore, building or real estate standardization is not only a concept of industrialization but also a more direct and perceptual concept of economics with the core element art. With the rapid development of the integration of urban and rural areas, the new rural agricultural modernization, informationization, modernization and internationalization, rural residential standards of value-added will become a poverty of rural development as an important research and practice hotspot [1].

2. Implication, Characteristics and Significance of Rural Residential Standardization Design

Standardization is one of the significant signs of modern industrialized and commercialized value system. In residential areas, people all know that city commercial housing prices, land prices have been rising steadily. In fact, either commercial housing price per square meter or the land price per mu all reflects the standard basis of valuation. However, the standardization of rural residence was ignored and with the development of regional, society economic, traditional culture, more and more differences have been appeared.

Why is the price of urban residential district which built by developers is several times more than rural residence even they are just separated by a block? The main reason is that the development of the urban residential district was after the standardization design and construction with perfect functions and environmental facilities [2]. It already has a well-established standard base and technique for valuation in cities, even in the whole country. Their value was relatively stable without the changeable nature of land utilization.

The general concept of standardization is to obtain the best order in a certain range and the development of common used, reused for the realistic or the potential problems which contents simplification, unification, generalization, serialization, modularization and so on.

Rural residential construction standardization is not only a basic technical work to urge the rural construction industry but also a basic of industrialization and modernization of rural construction. It is necessary to applicant rural construction standardization in the fields of advanced production experience, scientific and technological achievements that also promote the technical progress of the construction industry and make the rural construction industry to obtain the best comprehensive economic benefit and social benefit. More importantly, farmers who are the housing investors can achieve maximizing economic, social, cultural and environmental benefit that is an important way to realize the commercialization. Meanwhile, improve the utilization of rural land, human resources and materials can also improving the life cycle and value level of rural residence that improve the level of the commercialization of the farmers' property at the same time which has significance influence to the development of new rural agricultural modernization.

With the deepening of urbanization, China's rural residential construction is about to enter the era of commercialization. People have changed from pay attention to pursuit the external environment and casual quality to seek inner residential living quality and safety that requires smart growth and sophisticated construction. Rural residential have to advocate high quality of living environment, improve land use efficiency and reduce the cost that need people to make the right decisions in environment protection and rational use of resources aspects. Time opportunity requires standardization and modular design that this rule has verified by the developed countries as a valuable experience. Pay high attention to standardization and industrialization of residence is the inevitable trend of development. Advocate modulus-matching system that use customized products to form the concept of standardization to guide our construction standards. Lead the rural real estate development to reach a new level and has broad prospects.

China's traditional rural residential modular standard has the following characteristics:

- The standards have no enough sufficient theories and practices due to the limitations of the conditions and have not elaborate on the principle base, application method and coordination principle that leads to actual effect is not good and irregular construction occur frequently.
- Traditional rural housing has various standard map forms and few normative standards. Meanwhile, it still short of practical research combined with local cultural and innovation that cause regional practical application come apart with research. In addition, with the limitation of materials, technology and environment that results in the lack of standard value consciousness and great waste of irrationality construction.
- The traditional standard contents, definition and methods have not adapted to the needs of the development for new rural residence. Most of the standards were drew up decade ago that can hardly adapted to the current society.

Rural residential architectural standardization is not only mutual restriction but also mutual promotion with economic benefit. Solve the problem of formers' inherent concept to accept adopt standardized industrial production with other new technologies to build a new type of rural residential can bring leap forward developments for rural residence.

3. Main Problems, Existing Hazards and Causes of Standardization in China

With the acceleration of the process of rural modernization, China's farmers began to pay more attention in living environment requirements. Due to the needs of large residential area, more room numbers and high quality by modern formers result in rural residence numbers booming rapidly [4]. According to the statistics, China's rural construction area has more than 30 billion square meters in 2011, accounting for 60% of the country's total [5]. Seemingly that both economic development and land manage system have restricted the development of rural housing construction, which severely restricted the improvement of the living standard for peasants. However, the essence reason is that non-standardized design and non-standardized planning result in low value, functional deficiency, low land use efficiency and poor living environment rural residence. The main problems and hazards are as follows:

3.1. *Non-standardized building layout leads to the waste of land and resources*

Most rural residential sites lack of planning and unified standards in China. Irregular layout makes the residential located in optional location. Moreover, some residence even occupies the cultivated land. These actions not only causes the waste of land, but also destroyed the cultivated land that people depend on for existence. Some residential continuous expansion, occupy a lot of public spaces and public facilities; some residential frequently reconstruction, but has low utilization due to functional deficiency; some residence were abandoned lead to the Inner Decaying Village. All the phenomenon make China's rural residence development out-of-order and waste serious.

3.2. *Non-standardized building style are all in a muddle*

With the rapid development of China's economy, the formers' have a strong desire to improve their living environment. At present age, due to the lack of architecture design and urban planning knowledge, China's rural residential have face the problems of appearance confusion, appearance unattractive and the quality difference is large. Non-standardized architectural style also result in poor residential integrity, poor uniformity, poor planning and poor stability which not only affects the appearance of the building, but also bring about the low utilization rate of the building space and the hidden danger of the building structure.

3.3. *Non-standardized construction has low-level quality*

In recent years, lots of new materials and technologies are into use in cities while the rural area has just started with new construction mode and construction method. It is unable to reach the safety standards of construction with the lack of professionalism, less technology and equipment and non-uniform building materials. Moreover, because of poverty, some formers select the building materials with low quality that could lead to low reliability, short life cycle and hidden danger. From the perspective of the development trend of urban and rural integration and social equity, rural buildings not only have to maintain a beautiful natural scenery, but also have to fulfill the requirements of high safety and quality as same as cities buildings and should be convenience, comfortable, hygienic, multi-functional.

3.4. *Non-standardized buildings are lack of public facilities*

With the disappearance of the mess of rural collective mode of production, peasants' production and living change to home as a unit. The difference between each home courtyard makes the public functional facilities can hardly be matched with their self-built buildings that lead to a slow process of install public facilities and big difficulties of living condition improvement. Such as road system, underground engineering, water supply, heating, gas supply, garbage disposal system, etc. In spite of the constant improvement of the economic conditions in our countryside, it is still hard to meet the farmers' demands in Public Service Facilities and the construction of Environmental Hygiene. Most rural area's living environment remained at the end of the last century's state that only electricity and groundwater was supply to villages. In addition, because of the mismanagement, incomplete matching and without maintenance, it is difficult for the existing public facilities work properly.

Main problems caused by non-standardization of rural housing construction in China is that low construction quality, poor indoor environment, lack of living facilities, no ability to resist natural disasters and can hardly fulfill the requirements of living. There are few architectural design theory and technical support the rural housing construction[3], which only rely on the village mechanic's experience to keep construction technology evolving. Resulting in Brick-concrete Buildings with simple form, construction rough, low quality, high-energy consumption, low disaster prevention performance, lack of regional cultural characteristics everywhere. There is huge contradiction between the rural Brick-concrete Buildings with the harmonious society of human and nature. Therefore, the traditional rural residence will soon face the challenge of standardization design.

The harm of the above problem is huge. The key point is that governments and the architects have to realize the problem and take action to practice standardized design and construction in the Construction of New Countryside. It is also a general problem that worth further research and exploration.

4. Useful Suggestions for Standardization Design of Rural Residential

4.1. *Attach great importance to the necessity of standardization of rural housing construction*

In the first place, all-level governments should widely advertise to change people's mind. Countryside of China has vast mountains, rivers and forests; therefore, how to realize urbanization for rural areas under the guidance of standardization has important practical and historical significance. Especially in scientific planning and designing; land resources saving and better implementation of the future urbanization strategies. Along with the progress of the society, the farmers' demand for rural residents living environment is also increasing, so both governments and farmers should be aware of standardization design and construction's significance.

4.2. *Further study to develop strategies and action plans as soon as possible*

The key point is that governments at all levels and the relevant departments have to formulate strong guidance and scientific response. Investment in research, which promotes the development of rural residential measures, policies and problems in standardized planning, standardized design, standardized construction, and standardized function. Also, study on the relationship between commercialization and urbanization of rural residence, the value-added rural real estate and the standardization, which means that wc need both standardization design and adaptation to local conditions to deal with the diversity and unity of Art. Against the "the image project" which squanders resources under the guise of the standardization of rural housing. It is helpful to solving the relationship between inheritance and innovation development to prominent local characteristics of rural residential areas and the individual features of architectural art that could continue the regional architectural context as the root of regional economic and culture. It can truly improve the quality of farmers' living environment and lay a good foundation to achieve the well-off Society.

4.3. *Generalize successful experience play an exemplary role*

(1) Guarantee the planning and perfect system first

The design should follow the philosophy that top-level design, planning first and institutional Guarantee. Seize the opportunity of New Rural Construction, resettlement and rural transformation to combine with the development of economic, society, culture, ecology, environment, marketplace and urbanization. Meanwhile, standardization design also has to unite by regional culture, construction mode, science and technology, management methods and other multiple elements.

(2) Provide successful models of rural standardization residential

Guide the standardization of rural housing by "Four Comprehensive" and "Five Development Concept" and adopt high technology, appropriate technology and traditional skills to adaptation local conditions. Seeking the practical approaches to integrate the rural residential building and high technology could solve the current problems of rural residential construction.

(3) Popularize innovation of new technology in rural areas

In order to improve the building function and safety performance, reduce the construction cost and increase the value of rural residential. At the same time, it can also promote the modernization of residential transformation and improve the rural residential market level. Standardization design can help ease the practical problem brought by the rapid urbanization of rural labor shortage and integrate the advanced productive forces brought by the rapid development of economy and technology, which brings new development about 3D printing, standardized manufacturing and customized residential buildings technology.

(4) Establish adaptive local standards

The modular design based on the living habits of the local residents, spatial usage rules, building horizontal layout forms and local construction skills. The standards have to establish according to comprehensive analysis on various influence factors in all aspects to ensure the design could fulfill local conditions and fit the actual situation in rural areas.

5. Conclusion

In conclusion, researches and practices of our country's rural residential standards are still in the exploration stage. Under such conditions as lack of awareness and theory, weak foundation, realistic complexity, regional differences huge, imbalance development of the economy and society, all levels of government and architects should get together and try to brainstorm a better idea to promoting the standardization of rural residence and a way to increase in value of rural residence for farmers. Therefore, the standardization of rural residence is able to make an irreplaceable contribution for integrating Urban-Rural development and improve the living standards of farmers that help them to get rid of poverty and realize well-off standard of living.

References

1. CPC Central Committee and State Council, *Central Document No. 1* (2014), http://www.gov.cn/jrzg/2014-01/19/content_2570484.htm.
2. Liangyong Wu, *Introduction to human settlements*, (Beijing: China Building Industry Press, 2001).
3. Jun Wang, *Residence of the northwest*, (Beijing: China Building Industry Press, 2009).
4. Kecheng Liu and Zengfeng Yan, *Western Regional Architecture Research -- the construction of the National Key Laboratory of architecture science and technology in Western China*, Vol. 05 (Southern architecture, 2011), pp. 07-09.
5. Fuzhong Li, *Theory of housing industry — Economy, Technology and Management of housing industrialization*, (Beijing: Science Press, 2003).

The CNC machining process design and machining simulation of gear pump edge cover

Xiu-Rong Zhu*, Yang Liu, Zhi-Jie Sun

College of Mechanical Engineering, Jilin Teacher's Institute of Engineering and Technology,
Changchun, Jilin, 130052, China
**Email: zxr67811@163.com*

Through the analysis of CNC machining process of gear pump edge cover, gives a brief introduction about design parts processing route, and expounded the main processing contents of edge cover -milling. At the same time, describe the process characteristics of the machining of parts, and analysis the formulation of the method and steps of the parts processing procedure. Use the test method, to adjust and optimize the processing parameters, Through the numerical control machining simulation to verify the rationality of the simulation processing parameters and processing procedures.

Keywords: The Gear Pump Edge Cover; The CNC Machining; The Process Design; The Process Simulation.

1. Introduction

Manufacturing industry is the pillar industry of our national economy, its added value accounted for more than 40% of GDP in China. The advanced manufacturing technology is an important part of the revitalization of the manufacturing system engineering. The CNC machining is the core technology, its appearance and brought the huge benefit to arouse the attention of world science and technology with industry. At present, The CNC machining tool can machine disc parts, cover, box structure, the shaft and other parts. The gear pump edge cover is common milling parts. The processing of the disc parts milling CNC machining tools, to meet the needs of machining accuracy and surface roughness requirements. Therefore, the study of disc parts of CNC machining process design and simulation process is very necessary [1].

2. The CNC Machining Process Design of Gear Pump Edge Cover

2.1. *The analysis of parts of the drawings*

The dimension tolerance of the parts is between 0.05-0.1mm, and the convex thin wall thickness is 4mm, the area was large, and the surface roughness is relatively high, reaching Ra3.2um, relatively difficult processing, processing prone to deformation, not handled properly may lead to the wall thickness tolerance and surface roughness is difficult to meet the requirements.

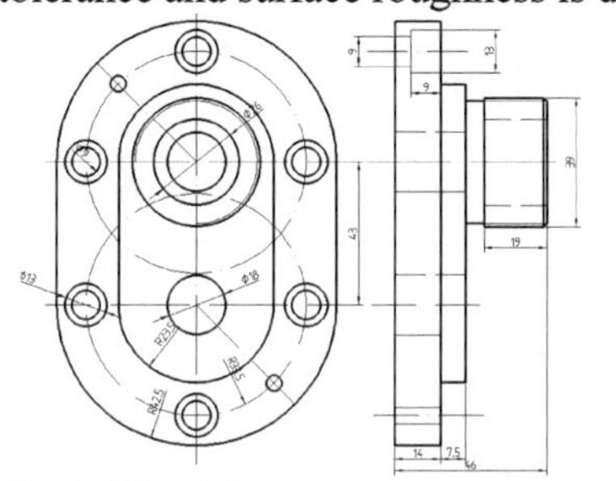

Fig. 1. The right gear pump cover parts diagram.

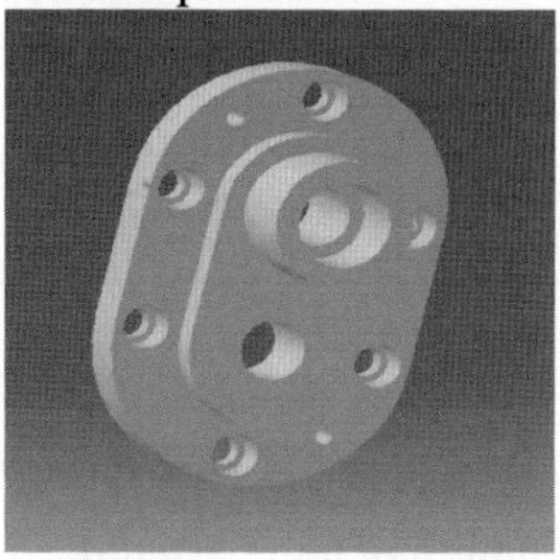

Fig. 2. Gear pump right end cover three-dimensional diagram.

The locating datum is based on the workpiece positioning clamping datum. First, Take a plane of the blank part for the coarse positioning. Second, Cut out the blank part's precision positioning surface, and reach to the provisions of the requirements and quality, as the clamping surface, the clamping surface as precise basis for the machining fixture and finished parts. Finally, the coarse datum processing to the size requirements[2]. Choice the dimension of blank: 128mm×85mm×46mm, profile of aluminum.

2.2. *Determine the processing route*

The processing sequence is first the inside after the outside, first the coarse after the fine, first the surface after the hole, use this method to divide the processing steps, the process division to consider comprehensive, part of the power of the first processing, the rest of the rough milling began finishing processing. Due to the rough and finish machining with a difference tool, so the processing scheme selection to consider [3].

The edge cover main processing surface is hole, plane and curved surface. Part with the hole φ18mm, size Tolerance H7 can choose the vertical milling through hole-boring. stepped hole Φ13mm surface rough degree is not high, directly with the milling through hole can be reached; plane and the outer surface roughness value Ra3.2um can be used rough milling--fine milling scheme. Above the holes diameter enlargement to the φ26mm. Rough milling outer contour. Finish milling outside contour. drill countersink counterbore the hole. Drill the hole diameter is φ5mm [4].

2.3. *Parts of the machine tools, fixture, tools, measuring tools and cutting parameters selection*

Table 1. Parallel jaw clamp parameters (mm).

Product name	Model	Width	height	Maximum opening	The positioning key width	Size length× width ×height
parallel jaw	Q12160	160	50	160	18	500×300×200

2.3.1. *The choice of machine tool*

Choose the FAUNC-21i CNC vertical machining center, the specification is 1300mm×850 mm×710 mm.

The FAUNC-21i CNC vertical machining center is superior flexibility than ordinary CNC machining center milling machine, there is a lathe tools library of 20 lathe tools and automatic lathe tools changer, to process complex parts need more lathe tools processing, change lathe tools can reduce a lot of auxiliary time, very convenient, and can process more complex surface workpiece. Therefore, to improve the efficiency of the machining center, and a reasonable use of programming techniques, the processing procedure of the preparation of high efficiency, to improve the efficiency of the machining center is more significant.

2.3.2. *Fixture selection*

The CNC milling machine and machining center fixture should meet the convenient installation and adjustment, good rigidity, high precision, durable degree good, so according to the shape of the parts consider V type block or machine with parallel jaw. meanwhile, the main consideration of the fixture structure is simple, which can meet the gear pump edge cover clamping requirements and selection of flat mouth vice specifications are as follows [5].

2.3.3. *The choice of tools*

Combined with the analysis of gear pump edge cover parts diagram, parts processing is the main content of milling, drilling, and have to be rough, finishing separately, therefore, the processing used in tool processing according to different content, is a tool with different requirements, generally pay attention to the efficiency performance of the cutter in rough machining, finish machining and more emphasize precision tool. First understand the ability of processing content, processing order and processing content accuracy, and then select the tool necessary. This part is flat and the cavity surface and pore characteristics, process more complex. In order to reduce the tool change and tool time, reduce the error caused by tool change, improve processing efficiency the rough, finishing as far as possible to choose the same tool, to ensure good accuracy [6].

Table 2. The choice of tools (mm).

No.	The content and requirement of processing precision	The processing method and tool selection			
		Processing method	The contents of the tool		
			Tool name	Specifications	Tool number
1	Rough milling edge face	Milling	Disc cutter of φ80	80	T01
2	Fine milling edge face	Milling	Vertical milling tool of φ16	16	T02
3	Milling Deep 7.5 outline Ra 3.2um φ39	Milling	Vertical milling tool of φ16	16	T02
4	Drill hole φ4.8, hinge φ5 H7	Drill, Hinge	The drill bit of φ4.8	4.8	T03
			The reamer of φ5	5	T04
5	Drill hole 6-φ9	Drill	The drill bit of φ9	9	T05
6	Reaming 6-φ13	Reaming	The drill bit of φ13	13	T06
7	Drill hole φ13, reaming hole φ18	Reaming	The drill bit of φ18	18	T07
8	Drill φ13 Hole, reaming φ26 Hole	Reaming	The drill bit of φ26	26	T08
9	Turning thread M39	Milling	Thread Turning tool		T09

2.3.4. *The selection of measuring instrument*

The inspection is the part of the inside and outside diameter precision whether meet the requirements of the drawings, first of all, we should choose reasonable measuring. If parts have been measured and the aperture size, you should choose vernier caliper, micrometer diameter or diameter dial indicator. It should also be looking at parts size precision, tolerance is relatively small, should select micrometer; tolerance is large, for the convenience of measurement, fast selection of vernier caliper. Choose reasonable measuring to consider the measurement of economic. Similarly a size can sometimes choose different instruments to achieve [6][7].

Table 3. The selection of measuring instruments (mm).

No	The content and requirement of processing precision	Gage name	Specifications	Accuracy
1	Milling edge face, milling deep 7.5 outline Ra 3.2um, milling φ39 outer circle .	Outside micrometer	50-75 25-50	0.001
		Vernier caliper	0 - 200	0.02
		Surface roughness tester		
2	The thread M39X1.5-6g.	Thread gage		
3	The φ13 deep 9 hole, the blind hole φ26 hole deep 19.	Vernier caliper	0 - 200	0.02

2.3.5. *Cutting the three elements of the selection*

Due to the processing of parts and materials for aluminum and its cutting performance good, and the amount of feed rate and depth are relatively small, so the investigation table reference, the cutting speed can get bigger, so the selection of the cutting speed for 160m/min. so when choose φ16mm cutter milling of parts. The principal axis speed [8].

$$n = \frac{1000 \times v}{\pi \times d} = 3000r / min \tag{1}$$

The results show that the principal axis speed can choice the roughing machining finishing is 3000r/min, the principal axis speed can choice the finishing machining is 3000r/min. Selection of cutting the three elements of the table below[9].

Table 4. Cutting the three elements of the selection.

No.	Process content	Three elements of cutting		
		Cutting depth (mm)	Feed (mm/min)	Cutting speed (m/min)
1	Rough machining	5	100	160
2	Finish machining	0.2	80	150

3. CNC Machining Simulation of Gear Pump Edge Cover

Through drawing analysis, determine the rough and finish machining route, choosing the suitable processing tools, rough machining and finish machining. In processing prior to machining simulation, repeated machining simulation, optimization of machining parameters, determined the final processing scheme. After simulation and optimization of processing parameters, using the data or IF card, the automatically generated program transmission to the CNC milling machine, processing to meet the precision requirements of the parts. At the same time, can effective use of CNC machine tools, improve productivity[10].

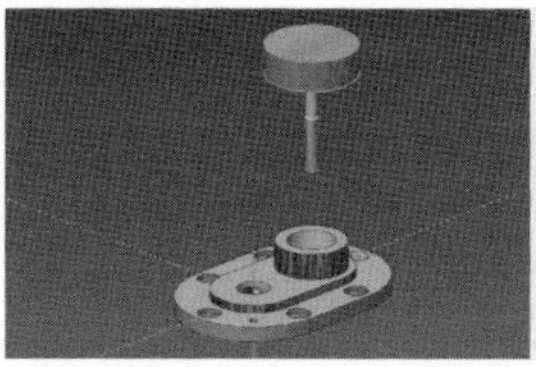

Fig. 3. The simulation processing diagram.

4. Conclusion

Through the gear pump edge cover CNC machining process analysis, design parts processing route, technical requirements of parts, parts of the material and the blank forms were briefly, describes the edge cover machining quality of the parts of the machine precision, performance and life has a direct impact on. Detailed introduces the edge cover of the main content of the processing-milling. And expounds the technological characteristics of machining, and the analysis of the formulation of the method and steps of the machining process planning. The test method, the processing parameters were adjusted to optimize, after CNC machining simulation to verify the rationality of machining parameters and the machining process planning.

References

1. Song Shushan, Numerical Control Machining Technology, Electronic Science Publishing House, Beijing (2008).
2. Zhao Xianzhong and Cheng Junlan, Numerical Control Processing Technology and Programming, Publishing House of electronics industry, BeiJing (2010).
3. Wang Kai and Feng Juan, J. Selection of Cutting Tools and Determination of Cutting Parameters. Journal of Changsha Aeronautical Vocational and Technical College. 8 (3): 65-67 (2008).
4. Wu Guoli, J. Analysis of Positioning in Machining Datum of the Selection and Application of Chinese Journal. (10):58 (published in late 2011).
5. Jess Zhang and Wang Hui, J. A Brief Discussion on the Technical Standard and its Application in the Machining Process of the Parts, The Market of CNC Machine Tools. (12): 53-54 (2010).

6. Zhang Shengwen, Zhang Ruiguo and Jia Wei, J. Research on NC Automatic Programming System For Complex Parts Based on Group Technology Manufacturing Technology and Machine Tool. (3): 92-95 (2010).
7. Song Hongming and Yang Feng, Numerical Control Processing Technology. Mechanical Industry Press, Beijing (2010).
8. Lu Dong, Research and Development of Intelligent Programming System CNC Machining Based on Template, Huazhong University of Science and Technology, WuHan, China (2011).
9. Li Zhongli, Li Xiaoli and Yang Hailiang, The NC Machining Process, Tsinghua University Press, Beijing (2015).
10. Xu Heng, Technology and Programming of NC Milling Machine and Machining Center, Chemical Industry Press, Beijing (2015).

Research of energy-saving mode of existing residential buildings in hot summer and warm winter zones- a case study of xiamen city

Jian-Xun Deng*, Zhi-Zhou Shen†

1, 2 *School of Architecture & Civil Engineering, Xiamen University, Xiamen, 361005, Fujian, China*

*Email: *djxun21cn@sina.com, †uniquevector@163.com*

In this paper, energy-saving mode of existing buildings of subtropical region has been studied, it analyzes a number of technology, management and financing for energy-saving and other issues of subtropical region. Drawing on experience at home and abroad to study the transformation of energy efficiency experience. Regarding Xiamen as the representative of the region, for the construction of referred energy-saving sector, to provide relevant recommendations to promote energy saving in similar areas of the existing problems.

Keywords: Hot Summer and Warm Winter Zones; Existing Residential Building; Energy-Saving; Financing Management.

1. Introduction

For a long time, people always regard the size of space, the rationality of layout, the aesthetic of shape and other factors as criteria of the building, while ignoring the connotation of quality, such as the thermal environment, acoustic environment and the air quality of indoor space and other quality, which directly leads to a large amount of energy during the process of using and maintaining building. Statistics show that China's urban and rural areas have more than 90% high-energy buildings among 430 billion square meters of building stock, and residential buildings account for about 70%, causing a huge waste of energy [1]. For the status quo of existing residential buildings in Xiamen, which faces the problem of large scale, dispersed ownership and inflexible mechanism, as well as it has difficulty in raising reconstruction funds, unifying the idea of transformation of the owner and the implementing management of transformation process of the owner in the reconstruction of energy saving. It is high time that we should research for suitable renovation mode of existing residential building and accelerate the pace of renovation of existing residential building in hot summer and warm winter zones.

2. The Current Problems of Renovation of Existing Residential Buildings in Xiamen City

2.1. *The energy saving is unfavorable to the owners because of the difficulty of financing*

Although the basic standard of energy-saving is 50% for existing residential building, but compared to developed countries, the absolute energy consumption of residential buildings is still at a relatively low level, which will inevitably lead to a long payback time of energy saving, generally a decade or so. Consider the opportunity cost of capital, from the point of view of energy-saving renovation investment costs, owners are somewhat unwilling to invest in it. In October 2005, the questionnaire of energy-saving buildings made by Ministry of Construction, in the whole country, showed that the number of residents who are willing to invest in residential energy-saving accounts for 58%, the unwilling group accounts for 8%, about 15% people depend on the energy-saving effect and 19% people depend on the national policies. People who are willing to bear more than 20% of the cost are only 6%, 74% people would bear 10% and 20% people can bear 10%-20%. In a word, the residents are willing to reconstruct, but unwilling to bear the cost of transformation.

2.2. *The low level of willingness to transform because of housing property*

Currently, when the departments of construction select energy-saving objects to offer subsidy, requiring the application before the transformation firstly, and then subsidize them based on the results of transformation. For commercial residential districts, the contact of the owners is not close, while the owners have unequal discretion of energy saving and family income and expense may also have a great difference, which will lead to the consensus of the owners hard to achieve that about the need for energy-saving and which program should be chosen. If there is no professional organization making intensive advocacy and mobilization at home, only rely on consciousness of the owners or rely on owner committee and property companies, which is loose and non-professional organization, it is hard to unify the views of residents and difficult to achieve satisfactory results.

2.3. *The implementation of management is difficult for lack of understanding of energy-saving program*

At president, energy conservation in existing residential buildings promoted by the administrative department of construction focus on a number of the courtyards belong to enterprises and institutions. During the transformation, design, selection of construction units, process of construction management and application for incentives after the completion of the transformation, which infrastructure departments take charge. In the residential district, there is still no mature approach on energy-saving project, the supervision of transformation process and the regulatory incentives of transformation. Owners not only stay at the loose situation but also not familiar with the construction program. If merely rely on the enthusiasm and voluntary of owners, it is difficult to effectively manage the process of transformation. The government should actively promote the professional energy-saving reconstruction company, and residents can reduce obstruction of energy-saving process because of information asymmetry.

3. Energy-Saving Reconstruction Mode of Existing Residential Building from Home and Abroad

3.1. *Financing mode of energy-saving reconstruction of existing buildings in Poland*

The forms of residential property are diversified and private ownership is nearly 50%. About 90% old houses belong to the concrete block building, resulting poor insulation, serious leakage and serious heat loss, in addition that Poland apply to join the EU, they must strictly implement EU standards of energy-saving buildings [2]. Therefore Poland requires for energy saving of existing residential buildings in large area.

Poland is financed by special funds, bank loans and investment of owners. Poland takes long-term financing model, which requires to use subsidies and loans in conjunction. Therefore the Poland government began to implement modern energy bill in December 1998, establish energy fund and subsidize energy conservation of existing buildings by the state budget. Poland government transfer the fund to the Poland state-owned bank BGK (Poland Residential Development Bank) to manage, and entrust a special body to review the application for funding of the project. After the approval of reconstruction project, 20% of self-transformation fee will be raised by owners, the rest is provided by BGK, payback period should not exceed 7 years, users only need to repay the 75% loan, and the other 25% will be regarded as incentives to users from the state. At the same time, the audit agency reviews the completion of transformation project to make sure that the predetermined energy-saving targets are achieved. In addition, the state also introduced a series of subsidies and preferential treatment for the installation of heat metering device, for an example, the state subsidizes the 50% cost of the heat meter to who installed it and 10% cost of the water meter to who installed it, subsidy and so on. Such a policy mechanism model not only conform to market economy status, but also promote the development of energy efficiency of related buildings and metering charges.

The relevant supporting policies, from 1990, Poland has made a series of mandatory laws and regulations and incentive policy in respect of heating companies, heat metering, charging system and heat pricing mechanism, covering the heating system of property rights, licensing system, energy use system, charging system and heat pricing mechanism. On the one hand, it helps regulate the relevant authorities in the transformation process, on the other hand, it encourages the transformation of the subject, to ensure the smooth progress of reconstruction work.

3.2. *Reconstruction mode of energy saving in Urumqi*

Municipal government of Urumqi attach great importance to energy efficiency in buildings. Currently, the policy of energy-saving reconstruction of existing buildings in Urumqi areas are mainly the type of forceful policy, supplemented by economic incentives. In 2005, Urumqi begin the energy-saving work of pilot residential building, the main source of funds comes from special funds supported by the government. The principle of reconstruction work adheres to" who owns, who reconstructs, who invests, who gets", adopted the program of maintaining and implementing existing buildings and construction of energy-saving heat metering transformation simultaneously. Focusing on the construction built from 1980 to 2002, which have poor insulation performance of existing energy-saving residential buildings. By the end of 2010, the area of existing energy-saving buildings in Urumqi reached 10 million square meters (residential buildings have 8.02 million square meters, public buildings have 1.98 million square meters). According to preliminary statistics, the reconstruction of existing buildings which the residents take the initiative for energy saving is about 5 million square meters, the area of seismic strengthening energy-saving insulation is about 93 million square meters. Saving 240,000 tons of standard coal per year, reducing 45 million tons carbon dioxide emissions per year, reducing 5,950 tons soot powder per year and 4,080 tons of sulfur dioxide per year [3].

The reconstruction of existing buildings for energy saving in Urumqi is still in infancy, which needs support from government. At present, there is no ESCO offering professional energy-saving service, during the government-led financing, financing is still mainly based on the mode of the content of transformation, energy-saving financing type of EMC is a supplement. The financing model of transformation which is based on the content is relatively mature, which accordingly can be proceeded by the appropriate program or similar projects. At the same time, we should vigorously support the Energy Service Company (ESCO) for EMC financing. Since one of the characteristics of energy-saving reconstruction in Urumqi is proceeding the reconstruction of existing residential building and the heating system simultaneously. In the absence of professional ESCO, we can support heating companies, with reference to the similar ESCO companies to establish a company, in order to ensure financing model of EMC energy-saving projects. Overall, the main choice of financing mode of existing residential buildings in Urumqi at this stage is still government-led transformation, and the government vigorously supports ESCO companies as the basis for heating enterprises, and gradually financing will transform into ESCO relying on heating companies.

The financing model of government-led energy-saving assurance of EMC is already applied at this stage, providing EMC the basis of development and foundation. At this time, financing model of energy saving of assured EMC, sharing EMC and the comprehensive EMC will become mainstream energy-saving work. When the initial use of energy saving of sharing EMC financing, we can cooperate with relatively mature ESCO from foreign or domestic country.

4. Energy-Saving Mode of Existing Residential Buildings in Xiamen

According to the actual situation of Xiamen, you can refer to the following energy-saving mode, and it is desirable to provide useful experiences:

4.1. *The mode of return on investment*

Since the dispersion of residential property, the government can not pay alone, and the role who should bear ought to be diverse. First, it should identify the energy consumption of buildings, and the value of reconstruction of existing housing will be rated, and then follow the number of inputs to determine the body who may participate in, to maximize the number of participants who may include government, real estate, utilities, heating company or specialized energy companies and individuals. For an example, the mode of investment of heating company can be direct investments, such as installing regulate indoor heating system renovation and reconstruction apparatus. Its sources of funding may include: absorbing direct investment, issuing shares, bank loans, issuance of corporate bonds, financial leasing, commercial credit and so on. But it also can apply for state loans, interest subsidies and direct financial subsidies. Resident individuals mainly responsible for the relevant cost of the transformation of their own property and buildings, appropriately, local finance department can offer a certain amount of subsidies. For low-income family which is going to take energy-saving renovation of existing residential buildings, local financial department should subsidizes all transformation costs [4]. A number of years after the transformation, the government unified distribution of energy-saving benefits in return by the ratio of investment. After the recovery of the cost of renovation, there will be new standards of fees in order to attract a wide range of money into energy-saving work. To return the energy-saving benefits to investors, and to take different policy of classification on existing residential buildings and charge different fees for residential between non-energy-saving buildings and energy-saving buildings can mobilize retrofit initiative of energy efficiency of existing residential buildings [5].

4.2. *The mode of energy management with contract*

The mode of energy management with contract is a new investment and financing, energy management contract is signed between the implemented companies (users) of energy projects and specialized energy service companies (EMC Corporation), and it helps to promote the implementation of energy conservation projects. Energy-saving service companies is also known as energy management company, which is based on contract energy management mechanism of specialized companies operated for profit. EMC company signs the contract with the customer who is willing to carry out energy-saving service, providing services like energy audits, feasibility studies, project design, project financing, equipment and materials procurement, construction, personnel training, joint energy detection system transformation, maintenance and management, etc. During the contract, energy-saving service companies share the profits with enterprise and customers, and obtain investment and a reasonable profit. After the end of the contract, efficient equipment and energy-saving profits are all owned by all corporate customers [6].

4.3. *The mode of rolling-type transformation*

Rolling transformation model is government firstly build a number of government-funded resettlement houses, we will relocate households for the implementation of the transformation, during the process of relocation, the government collects money for the spread between new buildings and old buildings and increased areas from households. Then the original building will be implemented with renovation of apartment and energy-saving reconstruction, after the transformation is completed, the building will be used as the new resettlement houses for other tenants.

The mode of rolling-type transformation will upgrade the value which is greater than the cost of the transformation. It is entirely feasible to required the payment of an appropriate spread. Thus, it not only can solve the problem of funding but also obtain substantial benefits from a one-time investment, which is a long-term development. But this transformation model requires organization and implementation by government departments, if it entirely relies on operation of social enterprises, the spread will be set too high, and then the problems of people's mentality and affordability may arise.

4.4. *The mode of encouraging transformation with financing incentives*

Economic incentives for building retrofits have a direct effect in promoting reconstruction, and therefore it should further improve fiscal policy, using the market mechanism to promote energy-saving buildings.

There are many methods for incentives like establishing an effective incentive mechanism, using economic leverage effect, considering to create financial policy system of incentives and disincentives, establishing and improving tax system related to energy-saving buildings. We should prompt consumers to conserve energy and improve energy efficiency by the tax system. For example, you can use financial interest, financial assistance, government procurement, tax incentives, low-interest loans and other measures to encourage building energy-saving initiative body and take measures of energy-related taxes and fees to increase the cost of construction of energy use for industry, but we need to control the cost of living due to increasing energy costs. It's time to establish related tax system and improve building energy efficiency. To encourages consumers to save energy and improve energy efficiency through tax system. Practical manifestations are: First, to restore the collection of investment direction adjustment tax of fixed asset and zero tax rate for energy-saving buildings, the second is the tax relief of the implementation of the policy to the users who purchase energy-efficient building products, and the third is to give tax breaks and other policies of the low-energy residential building appropriately. focusing on combining the relationship of central and local government. In classifying financial institutional framework, the central and local fiscal expenditure have their own scope and means of tax. Promoting building energy efficiency should fully mobilize the central government and local government, specifically confer their respective responsibilities and mission requirements.

5. The Relevant Measures and Suggestions

5.1. *To enhance the supervision of market access*

The supervision includes related technologies and products and businesses or institutions for retrofitting of existing buildings to provide services. For technology and product, we should take a filing mechanism for technology and product in line with our current standards of market access. For transformation of enterprises or institutions providing service, we should establish review system of qualification record. In addition to the market access regulation, we should also strengthen supervision of the implementation of the verification system.

5.2. *To strengthen supervision of the quality of renovation project*

We should not only control the quality of reconstruction, but also supervise whether relevant regulatory technical products consistent with existing national standards, including compliance with the standard of applicability, such as technology and products recommended directory or guides.

5.3. *To strengthen public awareness and information governance*

To build information governance platform, which can publish relevant information to the community and inform the relevant policies and regulations to each body of renovation market, facilitating timely understanding of the whole society and other relevant policies and regulations, information of engineering, business intelligence, and private practice qualifications and other information.

5.4. *To build a multi-user platform*

With the use of internet technology, we can put government, energy-saving enterprises, power sector, building material suppliers, equipment manufacturers, energy companies, banks and individuals into a unified management platform, providing real-time exchange of information and integrating energy-saving resources. In the early time of investment, costs of renovation and operational expenses are in the overall planning, in order that the platform will maximize the long-term interests.

5.5. *To establish technical support and evaluation system*

To provide technical support for energy conservation in existing buildings, the relevant departments should assist the building department, collaborating design departments to strengthen supervision and management of the construction process of existing buildings to conduct a comprehensive quality, strictly according to the bidding way to optimize the construction program selection and select construction unit. Finally, the relevant departments should carry out appraisal of energy-saving effect of the renovation project.

6. Conclusion

The thesis of this paper focus on the problem of reconstruction of existing residential building in Xiamen, introducing a number of domestic and foreign typical modes of existing energy-saving buildings. For characteristic of the subtropical region of Xiamen, the paper proposes the corresponding transformation patterns and related policy recommendations. Currently, energy-saving public buildings in Xiamen have gained the attention by the government departments, Xiamen has been implemented the plan of universal LED lamp lighting for existing public buildings by the EMC model[7], however, the plan of the energy saving of existing residential buildings has not yet widely formed, and there is no specific implementation details. The environmental policy of renovation of existing residential buildings still requires more improvement.

Acknowledgment

Foundation: Xiamen Construction and Technology Projects (Item Code: XJK2013-1-3), Student Innovation and Entrepreneurship Training Program of Xiamen University.

Jian-Xun DENG, male, Title: assistant professor, Research Field: engineering management, green building management, property development and investment decision.

Zhi-Zhou Shen, male, Title: master student, Research Field: civil engineering construction and management.

References

1. Li Lijun, Practical Design of Energy-efficient Building [M], Beijing: China Electric Power Press, 2008.

2. Kang Yanbing, The Market of Reconstruction of Energy-saving Building and Project Finance[M], Beijing: China Building Industry Press, 2011: 110-112.
3. Zhong Chen, Financing Study of Existing Energy-saving Building[J], MATERIALS INNOVATION & ENERGY SAVING, 2011 (6): 34-36.
4. Jin Zhanyong, Hao Youzhi, Ma Chongfang, Liu Changbin, Financing Model Design of Northern Areas of Existing Residential Energy-saving Buildings [J], Construction Economy (2008) (5): 56-59.
5. Guo Jian, Qin Li, Guo Mao, Modes of Reconstruction Technology of Existing Energy-saving Residential Buildings[J], 2009 (11):74-75.
6. Baidu Encyclopedia [EB / OL]. Http://baike.baidu.com/view/3679404.htm?fr=aladdin.
7. Xiamen City encourages the mode of EMC and strives to make LED lighting popular in three years, http://ecep.ofweek.com/2014-06/ART-93003- 8120-28836651.html.

Study on strength of upper filter system of mechanical water treatment equipment

De-Zhen Feng*, Jian-Hua Sun
School of Mechanical Engineering, University of Jinan,
Jinan, 250022, China
**E-mail: fengdzh@163.com*

Now mechanical water treatment technology is getting wide use gradually. This new technology is changing the work way of conventional water plants in practice. In the process of water treatment, the filter process is a very important step. In this paper, the strength and deformation of upper filter system which includes upper filter frames and upper filter plates were analyzed with ANSYS and useful results are got. Finally, according to the analysis results, the paper combined theory with practice, pointed out the disadvantage of design and provided the optimized advice about the upper filter system design.

Keywords: Water Treatment; Filter; Strength; Deformation.

1. Introduction

China has a vast territory and rich water resources. But the average water resource of every person is very low. Furthermore, the most water of our country is surface water and the components of the water are very complicated. This will threaten the safety of the water for life especially for rural water for life. So it is a very important to protect and apply the surface water scientifically. On the other hand, most parts of our country are mountains and desert. The available land resource is not abundant. With the increase of the population, the development of the economy and the quickening of urbanization process, the land resources will get more and more precious. So how to save in water treatment project has become an important subject.

Now the conventional water plants of our country use steel concrete structures as the main constructions which need large land and large building project which will cost long time and much money. On the other hand, the conventional water plants are not flexible in work. So the modular mechanical water treatment plants are getting wider and wider use because they need low invest, short time limit, small land and have the advantages of flexibility, high automation and easy operation.

According to the characteristics of modular mechanical water treatment equipment, this paper has analyzed the strength and stiffness of the upper filter system, discussed the stress and the deformation of the filter frame and plate. At last, combining theory with practice, the paper has improved and optimized the design of filter system.

2. Structure of Filter Systems

According to the requirement of practical production, the filter systems should be changed in time to adapt to the need of production process. The filter systems are composed of upper and lower filter systems. In the upper filter system, filter plate uses 30mm PVC plate on which a number of filters are mounted, the filter frame uses 3*9 layout plan, i.e. three 120*80*5mm longitudinal steel tubes and nine 120*80*5mm transverse steel tubes which all are welded to the longitudinal tubes. Under the middle longitudinal beam, there are four uniformly distributed 80*80*5mm upright columns, as shown in Figure 1. The lower filter frame uses 3*9 layout plan, i.e. three 80*80*5mm longitudinal beam and nine 80*80*5mm transverse beams. The supports columns are twelve (3*4) uniformly distributed 80*80*5mm upright steel tubes. The paper only studies the upper filter system.

In practice, the water treatment plant uses one layer or two layers homogeneous filter materials in one filter chamber which includes frames, filter plates, filter materials etc. Filter frames support the filter plates and filter plates bear the gravity of filter materials. The upper layer filter materials are quartz sands whose density is about 2.6g/ cm3 and stacking density is 1.6 g/ cm3. The filter plates are rigid plastic PVC plates whose thickness is 30mm. The filter frames are composed of Q235B steel tubes whose size is 120*80*5mm and the supporting tubes' size is 80*80*5mm.

3. Strength Analysis of Upper Filter System under Normal Work Load

In the upper filter system, the filter plate uses 30mm PVC plate in which a number of filters are mounted. The filter frame uses 3*9 layout plan, i.e. three 120*80*5mm longitudinal tubes and nine 120*80*5mm transverse steel tubes. There are four uniformly distributed 80*80*5mm upright square tubes that support the longitudinal beam, as shown in figure 1. In transverse direction, the frame uses nine 120x80x5mm rectangular steel tubes which all are welded to the longitudinal beam. The load of quartz sand is about 0.02785 MPa.

For the rectangular sheets with usual constraints, we can use theoretical formulae to compute [1]. But for the complicated combination of beams and sheets, the theoretical formulae can not be used to analyze the strength. In this case, numerical calculation must be used. In this paper, the FEM software ANSYS is used to calculate the stress and deformation of complicated structures [2,3]. Figure 2 ~ 8 are the analytical results of upper filter system.

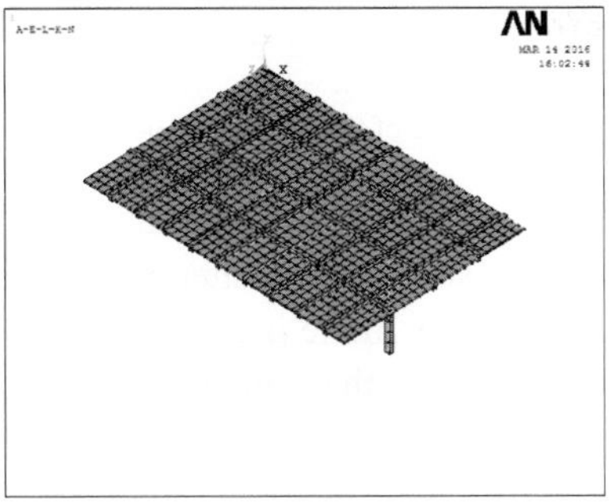

Fig. 1. Model of upper filter system.

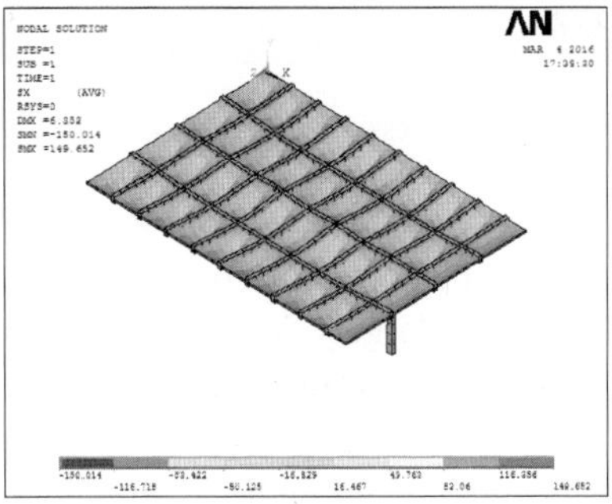

Fig. 2. Stress of upper filter system.

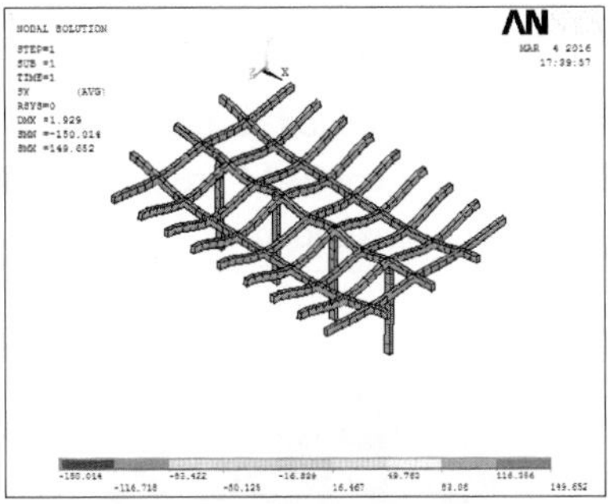

Fig. 3. Stress of upper frame.

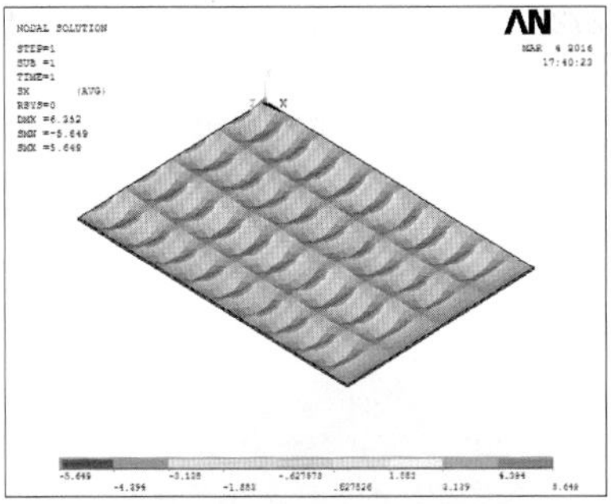

Fig. 4. Stress of upper filter plate.

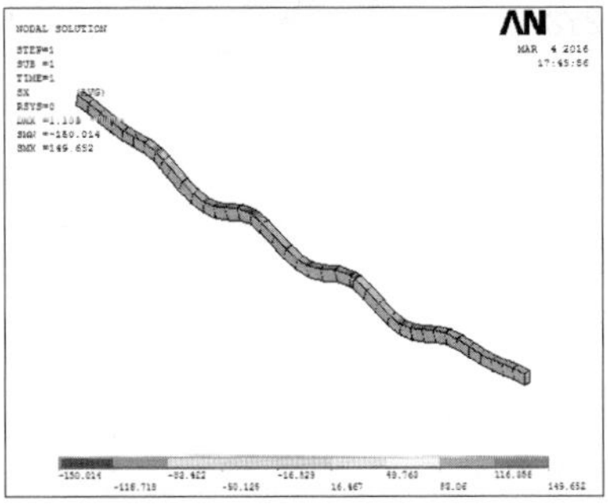

Fig. 5. Stress of the middle longitudinal beam.

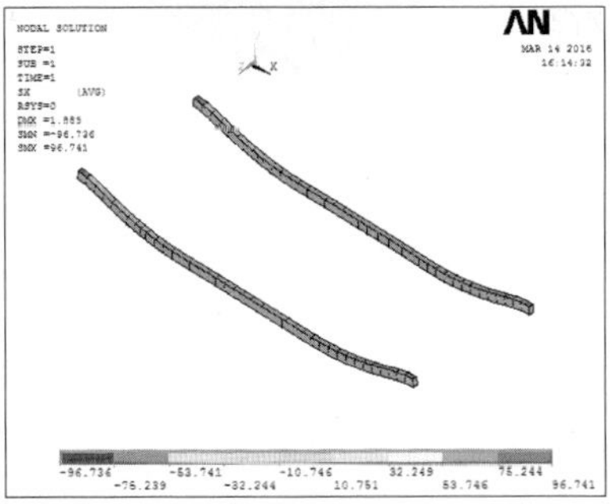

Fig. 6. Stress of the flank longitudinal beams.

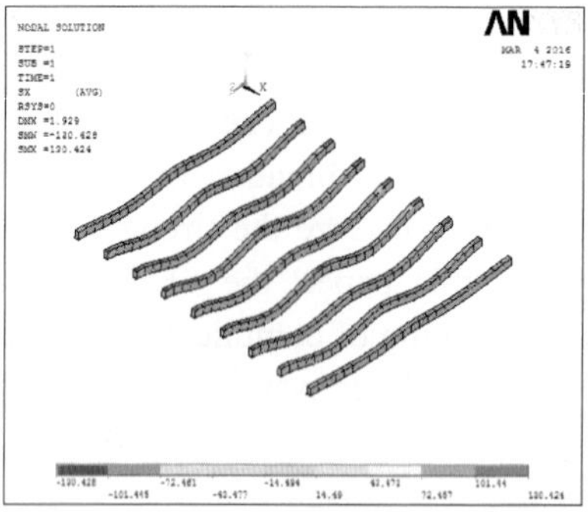

Fig. 7. Stress of the transverse beams.

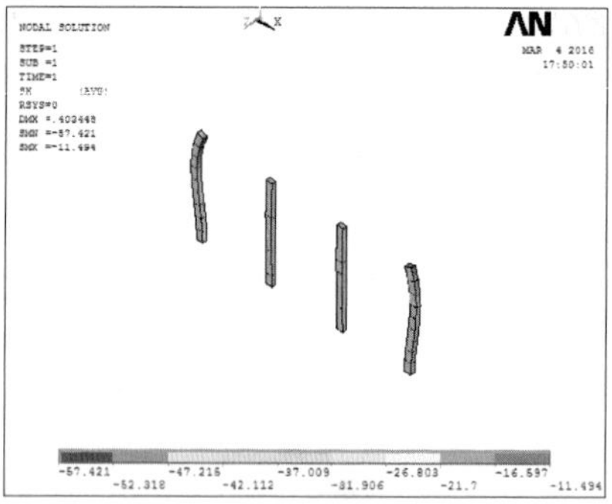

Fig. 8. Stress of the supporting beams.

From the FEM analytical results we find that the largest deformation of the upper filter plate is about 6.4mm and the largest stress of the filter plate is about 5.6 MPa. The largest deformation of the upper filter frame is about 1.9mm, and the largest stress of the frame is about 150.0 MPa. At the same time, the results show that the middle longitudinal beam is most dangerous because the largest stress occurs in the middle longitudinal beam. But they all are in safety.

4. Strength Analysis of Upper Filter System under Back Wash Load

In the process of water treatment, the back wash is needed to regularly. When back wash, the back wash, load is about 0.05MPa. Figure 9~14 are the analytical results of upper filter system under the regular back wash load.

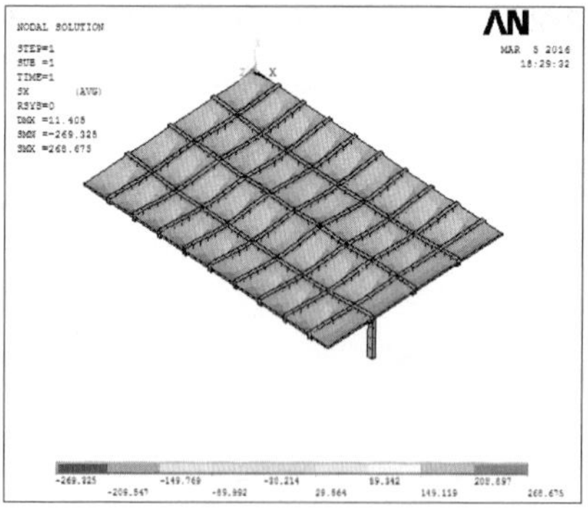

Fig. 9. Stress of upper filter system.

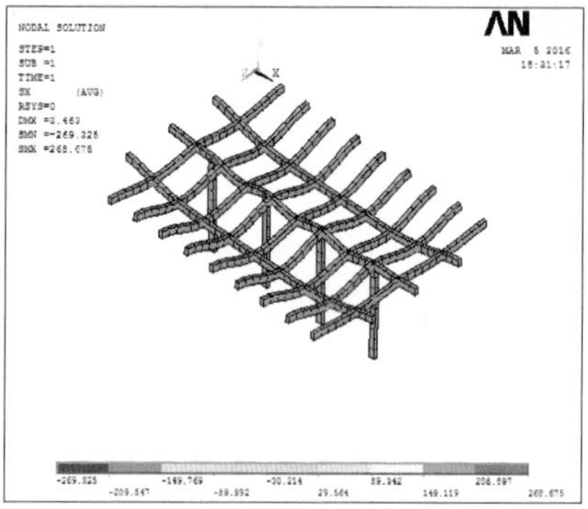

Fig. 10. Stress of upper frame.

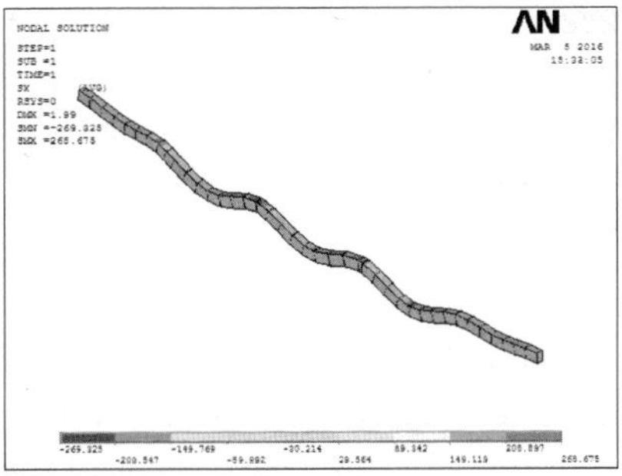

Fig. 11. Stress of the middle longitudinal beam.

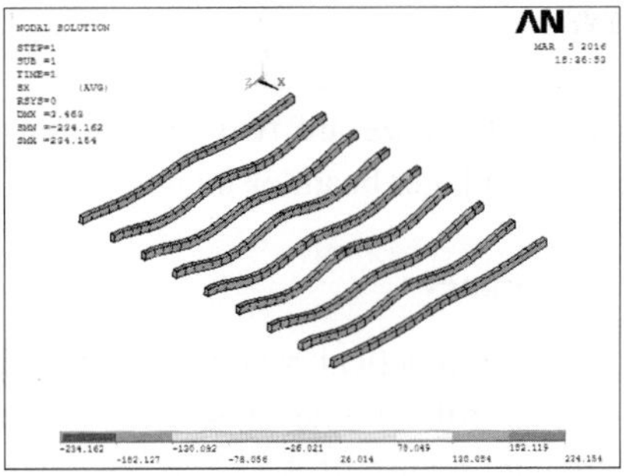

Fig. 12. Stress of the transverse beams.

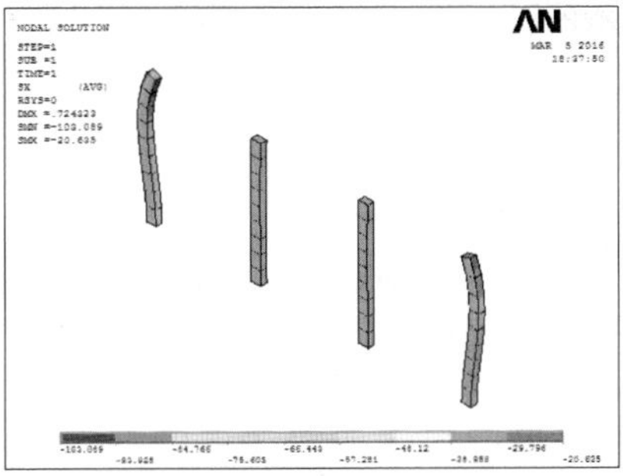

Fig. 13. Stress of the supporting beams.

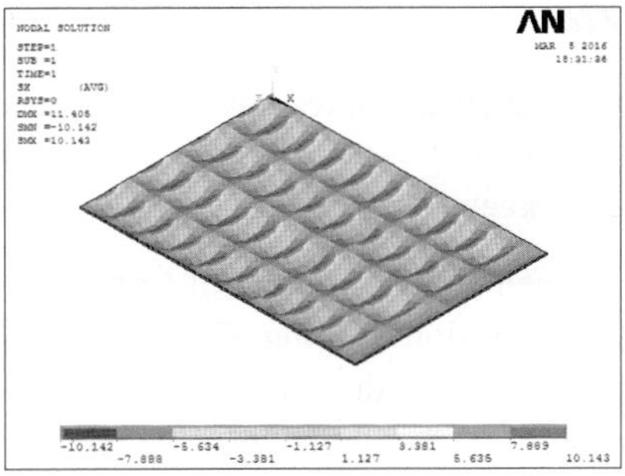

Fig. 14. Stress of upper filter plate.

From the FEM analytical results we find that the largest deformation of the upper filter plate is about 11.4mm and the largest stress of the filter plate is about 10.1 MPa. The largest deformation of the upper filter frame is about 3.5mm, and the largest stress of the frame is about 269.3 MPa. Just like the case in the normal work load, when back wash, the middle longitudinal beam is most dangerous too.

5. Analysis of Results

The deformations and stresses of the upper filter system under the normal work load and back wash load are shown in Table 1.

Table 1. Deformations and stresses of the upper filter system.

	Work load		Back wash load	
	Deformation (mm)	Stress (MPa)	Deformation (mm)	Stress (MPa)
Filter frame	1.9	150.0	3.5	269.3
Filter plate	6.4	5.6	11.4	10.1

From Table 1 we find that under the normal load and the back wash load the stress of the upper filter plates all is in the safety range. But under the back wash load, the stress of upper frames is larger than the yield limit of Q235B steel and not safe. This is a serious problem. So, the upper filter frames must be reinforced. There are two reinforcing ways: changing the middle longitudinal beam with lager beam or increasing the number of the supporting tubes.

6. Summary

Combining practical production, the paper analyzed the strength and stiffness of the upper filter frames and filter plates of modular mechanical water treatment equipment by use of ANSYS. The analysis results show that the designs of filter system can satisfy the requirements of strength and stiffness. But when back wash, the stress of upper frames is larger than the yield limit of Q235B steel and the upper filter frames must be reinforced.

References

1. Chen Jingyi. Manual of Computer Aided Engineering Analysis of ANSYS. 1st Edition. Beijing: China Railway Press, 2001.
2. T. R. Chandrupatla. Introduction to Finete Elements in Engineering. 1st Edition. Beijing: China Machine Press, 2008.
3. ANSYS Inc. ANSYS Structural Analysis Guide Release 5.4 Third Edition. SAS, IP Inc., 1997.

Study on piston wind effect on the smoke control in a cross transfer station fire

Yan-Qiang Cui*, Yan-Feng Li, Yi-Xin Zhang, Hong-Yi Wang
University Department, Beijing University of Technology, Beijing, 100124,China
**E-mail: liyanfeng@emails.bjut.edu.cn*

A cross type interchange subway station was selected as research object. Fire happened in the 3rd underground subway platform while the metro train in the neighbor line passed the 2nd underground subway platform. Piston effect by passing train would be produced. It would influence the smoke propagation in the bottom platform. Fire dynamics simulator(FDS) software was used to simulate the piston effect by comparing two fire conditions where trains keep passing by or not on the neighbor platform. Smoke concentration, temperature and visibility on 3rd underground platform were calculated. Results show that piston wind effect would enhance destabilization on smoke layer and do harm to passenger evacuation.

Keyword: Interchange Subway Station; Piston Wind; Smoke Numerical Simulation.

1. Introduction

In case of subway train fire large quantities of smoke are likely to spread rapidly to entire subway station due to stack effect and confinement of the tunnel wall [1]. Especially, because smoke spread path usually coincide with passenger's evacuation path it will reduce visibility and can cause fatalities by asphyxiation. On 18 February 2003, as an example, a subway train was set on fire with gasoline by a mentally ill patient at Jungangno Station in Daegu, South Korea. The fire quickly spread to all six coaches of the train within 2 min due to the highly flammable interior of the train, destroying two trains and causing large casualties of 192 deaths and 148 injuries. Many scholars have conducted much research a on the smoke propagation of subway stations. Park, Rie et studied ventilation and smoke flow characteristics of accidents subway fire [2]. Chongling Shi et through numerical simulation method to study smoke horizontal and vertical spread rules of the deep Island subway station, proposed to open the screen door when the screen door is fitted with a fire station and use station tunnel ventilation systems assistant exhaust smoke [3].

In China, more and more interchange stations have been built with the increase of subway lines in city such as Beijing, Shanghai, Shenzhen and Guangzhou. The importance and necessity of the interchange station fire research has been get a consensus. However, fire safety study on subway interchange station started lately , especially smoke propagate in the complex space and the piston wind effects lack of in-depth study.

In this paper, FDS software is used to explore the piston wind effect on the underlying platform smoke spread rules of subway interchange station. The smoke concentration and temperature and visibility are calculated for analyzing the smoke control effect.

2. CFD Modeling

FDS simulation program was developed by the US NIST, using a computational fluid dynamics to solve the issue of the flame-driven fluid motion (CFD) models [4]. A series of tests were used to validate the FDS analysis effectiveness of fire safety [5].

2.1. *Physical model of subway interchange station*

The study was conducted in a typical subway interchange station with three stories below the ground level as shown in Fig.1. The interchange station is used side - side transfer mode. Simplified, the station basement 1 is two separate concourse space, two separate spaces, respectively, as two smoke partitions. Each partition has four communicating outside entrance. There were two lines each in basements 2 and 3. Fire source located in the central area of the rail line.

2.2. *Grid analysis*

The FDS use large eddy simulation (LES) to deal with turbulent flow problem. The numerical simulation result is quite sensitive to the mesh size used [6]. Therefore, when using a large eddy simulation (LES), the appropriate grid sizes is necessary [7]. The mesh size near fire source is generally determined by the characteristics of the fire source diameter. Fire source characteristic diameter is determined by

$$D^* = \left(\frac{Q}{\rho_\infty c_p T_\infty g^{1/2}} \right)^{2/5} \tag{1}$$

D* for fire source characteristic diameter, m; Q for the heat release rate, kw; $\rho_\infty$for the environmental density, kg/$_{m^3}$;T_∞ for the ambient temperature, K; C_p for the air specific heat capacity, kJ/(kg·k); g for the gravitational acceleration, m/s^2.

Studies shown that when the grid size of FDS model d=0.1D* simulation results and experimental results can be better consistency[8]. In this paper, heat release rate (HRR) is 7.5MW and equation (1) is calculated D*= 2.1. Therefore, the grid near fire source use 0.25m × 0.25m × 0.25m ,basement 2 and 3 using 0.5m × 0.5m × 0.5m grid, basement 1 using 1m × 1m × 1m grid.

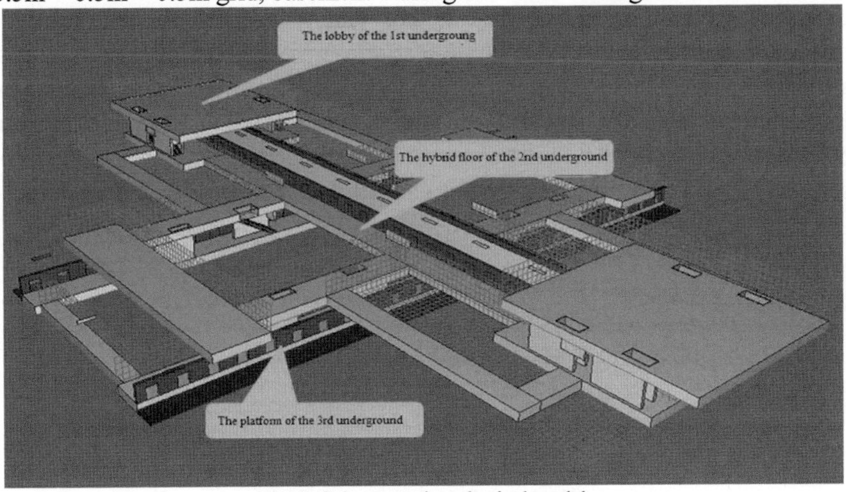

Fig. 1. Subway station physical model

2.3. *Boundary conditions and fire sources*

Take the ambient temperature 20°C, ambient pressure 101.325kPa. In the absence of piston wind simulation conditions---condition 1, the boundary physical model such as entrances, the rail line connecting the outside world with the model was set up for the free surface, so that the model is connected with the atmosphere. In a piston wind conditions---condition 2, the basement 2 area at both ends of the rail line additional piston wind speed, others same to condition 1.

The fire occurs in the central area of rail line of train electrical equipment. Fire development curve is $Q=at^2$. In this paper, relatively conservative assumptions, the fire source development is set to super- fast fire, a = 0.1876. And the fire source development process ignored the attenuation of the fire. When the fire grew to the maximum, still maintain this maximum until the end of the simulation.

3. Results and Discussion

When the basement 3 rail line area fire, after 60s basement 3 open security doors and mechanical exhaust systems, basement 2 open air supply systems, but not open the security door. Condition 1, basement 2 no piston winds, while in condition 2 basement 2 with piston winds.

3.1. *Smoke propagation simulation*

Figure 2 is a condition 1 and conditions 2 on fire every 60 s of the smoke diffusion. As can be seen from Fig(a), the smoke propagated gradually from the top of the rail line area to the whole platform area. Before the smoke exhaust system response smoke has not yet spread to the nearest stairs, with the increase in the amount of smoke, smoke gradual propagation to the platform area. As can be seen from Fig.(b), due to the disturbance of the piston wind, the smoke layer settlement quickly. When 360s the fire has diffused to the entire platform area.

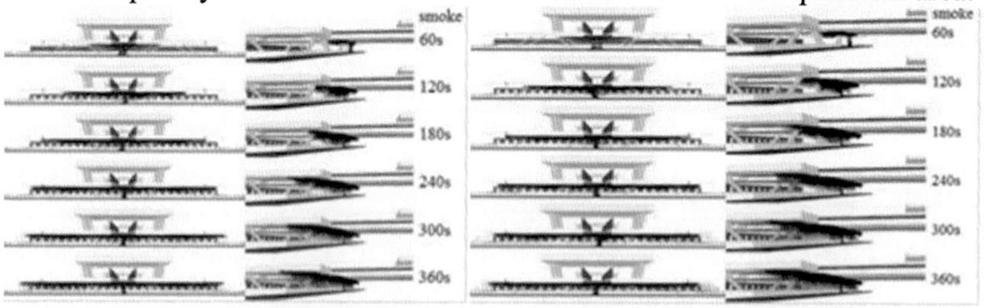

(a) no piston wind (b) With a piston wind

Fig. 2. Smoke propagation condition

3.2. *Smoke temperature simulation*

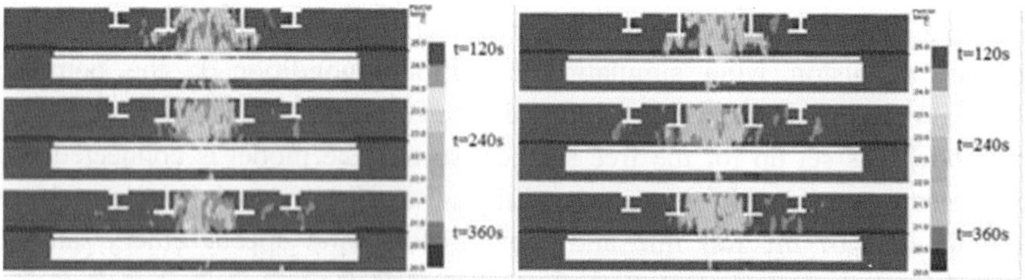

(a) no piston wind (b) With a piston wind

Fig. 3. Temperature contour of the plane Z=1.8m

Figure 3 is conditionl and condition 2 at 1.8m height of smoke layer temperature field at intervals of 120s. As it can be seen from Figure (a), in the 360s, smoke layer temperature field changed. Because of the fire source disposed in the central rail line area, so the temperature of the central public areas is higher than other places, it will bring some difficulties for evacuation. From Figure (b), the smoke layer disturbed by piston wind, the settling velocity to accelerate, with an average temperature rise of 0.5-1°C , heat radiating effect of the smoke layer has been enhanced at 1.8m height.

3.3. *Visibility simulation analysis*

Figure 4 is condition t 1 and condition 2 at 1.8m height visibility at intervals of 120s. Figure (a) can be seen as the smoke spread to the top of the station, visibility are generally affected at a height of 1.8m, but visibility still can guarantee more than 10m. Figure (b) due to the effect of piston wind, the smoke layer of the stairs around rapidly decline and reduce visibility in this area.

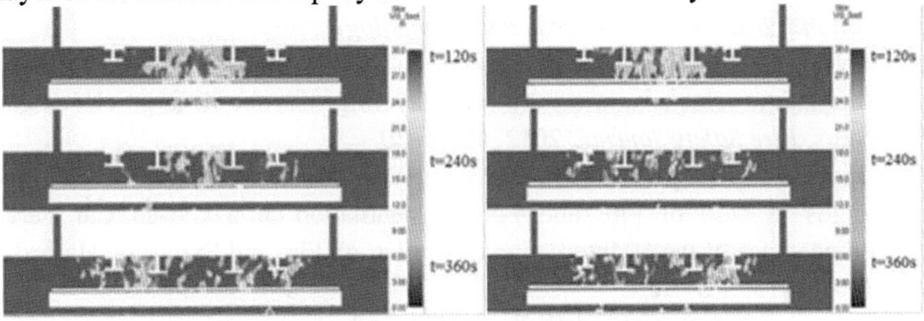

(a) no piston wind (b) With a piston wind

Fig. 4. Visibility contour of the plane Z=1.8m

4. Conclusion

In this paper, fire in a typical cross interchange subway station is studied by FDS .The piston wind effect on a bottom platform smoke is analyzed Through analyses on the simulated results we can get the following conclusions. Because of the disturbance of piston wind, airflow temperature on platform significantly increase by 0.5-1°C. Since the piston wind effect is sufficient, smoke layer of underlying platform decline rapidly. The critical value of visibility at safety height reduced to less than 10m. It would take disadvantages for passenger evacuation .

Acknowledgements

This work was supported by Beijing Joint Science Foundation(Grant No. 14L00001) and Natural Science Foundation (Grant No. 51378040).

References

1. Zhang Yi-xin. Study on smoke control system in cross-typed interchange subway station. *Master's degree thesis of Beijing University of technology*.2015.
2. Park W. H, Kim D. H, Chang H. C. Numerical predictions of smoke movement in a subway station under ventilation. *Tunnelling and Underground Space Technology*, 21(3): 304-309, 2006.
3. Shi Ling-cong, Zhong Mao-hua, Tu Xu-wei, et al. Deep subway station fire experiment and numerical analysis. *Science and Technology Press*,2009.
4. MCGRATTAN K, HOSTIKKA S, BAUM H, et al. Fire dynamics simulator(version5.3) technical reference guide NIST 1018-5.*Gaithersburg: national Institute of Standards and Technology*,2009.
5. Nele Tilley, Xavier Deckers, Bart Merci. CFD study of relation between ventilation velocity and smoke back layering distance in large closed car parks. *Fire Safety Journal*, 2012, 48:11–20.
6. Deckers, X, Jangi, M, Haga, S, Merci, B. Numerical Simulations of Flow Fields in case of Fire and Forced Ventilation in a Closed Car Park. Proceedings of the 6th International Seminar on Fire and Explosion Hazards, 1104-1115.
7. HU L H,HUO R, CHOW W K. Studies on buoyancy-driven back layering flow in tunnel fires.*Experimental Thermal and Fluid Science*, 2008, 32(8):1468-1483.
8. Zhang Hui-bin. Different boundary conditions on the influence of tunnel fire simulation results. *Master's degree thesis of Southwest Jiaotong University.2007.*

Study on the statistics — Infiltration coupling forecast model for rainfall-induced landslide in Chongqing area

Hong-Kai Chen*, Si-Qi Chen, Ming Yang, Tao Chen, Ke-Xuan Guo
College of Hydraulic & environmental Engineering, China Three Gorges University, Yichang, Hubei, 443002, China
**E-mail: chkcxf@163.com*

Considering the maximum infiltration depth in the role of heavy rainfall and taking into account the water pressure in tension cracks located in trailing edge of landslide, this paper revises the transfer coefficient method traditionally used to determine landslide stability factor and thrust force. What's more, based on integrating the effective cumulated rainfall and the daily maximum hourly rainfall, this article establishes the statistics-infiltration coupling forecast model for rainfall-induced landslide in Chongqing territory.

Keywords: Statistics-Infiltration Coupling Forecast Model; Index of Landslide Occurrence; Rainfall-Induced Landslide; Weather Forecasting; Chongqing Territory.

1. Introduction

Landslide is one of the major types of geological disaster in China, which is particularly sensitive to the process of response to rainfall [1]. Khan Y.A. et al. had made a statistics about the landslide rainfall occurred in Bangladesh Chittagong, and studied the response relationship for landslide with intensity and duration of rainfall [2]; Montrasio L. et al. had analyzed the response relationship between safety factor of rainfall-induced landslide and rainfall intensity from the geological and climatic perspective surrounding Emilia-Romagna Apennines [3]; Gattinoni P. et al. through field investigation into landslide which occurred in southern Italy Maierato, had studied the before-and-after rainfall characteristics of landslide failure and instability mechanism under simulated rainfall condition [4]; Capparelli G. et al. had built his landslide forecasting model (FLAIR) [5]; Enrico C and Antonello T had analyzed the mechanical mechanism about rainfall-induced shallow landslide [6]; Li Liu et al. found that the presence of cracks on landslide surface has a greater effect on the stability of landslide [7]. Jian Chen et al. discovered that the largest rainfall intensity within 24h before landslide occurred could be used as an effective predictor of landslide [8]; Zhen

Zhang et al. suggested that landslides activity, behind the border of Chongqing, was proportional to rainfall intensity and landslides mostly occurred in 4 days after the rainfall [9].

However, at present, the meteorological department adopts landslide hazard weather forecasting model part of statistical models, which is only suitable for regional forecasting and can't apply to the monomer landslide. In view of this point, this text comes up with landslide forecasting statistical model suitable for regional weather forecasting. At the same time, considering the effect of rainfall infiltration, this article also built statistics-infiltration coupling forecast model for rainfall-induced landslide aiming at monomer landslide.

2. Statistics Forecast Model for Rainfall-Induced Landslide in Chongqing

Upon the 285 rainfall-induced landslides of Chongqing area, this article proposes a corrected rainfall-induced landslides forecasting model:

$$P = 0.8P(R') + 0.2P(R_h)\,. \tag{1}$$

Where P is the probability of occurrence for rainfall-induced landslide ranging from 0 to 1; $P(R')$ is the probability of occurrence for rainfall-induced landslide by the effective cumulated rainfall forecasting model; $P(R_h)$ is the probability of occurrence for rainfall-induced landslide by maximum hourly rainfall predicting model, respectively; The fitting formulas are (2) and (3):

$$P(R') = -6\times10^{-5}\times R'^3 + 0.016\times R'^2 - 0.3909R' + 1.0347\,. \tag{2}$$

$$P(R_h) = -2\times10^{-6}R_h^5 + 4\times10^{-4}R_h^4 - 0.0252R_h^3 + 0.5602R_h^2 - 0.7039R_h + 3.5157\,. \tag{3}$$

Where R' is the effective cumulated rainfall (mm); R_h is maximum hourly rainfall (mm).

3. Stability Coefficient Calculation Method for Landslide Considering Rainfall in Filtration

We should regard the rainfall infiltration area controlled by maximum rainfall infiltration depth to be saturated. When $h_{\mathrm{inf}} < H$ (h_{inf} represents maximum rainfall infiltration depth (m), H represents the average thickness of landslide (m)), sliding surface is in natural state, and the impact of rainfall infiltration on slope stability is mainly reflected on gaining self-weight of infiltration-zone soil ; When $h_{\mathrm{inf}} \geq H$, the whole landslide including sliding surface is saturated and the related strength parameters should take saturated state parameters. When there exists tension cracks on the trailing edge of landslides, we can define the area as seepage-affected zone. In the area, mechanical parameters of landslide body and sliding surface take saturated parameters.

Based on the rainfall infiltration experiment study on typical strongly weathered mudstone in Chongqing area, writer proposed empirical formula to determine the maximum rainfall infiltration depth

$$h_{\inf} = 139.8219I - 1.7848t + 0.0919t^2 - 569.9429I^2 . \tag{4}$$

Where I is rainfall intensity (mm / min); t is rainfall duration (min); $h_{\inf}$ is the maximum infiltration depth (mm).Depth of soil tensile cracks can be determined

$$Z = 2c \Big/ \gamma\sqrt{\tan(45^\circ - \frac{\varphi}{2})} \ . \tag{5}$$

Where Z is depth of soil tensile cracks (m); c is soil cohesion force in natural state (kPa); γ is bulk density of soil in natural state (kN/m³); φ is internal friction angle of soil in natural state (°). Rainfalls' water pressure in fissure, the pressure value can be calculated by following formula

$$P_w = \frac{1}{2}\gamma_w Z^2 \ . \tag{6}$$

Where P_w is the water pressure in fissure, γ_w is bulk density of water, γ_w =10kN/m³; Z is depth of tensile cracks (m); Currently, the transfer coefficient is often used in civil engineering to finger out landslide stability coefficient[10]. This paper introduces the role of rainfall infiltration to correct transfer coefficient method. Calculation formula of landslide stability coefficient is

$$F_s = \sum_{i=1}^{n-1}(R_i \prod_{j=i}^{n-1}\psi_j) + R_n \Big/ \sum_{i=1}^{n-1}(T_i \prod_{j=i}^{n-1}\psi_j) + T_n \ . \tag{7}$$

Where R_i means skid-resistance force acting on the ith slice (kN/m); R_h means skid-resistance force acting on the nth slices (kN/m); T_i is component force acting on the ith slice along the landslide surface (kN/m); T_i is component force acting on the nth slice along the landslide surface (kN/m); φ_j is the transfer coefficient from i^{th} slice to $(i+1)^{th}$, $j = i$; W_i is the self-weight of the ith slice after rainfall infiltration (kN/m); n is the number of slices. Calculation formulas about related variables are as follows

$$\psi_i = \cos(\alpha_i - \alpha_{i+1}) - \sin(\alpha_i - \alpha_{i+1})\tan\varphi. \tag{8}$$

$$\prod_{j=i}^{n-1}\psi_j = \psi_i \cdot \psi_{i+1} \cdot \psi_{i+2} \cdots \psi_{n-1} . \tag{9}$$

$$R_i = W_i \cos\alpha_i \tan\varphi + cl_i . \tag{10}$$

$$T_i = W_i \sin\alpha_i . \tag{11}$$

$$W_{\rm i} = W_{1i} + W_{2i} = [\gamma_{sat} h_{i^{\rm th}\rm inf} + \gamma(H_i - h_{i^{th}\rm inf})] l_{\rm i} \cos\alpha_i . \tag{12}$$

Where φ_j is the friction angle on sliding surface of the ith slice (°); c_i is the cohesion on sliding surface of the ith slice (kPa); α_i is the angle of sliding surface of the ith slice (°); l_i is the length of sliding surface of the ith slice (m).Further, taking the hydrostatic water pressure in tension cracks into consideration, the formula (7) changes to another one, like formula (13)

$$F_s = \sum_{i=1}^{n-1}(R_i \prod_{j=i}^{n-1} \psi_j) + R_n \Bigg/ \sum_{i=1}^{n-1}(T_i \prod_{j=i}^{n-1} \psi_j) + T_n + P_w . \tag{13}$$

4. The Statistics-Infiltration Coupling Forecast Model in Chongqing

Based on formula (1) and (13), the paper comes up with a statistics-infiltration coupling forecast model for rainfall-induced landslide which is only suitable for monomer landslide in territory of Chongqing.

$$\eta = p/F_s . \tag{14}$$

Where η is coupling occurrence index of landslide statistics-rainfall infiltration. Aiming at monomer rainfall-induced landslide, writer puts forward related prediction grades shown in table 1.

Table 1. Forecast grades for rainfall-induced landslide.

Forecasting grades	η / %	P / %	F_s	Possibility of occurrence for landslide
Grade 1	[0, 8.7)	[0, 10)	≥1.15	Minimal
Grade 2	[8.7, 22.7)	[10, 25)	[1.10, 1.15)	Little
Grade 3	[22.7, 47.6)	[25, 50)	[1.05, 1.10)	Bigger
Grade 4	[47.6, 75)	[50, 75)	[1.00, 1.05)	Large
Grade 5	[75, +∞)	[75, 95)	<1	Maximal

5. Model Experiment

The area is 460cm × 400cm. There is uniformly filled with 60cm-thick silty clay and its mainly physical and mechanical parameters see table 2. Rainfall device

(Patent No.ZL201110113318.7) selects 7.5mm/10min for 40 minutes first, after 120 minutes then continue raining. According to macroscopic judgment method on landslide stability, the stability factor for the landslide model is 0.85~0.9 [10] .

Table 2. Physical and mechanical parameters of experimental soil.

	Natural state	Saturated state
Water content (%)	12	20
Bulk density (kN/m³)	21.2	21.9
Cohesion (kPa)	16.1	5.4
Internal friction angle (°)	15.94	12.68
Permeability coefficient (cm/s)	/	4.5×10^{-6}

In the process of model experiment, there were two kinds of rainfall modes: long-time continual rainfall and short-time heavy rainfall. The former mode, on 12 November, 2012, had a rainfall of 60mm, which was divided into twice (30mm each time); the later, on 13 November, 2012 had a rainfall of 90mm and lasted two hours that was equal to average rainfall intensity of 45mm/h. In the whole process, it took a total of 2 days (2880 minutes), so the average rainfall intensity was 0.052mm/min. Finally, we can use formula (4) to determine the maximum depth of rainfall infiltration:

$$h_{\inf} = 139.8219I - 1.7848t + 0.0919t^2 - 569.9429I^2 = 75.19cm.$$

Through formula (5), we can calculate the tensile crack depth generated on the trailing edge of landslide:

$$Z = \frac{2c}{\gamma\sqrt{K_a}} = \frac{2c}{\gamma \times \tan(45^\circ - \frac{\varphi}{2})} = \frac{2\times 16.1}{21.2\times\tan(45^\circ - \frac{15.94}{2})} = 201cm.$$

Considering the most unfavorable conditions, we took value of 60cm for the depth of tension crack, then

$$P_w = \frac{1}{2}\gamma_w Z^2 = 0.5\times 10\times 0.6^2 = 1.8\text{kN}.$$

$$W = W_1 + W_2 = [\gamma_{sat} h_{\inf} + \gamma(H - h_{\inf})]l\cos 30^\circ = 21.9\times 0.6\times 4\times\cos 30^\circ = 45.52\text{kN}$$

$$R = \left(45.52\times\cos 30^\circ \text{-} 1.8\times\sin 30^\circ\right)\tan 12.68^\circ + 5.4\times 4 = 30.27\text{kN}.$$

$$T = W\sin\alpha + P_w\cos\alpha = 45.52\times\sin 30^\circ + 1.8\times\cos 30^\circ = 24.32\text{kN}.$$

The sliding surface was flat, so the transfer coefficient is 1, then

$$F_s = \frac{\sum_{i=1}^{n-1}(R_i\prod_{j=i}^{n-1}\psi_j)+R_n}{\sum_{i=1}^{n-1}(T_i\prod_{j=i}^{n-1}\psi_j)+T_n} = \frac{\sum_{i=1}^{n-1}(R_i\times 1)+R_n}{\sum_{i=1}^{n-1}(T_i\times 1)+T_n} = \frac{\sum R}{\sum T} = \frac{24.32}{30.27} = 0.80\,.$$

We can confirm rainfall parameters $R' = 150mm$, $R_h = 45mm$, put them into formula (1), we have flowing result

$$P = 0.8\times P(R') + 0.2\times P(R_h) = 0.8\times 0.86 + 0.2\times 0.81 = 0.85.$$

Finally, based on formula (14), the failure probability of landslide model experiment can be obtained

$$\eta = \frac{p}{F_s} = \frac{0.85}{0.80} = 1.0625\,.$$

The results indicated that the landslide model instability index was 106.25% and the corresponding landslide forecasting level was Grade 5, namely, it was highly likely to landslide. All the results are consistent with the observations.

6. Conclusions

In summary, the paper has revised the transfer coefficient method traditionally used to determine landslide stability factor and thrust force and has put forward forecasting grades for landslide and built statistics-infiltration coupling forecast model for rainfall-induced landslide which is suitable for the monomer landslide forecasting in territory of Chongqing area. By indoor model experiment, we have proved the statistics-infiltration coupling forecast model for rainfall-induced landslide in Chongqing area to be reasonable. It can effectively make weather forecast for the typical monomer landslide potentially occurring under the condition of heavy rainfall in Chongqing.

References

1. Ma Li, Zeng Xiang-ping, Xiang Ping. The conditions analysis of landslides occur precipitation in Chongqing[J]. Journal of Mountain Research, 2002, 20(2): 246-249.
2. Khan Y. A, Lateh M. H, Baton A, et al. Critical antecedent rainfall conditions for shallow landslide in Chittagong City of Bangladesh [J]. Environmental Earth Sciences, 2012 (67): 97- 106.
3. Montrasio L, Valentino R, Lois G L. Shallow landslides triggered by rainfalls: modeling of some case histories in the Reggiano Apennine (Emilia Romagna Region, Northern Italy) [J]. Natural Hazards, 2012 (60): 1231-1254.

4. Gattinoni P, Scesi L, Arieni L, et al. The February 2010 large landslide at Maierato, Vibo Valentia, Southern Italy [J] Landslides, 2012(9): 255-261.
5. Capparelli G, Pasquale I, Giulio G R, et al. Modeling the rainfall-induced mobilization of a large slope movement in northern Calabria [J] . Natural Hazards, 2012(61): 247-256.
6. Enrico C, Antonello T. Simplified Approach for the Analysis of Rainfall-Induced Shallow Landslides[J]. Journal of Geotechnical and Geoenvironmental Engineering, 2012, 138(3): 398-406.
7. Liu Li-ling, Yin Kun-long. Analysis of rainfall infiltration mechanism of rainstorm landslide [J]. Rock and Soil Mechanics, 2008, 29(4): 1061-1066.
8. Chen Jian, Yang Zhifa, Li Xiao. Relationship between landslide probability and rainfall in Three Gorges Reservoir Area[J]. Chinese Journal of Rock Mechanics and Engineering, 2005, 24(17): 3052-3056.
9. Zhang Zhen, Li Shihai, Ma Li. Probability Analysis of Relationship Between Landslide and Rainfall in Chongqing Area[J]. Chinese Journal of Rock Mechanics and Engineering, 2005, 24(17): 3185-3191.
10. The Chongqing Provincial standards. Design standard of Geological disasters prevention and control, DB 50/5029-2004.

New type element for elastic foundation beam

Jun-Xing Zhang, Min Ding*, Shi-Ping Li, Xiao Li, Shi-Ping Meng

College of Water Resources and Civil Engineering, China Agricultural University, Beijing, 100083, China

**E-mail: dingmin@cau.edu.cn*

According to the general analytical solution of deflection equation deduced from Winkler elastic foundation beam theory, the element displacement shape function was determined. The analytical element stiffness matrixes and node equivalent load arrays were established via the principle of minimum potential energy. The analytical Euler beam element AWFB-E and Timoshenko beam element AWFB-T were obtained. Meanwhile, the Euler beam element RWFB-E and Timoshenko beam element RWFB-T deduced from Ritz method were received. Finally, the comparisons between the calculation results and theoretical solutions were conducted. The results show that the precision and efficiency of existing finite element models are insufficient. The results of RWFB-E do not match well with theoretical solution, but the difference becomes smaller with element number increasing. The results of AWFB-E are almost identical with theoretical solution irrespective of element number. This means the displacement shape function is much more accurate than that of general polynomial. The corresponding elastic foundation beam elements are accurate and efficient and can be applied in practice.

Keywords: Elastic Foundation Beam; Finite Element; Principle of Minimum Potential Energy; High-Precision Element.

1. Introduction

Elastic foundation beams are mainly employed in harbor engineering, hydraulic engineering, and civil engineering. Foundation calculation model in accord with the characteristics of beam-foundation system was firstly put forward by Winkler. After that, double parameter foundation model was presented [1, 2]. Nowadays, a large number of theories and methods for internal force and deformation calculation of beam have been suggested [3-7].

Finite element method was firstly proposed by Clough [8]. The basic analytical solution was used as the element trial function in analytical trial function method. Discrete finite element method takes advantage of analytical solution [9]. It embodies the superiority of trial function using basic analytical solution. A simple finite element model which was used to evaluate rectangular plate behavior was developed after thermal buckling [10]. A new finite element method based on elastic foundation beam theory was deduced by applying variational principle [11].

This paper focuses on constructing new type elastic foundation beam element using analytical trial function method based on Winkler elastic foundation beam theory by applying principle of minimum potential, which will be useful to the calculation of internal force and deformation of elastic foundation beam.

2. Construction Method of Beam Element

Fig. 1 presents load condition of Winkler foundation model. The displacement of beam and foundation loaded with $q(x)$ is $w(x)$, and the corresponding pressure between beam and foundation is $p(x)$.

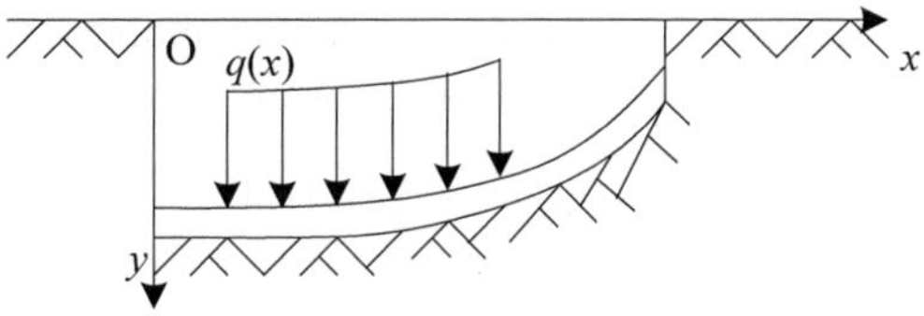

Fig. 1. Load condition of Winkler foundation mode.

2.1. *Element displacement shape function*

Element displacement and node displacement are defined as $\{\Delta\}=\{u(x)\ w(x)\}^T$ and $\{\delta\}^e=\{u_1\ \ w_1\ \ \theta_1\ \ u_2\ \ w_2\ \ \theta_2\}$ respectively. The relationship between element displacement and node displacement is $\{\Delta\}=[N]\{\delta\}^e$.

Axial displacement shape function $[N_u]$ and flexural displacement shape function $[N_W]$ are calculated respectively irrespective of the interaction of axial deformation and flexural deformation. The corresponding element displacement shape function matrix can be expressed as

$$[N]=\{N_{\mathrm{u}}\quad N_{\mathrm{w}}\}^{\mathrm{T}}. \tag{1}$$

In Eq. (1), $[N]_u$ determined by Lagrange linear interpolation function is expressed as

$$[N_{\mathrm{u}}]=[1-x/l\quad 0\quad 0\quad x/l\quad 0\quad 0]. \tag{2}$$

$[N_W]$ deduced from Hermite linear interpolation function is expressed as

$$[N_{\mathrm{w}}]=[0\quad N_1\quad N_2\quad 0\quad N_3\quad N_4]. \tag{3}$$

where $N_1 = 1 - 3(x/l)^2 + 2(x/l)^3$, $N_2 = l\{x/l - 2(x/l)^2 + (x/l)^3\}$,
$N_3 = 3(x/l)^2 - 2(x/l)^3$, $N_4 = l\{-(x/l)^2 + (x/l)^3\}$.

2.2. *Establishment of potential energy functional*

According to energy principle, potential energy functional of Euler beam is given by

$$\Pi_p = U + V = \frac{1}{2}\int_0^l EAu'^2(x)dx + \frac{1}{2}\int_0^l EIw''^2(x)dx - \sum_{i=1}^{n} P(x_i)w(x_i) - \int_0^l q(x)w(x)dx + \frac{1}{2}\int_0^1 kw^2(x)dx. \quad (4)$$

Where U is element strain energy, V is element load potential energy, $q(x)$ is uniformly distributed load on the element, k is Winkle foundation coefficient, E is elastic modulus, I is cross-sectional moment of inertia.

Potential energy function of Timoshenko beam is expressed as

$$\Pi_p = U + V = \frac{1}{2}\int_0^l EAu'^2(x)dx + \frac{1}{2}\int_0^l EIw''^2(x)dx + \frac{a_s(EI)^2}{2GA}\int_0^l w'''^2(x)dx + \frac{1}{2}\int_0^1 kw^2(x)dx$$
$$-\sum_{i=1}^{n} P(x_i)w(x_i) - \int_0^l q(x)w(x)dx. \quad (5)$$

where A is cross-sectional area, a_s is shear coefficient, G is elastic shear modulus.

2.3. *Establishment of stiffness matrix*

In the light of principle of minimum potential energy, which is $\frac{\partial \Pi_p}{\partial \{\delta\}^e} = 0$, element stiffness equation is obtained:

$$[K]^e\{\delta\}^e = [P]^e. \quad (6)$$

where element stiffness matrix of Euler beam is expressed as

$$[K]^e = \int_0^l EA[N'_u]^T[N'_u]dx + \int_0^l k[N_w]^T[N_w]dx + \int_0^l EI[N''_w]^T[N''_w]dx. \quad (7)$$

Element stiffness matrix of Timoshenko beam is expressed as

$$[K]^e = \int_0^l EA[N'_u]^T[N'_u]dx + \int_0^l k[N_w]^T[N_w]dx + \int_0^l EI[N''_w]^T[N''_w]dx + \frac{a_s(EI)^2}{GA}\int_0^l [N'''_w]^T[N'''_w]dx. \quad (8)$$

Element node equivalent load array is expressed as

$$[P]^{e}=\sum_{i=1}^{n}P(x_i)[N_w]^T+\int_0^l q(x)[N_w]^T dx. \tag{9}$$

3. Elastic Foundation Beam Element Via Ritz Method

By applying principle of minimum potential energy, after matrix operation, Euler beam element stiffness matrix $[K]^e$ can be expressed as

$$[K]^e=\begin{bmatrix} EA/l & 0 & 0 & -EA/l & 0 & 0 \\ 0 & k_{11} & k_{12} & 0 & k_{13} & k_{14} \\ 0 & k_{21} & k_{22} & 0 & k_{23} & k_{24} \\ -EA/l & 0 & 0 & EA/l & 0 & 0 \\ 0 & k_{31} & k_{32} & 0 & k_{33} & k_{34} \\ 0 & k_{41} & k_{42} & 0 & k_{43} & k_{44} \end{bmatrix}. \tag{10}$$

where

$k_{11}=k_{33}=\frac{12EI}{l^3}+\frac{13kl}{35}$, $k_{12}=k_{21}=\frac{6EI}{l^2}+\frac{11kl^2}{210}$, $k_{13}=k_{31}=-\frac{12EI}{l^3}+\frac{9kl}{70}$,

$k_{14}=k_{41}=\frac{6EI}{l^2}-\frac{13kl^2}{420}$, $k_{22}=k_{44}=\frac{4EI}{l}+\frac{kl^3}{105}$, $k_{23}=k_{32}=-\frac{6EI}{l^2}+\frac{13kl^2}{420}$,

$k_{24}=k_{42}=\frac{2EI}{l}-\frac{kl^3}{140}$, $k_{34}=k_{43}=-\frac{6EI}{l^2}-\frac{11kl^2}{210}$.

Finally, load array can be obtained with $[P]^{e}=\sum_{i=1}^{n}P(x_i)[N_w]^T+\int_0^l q(x)[N_w]^T dx$, and it is identical with node load array of general two-node beam element.

Similarly, Timoshenko beam element stiffness matrix $[K]^e$ can be expressed as Eq. (10), and the element value in the matrix is given by

$k_{11}=k_{33}=12\frac{EI}{l^3}+\frac{13}{35}k+\frac{144a_s(EI)^2}{GAl^5}$, $k_{12}=k_{21}=6\frac{EI}{l^2}+\frac{11}{210}kl+\frac{72a_s(EI)^2}{GAl^4}$,

$k_{13}=k_{31}=-12\frac{EI}{l^3}+\frac{9}{70}k+\frac{144a_s(EI)^2}{GAl^5}$, $k_{14}=k_{41}=6\frac{EI}{l^2}-\frac{13}{420}kl+\frac{72a_s(EI)^2}{GAl^4}$,

$k_{22}=k_{44}=4\frac{EI}{l}+\frac{k}{105}l^2+\frac{36a_s(EI)^2}{GAl^3}$, $k_{23}=k_{32}=-6\frac{EI}{l^2}+\frac{13}{420}kl+\frac{72a_s(EI)^2}{GAl^4}$,

$k_{24}=k_{42}=2\frac{EI}{l}-\frac{k}{140}l^2+\frac{36a_s(EI)^2}{GAl^3}$, $k_{34}=k_{43}=-6\frac{EI}{l^2}-\frac{11}{210}kl+\frac{72a_s(EI)^2}{GAl^4}$.

The calculation method of load array $[P]^e$ is identical with the previous section.

4. Analytical Elastic Foundation Beam Element

4.1. *Element displacement shape function*

Flexural deflection function $w(x)$ is the basic unknown quantity of elastic foundation beam, and its basic control equation is $EI\,d^4w(x)/dx^4 + kw(x) = q(x)$.

Analytical solution of deflection equation of Euler beam is expressed as

$$w(x) = C_1e^{\lambda x}\cos\lambda x + C_2e^{\lambda x}\sin\lambda x + C_3e^{-\lambda x}\cos\lambda x + C_4e^{-\lambda x}\sin\lambda x = \sum_{i=1}^{4} f_i(x)c_i = \{F\}\{C\}. \quad (11)$$

where $\{F\} = \{f_1(x) \quad f_2(x) \quad f_3(x) \quad f_4(x)\}$,

$f_1(x) = e^{\lambda x}\cos\lambda x$, $f_2(x) = e^{\lambda x}\sin\lambda x$, $f_3(x) = e^{-\lambda x}\cos\lambda x$, $f_4(x) = e^{-\lambda x}\sin\lambda x$.

If displacement vector of beam end node is defined as

$$\{\delta\} = \{w_1 \quad \theta_1 \quad w_2 \quad \theta_2\}^{\mathrm{T}}. \quad (12)$$

According to Eq.(11), Eq.(12) becomes

$$\{\delta\} = \begin{bmatrix} f_1(0) & f_2(0) & f_3(0) & f_4(0) \\ f_1'(0) & f_2'(0) & f_3'(0) & f_4'(0) \\ f_1(l) & f_2(l) & f_3(l) & f_4(l) \\ f_1'(l) & f_2'(l) & f_3'(l) & f_4'(l) \end{bmatrix} \{C\} = [A]\{C\}. \quad (13)$$

where $[A] = \begin{bmatrix} 1 & 0 & 1 & 0 \\ \lambda & \lambda & -\lambda & \lambda \\ e^{\alpha}\cos\alpha & e^{\alpha}\sin\alpha & e^{-\alpha}\cos\alpha & e^{-\alpha}\sin\alpha \\ \lambda e^{\alpha}(\cos\alpha - \sin\alpha) & \lambda e^{\alpha}(\cos\alpha + \sin\alpha) & -\lambda e^{-\alpha}(\cos\alpha + \sin\alpha) & \lambda e^{-\alpha}(\cos\alpha - \sin\alpha) \end{bmatrix}$,

$\alpha = \lambda l$.

The undetermined parameter can be expressed as

$$\{C\} = [A]^{-1}\{\delta\}. \quad (14)$$

With Eq. (14), Eq. (11) becomes

$$w(x) = \{F\}\{C\} = \{F\}[A]^{-1}\{\delta\} = [\tilde{N}]\{\delta\}. \quad (15)$$

Element deflection shape function can be expressed as

$$[\tilde{N}] = \{F\}[A]^{-1} = [N_1 \quad N_2 \quad N_3 \quad N_4]. \quad (16)$$

where N_1, N_2, N_3, N_4 are deflection shape functions. And after matrix operation, they can be given by

$$N_1 = \frac{1}{S}\begin{bmatrix} \left(e^{\lambda x-2\alpha} + e^{-\lambda x+2\alpha} - 2e^{-\lambda x} - 2e^{\lambda x}\right)\cos\lambda x \\ +\left(e^{-\lambda x} - e^{\lambda x}\right)\sin\left(-\lambda x + 2\alpha\right) \\ +\left(e^{-\lambda x} + e^{\lambda x}\right)\cos\left(-\lambda x + 2\alpha\right) \end{bmatrix},\ N_2 = \frac{1}{\lambda S}\begin{bmatrix} 2\left(e^{-\lambda x} - e^{\lambda x}\right)\cos\lambda x \\ +2\left(e^{\lambda x} - e^{-\lambda x}\right)\cos\left(-\lambda x + 2\alpha\right) \\ +2\left(e^{\lambda x-2\alpha} + e^{-\lambda x+2\alpha} + e^{\lambda x} + e^{-\lambda x}\right)\sin\lambda x \end{bmatrix},$$

$$N_3 = \frac{1}{S}\begin{bmatrix} \begin{pmatrix} e^{-\lambda x-\alpha} + e^{\lambda x+\alpha} \\ -2e^{\lambda x-\alpha} - 2e^{-\lambda x+\alpha} \end{pmatrix}\cos\left(-\lambda x + \alpha\right) \\ +\left(e^{\lambda x-\alpha} + e^{-\lambda x+\alpha}\right)\cos\left(\lambda x + \alpha\right) \\ +\left(e^{\lambda x-\alpha} - e^{-\lambda x+\alpha}\right)\sin\left(\lambda x + \alpha\right) \\ +\left(e^{\lambda x+\alpha} - e^{-\lambda x-\alpha}\right)\sin\left(-\lambda x + \alpha\right) \end{bmatrix},\ N_4 = \frac{1}{\lambda S}\begin{bmatrix} \left(e^{-\lambda x+\alpha} - e^{\lambda x-\alpha}\right)\cos\left(-\lambda x + \alpha\right) \\ +\begin{pmatrix} e^{\lambda x-\alpha} - e^{\lambda x+\alpha} \\ -e^{-\lambda x-\alpha} + e^{-\lambda x+\alpha} \end{pmatrix}\sin\left(-\lambda x + \alpha\right) \\ +\left(e^{\lambda x-\alpha} - e^{-\lambda x+\alpha}\right)\cos\left(\lambda x + \alpha\right) \end{bmatrix}.$$

where $S = e^{-2\lambda l} + e^{2\lambda l} - 4\sin^2\lambda l - 2$.

With element node displacement $\{\delta\}^e$, element deflection becomes

$$w = [N_w]\{\delta\}^e. \tag{17}$$

Element deflection shape function matrix is given by

$$[N_w] = [0\quad N_1\quad N_2\quad 0\quad N_3\quad N_4]. \tag{18}$$

4.2. *Euler beam element and Timoshenko beam element*

By applying principle of minimum potential energy, after matrix operation, analytical Euler beam element stiffness matrix $[K]^e$ can be expressed as Eq.(10), and the element value in the matrix is given by

$$k_{11} = -\frac{\lambda A}{4\left(1 - 6e^{2\lambda l} + 4e^{2\lambda l}\cos^2\lambda l + e^{4\lambda l}\right)^2}. \tag{19}$$

where

$$\begin{aligned} A = {} & 4EI\lambda^4\left(1 + 6e^{6\lambda l} - e^{8\lambda l}\right) - 32\lambda lke^{4\lambda l}\sin^2\lambda l - 6k\left(e^{2\lambda l} + e^{6\lambda l}\right)\sin 2\lambda l + 3k\left(1 - e^{8\lambda l}\right) \\ & -16EI\lambda^4\cos^2\lambda l\left[e^{6\lambda l}\left(\tan\lambda l + 1\right) - e^{2\lambda l}\right] + \left(12 - 18k\right)\left(e^{2\lambda l} - ke^{6\lambda l}\right)\cos^2\lambda l \\ & -16\lambda lk\left(e^{6\lambda l} + e^{2\lambda l}\right)\left(\cos^2\lambda l + 1\right) + 4e^{4\lambda l}\left(3k + 4EI\lambda^4\right)\left(1 + 2\sin^2\lambda l\right)\sin 2\lambda l \\ & -64EIl\lambda^5\left(e^{2\lambda l} + e^{6\lambda l} + 2e^{4\lambda l}\right)\sin^2\lambda l. \end{aligned}$$

Other element values are not listed here due to the space limitation, and more details can be found in reference [6].

When element is loaded with concentrated load, equivalent load array can be expressed as

$$P(x_i)[N]^{T} = P\begin{bmatrix} 0 \\ N_1(x_p) \\ N_2(x_p) \\ 0 \\ N_3(x_p) \\ N_4(x_p) \end{bmatrix}. \tag{20}$$

where x_p is the absolute coordinate of force P point in the element.

If element is loaded with uniformly force $q(x) = q_0$, then

$$\int_0^l q(x)[N_w]^T dx = q_0\{0 \quad P_1 \quad P_2 \quad 0 \quad P_3 \quad P_4\}^T. \tag{21}$$

where

$$P_1 = P_3 = \frac{-\tilde{P}}{\alpha - 6\alpha e^{2\alpha l} + \alpha e^{4\alpha l} + 4\alpha e^{2\alpha l}\cos^2(\alpha l)}, P_2 = -P_4 = -\frac{1 - e^{2\alpha l} + 2e^{\alpha l}\sin(\alpha l)}{2\alpha^2 e^{2\alpha l} - 2\alpha^2 + 4\alpha^2 e^{\alpha l}\sin(\alpha l)},$$

$$\tilde{P} = 1 - \cos(\alpha l)\left[2e^{\alpha l} - 2e^{3\alpha l} + 4e^{2\alpha l}\sin(\alpha l)\right] - e^{4\alpha l} + 2e^{\alpha l}\sin(\alpha l) + 2e^{3\alpha l}\sin(\alpha l).$$

Similarly, analytical Timoshenko beam element stiffness matrix $[K]^e$ can be expressed as Eq.(10). Element values in the matrix are not listed here due to the space limitation, and more details can be found in reference [6].

The calculation method of equivalent load array is identical with the previous section.

5. Comparison and Verification

5.1. *Calculation model*

The parameters of calculation model are as follows: beam dimension is *4m×0.5m×0.5m*, elastic modulus $E_C = 1.35\times10^4 MPa$, poisson ratio $\rho = 0.3$, shear modulus $G = 5.19\times10^9 MPa$; foundation bed coefficient $k_0 = 1.5\times10^8 N/m^3$, foundation poisson ratio $\rho_s = 0.35$, foundation elastic modulus $E_S = 45MPa$.

5.2. *Element precision* and *element efficiency*

The beam is divided into 2 elements and 3 elements respectively according to force point. The calculation results are shown in Table 1-2.

The beam is divided into 4 elements, 8 elements and 12 elements respectively. The calculation results are shown in Table 3-4.

Table 1. Result comparisons with concentrated force at mid-span.

Element number	Location		0	$l/2$	l
2 elements	Theoretical solution	Value (mm)	0.18248	0.43605	0.18248
	Chain-bar method	Value (mm)	0.13701	0.39168	0.13701
		Relative error (%)	-24.9178	-10.1754	-24.9178
	Coupling method	Value (mm)	0.04751	0.09351	0.04751
		Relative error (%)	-73.9637	-78.5552	-73.9637
	RWFB-E	Value (mm)	0.17912	0.43239	0.17912
		Relative error (%)	-1.8413	-0.8394	-1.8413
	AWFB-E	Value (mm)	0.18248	0.43605	0.18248
		Relative error (%)	0.0000	0.0000	0.0000

Table 2. Result comparisons with symmetrical concentrated force.

Element number	Location		0	$l/4$	$3l/4$	l
3 elements	Theoretical solution	Value (mm)	0.60313	0.68110	0.68110	0.60313
	Chain-bar method	Value (mm)	0.52130	0.80739	0.80739	0.52130
		Relative error (%)	-13.5676	18.5421	18.5421	-13.5676
	RWFB-E	Value (mm)	0.42875	0.71527	0.71527	0.42875
		Relative error (%)	-28.9125	5.0169	5.0169	-28.9125
	AWFB-E	Value (mm)	0.60313	0.68110	0.68110	0.60313
		Relative error (%)	0.0000	0.0000	0.0000	0.0000

Table 3. Result comparisons with concentrated load at mid-span (relative error, %).

Element number	Chain-bar method	Coupling method	RWFB-E	AWFB-E
2 elements	-10.1754	-78.5552	-0.8394	0.0000
4 elements	-6.0796	-78.2431	-0.0527	0.0000
8 elements	-0.2041	-77.7638	-0.0023	0.0000
12 elements	-0.6398	-77.6425	1.0297	0.0000

Table 4. Result comparisons with symmetrical concentrated load (relative error, %).

Element number	Chain-bar method	Coupling method	RWFB-E	AWFB-E
4 elements	-7.8933	-79.4797	-0.0639	0.0000
8 elements	-4.7308	-79.8329	-0.0044	0.0000
12 elements	-3.6698	-79.8619	1.0479	0.0000

5.3. *Result analysis*

(1) The results of AWFB-E are almost identical with theoretical solution irrespective of element number. Then just divide beam element at force point to meet the precision requirements, and have a better match with theoretical solution. It indicates that AWFB-E is analytical.

(2) The deflection value at mid-span of RWFB-E is close to theoretical solution, while at the end, it is far away from theoretical solution. It indicates that the actual beam deflection curve is not polynomial function. Meanwhile, with the beam element number increases, the difference of deflection value at

mid-span and theoretical solution is larger. It indicates that the precision of elastic foundation beam element via Ritz method is far below that of analytical element.

(3) Chain-bar method result for simulating elastic foundation beam is not in accordance with theoretical solution. Its relative error is about 10%. The larger relative error leads to precision insufficiency of the method. But the relative error becomes smaller with element number increasing. It indicates that the efficiency of the method is insufficiency.

6. Conclusions

Based on the results of this investigation, the following conclusions can be drawn:

(1) Chain-bar method result and coupling method result for simulating elastic foundation beam are not in accordance with theoretical solution. It indicates that the precision and efficiency of existing finite element models are insufficient.

(2) Elastic foundation beam element stiffness matrix is obtained via Ritz method. The corresponding deformation is calculated by establishing elastic foundation beam model, and is compared with theoretical solution. The result indicates that polynomial function used for displacement trial function of elastic foundation beam can't meet the requirements of precision and efficiency.

(3) Elastic foundation beam element stiffness matrix is obtained by applying analytical trial function put forward in this paper. The corresponding deformation is calculated by establishing elastic foundation beam model, and is compared with theoretical solution. The result indicates that elastic foundation beam element in the light of analytical trial function put forward in this paper is high-precision element.

Acknowledgments

The research is supported by National Natural Science Funds (Grant No. 11272340), Beijing Natural Science Foundation (Grant No.3144029), Key Laboratory of Farm Building in Structure and Construction, Ministry of Agriculture, P.R. China (Grant No.201502), and Chinese Universities Scientific Fund (Grant No. 2015SYL004).

References

1. Filonenko M., Borodich M, Some approximate theories of the elastic foundation, *Uchenyie Zapiski Moskovskogo Gosudarstvennoho Universiteta Mekhanica*, 46, 3-18, (1940).

2. Hetenyi M, Beams on elastic foundations, *The University of Michigan Press*, Ann Arbor, Michingan, (1946).
3. Chen Y., Qian J, Fuzzy finite element analysis of beams and piles on Winkler foundation, *Engineering Mechanics*, 21(1), 125-173, (2004).
4. Sun F. X., Cai X. H., Zhu Y. H, Analytical solution of internal force and displacement in multi-center circular arc tunnel lining based on initial parameter method, *Rock and Soil Mechanics*, 30(4), 1127-1130, (2009).
5. Li X., Ding M., Jiang X. G, Theoretical analysis of the sole plate of semi-rigid light steel column footings on the basis of Winkler model of elastic foundation beam, *Advanced Materials Research*, 660, 105-110, (2013).
6. Li S. P, The structure of the analytical model of elastic foundation beam element, Master's thesis, China Agricultural University, (2013).
7. Shi X. D, Development of plate elements on elastic foundation based on analytical trial function, Master's thesis, China Agricultural University, (2013).
8. Clough R.W, The finite element method in plane stress analysis, *Proceedings of 2nd ASCE Conference on Electronic Computation*, USA, Pittsburgh, (1960).
9. Ma F., Ai Z. Y, The finite element method of elastic subgrade beam with the contact friction effect considered, *Rock and Soil Mechanics*, 23(1), 93-96, (2002).
10. Raju K.K., Rao G. V, Thermal post-buckling of a square plate resting on an elastic foundation by finite element method, *Computers and Structures*, 28(2), 151-156, (1988).
11. Huang F. Y., Shi G. L, Finite element analysis of pressure vessel using beam on elastic foundation analysis, *Finite Elements in Analysis and Design*. 28(4), 293-302, (1998).

Space robustness of greenhouse structure based on stable bearing capacity

D. Zhu, M. Ding*, M. M. Li, Z. M. Liang, J. Q. Lei
College of Water Resources and Civil Engineering, China Agricultural University, Beijing, 100083, China
E-mail: dingmin@cau.edu.cn

In order to investigate the space robustness of greenhouse structure, three types of greenhouse including huabei-type greenhouse, venlo greenhouse and sunlight greenhouse, were chosen to be prototypes, and the numerical simulations on the overall collapse process of single hinged frame, whole frame structure, whole frame structure with plate glass covering, whole frame structure with PC sun board covering. The results show that buckling failure mode happens easily for greenhouse structure under vertical snow load, the whole structure bearing capacity is higher for greenhouse structure under horizontal wind load. Plate glass covering is the most favorable to the stable bearing capacity of greenhouse structure, PC sun board is the second and membrane is the last.

Keywords: Greenhouse Structure; Space Robustness; Numerical Simulation; Stable Bearing Capacity; Space Robustness Coefficient.

1. Introduction

In 2007, large snowstorm with average snow depth reaching about 40cm in Anshan of Liaoning province leaded to damages at different degree of greenhouses and great financial losses [1]. Strong snow resulted in collapse of 2681 sunlight greenhouses and damage of 1180 in Chengde [2]. In 2009, the largest heavy snowfall for 54 years in Shijiazhuang leaded to collapse of 237 greenhouses in Luancheng [3]. Greenhouse is both building facility and production facility [4]. As special agricultural building, greenhouse structure should have certain carrying capacity to bear all kinds of loads, such as wind load, snow load, and so on [5].

At present, studies on greenhouse have focused on environmental factors, such as illumination, temperature, gas, humidity, and so on, and there are few researches on greenhouse structure. Greenhouse structure design is short of national standard and industry standard [6]. Researches on mechanical behavior of greenhouse structure are including as follows: Li [7] conducted static linear finite element analysis on greenhouse with abnormity structure; Sun [8] suggested partial space robustness calculation model should be used for greenhouse structure needing accurate analysis and optimal design, Wang [9] suggested the larger influence of fluid-structure interaction by using CFD; Wang [10] simulated the deformation and bearing capacity of Ningxia sunlight greenhouse structure, Zhang [11] analyzed the mechanical behavior of greenhouse structure under three typical unfavorable load combination; Luis Iribarne [12,13] built a model to analyze multi-span greenhouse based on UML and XML. There are few researches on greenhouse by taking cover material and skeleton of as a whole. And there are few researches on greenhouse structure design theory considering supporting role of cover material. In order to make design working condition of greenhouse structure consistent with its actual working condition, it is necessary to conduct this research.

Five types of structure calculation model including single hinged frame, whole frame structure, whole frame structure with plate glass covering, whole frame structure with PC sun board covering, and whole frame structure with membrane covering were built by using finite element analysis software ANSYS. The whole damage process of greenhouse structure under wind disaster or snow disaster was simulated. Stable bearing capacity and space robustness of three types of greenhouse including huabei-type greenhouse, venlo greenhouse and sunlight greenhouse, were analyzed.

2. Greenhouse Structure and Finite Element Model

2.1. *Greenhouse structure prototype and model dimension parameter*

Huabei-type greenhouse, venlo greenhouse and sunlight greenhouse are selected as greenhouse structure prototype. Huabei-type greenhouse and venlo greenhouse are more multi-span and multi-bay structure, and whether its overall unstability or member buckling is bent bifurcation buckling in-plane or out-plane. Their stable bearing capacity depends on member end restraint stiffness, member out-plane brace distance and member stiffness.

Huabei-type Greenhouse: The dimension parameters of huabei-type greenhouse structure are as follows: span is 8m, roof heigth is 4.5m, eaves height is 3m, and bay is 3m. In order to take greenhouse structure space robustness into account, the simulation model of whole frame structure with 3 bays was built as shown in Fig.1a

Venlo Greenhouse: The dimension parameters of venlo greenhouse structure are as follows: span is 9.6m, roof heigth is 5.8m, eaves height is 5m, bay is 4m, and roof angle is 26.5°. In order to take greenhouse structure space robustness into account, the simulation model of whole frame structure with 3 bays was built as shown in Fig. 1b.

Sunlight Greenhouse: The dimension parameters of sunlight greenhouse structure are as follows: span is 7.5m, roof heigth is 3.6m, backwall height is 2.6m, front roof angle is 30°, back roof angle is 45°, arch structure distance is 1m. In order to take greenhouse structure space robustness into account, the simulation model of whole frame structure with 8 bays was built as shown in Fig.1c.

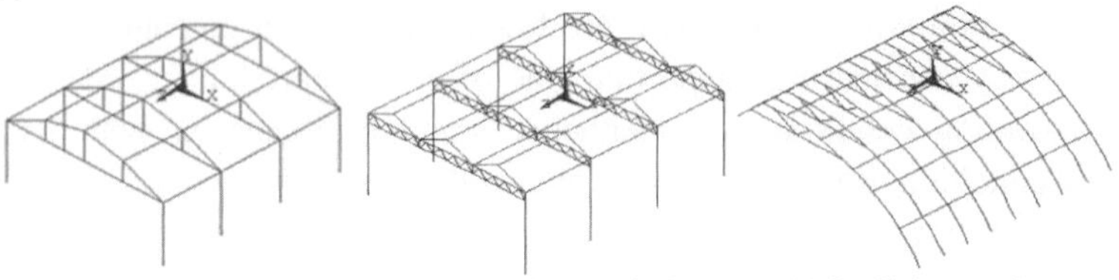

(a)Huabei-type greenhouse (b) Venlo greenhouse (c) Sunlight greenhouse

Fig. 1.Whole frame structure of the three greenhouses.

2.2. *Material model and element type*

Greenhouse skeleton is light steel structure. Steel is Q235. Cover material is plate glass, PC board and membrane, respectively, and there parameters are as shown in Table 1.

Beam188 is used for the column and roof beam of huabei-type greenhouse and venlo greenhouse, Link180 is used for purline and brace of them. For sunlight greenhouse, Beam188 is used for arch bar and Link180 is used for purline and diagonal brace. Shell181 is used for greenhouse cover material.

Table 1. Parameters ofcovering materials.

Covering material	Thickness (mm)	Density (kg·m^3)	Poisson ratio	Elastic modulus GPa)	Yield stress (MPa)
Steel	—	7850	0.3	206	200
Membrane	0.12	940	0.439	0.172	/
Plate glass	5	2400	0.25	70	58.8
PC board (single layer)	10	1200	0.3	2.4	63

2.3. *Load case*

Three live load cases used in this paper are only wind load, only snow load, both wind load and snow load, calculated according to Greenhouse Structure Design Load (GB/T 18622-2002) load combination controlled by live load effect is also in accordance with the Code.

2.4. *Restraint*

The column base of greenhouse structure is all simple steel column base and it is different from ideal clamped base and ideal simple-supported base. The column base of greenhouse structure is usually simplified for easy to structural analysis and processing installation.

Column, roof beam and purline, is hinged with each other. Five degrees of freedom at each node are all restrained and only z axis rotation degree of freedom is released to simulate hinged joint.

3. Results and Analysis

In order to ensure load consistency during the process of model calculation, wind load and snow load are imposed on the greenhouse skeleton in the shape of linear load. When analyzing the results, linear load is conversed to area load defined as equivalent wind load or snow load.

3.1. *Huabei-type greenhouse*

Table 2. Space robustness of huabei-type greenhouse structure under the three load cases.

Load case	Model	wind load (kN·m^2)	snow load (kN·m^2)	Structure safety coefficient	Space robustness coefficient
Only wind load	Single hinged frame	8.71	-	19.36	1.00
	Whole frame structure	8.89	-	19.76	1.02
	Whole frame structure with membrane covering	4.74	-	10.53	0.54
	Whole frame structure with plate glass covering	7.02	-	15.60	0.81
	Whole frame structure with PC board covering	4.20	-	9.33	0.48
Only snow load	Single hinged frame	-	1.89	4.73	1.00
	Whole frame structure	-	2.52	6.30	1.34
	Whole frame structure with membrane covering	-	3.55	8.88	1.88
	Whole frame structure with plate glass covering	-	5.40	13.5	2.86
	Whole frame structure with PC board covering	-	3.88	9.70	2.05
Both wind load and snow load	Single hinged frame	0.77	0.69	1.71	1.00
	Whole frame structure	0.78	0.69	1.73	1.00
	Whole frame structure with membrane covering	0.99	0.88	2.20	1.28
	Whole frame structure with plate glass covering	1.13	1.00	2.51	1.46
	Whole frame structure with PC board covering	1.05	0.93	2.33	1.35

The load-displacement curve of the whole damage process of huabei-type greenhouse under wind load is shown in Fig. 2. Equivalent wind load of ultimate bearing capacity, structure safety coefficient and space robustness coefficient of every structure are shown in Table 2.

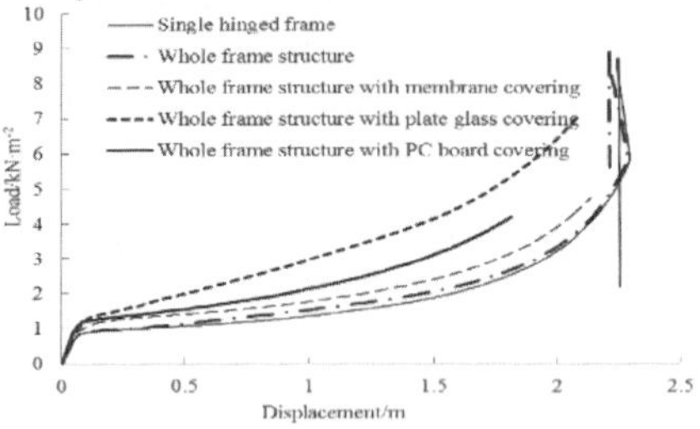

Fig. 2. Load-displacement curves of huabei-type greenhouse structure under only sonw load.

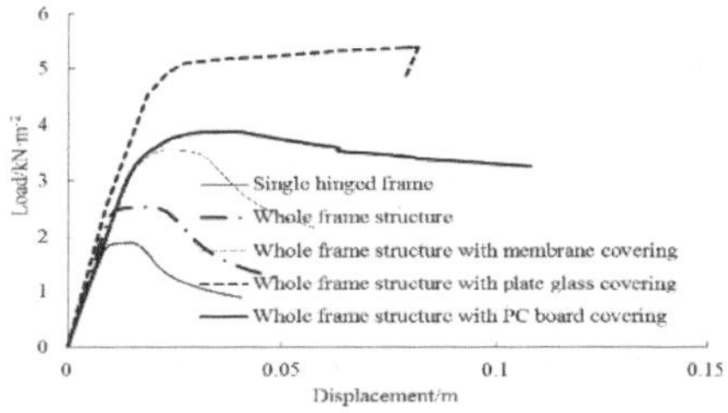

Fig. 3. Load-displacement curves of huabei-type greenhouse structure under both wind load and snow load.

The load-displacement curve of the whole damage process of huabei-type greenhouse under both wind load and snow load is shown in Fig. 3. Equivalent wind load and snow load of ultimate bearing capacity, structure safety coefficient and space robustness coefficient of every structure are shown in Table 2.

3.2. *Venlo greenhouse*

The load-displacement curve of the whole damage process of venlo greenhouse under wind load is shown in Fig. 4. Equivalent wind load of ultimate bearing capacity, structure safety coefficient and space robustness coefficient of every structure are shown in Table 3.

The load-displacement curve of the whole damage process of venlo greenhouse under snow load is shown in Fig. 5. Equivalent snow load of ultimate bearing capacity, structure safety coefficient and space robustness coefficient of every structure are shown in Table 3.

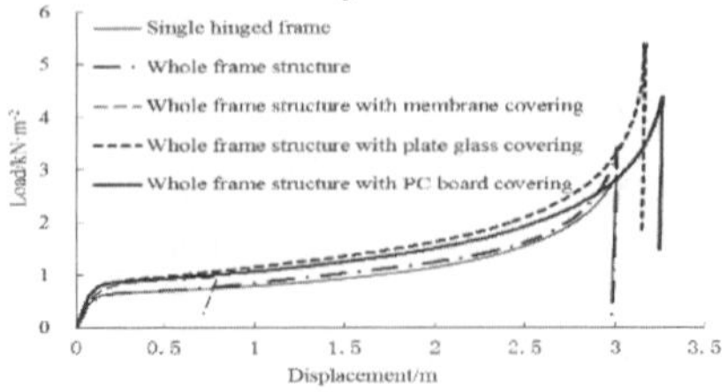

Fig. 4. Load-displacement curves of venlo greenhouse structure under only wind load.

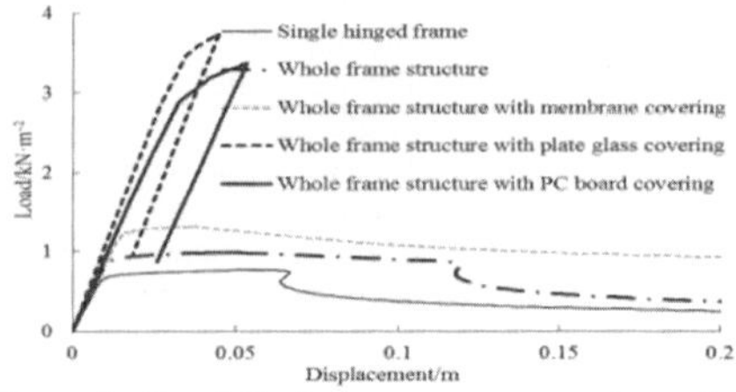

Fig. 5. Load-displacement curves of venlo greenhouse structure under only sonw load.

The load-displacement curve of the whole damage process of venlo greenhouse under both wind load and snow load is shown in Fig. 6. Equivalent wind load and snow load of ultimate bearing capacity, structure safety coefficient and space robustness coefficient of every structure are shown in Table 3.

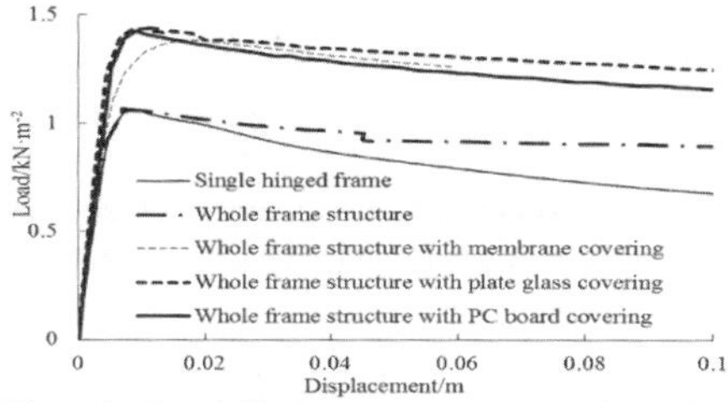

Fig. 6. Load-displacement curves of venlo greenhouse structure under only wind load.

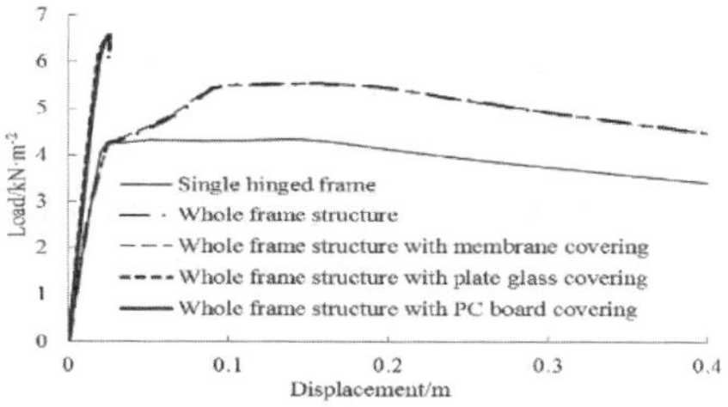

Fig. 7. Load-displacement curves of sunlight greenhouse structure under both wind load and snow load.

Table 3. Space robustness of venlogreenhouse structure under the three load cases.

Load case	Model	Wind load (kN·m^2)	Snow load (kN·m^2)	Structure safety coefficient	Space robustness coefficient
Only wind load	Single hinged frame	2.89	-	6.42	1.00
	Whole frame structure	3.43	-	7.62	1.18
	Whole frame structure with membrane covering	0.96	-	2.13	0.33
	Whole frame structure with plate glass covering	5.40	-	12.00	1.87
	Whole frame structure with PC board covering	4.39	-	9.76	1.52
Only snow load	Single hinged frame	-	0.77	1.93	1.00
	Whole frame structure	-	0.99	2.48	1.28
	Whole frame structure with membrane covering	-	1.31	3.28	1.71
	Whole frame structure with plate glass covering	-	3.73	9.33	4.86
	Whole frame structure with PC board covering	-	3.37	8.43	4.39
Both wind load and snow load	Single hinged frame	0.56	0.50	1.24	1.00
	Whole frame structure	0.56	0.50	1.24	1.00
	Whole frame structure with membrane covering	0.73	0.65	1.62	1.30
	Whole frame structure with plate glass covering	0.76	0.68	1.69	1.35
	Whole frame structure with PC board covering	0.75	0.67	1.67	1.34

3.3. *Sunlight greenhouse*

The load-displacement curve of the whole damage process of sunlight greenhouse under wind load is shown in Fig. 7. Equivalent wind load of ultimate bearing capacity, structure safety coefficient and space robustness coefficient of every structure are shown in Table 4.

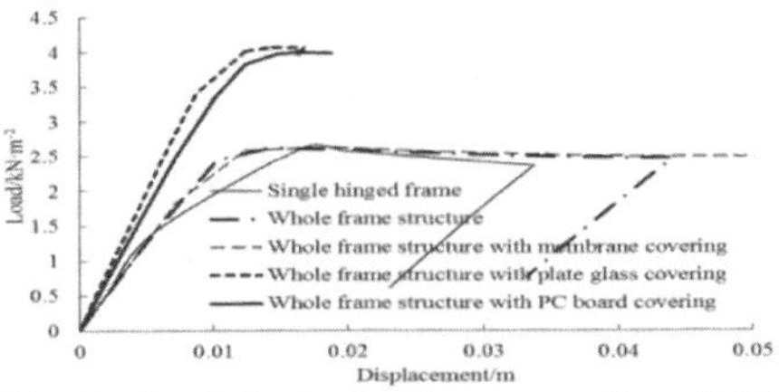

Fig. 8. Load-displacement curves of sunlight greenhouse structure under only snow load.

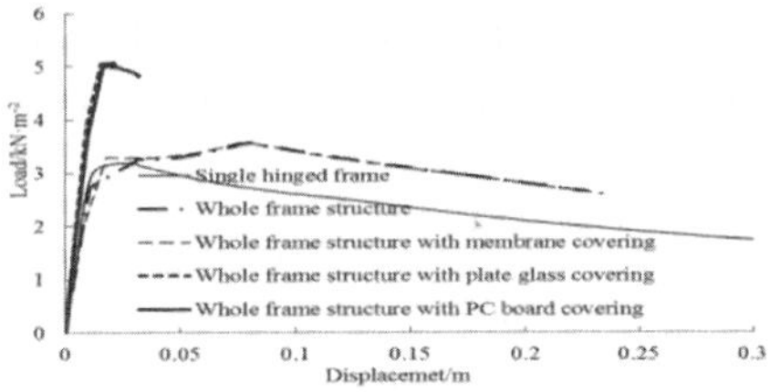

Fig. 9. Load-displacement curves of sunlight greenhouse structure under both wind load and snow load.

Table 4. Space robustness of sunlight greenhouse structure under the three load cases.

Load case	Model	Wind load ($kN{\cdot}m^2$)	Snowload ($kN{\cdot}m^2$)	Structure safety coefficient	Space robustness coefficient
Only wind load	Single hinged frame	4.35	-	9.67	1.00
	Whole frame structure	5.54	-	12.31	1.27
	Whole frame structure with membrane covering	5.54	-	12.31	1.27
	Whole frame structure with plate glass covering	6.57	-	14.60	1.51
	Whole frame structure with PC board covering	6.55	-	14.56	1.51
Only snow load	Single hinged frame	-	2.62	6.55	1.00
	Whole frame structure	-	2.62	6.55	1.00
	Whole frame structure with membrane covering	-	2.63	6.58	1.00
	Whole frame structure with plate glass covering	-	4.06	10.15	1.55
	Whole frame structure with PC board covering	-	3.99	9.98	1.52
Both wind load and snow load	Single hinged frame	1.70	1.50	3.78	1.00
	Whole frame structure	1.88	1.68	4.18	1.11
	Whole frame structure with membrane covering	1.90	1.68	4.22	1.12
	Whole frame structure with plate glass covering	2.69	2.39	5.98	1.59
	Whole frame structure with PC board covering	2.66	2.37	5.91	1.57

The load-displacement curve of the whole damage process of sunlight greenhouse under snow load is shown in Fig. 8. Equivalent snow load of ultimate bearing capacity, structure safety coefficient and space robustness coefficient of every structure are shown in Table 4.

The load-displacement curve of the whole damage process of sunlight greenhouse under both wind load and snow load is shown in Fig. 9. Equivalent wind load and snow load of ultimate bearing capacity, structure safety coefficient and space robustness coefficient of every structure are shown in Table 4.

3.4. *Brief summary*

Table 5 presents the suggested value of space robustness coefficient of the three type greenhouse under three load cases.

Table 5. Suggested value of space robustness coefficient.

Greenhouse type	Structure type		Space robustness coefficient: Wind load	Snow load	Both wind load and snow load
Huabei-type greenhouse	Single hinged frame		1.0	1.0	1.0
	Whole frame structure		1.0	1.3	1.0
	Whole frame structure with cover material	Membrane	1.0	1.8	1.2
		Plate glass	1.0	2.0	1.4
		PC board	1.0	2.0	1.3
Venlo greenhouse	Single hinged frame		1.0	1.0	1.0
	Whole frame structure		1.1	1.2	1.0
	Whole frame structure with cover material	Membrane	1.0	1.7	1.3
		Plate glass	1.8	2.0	1.3
		PC board	1.5	2.0	1.3
Sunlight greenhouse	Single hinged frame		1.0	1.0	1.0
	Whole frame structure		1.2	1.0	1.1
	Whole frame structure with cover material	Membrane	1.2	1.0	1.1
		Plate glass	1.5	1.5	1.5
		PC board	1.5	1.5	1.5

4. Conclusions

Based on the results of this investigation, the following conclusions can be drawn:

(1) The improve effect of space robustness on integral stability of sunlight greenhouse structure can be taken into account for load combination controlled by wind load.

(2) Space robustness effect of huabei-type greenhouse structure and venlo greenhouse structure can be taken into account for load combination controlled by snow load.

(3) Among the three cover materials, such as plate glass, PC board and membrane, plate glass has the biggest contribution to the improvement of structure stable bearing capacity, between 20%~280%; PC board is the second, between 20%~240%; membrane is the third, between 0~40%.

Acknowledgments

The research is supported by National Natural Science Funds (Grant No. 51279206), Beijing Natural Science Foundation (Grant No.3144029), Key Laboratory of Farm Building in Structure and Construction, Ministry of Agriculture, P.R. China (Grant No.201502), and Chinese Universities Scientific Fund (Grant No. 2015SYL004).

References

1. Changji Zhou, Revelation of multi-span greenhouse collapse induced by snowstorm[J]. *Agricultural Engineering Technology*, (12): 13-15 (2007).
2. Lianhe Fang, Reason analysis of sunlight greenhouse collapse induced by snow disaster[J]. *Inner Mongolia Agricultural Science and Technology*, (5): 98, 102 (2007).
3. Zihang Shi, Defense disastrous weather for facility agriculture development[J]. *New Agriculture*, (8): 35-36 (2010).
4. Changji Zhou, Design standards on greenhouse safety and theirs essential items[J]. *Transactions of the Chinese Society of Agricultural Engineering*, 18(sup.): 16-19 (2002).
5. Changji Zhou, Wind and snow load classification for greenhouse structure design[J]. *Transactions of the Chinese Society of Agricultural Engineering*, 16(4): 103-105 (2000).
6. Ruigang Fang, Hua Lu, Discussion on glass greenhouse structure design[J]. *Rural Practical Engineering Technology*, (9): 32-33 (2005).
7. Chengzhi Li, Zongmin Liang, Jinsan Ju, Spatial finite element analysis of heteromorp greenhouse[J]. *Journal of China Agricultural University*, 12(2): 84-87 (2007).
8. Defa Sun, Study on the design theory of multi-span plastic greenhouse structure and engineering applications[D]. Hangzhou: Zhejiang University (2002).
9. Dongxia Wang, Study on the wind vibration effect on greenhouse structures[D]. Beijing: China Agricultural University (2006).
10. Bin Wang, Baohong Jin, Jianxia Song, Finite element analysis of load bearing capacity of steel skeletons for solar greenhouses under vertical loads[J]. Journal of Ningxia University, 30(4): 336-338 (2009).
11. Lianyong Zhang, Analysis on the steel frame structure of multi-span plastic greenhouse in North China[D]. Taian: Shandong Agricultural University (2008).
12. Luis Iribarne, Using computer modeling techniques to design tunnel greenhouse structures[J]. *Computers in Industry*, 58: 403-415 (2007).
13. F.D. Molina-Aiz, D.L. Valera, A.J. Alvarez, Measurement and simulation of climate inside Almerı´a type greenhouses using computational fluid dynamics[J]. *Agricultural and Forest Meteorology*, 125: 33-51(2004).
14. GB/T 18622-2002, Greenhouse structure design load[S]. (2002).

Analytical element for torsion bar based on vlasov theory

Wen-Zhong Xia[1], Xiu-Gen Jiang[1,*], Min Ding[1], Hong-Zhi Wang[1], Xing-Hua Chen[2], Xiao-Ke Cheng[3], Wei-Tong Guo[1]

[1]*College of Water Resources & Civil Engineering, China Agricultural University,*
[2]*China Machinery TDI International Engineering Co.,Ltd,*
[3]*China Space Civil & Building Engineering Design & Research Institute (Group), Beijing, 100071, China*
**E-mail: jiangxg@cau.edu.cn*

According to Vlasov restrained torsion theory, shape function for restrained torsion bar is determined based on torsional angular displacement equation to analyze restrained torsion of bar. Potential energy functional of torsion bar is established by applying potential energy principle. Element stiffness matrix is obtained via variation principle of potential energy stationary value, and analytical element formulations for restrained torsion are also received. The comparisons between the calculation results of ETA in this paper, analytical solution, ETL, and ETI were conducted, and the results show that the relative error between the solution of ETA by the least element number and the theoretical solution is less than 10^{-9}. And it is proved that ETA can satisfy the high accuracy and efficiency requirement and be applied in practice.

Keywords: Restrained Torsion; Analytical Function; Variatio N Principle; Finite Element; Element Stiffness Matrix.

1. Introduction

Torsion bar refers to the bar bearing moment in the same direction with bar axis [1]. For the bar bearing torque, not only its shape distortion in plane, but also non-uniform displacement out-plane, that is warpage [2], occur. Torsion bar without warpage on normal section, or with warpage unrestricted, is defined as free torsion bar. Torsion bar with restrained warpage is defined as restrained torsion bar [3].

Various theories for torsion analysis have been presented according to the characteristics of bar torsion deformation. Among them, Vlasov [4] provided stiffness equation about deflection and internal force. This model neglects shear deformation of middle plane and is only suitable for restrained torsion of open section. There are three methods [5,6] to analyze the deformation of torsion structure, including static method, energy method and numerical method. In numerical method, force control differential equation is directly or indirectly solved. Among them, displacement finite element method [7] is the most widely used structure analysis method. The efficiency of finite element method depends on element number and accuracy. The element accuracy depends on element displacement shape function. There are different torsion theories and torsion displacement shape functions of bar element to construct different torsion elements [8-10]. The results show that this analytical element is more accurate and efficient than traditional interpolation shape function.

This paper focuses on constructing analytical torsion element of bar (hereafter referred to as ETA) on the basis of Vlasov theory. The calculation result of ETA is compared with that of analytical solution, ETL, and ETI. And the accuracy of ETA is verified via dividing different element number.

2. Element Displacement Shape Function

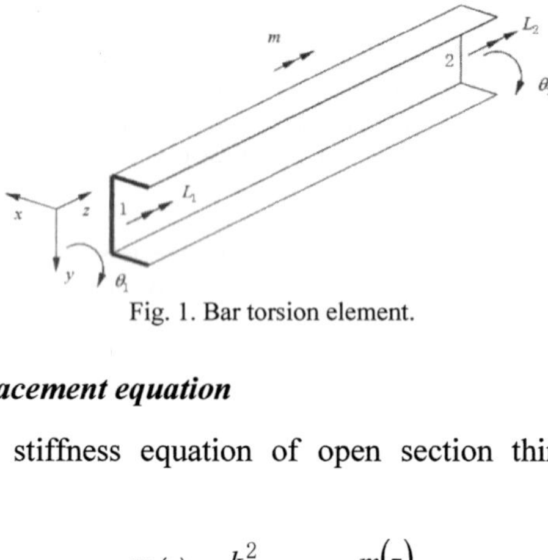

Fig. 1. Bar torsion element.

2.1. *Torsion displacement equation*

Restrained torsion stiffness equation of open section thin-walled bar[3] is expressed as

$$\theta^{\mathrm{IV}}(z) - \frac{k^2}{l^2}\theta'' = \frac{m(z)}{E_1 J_\omega}. \tag{1}$$

Where, E_1 is converted elastic modulus, G is shear modulus, θ is torsion angle, J_d is free torsion inertia moment of section, J_w is principle fan-shaped inertia moment, l is bar length, $m(z)$ is external torque intensity along bar length, k is characteristic coefficient of bar restrained torsion, and can be calculated by

$$k = \sqrt{\frac{GJ_d l^2}{E_1 J_\omega}}. \tag{2}$$

General solution of Eq. (1) can be expressed by matrix as

$$\theta(z) = \sum_{i=1}^{n} c_i f_i(z) = \{f\}\{c\}. \tag{3}$$

Where, $\{f\}$ is basis function vector of displacement, $\{c\}$ is undetermined coefficient vector, and both are n order vector. The expressions are as follow:

$$\{f\} = [1 \quad z \quad \sinh(kz/l) \quad \cosh(kz/l)]. \tag{4}$$

$$\{c\} = \{c_1 \quad c_2 \quad c_3 \quad c_4\}^{\mathrm{T}}. \tag{5}$$

2.2. *Element displacement shape function*

For torsion element as shown in Fig. 1, torsion displacement of any section can be expressed by the four end displacements of the two bar ends, that is the element four degrees of freedom. Its formula is

$$\{\delta\}^{(e)} = \{\theta_1 \quad \theta_1' \quad \theta_2 \quad \theta_2'\}^{\mathrm{T}} \tag{6}$$

where θ_1, θ_2 mean torsion angle of node 1 and node 2, θ_1', θ_2' mean warp rate of node 1 and node 2.

All node displacement can be expressed by matrix as

$$\{\delta\}^{(e)} = \begin{bmatrix} 1 & 0 & 0 & 1 \\ 0 & 1 & k/l & 0 \\ 1 & l & \sinh(k) & \cosh(k) \\ 0 & 1 & k/l\cosh(k) & k/l\sinh(k) \end{bmatrix} \begin{Bmatrix} c_1 \\ c_2 \\ c_3 \\ c_4 \end{Bmatrix}. \tag{7}$$

From Eq. (8), the determined coefficient vector can be obtained:

$$\begin{Bmatrix} c_1 \\ c_2 \\ c_3 \\ c_4 \end{Bmatrix} = \begin{bmatrix} 1 & 0 & 0 & 1 \\ 0 & 1 & k/l & 0 \\ 1 & l & \sinh(k) & \cosh(k) \\ 0 & 1 & k/l\cosh(k) & k/l\sinh(k) \end{bmatrix}^{-1} \{\delta\}^{(e)}. \tag{8}$$

By substituting Eq. (9) into Eq. (8), the following can be obtained:

$$\theta(z) = [1 \quad z \quad \sinh(kz/l) \quad \cosh(kz/l)] \begin{bmatrix} 1 & 0 & 0 & 1 \\ 0 & 1 & k/l & 0 \\ 1 & l & \sinh(k) & \cosh(k) \\ 0 & 1 & k/l\cosh(k) & k/l\sinh(k) \end{bmatrix}^{-1} \{\delta\}^{(e)} = [N(z)]\{\delta\}^{(e)}. \tag{9}$$

Where $[N(z)]$ is displacement shape function matrix of restrained torsion bar, and can be expressed after operation as

$$[N(z)] = [N_1(z)\ \ N_2(z)\ \ N_3(z)\ \ N_4(z)],\ N_1(z) = \frac{\vartheta}{l\alpha},\ N_2(z) = \frac{\varpi}{k\alpha},\ N_3(z) = \frac{\iota}{l\alpha}\ N_4(z) = \frac{\ell}{k\alpha}.$$

where, $\alpha = \cosh(k^2) - 2\cosh(k) - \sinh(k^2) + k\sinh(k) + 1$,

$\vartheta = l\sinh(k)\sinh(kz/l) - l[\cosh(k) - 1]\cosh(kz/l) - kz\sinh(k) + l\cosh(k^2) - l\sinh(k^2) + kl\sinh(k) - l\cosh(k)$

$\varpi = l[1 - \cosh(k) + k\sinh(k)]\sinh(kz/l) + l[\sinh(k) - k\cosh(k)]\cosh(kz/l) - k[\cosh(k) - \cosh(k^2) + \sinh(k^2)]z$
$- l[\sinh(k) + k\cosh(k)]$

$\iota = -l\sinh(k)\sinh(kz/l) + l[\cosh(k) - 1]\cosh(kz/l) + k\sinh(kz) - l[\cosh(k) - 1]$,

$\ell = l[\cosh(k) - 1]\sinh(kz/l) + l[k - \sinh(k)]\cosh(kz/l) - k[\cosh(k) - 1]z + l\sinh(k) - kl$.

3. Element Potential Energy

Strain energy due to restrained torsion of open section thin-walled bar is composed of strain energy by normal stress of restrained torsion and that by shear stress[3], and can be expressed as

$$U^{(e)} = \frac{1}{2}\int_0^l \left[E_1 J_\omega \theta''^2 + GJ_d \theta'^2\right] dz. \tag{10}$$

With Eq. (10), Eq. (11) becomes

$$U^{(e)} = \frac{1}{2}\{\delta\}^{(e)T}\left[\int_0^l [N''(z)]^T E_1 J_\omega [N''(z)] dz + \int_0^l [N'(z)]^T GJ_d [N'(z)] dz\right]\{\delta\}^{(e)}. \tag{11}$$

If external loads on thin-walled bar include uniformly distributed torque m, concentrated torque L_p at z_p, torque L and B at bar ends. External force potential energy is defined as the negative product of load and the corresponding displacement. Then the external force potential energy of thin-walled bar can be expressed as

$$U_p^{(e)} = -\int_l m\theta dz - [L\theta - B\theta']_0^l - L_p\theta(z_p). \tag{12}$$

With Eq. (10), Eq. (13) becomes

$$U_p^{(e)} = -\{\delta\}^{(e)T}\left(\int_0^l m[N(z)]^T dz + L_p[N(z)]^T + \{F\}^{(e)}\right) \tag{13}$$

where, $\{F\}^{(e)} = [L_1\ \ B_1\ \ L_2\ \ B_2]^T$.

The bar total potential energy Π is composed of strain energy $U^{(e)}$ and external force potential energy $U_p^{(e)}$, and can be expressed as

$$\Pi = U^{(e)} + U_p^{(e)}$$

$$= \frac{1}{2}\{\delta\}^{(e)T}\left[\int_0^l [N''(z)]^T E_1 J_\omega [N''(z)]dz + \int_0^l [N'(z)]^T GJ_d [N'(z)]dz\right]\{\delta\}^{(e)} \tag{14}$$

$$-\{\delta\}^{(e)T}\left(\int_0^l m[N(z)]^T dz + L_p[N(z)]^T + \{F\}^{(e)}\right)\Big|_0^l .$$

4. Element Formulation of ETA

4.1. *Element stiffness matrix*

According to the structure actual displacement, the following can be obtained

$$\frac{\partial \Pi}{\partial \{\delta\}^{(e)}} = \frac{\partial U^{(e)}}{\partial \{\delta\}^{(e)}} + \frac{\partial U_p^{(e)}}{\partial \{\delta\}^{(e)}} = 0 . \tag{15}$$

By substitute Eq. (12) and Eq. (14) into Eq. (16), it can be expressed as

$$\{F_\theta\}^{(e)} = \{F\}^{(e)} + \{F_0\}^{(e)} + \{F_{z_p}\}^{(e)} = [k]^{(e)}\{\delta\}^{(e)} \tag{16}$$

where, $\{F\}^{(e)}$ is bar end force, $\{F_0\}^{(e)}$ is equivalent node force, $\{F_{z_p}\}^{(e)}$ is equivalent node force of concentrated torque, $[k]^{(e)}$ is element stiffness matrix.

From the comparison of Eq. (16) and Eq. (17), the following can be obtained:

$$\frac{\partial U^{(e)}}{\partial \{\delta\}^{(e)}} = [k]^{(e)}\{\delta\}^{(e)} \tag{17}$$

$$\frac{\partial U_p^{(e)}}{\partial \{\delta\}^{(e)}} = -\{F\}^{(e)} - \{F_0\}^{(e)} \tag{18}$$

where, $[k]^{(e)} = \int_0^l [N''(z)]^T E_1 J_\omega [N''(z)]dz + \int_0^l [N'(z)]^T GJ_d [N'(z)]dz \cdot$

From Eq. (18), clement stiffncss matrix can bc obtaincd:

$$[k]^{(e)} = \begin{bmatrix} k_{11} & k_{12} & k_{13} & k_{14} \\ k_{21} & k_{22} & k_{23} & k_{24} \\ k_{31} & k_{32} & k_{33} & k_{34} \\ k_{41} & k_{42} & k_{43} & k_{44} \end{bmatrix} \tag{19}$$

where, $[k]^{(e)}$ is 4×4 order symmetric matrix, and

$$k_{ij} = E_1 J_\omega \int_0^l N_i''(z) N_j''(z) dz + GJ_d \int_0^l N_i'(z) N_j'(z) dz .$$

By substituting shape function value into the above formula, the following can be obtained:

$$k_{11} = -k_{13} = k_{33} = \frac{\gamma(1 + k\zeta\mu')}{l^3\beta},\ k_{12} = k_{14} = -k_{23} = -k_{34} = \frac{\gamma(1 + \zeta\mu)}{2l^2\beta},$$

$$k_{22} = k_{44} = \frac{\gamma(\lambda + \zeta\xi)}{kl\chi},\ \ k_{24} = \frac{\gamma(\lambda + \zeta\eta)}{kl\chi}.$$

where, $\beta = k[k - 4\sinh(k)] + (k^2 + 4)\cosh(k) - 4$,

$$\eta = -\sinh(2k) + (2 - 3k^2)\sinh(k) + k[(k^2 + 6)\cosh(k) - 6],$$

$$\chi = 8k\sinh(k)[1 - \cosh(k)] + (4 + k^2)\cosh(2k) - 16\cosh(k) - k^2 + 12,$$

$$\gamma = EJ_\omega k^3[\sinh(k) - k],$$

$$\xi = \sinh(k)[(2 + k^2)\cosh(k) + 2k^2 - 2] + k[2\cosh(k) - 2\cosh(2k) - k^2],$$

$$\varsigma = -\sinh(2k) + k[\cosh(2k) + 1] + (k^2 + 2)[\sinh(k) - k\cosh(k)],$$

$$\mu = k[\sinh(k) + k] + 4[1 - \cosh(k)],$$

$$\lambda = \frac{\sinh(k)[(2 + k^2)\cosh(k) - 2k^2 - 2] + k[2\cosh(k) - \cosh(2k) + k^2 - 1]}{k[\sinh(k) - k]},$$

$$\zeta = \frac{\sinh(k)[(2 + k^2)\cosh(k) - 2k^2 - 2] + k[2\cosh(k) - \cosh(2k) + k^2 - 1]}{\lambda}.$$

4.2. *Equivalent node force vector*

External load total potential energy U_p can be expressed as

$$U_\mathrm{p} = -\sum_e \{\delta\}^\mathrm{T}\left(\left[A\right]^{(\mathrm{e})}\right)^\mathrm{T}\left(\int_0^l m[N(z)]^\mathrm{T}\mathrm{d}z + L_\mathrm{p}[N(z)]^\mathrm{T} + \{F\}^{(\mathrm{e})}\right). \tag{20}$$

where, $[A]^{(\mathrm{e})}$ is the transfer matrix between structure node displacement and element bar end displacement.

For uniform torque, its equivalent node force vector can be expressed as

$$\{F_0\}^{(\mathrm{e})} = [D]^\mathrm{T} = -\int_0^l m[N(z)]^\mathrm{T}\mathrm{d}z. \tag{21}$$

where, $[D] = \{D_1\ \ D_2\ \ D_3\ \ D_4\}$, $D_1 = D_3 = -ml/2$,

$$D_2 = -D_4 = m[kl^2(1 + e^k) + 2l^2(1 - e^k)]/[2k^2(1 - e^k)].$$

For concentrated torque L_p at z_p, its work can be expressed as

$$U_\mathrm{p} = -L\{\delta\}^{(\mathrm{e})\mathrm{T}}[N(z)]^\mathrm{T}. \tag{22}$$

Its equivalent node force vector is as follow:

$$\{F_{z_p}\}^{(e)} = \frac{\partial \Pi}{\partial \{\delta\}^{(e)}} = [Y]^T . \tag{23}$$

where, $[Y]=[Y_1 \; Y_2 \; Y_3 \; Y_4]$, $Y_1 = -L\vartheta(z_p)/l\alpha$, $Y_2 = -L\varpi(z_p)/k\alpha$, $Y_3 = -L\iota(z_p)/l\alpha$, $Y_4 = -L\ell(z_p)/k\alpha$, where, α is the same with that in Eq. (10).

5. Example and Comparison

5.1. *Example*

Restrained torsion of thin-walled C-shape bar as shown in Fig.2 is analyzed by finite element method. Both ends of the bar are hinged, the bar length *L*=2000mm, the width of C-section flange d_1=100mm, the height of C-section d_2=200mm, the thickness of C-section *t*=10mm, material elastic modulus *E*=2.1×105MPa, material shear modulus *G*=8.4×104MPa, Poisson's ratio $\mu = 0.25$. Concentrated torque at mid-span *M*=87500N·mm.

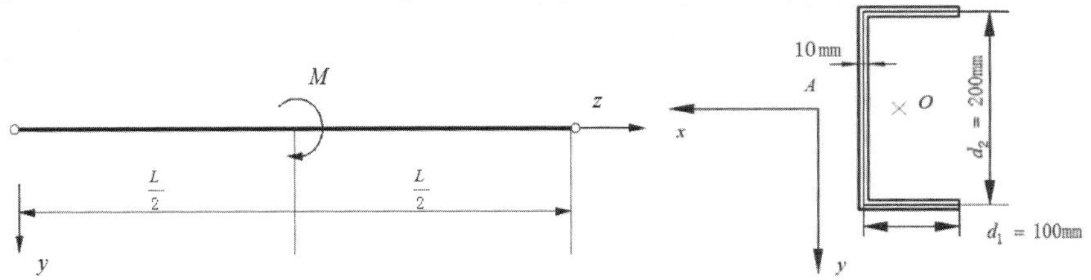

(*a*) (*b*)

Fig. 2. Bar and its section dimension.

Table 1 presents the main parameters.

Table 1. Main parameters.

Parameter	J_w (mm^6)	J_d (mm^4)	E_1 (N·mm^{-2})	*k*
Result	1.79167×10^{11}	1.333×10^5	2.24×10^5	1.056408

5.2. *Result and comparison*

The calculation results by initial parameter method, ETL, ETI, and ETA in this paper are presented and compared in Table 2, Fig. 3 and Fig. 4.

As shown in Table 2, Fig. 3 and Fig. 4, with the same element number, the calculation accuracy of torsion angle and warp rate by ETA is much higher than that by ETI, and that by ETL is so poor that it can't be shown in the Fig. 3 and Fig. 4. With the increasing of element number, the accuracy improves. And the calculation accuracy by ETA reaches 5.1×10^{-9} when element number is 2.

The reasons for these results are as follow: linear shape function model does not take the DOF of bar warp rate into account. With the consideration of warp rate, ETA is based on analytical function and the effect of model error is eliminated to a large extent. The relative error is reduced to 10^{-9}.

Table 2. Comparison of different element results.

Calculation method				Section at supports: Torsion angle ($\times 10^{-6}$rad)	Section at supports: Warp rate ($\times 10^{-7}$mm^{-1})	Section at mid-span: Torsion angle ($\times 10^{-6}$rad)	Section at mid-span: Warp rate ($\times 10^{-7}$ mm^{-1})
Analytical solution				0	4.88152870132	326.9335782	0
Finite element solution	ETL	2 elements	Calculation value	0	78.1445361340	0.3907×10^{6}	78.1445361340
			Error(%)	0	1500.8210	1.1940×10^{5}	∞
		4 elements	Calculation value	0	78.1445361340	1.9536×10^{6}	78.1445361340
			Error(%)	0	1500.8210	5.9745×10^{5}	∞
		6 elements	Calculation value	0	78.1445361340	1.3024×10^{6}	78.1445361340
			Error(%)	0	1500.8210	3.9827×10^{5}	∞
	ETI	2 elements	Calculation value	0	4.88103267714	326.9082888	0
			Error(%)	0	1.0×10^{-2}	7.7×10^{-3}	0
		4 elements	Calculation value	0	4.88082922608	326.8892714	0
			Error(%)	0	1.4×10^{-2}	1.4×10^{-2}	0
		6 elements	Calculation value	0	4.88207432343	326.9684702	0
			Error(%)	0	1.1×10^{-3}	1.1×10^{-2}	0
	ETA	2 elements	Calculation value	0	4.88152870157	326.9335779	0
			Error(%)	0	5.1×10^{-9}	9.2×10^{-8}	0

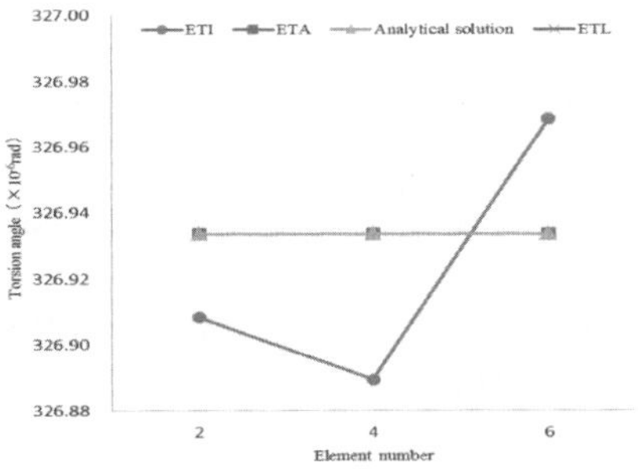

Fig. 3. Torsion angle of different element.

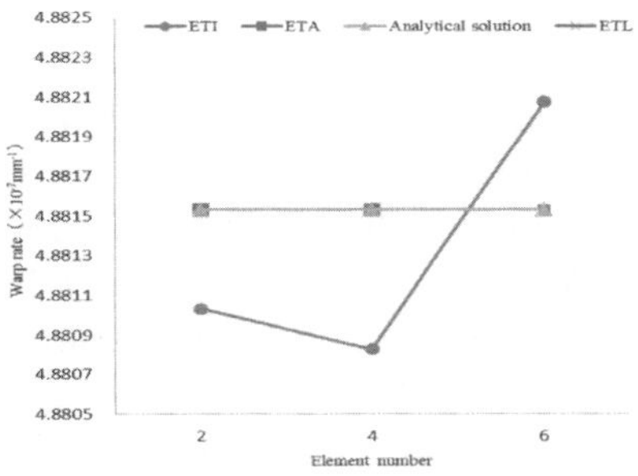

Fig. 4. Warp rate of different element.

6. Conclusions

Based on the results of this investigation, the following conclusions can be drawn:

(1) Analytical shape function for restrained torsion displacement of bar is determined based on Vlasov restrained torsion theory and torsional angular displacement equation.

(2) Potential energy functional of torsion bar is established by applying potential energy principle. Element ETA is constructed via variation principle of potential energy stationary value, and analytical element formulations for restrained torsion are also received.

(3) The calculation accuracy of torsion angle and warp rate by ETA constructed in this paper is much higher than that by ETI and ETL. The relative error between the solution of ETA by the least element number and the theoretical solution is less than 10^{-9}. And it is proved that ETA can satisfy the high accuracy and efficiency requirement.

Acknowledgments

The research is supported by Chinese Universities Scientific Fund (Grant No. 2015SYL004), Beijing Natural Science Foundation (Grant No.3144029), Key Laboratory of Farm Building in Structure and Construction, Ministry of Agriculture, P.R. China (Grant No.201502),National Natural Science Funds (Grant No. 11272340), and Foundation of National Science of China (51279206).

References

1. Sun Xunfang, Fang Xiaoshu, Guan Laitai. Material mechanics(I)[M]. Beijing: *Higher Education Press*, 51-76, (2001).
2. Gere J M, Timoshenko S P. Mechanics of materials[M]. SI Edition. New York: *Van Nostrand Reinhold*, 10-30, (1984).
3. Bao Shihua, Zhou Jian. Structural mechanics of thin walled bar[M]. Beijing: *China Building Industry Press*,38-69,(1991).
4. V.Z.Vlasov. Thin-walled elastic beams[M]. Jerusalem: *Israel Program for Scientific Translation*, 1-35, (1961).
5. Chen Chuanxu, Cao Fengqi. Calculation theory of thin walled truss structure[M]. Beijing: *People's Railway publishing house*,22-56,(1958).
6. Long Yuqiu, Bao Shihua. Structural Engineering[M]. Beijing: *Higher Education Press*, 168-257, (1979).
7. O.C. Zienkiewicz, Taylor RL. The finite element method[M]. 5th edition. Oxford: *Butterworth Heinemann*, 5-21, (2000).
8. Fu Xiangrong, Tian Ge. Advances in the finite element method based on the analytical trial functions[J]. Engineering Mechanics, s2,78-84, (2012).
9. Cao Qiongqiong, Xu Jing, Cheng Xiaoke, Zhang Qingxia, Zhang Peng, Jiang Xiugen. Element for Beam Analysis based on Parabolic Distribution Stiffness[J]. Architectural Engineering and New materials, 1-6, (2015).
10. Tang Lifeng, Xu Jing, Wang Hongzhi, Chen Xinghua. Accurate element of compressive bar considering the effect of displacement[J]. Mathematical Problems in Engineering, 1-8, (2015).

Research and development of similar material for the expansive soil tunnel in model test

Ke Wu

School of Civil Engineering, Shandong University,
Jinan, Shandong 250061, China
E-mail: wk4223@163.com

One new similar materical for the expansive soil tunnel in model test is researched and defined which can simulate the most features of expansive soil during wetting and drying cycles. Then, the failure mechanism and deformation rules of expansive soil tunnel are investigated based on the new similar material. Through the research and development of new similar material, those are able to obtain that the new similar material is good at simulate the expansive soil and is able to use for model test to investigate the features of expansive soil tunnel subjected to wetting and drying cycles.

Keywords: Similar Material; Expansive Soil; Tunnel; Model Test; Deformation.

1. Introduction

The expansive soil is one medium of geological hazard in civil engineering, which is widely distributing in the south west of China [1]. With the development of highway and high-speed railway, a lot of tunnels in civil engineering have to be built in those areas. As we all know, the expansive soil has many features such as water swelling and dry shrinking, which is easy to cause the surrounding rock of tunnel to bring the large deformation [2]. So, many researches are used the theoretical deduction, numerical simulation and basic experiments of soil mechanics to investigate the failure mechanism of tunnel in expansive soil area [3-6]. However, the model test, as one research method, is less used to study the characteristics of expansive soil [7, 8]. One reason is difficult for similar material to simulate the expansive soil [9]. For this reason, one new model test similar material of expansive soil is researched and defined which can simulate the most features of expansive soil during wetting and drying cycles. Then, the failure mechanism and deformation rules of expansive soil tunnel are investigated based on the new similar material.

2. Simlar Material of Expansive Soil

The model test similar material of expansive soil is mainly comprised of clay and swelling soil which are screening from 4.75m cheese. For different expansive level, it is made by weight ratio of ingredients. For weak expansive soil, the similar material is comprised of 100 parts of clay and greater than or equal to 2 partsbut less than 10 parts of swelling soil. For medium expansive soil, the similar material is comprised of 100 parts of clay and greater than or equal to 10 parts but less than 22 parts of swelling soil. For strong expansive soil, the similar material is comprised of 100 parts of clay and greater than or equal to 22 parts but less than 30 parts of swelling soil. And the mechanical properties of similar material for medium expansive soil is shown as table 1.

Table 1. The composition and physical mechanical properties of the materials of medium expansive soil.

Level	Proportion of raw material		Free expanion ratio /%	Load expansion ratio /%			Expanion force /KPa	Plasticity index /%	Compaction degree /%	Initial water content /%
	Clay	Swelling Soil		10KPa	25KPa	50KPa				
medium	100	10	60	5.41	3.89	3.14	85.38	26.2	0.96	25.7

3. Model Test For Expansive Soil Tunnel

According to the Nahe tunnel in the high-speed railway of YunGui in Guangxi province, the proportion of model test is 1:50 and the new similar material of expansive soil is filling in the model test which is used the medium expansive level. The size of model test is shown in figure 1. Then, according to the propotion of model test and medium expansive level, the similar material of expansive soil is compounded, which also mixes a little of quartz sand and barite powder for auxiliary material. The purpose of adding quartz sand is to reduce the strength and elastic modulus of material. When the certain content of barite powder and bulk density of the material will be increased with the content of quartz sand decreased. When the barite powder, quartz sand, a certain amount of water, the material elastic modulus and compressive strength will increase with the increase of the content of gypsum. And, when a certain amount of quartz sand, the material bulk density, compressive strength, elastic model is also increasing with the increase of barite powder content.

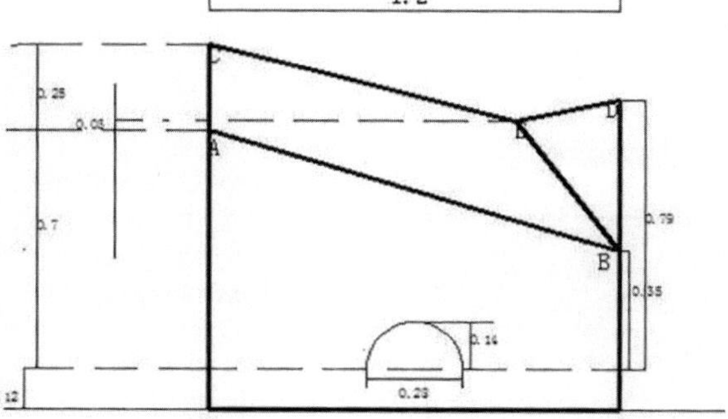

Fig. 1. Size of model test.

The materials of model test are divided into three categories, which are the main body of rock, the initial support or two lining similar materials and the bolt similar materials. Among them, the initial support similar material is used the gypsum to simulation and applied in the outside of two lining in model test, which can make the lining and surrounding rock fully pressuring. The two lining support is composed of lime and water in a certain porportion and it is brittle materials, which has only the mechanical properties of the compressive and tensile properties. And the anchor is used aluminum wire to simulation. So, the mechanical parameters of prototype and model is given in table 2.

Table 2. Mechanical parameters of prototype and model test.

Structure type	Modulus of elasticity E/GPa		Cohesional strength c/kPa		Poisson's ratio u		Friction angle φ/°		Bulk density r/kNm^{-3}	
	Prototype	Model test	Prototype	Model test	Prototype	Model test	Prototype	Model test	Prototype	Model test
IV level of surrounding rock	9.9	0.198	200	4.0	0.17	0.17	30	30	16.8	16.8
V level of surrounding rock	5.2	0.104	50	1.0	0.17	0.17	25	25	16.6	16.6
Two lining support	31	0.62			0.2	0.2			25	25
Initial support	23	0.46			0.2	0.2			23	23
Anchor	210	4.2			0.2	0.2			23	23

The deformation of surrounding rock in expansive soil tunnel is measuring by micro displacement meters. And, the stresses of surrounding rock and supports are obtained by miniature earth pressure box. All data from sensors are collected in the data acquisition instrument YE2539. Based on the similar materials and monitoring instruments, the model test is found in figure 2. Among them, the tunnel's excavation in the model test is used artificial excavation by upper and lower steps excavation, likely as figure 3. And, the rainfall is realized through the artificial rainfall which is used one plastic bucket with lots of 2mm holes. The time of rainfall is converted by the actual rainfall time.

Fig. 2. Model test.

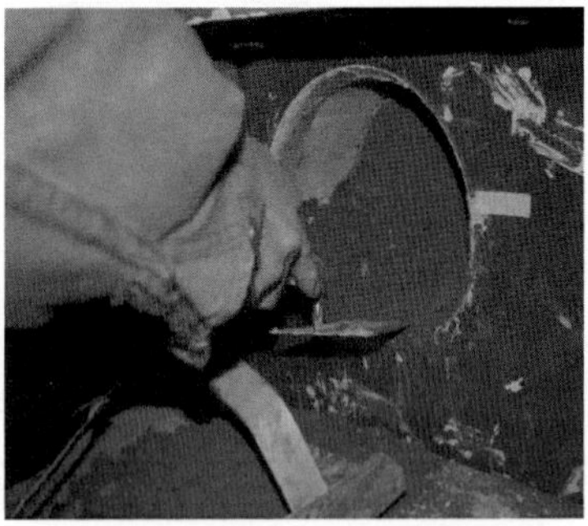

Fig. 3. Artificial excavation.

4. Model Test For Expansive Soil Tunnel

The layout of minitoring points is shown in figure 4. After the artificial rainfall, the tunnel in model test is beginning to be excaved. All monitoring instruments are recording the strain and stress of surrounding rock of expansive soil tunnel.

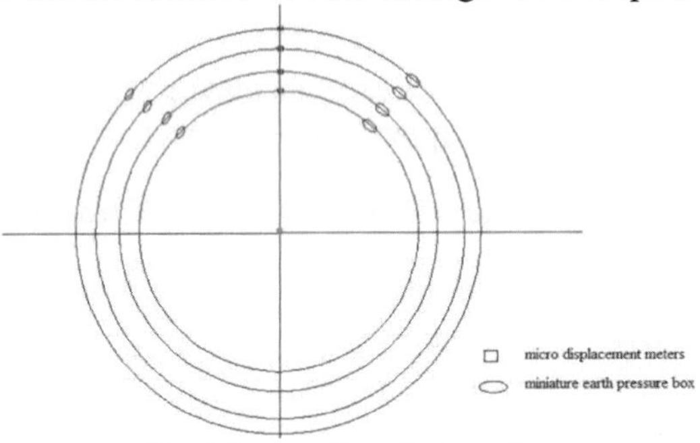

Fig. 4. Layout of monitoring points.

Figure 5 gives the displacement of different monitoring points. The monitoring points have 4 points which are from inside to outside of tunnel. It can be observed that: (1) the displacements of 4 points are increasing with the development of time of artificial rainfall and excavation. (2) The displacement of inside point of tunnel is larger than that of outside. (3) The curves of displacement have two notable changing. The first one is because that expansive soils are wetting and exposuring to the atmosphere, which cause the similar material volume expansion. The second one is because that the lower step is excaved which leads surrounding rock of tunnel to deform. (4) The displacements of expansive soil tunnel are increasing when they are wetting and exposuring to the atmosphere. It is because of the features of expansive soil.

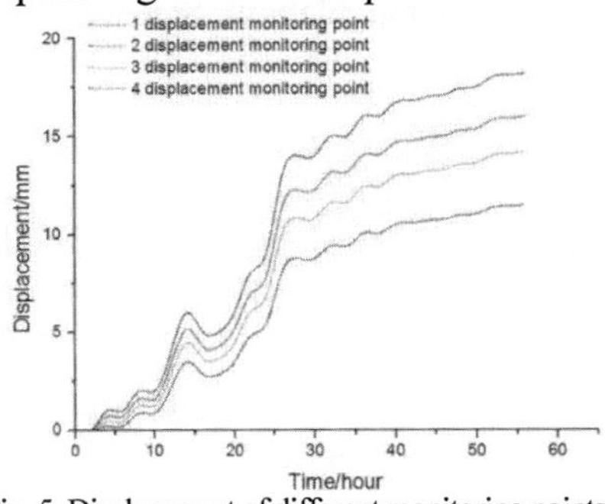

Fig. 5. Displacement of different monitoring points.

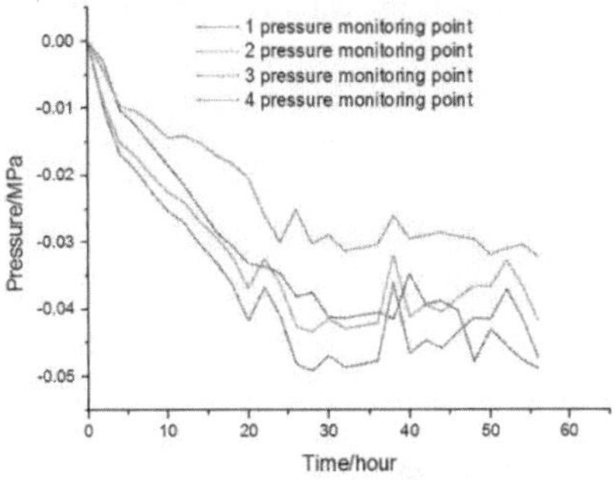

Fig. 6. Pressure of different monitoring points.

Figure 6 gives the pressure of different monitoring points. The monitoring points have 4 points which are from inside to outside of tunnel. It can be obtained that: (1) the earth pressures of different monitoring points are increasing with the development of artificial rainfall and excavation. (2) The pressure of inside point of tunnel is larger than that of outside. These features of curves are similar to the displacements of different monitoring points. (3) The curves of earth pressures have not two notable changing like as the curves of displacements. These are always increasing because the expansive soils are wetting and their volumes are expansion. Only in the lower step excavation, the curves of earth pressures are fluctuating because the surrounding rock are excavation unloading.

5. Conclusions

In order to investigate the features of expansive soil tunnel in model test during wetting and drying cycles, one new similar material of expansive soils is researching and defining which can simulate the most features of expansive soil during wetting and drying cycles. Then, the displacement and earth pressure rules of expansive soil tunnel are investigated based on the new similar material. Through the research and development of new similar material, those are able to obtain that the new similar material is good at simulate the expansive soil and is able to use for model test to investigate the features of expansive soil tunnel subjected to wetting and drying cycles.

Acknowledgments

This research was financially supported by National Natural Science Fund (51308323).

References

1. B. X. Shi, S. S. Chen, H. Q. Han and T. B. Wang, "Failure law of the expansive soil slope considering the suction changes", SHULI XUEBAO, vol. 45, pp. 1499-1506, Dec. 2014.
2. K. L. Cheng, "CBR test study on expansive soils and non-expansive soils of central China", Highway Engineering, vol. 40, pp. 209-211, Jan. 2015.
3. W. Z. Chen, G. L. Jiang, Z. H. Yuan, D. W. Wang, Z. M. Wang and A. H. Li, "Field soaking tests on expansive soil foundation under low embankment", Chinese Journal of Rock Mechanics and Engineering, vol. 33, pp. 1609-1618, Aug. 2014.
4. Khave, Ghassem Jalilian, Khave and G. J., "Engineering classification of soil masses for design of tunnel support based on observations in Tehran eastern main sewerage tunnel", Transactions of the Institutions of Mining and Metallurgy, vol. 123, pp. 1-8, Jan. 2014.
5. Tanaka, M. A., Maezaki, D., Ishii, T., Okubo, A., Hiramatsu, R., Ono and T., Mibu, K., "Interface magnetism of Co2FeGe Heusler alloy layers and magnetoresistance of Co2FeGe/MgO/Fe magnetic tunnel junctions", Journal of Applied Physics, vol. 116, pp. 243-245, Aug. 2014.
6. Y. J. Zhang, X. Wang, Q. G. Liang, D. J. Jiang and X. N. Ma, "Development of model test similar material of collapsible loess", Chinese Journal of Rock Mechanics and Engineering, vol.32, pp. 4019-4024, Feb. 2013.
7. K. Wu, "Study on the relationship between the law of crack propagation of tunnel surrounding rock's microscope structure in dilative soil tunnel with the engineering properties", International Conference on Mechanics and Civil Engineering, Chengdu, 2014, pp. 112-115.
8. Demirci, S., Erer, I. and Unaldi, N., "Multi-scale vector tunnel classification algorithm for hyperspectral images", The International Society for Optical Engineering, Wuhan, 2013, pp. 378-385.
9. W. Jin, W. Q. Ding, Q. W. Xu and L. Ye, "Model-testing technology for an extra-large span tunnel in soft rock", Modern Tunneling Technology, vol. 51(5), pp. 99-107, Oct. 2014.

The assessment of values and its effect on intervention in historic building conservation

Qiao Ning*, Yi Mou†
Civil Engineering School, Shandong University,
17922 Jingshi Road, Jinan, China
*Email: *yushou@sdu.edu.cn, †sdmouyi2014@hotmail.com*
www.sdu.edu.cn

In conservation practice, except for the basic maintenance to protect buildings from deterioration, other active actions of intervention have been applied as feasible alternatives to conserve historic buildings. However, it is undeniable that specific degrees of intervention, which vary from one case to another as a result of balancing and discussions, may have some adverse impact on historical buildings to some extent. Therefore, it is essential to understand the criteria of decision-making in terms of intervention degrees, which highly relies on value assessment of historic buildings.

Keywords: Conservation of Historic Buildings; Value Assessment; Intervention Degrees.

1. Introduction

In historic building conservation, there are several degrees of intervention which are identified by different scales and levels of intensity. Their application in practical cases is based on both physical conditions and the comprehension of the significance of a cultural heritage, between which the latter is generally decisive.

Understanding and assessment of values of a historical building is necessary before intervention for three reasons. Firstly, conservation of a cultural heritage originates intrinsically from the desire to preserve its significance and values. Secondly, this evaluation is useful to balance among conflicting values for a least bad alternative of intervention. Thirdly, it is easier and more effective to regulate conservation work and focus on limited values of a heritage. In practice, degrees of intervention differ from one case to another with reference to distinctive values of each heritage.

2. Values and Intervention in Historic Buildings

2.1. *Values in historic buildings*

When dealing with archaic possessions or objects, two easy actions are usually taken – to keep or to abandon. Similarly, the most direct and obvious attitude towards historic buildings is either to save or to change [1]. The first is associated with an aspiration of a pure history and no attempt can be tolerated to change it. The latter, however, is adopted as a belief that development is absolute in reality [2]. However, it cannot be as straightforward due to the fact that a building is always bearing much more than a personal item. Practically in most cases, the accessible approach to conserve a historic building is to find a balance between these two opposites – an intrinsic dichotomy upon which the philosophy of building conservation is literally established [1].

Neither of the two opposites can be applied separately in one practical case while any attempt to conserve a building will incorporate a difference. Therefore, it is crucial to make sure where the change is oriented and how much can be attempted – the degree of intervention [1]. Presumably there is a decayed column that hardly works structurally, which brings about a dilemma of whether to replace it or to add additional support. The decision should be made according to the anticipation of the building, which, in other words, highly relies on the understanding and assessment of its values.

Value is identified as the quality of an object which is desirable or treasured. According to Feilden [3], there are three principle categories of values in buildings – “emotional values”, “cultural values” and “use values”, and each of them contains several facets such as identity, aesthetics, functions, and so on.

2.2. *Historical background of values and intervention*

The philosophy of the dichotomy, to save or to change, originates from a conflict between two trends in the nineteenth century. One was restoration, to reinstate a building into a perfect form, while the other was preservation, to keep the object without any unnecessary interventions.

Both motives were generated due to the effect of multiple factors, of which apprehension of values in buildings was an indispensable one. Restoration was highly influenced by the Gothic Revival, which was characterised by the strong admiration of medieval architecture, in a way that certain styles – Gothic in this case – were believed to be superior to others [4]. This faith was intensively emphasised on the architectural value of a specific style with other values in a building completely neglected, which convinced restorers to reproduce and represent the perfect style through audacious interventions. By contrast, the attitude of its opponents was based on a thorough comprehension of architectural values that when contemplating a building. Ruskin argued that a historic building was a unique innovation by an artisan under certain historic context, which encompassed both individual contributions and the remarkable value of age [5]. Correspondingly, it should be conserved by limited intervention, such as daily maintenance to prevent deterioration.

Values and significance of a historic building were clearly identified for the first time by Ruskin in the history of conservation development while the anti-restoration movement and the foundation of SPAB set the threshold of modern conservation [5].

3. Value Assessment Before Intervention

As is generally accepted, modern conservation work is a process started by defining the objective including building itself as well as its context or environment [6]. The next is to analyse and evaluate its values and set them in order according to their significance. In other words, it is a necessity to assess values before any intervention is commenced in building conservation [1,3].

To start with, values stand for the very essence of historic buildings and conservation. In the Burra Charter [7], conservation is defined as "all the processes of looking after a place so as to retain its cultural significance" and the "cultural significance" is "aesthetic, historic, scientific, social or spiritual value for past, present or future generations". In Britain, conservation is also designated as a process of "managing change" that will sustain the heritage values by maximum while enabling development to reinforce those values [8]. It is the human nature in all societies through history to pursue the answer for where we came from and to understand the significance of a building satisfies the curiosity through the association with past people and events, which also serves as a study material for the present [9].

Various values are considered to be of both tangible and intangible characters [10] staying intrinsically in a historic building, which is confirmed as a living witness of the past times [11]. It is conceivable that a cultural heritage must have its distinctive values and to weigh and assess these values provides a feasible approach to a deep understanding of the heritage. To a large extent, conservation virtually means to maintain the significance. For this reason, it is the inescapable responsibility for anyone involved to thoroughly evaluate a historic building before interventions.

There is one more point. As mentioned before, any degree of intervention will bring along at least some changes or in other words, the loss of a previous state, which indicates the detriment of certain values. Insall [1] states that even the basic cleaning would eliminate evidence of history to some extent. The relative importance should be taken into consideration due to the nature of conflicting values [9].

Therefore, in a real project, as there are many options of intervention actions, a decision has to be made beforehand which is generally determined by the understanding of heritage significance that consists of a wide variety of values.

4. Minimum Intervention Principle

Given the fact that poor structural condition of a building can hardly provide a possible future, it is unavoidable to introduce minimum intervention [12]. In addition, it is also considered to be of vital importance to keep a building in use in terms of both financial and conservation perspectives [6,10]. Therefore, it is legitimate and practical to sacrifice some values for lengthening the life of a building. In other words, in order to preserve a building, compromise has to be made and there is no choice but to select some of the values over others, which means the least bad alternative of intervention. For example, a medieval church can embrace a large scale of values, such as a manipulation of the space; yet if its craftsmanship is appreciated as the most highly achieved after appraisal, then that should be the last to be removed [1].

It has been proved successful to focus on limited facets and values of the heritage. By applying such approach, the most treasured and outstanding features of domestic buildings can survive [1].

The conservation of Forty Hall is an actual case of showing how exactly value assessment affects the resulting intervention actions. As a Grade I listed building, Forty Hall was built between 1629 and 1632 for Nicholas Rainton (1569-1646) who was a wealthy merchant, with the architect unknown. The hall was bought and used by Enfield Borough Council as an open museum and gallery for the public in 1951 and the collections were removed to a new building in 2005. It was intended to transform Forty Hall into a house of great public attraction for its historical and cultural values.

Forty Hall is a house constructed with brick in three storeys on a semi-basement. In contrast with its contemporary buildings which tend to be in the artisan mannerist mode, the exterior of the house is plain with simplicity and rigorously symmetrical on a square plan, showing the character of an early

Georgian country house. However, room arrangement of the plan and its interior decoration is conservative and old-fashioned [13]. Therefore, there is a tremendous contrast within the house externally and internally, which identifies the house to be of great significance as a combination of both medieval and classical architecture.

The form of the building before conservation in 2010 was probably attributed to the work in 1897 by Henry Ferryman Bowles, who extended the house and inserted the main staircase [13].According to the conservation plan, this staircase was replaced with one based on the composition from an earlier stage in the history of the Hall.

The house, with existing elements of its 17th century design, was identified to be of great architectural value for the uniqueness as few buildings recording such transition had survived [14]. The Victorian staircase of 1897, however, was thought to be of much lower significance for the reason that it not only confused the original plan and the circulation which decreased value of the building, but also affected the early decoration schemes of the house. Meanwhile, there was an opportunity for the building to develop for future utility as a public attraction, which required the installation of a lift to provide access for the disabled. The new lift and staircase would restore the building to a more valued period and recreate the original circulation for the public to experience as the owner did. On account of this, the proposal was justified for the loss of significance [10]. In the meantime, with no evidence for the detailing of the previous staircase, the new one was designed partly by other 17th-century precedents and partly by conjectures [10], and was also turned 180° to reopen the doorway on the former half landings which was previously blocked. Traditional techniques and oak materials were also used to reduce the risk of unexpected consequences, while modern technology was introduced to allow the installation of the lift to be reversible if necessary [10].

The decision was actually relied on two essential conservation principles that it is vital to understand the significance of the place and that the change must sustain the values. In other words, the adverse effect of the outcome must be outweighed by the advantages to enable the development [8,10].

5. Conclusion

In historic building conservation, scales and levels of intervention are usually based on the understanding of heritage significance, which consists of various values. This evaluating process contains the arduous work taken by both conservators to assess different values with reference to the contemporary conditions as well as historical evolution of the building, and the public to make sure the rationality of the assessment. As a result, the most valuable aspects of an architectural heritage can be sustained and reinforced for future development by an application of proper intervention actions. Although this change will harm certain values, it can be justified for the building itself to survive with a satisfactory prospects as an object that is alive, and for the unique value or authenticity to be continually identified as a cultural heritage.

Still, it is worth mentioning that a historic building should be treated with great respect as a whole and all its values deserve protection. Therefore any action must follow the principle of minimum intervention with great sympathy.

References

1. D. W. Insall, *Living buildings: architectural conservation: philosophy, principles and practice* (Images Publishing Group, Mulgrave, Vic, 2008).
2. C. Erder, *Our architectural heritage: from consciousness to conservation* (the United Nations Educational, Scientific and Cultural Organization, Paris, 1986).
3. B. M. Feilden, *Conservation of historic buildings*, 3rd ed. (Architectural Press, London, 2003).
4. M. Hunter, Introduction: the fitful rise of British preservation, in *Preserving the past: the rise of heritage in modern Britain*, ed. M. Hunter (Alan Sutton Publishing Ltd, Stroud, 1996) pp.1-16.
5. J. Jokilehto, *A history of architectural conservation* (Butterworth-Heinemann, Oxford, 1999).
6. J. Earl, *Building conservation philosophy*, 2nd ed. (The College of Estate Management, Reading, 1997).
7. The Australia ICOMOS, *The Burra Charter: The Australia ICOMOS Charter for Places of Cultural Significance* (Australia ICOMOS Incorporated, Sydney, 2013), p.4.
8. English Heritage, *Conservation principles: policies and guidance for the sustainable management of the historic environment* (English Heritage, London, 2008), p.22.
9. P., Druty, 2009. *A sense of value. Conservation bulletin: conservation principles in practice*, **60**, pp.7-10. Http://www.english-heritage.org.uk/publications/conservation-bulletin-60/codifyingconservationpp1-14.pdf.
10. English Heritage, *Practical building conservation: conservation basics* (Ashgate Publishing, Farnham, 2013).
11. A. Dobby, *Conservation and planning* (Hutchinson & Co, London, 1978).

12. I. Hume, The philosophy of conservation engineering, In *Structures & construction in historic building conservation*, ed. M. Forsyth (Blackwell Publishing, Oxford, 2007) pp.12-18.
13. R. Peats, Forty Hall, Enfield: Continuity and innovation in a Carolean Gentry House. *Architectural history*, **33-62**, 51 (2008).
14. The Paul Drury Partnership, *Forty Hall and Estate, Enfield: conservation management plan: executive summary, introduction* (Enfield Council, Enfield, 2007),
15. https://governance.enfield.gov.uk/documents/s6190/Appendix%20-%20Conservation%20Management%20Plan%20for%20Forty%20Hall%20and%20Estate%20and%20Preparation%20for%20the%20bid%20to%20the.pdf.

Study on fracture stability analysis of toppling perilous rock

Xiao-Han Zhou, Hong-Kai Chen*, Hong-Mei Tang, He Wang
Institute of Geotechnical Engineering, Chongqing Jiaotong University, Chongqing, 400074, China
**E-mail: chk99@163.com*

The unstable rock aggregates with complex genetic relation is the basic condition to make large collapse disaster, the sudden destruction of any rock block in the aggregates can release energy which will spread around to produce excitation effect and the excitation signal can be characterized by excitation acceleration. For the falling perilous rock, the probability statistical features are analyzed through the indoor model test. The results show that the energy released by the unstable rock has the characteristic of point load, the probability density of excitation signal showing a single peak approximates a normal distribution, and the peak intensity in the horizontal is greater than that in the vertical; the average value, the valid value and the standard deviation value of the excitation signal is greater as it closer to the vibration source and the measuring value of the vertical direction is greater than that in the horizontal direction; excitation signal has obvious autocorrelation and it indicates that the dependence of the excitation signal to the time is more obvious as the autocorrelation factor gets bigger, and the autocorrelation factor in the vertical direction is larger than that in the horizontal direction; the completeness of the interface between perilous rocks significantly affect the excitation signal transmission, the peak of the statistical characteristics and the probability density of the excitation signal is greater as the completeness gets better, and the transfer effect of the excitation signal is better as the attenuation time gets longer. The study result has positive significance to the reveal of failure mechanisms of the unstable rock mass.

Keywords: Rock Mechanics; Stability Analysis; Toppling Perilous Rock; Rupture Model; Fracture Mechanics.

1. Introduction

Collapse of unstable rock is main types of geological disasters in our mountains, which seriously threatened the rail transportation of mountain road, urban residents and mining safety, the national geological disasters released by Ministry of Land and Resources in 2012 showed that, over the past decade there are an annual average of 2,000 times collapse hazard in our country, such as 5575 occurred in 2010 and 2319 occurred in 2011, which accounted for 30% to 40% of the total number of geological disasters, causing direct economic losses for more than 1.3 billion yuan each year. Unstable rock stability analysis is the core link to judge crag collapse disaster and govern unstable rock disasters effectively[1]. The publicly reported research results of unstable rock failure mechanism are less. Amoung the related researches, Xu Qiang and others think that the key block presence on the cliffs of Wulong chicken mountain collapse in Chongqing[2], Wang Jiachen and Tan Wenhui put forward a three-dimensional stochastic model of progressive failure of slope[3], Zhang Yongxing analyzed the formation and destruction caused by the stress and tension of slope and the rock cavity depth on the development of differential weathering perilous[4], Frayssines and Hantz analyzed the failure mechanism of limestone high slope[5], Riccardo Fanti and others got the related information of slope surface rock structure by using 3D laser scanning method, then accordingly analyzed rock collapse hazard analysis[6], B. Palma and others explored the dynamics of collapse hazard along the road in limestone area through field rockfall trials[7], Claudia Mignelli and others proposed a landslide disaster risk assessment method based on the acceptable threshold of the number of accidents and disasters[8]. Since 2000, Chen Hongkai and others established a dangerous rock stability analysis method based on limit equilibrium theory, which can determine stability state under natural state and rainstorms status conditions[9], however, the stability characterization of the state of unstable rock only related witn working conditions, but has nothing to do with the time factor, therefore, this cannot accordingly forecast collapse hazard safety. Chen Hongkai and Tang Hongmei studied the dangerous rock failure mechanism[10,11], which showed that although the unstable rock has a sudden damage[12] and a significant effect of excitation[13], the essence is the fracture propagation of master structure plane[14], they also explored the basic ideas of fracture mechanics method used in stability analysis of dangerous rock[15]. This paper focused on the emergency security of alert needs of landslide disaster after the destruction of unstable rock, systematically building fracture stability analysis of toppling perilous rock by application of fracture mechanics, which solved the core foundation theoretical issues of real-time monitoring and security alerts of toppling perilous rock landslide disasters.

2. Failure Mode and Load Calculation of Toppling Perilous Rock

2.1. *Geological model*

Geological model of toppling perilous rock is a key link to establish the mechanical model, which can be achieved through on-site investigation, measurement and exploration. Such as National Scenic Area Red Rock

Mountains of Simian Mountain in Jiangjin, where the toppling perilous rocks are widely developed, as shown in figure 1. Such perilous rocks usually have a certain depth of rock cavities at the bottom, the controlling structural plane is located in the base of perilous rocks, where the strength is high, the perilous rock always appeared dumping destruction under the conditions of no shear failure, the typical geologic model as is shown in Figure 2.

Fig. 1. Perilous rocks at Mt. Hongyan, Simianshan national scenic area in Jiangjin

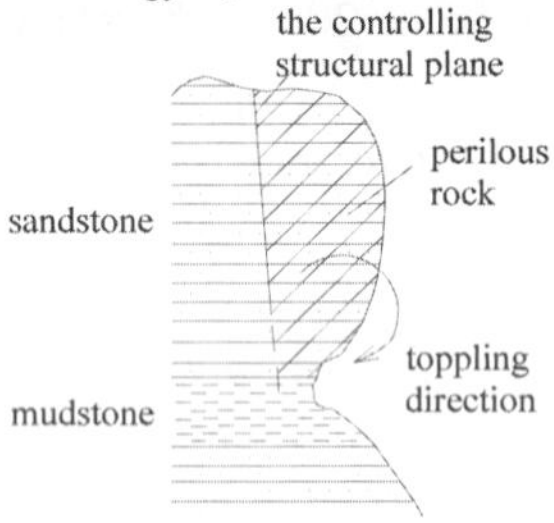

Fig. 2. Representative geological model of toppling perilous rock

2.2. *Mechanical model*

Toppling perilous rock exists tipping point in the bottom, which is always the weathered band endpoints in the base. According to the relationship between heart and tipping point crag weight, toppling perilous rock mechanical model can be divided into two categories-type. Firstly, the tipping point is located inside the center of gravity (Figure 3). Secondly, the tipping point is located outside the center of gravity (Figure 4). Variables in figure: W is the weight per unit width of perilous rock (kN); P_v is vertical seismic force centered on the heart of perilous rock (kN); P_h is the horizontal seismic force centered on the heart of perilous rock (kN); H is the average height of perilous rock (m); Q is the resultant of fissure water pressure in controlling structural plane (kN); e is penetrating segment length in controlling structural plane (m); e_w is the distance between fissure water pressure force acting point and controlling structural plane ends (m); c is the width of the bottom base of perilous rock (m); $[\sigma_c]$is the allowable compressive strength in the base of perilous rock (kPa); β is inclination of controlling structural plane (°); a_c is the Horizontal distance of intersection between controlling structural plane and bottom of perilous rock and center of gravity (m); b_c is the vertical distance between center of gravity and the bottom of perilous rock (m); O is the perforation segment endpoints of controlling structural plane; C is the center of gravity of perilous rock; B is the tipping point.

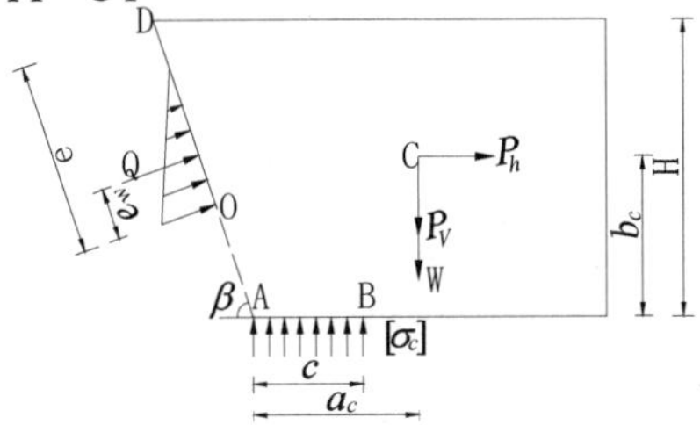

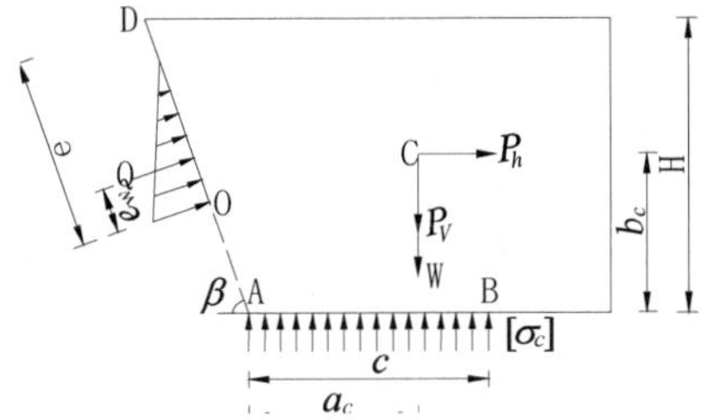

Fig. 3.1st mechanic model of toppling perilous rock. Fig. 4.2st mechanic model of toppling perilous rock

2.3. *Failure mode and load calculation*

Definiting the sinking state function in the base of toppling perilous rock as F:

$$F = P_v + W - c[\sigma_c] - \frac{1}{18}\gamma_w e^2 \sin 2\beta - (H - e\sin\beta)[\tau] \tag{1}$$

In the formula, [τ] is the allowable shear strength of non-perforation in controlling structural plane (kPa); γ_w is the Bulk density of water in controlling structural plane (kN/m^3); the remaining variables are the same as before.

(1)When F<0, there is no settlement in the base of perilous rock, toppling perilous rock destruction belong to tensile failure mode (Figure 5), the moment of all loads on the end point O in controlling structural plane acts on the perilous rock can be calculated as follows:

$$M = P_h(b_c - H + e\sin\beta) + (W + P_v)(a_c + H\text{ctan}\beta - e\cos\beta) - c[\sigma_c](0.5c + H\text{ctan}\beta - e\cos\beta) + \frac{1}{162}\gamma_w e^3 \sin^3\beta \tag{2}$$

(2)When F≥0, there is settlement in the base of perilous rock, normal stress σ acts on the controlling structural plane is negative, toppling perilous rock belongs to tensile shear failure mode (Figure 6); normal stress σ acts on the controlling structural plane is positive, toppling perilous rock belongs to compression and shear failure mode (Figure 7). In both of these failure modes, controlling structural plane of perilous rock are also subject to the master structure M (moment), σ (tensile stress or compressive stress) and τ (shear stress), calculation method as follows:

Carrying all the loads (Figure 3 and Figure 4) act on perilous rock on controlling structural plane, the calculation formula of all loads act on controlling structural plane ends O is as follows:

$$M = P_h(b_c - H + e\sin\beta) + (W + P_v)(a_c + H\text{ctan}\beta - e\cos\beta) + \frac{1}{162}\gamma_w e^3 \sin\beta \quad (3)$$

All the loads act on perilous rock can be decomposed along the normal and tangential direction of controlling structural plane, then getting the normal force and tangential force, assuming the normal force and tangential force are uniformly distributed along the controlling structural plane, finally getting the calculation formula of normal stress σ and tangential stress τ:

$$\sigma = \frac{[(P_v + W)\cos\beta - P_h \sin\beta - \frac{1}{18}\gamma_w e^2 \sin\beta]\sin\beta}{H} \quad (4)$$

$$\tau = \frac{[(P_v + W)\sin\beta + P_h \cos\beta]\sin\beta}{H} \quad (5)$$

In the formula 2 to 5, physical meaning of the variable are the same with the former, Vertical seismic force P_v and horizontal seismic force P_h can not be considered simultaneously[1].

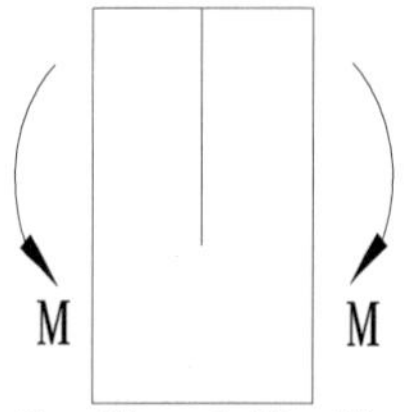

Fig.5. (the left) Tensile rupture model of toppling perilous rock

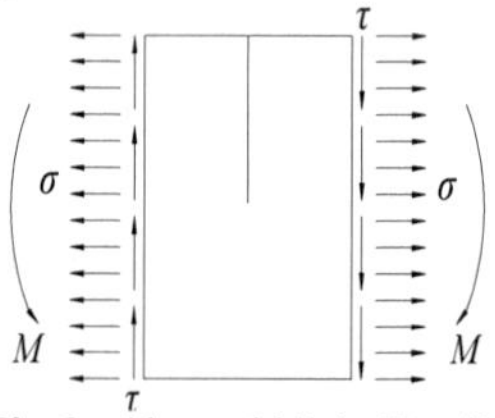

Fig.6. (the middle) Tensile shear rupture model of toppling perilous rock

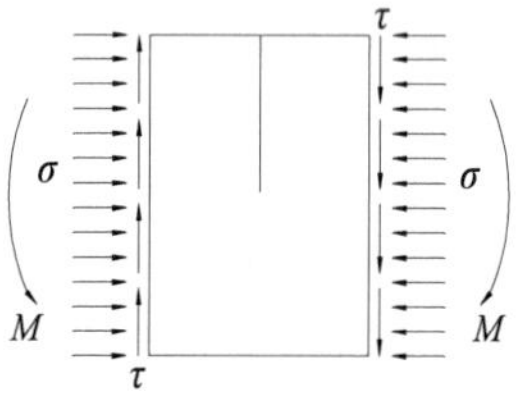

Fig.7. (the right) Compressive shear rupture model of toppling perilous rock

3. Fracture Model and Solution Formula of Toppling Perilous Rock

3.1. *Fracture model*

Decomposing the failure mode of toppling perilous rock (Figure 8 and Figure 9), building four basic fracture model: Bending model, shear stress model, tensile stress model and compressive stress model (Figure 10). Solving each basic fracture model, solution about fracture mechanics of toppling perilous rock can be gotten by basic fracture model solution portfolio.

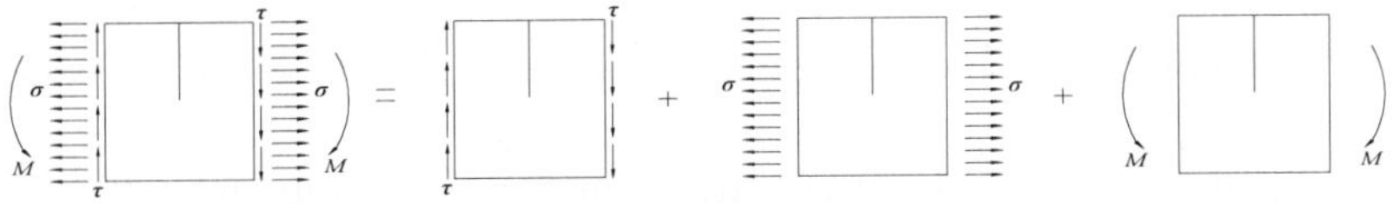

Fig. 8. Disintegration of tensile shear rupture model for toppling perilous rock

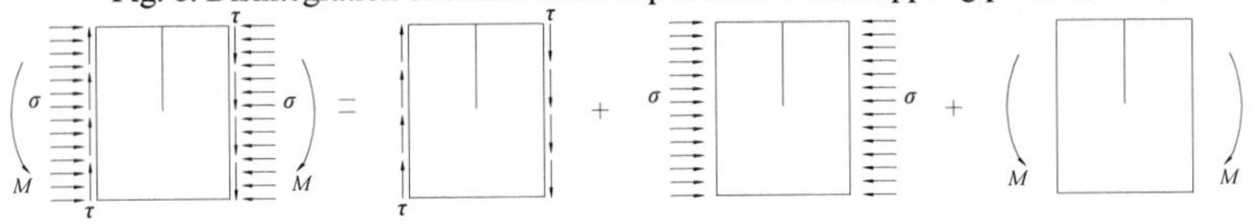

Fig. 9. Disintegration of compressive shear rupture model for toppling perilous rock

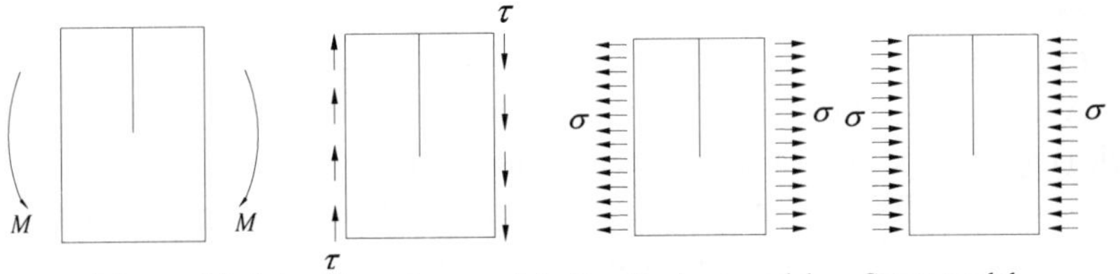

Moment Model Shear stress model Tensile stress model Stress model

Fig. 10. Elementary fracture models for toppling perilous rock rupture model

3.2. *Joint fracture stress intensity factor*

(1)Solution of basic fracture model

Moment Model: using integral transform method to obtain the stress intensity factor calculation[16]:

$$\mathrm{KI}=3.975M\left(\frac{H}{\sin\beta}-e\right)^{-\frac{3}{2}},\quad \mathrm{KII}=0 \tag{6}$$

In the formula, both units of K_{I} and K_{II} are $\mathrm{kPa}\sqrt{\mathrm{m}}$; the physical meaning of remaining variables are the same as before.

Shear stress model: using Fourier transformation method to obtain the stress intensity factor calculation[16]:

$$\mathrm{KI}=0,\quad \mathrm{KII}=\tau\sqrt{1.5B\sin\beta}\left(\frac{2H}{B\sin^2\beta}+0.2865\right) \tag{7}$$

In the formula, B is the average width of perilous rock (m); the remaining variables are the same as before.

Tensile stress model: using Westergard stress function method to obtain the stress intensity factor calculation[16]:

$$\mathrm{KI} = 1.5958\sigma\sqrt{\frac{H}{\sin\beta}},\quad \mathrm{KII} = 0 \tag{8}$$

In the formula, the physical meaning of remaining variables are the same as before.

Stress model: using Westergard stress function method to obtain the stress intensity factor calculation[16]:

$$\mathrm{KI}=-1.5958\sigma\sqrt{\frac{H}{\sin\beta}},\quad \mathrm{KII}=0 \tag{9}$$

In the formula, the physical meaning of remaining variables are the same as before.

(2)Fracture model solution of toppling perilous rock

Aiming at breaking angle θ_0 of any point Q in end area of toppling perilous rock (Figure 11), according to fracture mechanics, the calculation formula of expansion joint fracture stress intensity factor Ke along the diagonal direction θ_0 in controlling structural plane ends of perilous rock is:

$$K_e = \cos\frac{\theta_0}{2}[K_I\cos^2\frac{\theta_0}{2} - 1.5K_{II}\sin\theta_0] \tag{10}$$

In the formula, Ke is the joint fracture stress intensity factor in controlling structural plane ends of perilous rock ($\mathrm{kPa}\sqrt{\mathrm{m}}$); θ_0 is the fracture propagation angle in controlling structural plane ends of perilous rock (°); The physical meaning of remaining variables are the same as before.

When $K_{II}=0$ and $K_I\neq0$, $\theta_0=0°$, the controlling structural plane extending along the main direction of extension;

When $K_{II}\neq0$, then $\theta_0\neq0°$, assuming $k_0 = (K_I / K_{II})^2$, extension angle θ_0 can be determined by the following formula:

$$\theta_0 = \arccos[\frac{3+\sqrt{k_0^2+8k_0}}{k_0+9}] \tag{11}$$

Firstly, tensile failure joint fracture stress intensity factor of toppling perilous rock

$$\mathrm{Ke}=\mathrm{KI}=3.975M(\frac{H}{\sin\beta}-e)^{-\frac{3}{2}},\quad \theta0=0° \tag{12}$$

Secondly, tensile shear joint destruction fracture stress intensity factor K_e of toppling perilous rock and extension angle θ_0 in the controlling structural plane ends can be determined by formula (10) and formula (11):

$$\text{KI}=3.975M(\frac{H}{\sin\beta}-e)^{-\frac{3}{2}}+1.5958\sigma\sqrt{\frac{H}{\sin\beta}} \tag{13}$$

$$\text{KII}=\tau\sqrt{1.5B\sin\beta}(\frac{2H}{B\sin^2\beta}+0.2865) \tag{14}$$

Thirdly, compression and shear failure fracture stress intensity factor Ke of toppling perilous rock and extension angle θ_0 in the controlling structural plane ends can be determined by formula (10) and formula (11), in the formula, K_{II} can be determined by formula (14), K_I can be determined by the following formula:

$$\text{KI}=3.975M(\frac{H}{\sin\beta}-e)^{-\frac{3}{2}}-1.5958\sigma\sqrt{\frac{H}{\sin\beta}} \tag{15}$$

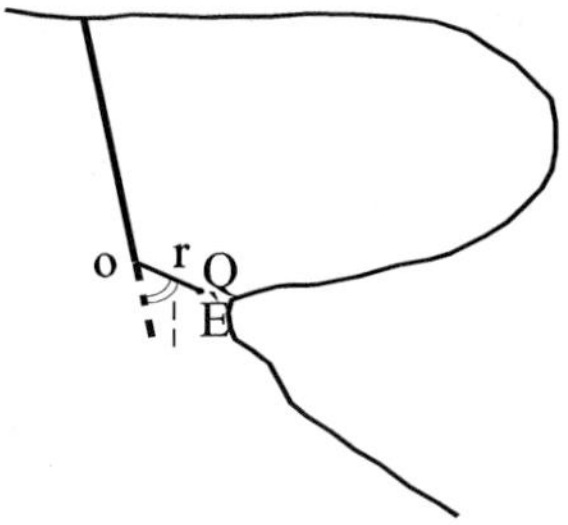

Fig. 11. Polar coordinates focusing on the end of dominant fissure in toppling perilous rock

4. Fracture Stability Analysis of Toppling Perilous Rock

4.1. *Evaluation criterion of stability for perilous rock*

Chen Hongkai divided the stability of perilous rock into three states of instability, the basic stability and stable according to stability factor of perilous rock[9] (Table 1). In general, perilous rock in a steady state cannot use engineering control measures, perilous rock in the basic steady state should be strengthen the monitoring and conduct the surface drainage and utilize local governance project in necessary part, perilous rock in an unstable state in addition to be used monitoring, surface and underground drainage, still need to adopt systematic engineering control measures.

Table 1. Evaluation criterion of stability for perilous rock

Failure mode of perilous rock	Instability	Basic stability	Stable
Sliding perilous rock	<1.0	1.0~1.3	>1.3
Toppling perilous rock	<1.0	1.0~1.5	>1.5
Falling perilous rock	<1.0	1.0~1.5	>1.5

Note: The data in the table is stability factor of perilous rock

4.2. *Fracture stability factor of perilous rock*

According to the joint fracture stress intensity factor K_e of toppling perilous rock, combined with the fracture toughness of intact rock K_{IC}, establishing fracture stability coefficient calculation formula of perilous rock:

$$F_s = \frac{K_{IC}}{K_e} \tag{16}$$

In the formula, K_{IC} is fracture toughness of the controlling structural plane ends, which can be determined by fracture test, the unit is $\mathrm{kPa}\sqrt{\mathrm{m}}$; The remaining variables are the same as before.

The steady state of perilous rock can be determined referring to table 1. As K_{IC} is the function of normal stress σ and shear stress τ, as long as collecting normal stress σ and shear stress τ of controlling structural plane ends by using corresponding sensor, this method can be used to judge the state of stability of toppling perilous rock.

5. Conclusions

(1) Building the geological model and the mechanical model of toppling perilous rock, definiting the sinking state function F in the base of toppling perilous rock, when $F<0$, the base does not occur settlement, when $F\geq 0$, the base will occur settlement.

(2) Failure mode of toppling perilous rock include three forms, that is tensile failure ($F<0$), the pull shear failure (when $F\geq 0$ and σ is negative) and compression and shear failure (when $F\geq 0$, and σ is positive).

(3) The controlling structural plane of tensile failure mode of toppling perilous rock undertake moment, both tensile shear and compressive shear failure modes and failure mode undertake the forces of moment, tensile stress

(or compressive stress) and the shear stress, then establishing the corresponding calculation method.

(4) Decomposing the failure mode of toppling perilous rock, then getting four basic fracture model: moment model, shear stress model, tensile stress and compressive stress model model. Based on Fracture Mechanics, getting the basic fracture model for each answer, then getting the fracture mechanics solutions of failure mode of toppling perilous rock and establishing calculation formula of fracture propagation angle and joint fracture stress intensity factor of controlling structural plane ends of perilous rock.

(5) Based on joint fracture stress intensity factor and fracture toughness of toppling perilous rock, establishing the fracture stability analysis of toppling perilous rock, which has been demonstrated reasonable through projects, this method has solved basic theoretical issues for emergency security alarm of toppling perilous rock.

Acknowledgement

This work is supported by the National Natural Science Foundation of China (No. 11272185, 50678182); Projects funded by innovative team building programs of Chongqing university in 2013 (No. KJTD201305); Funded by Par-Eu Scholars Program of Chongqing. Corresponding author: CHEN Hong-kai, (1964-), (Doctor, Processor, Doctoral tutor. Engaging in the research on geomorphological processes and their mitigation mechanics. Tel: 023-62652719, E-mail:chk99@163.com)

Work partially supported by grant 2-4570.5 of the Swiss National Science Foundation.

References

1. CHEN Hong-kai, TANG Hong-mei, WANG Lin-feng, YE Si qiao. *Evolution theories and its application for perilous rock and rockfall* [M]. Beijing: The Science Press, 2009
2. XU Qiang, HUANG Run-qiu, YIN Yue-ping, HOU Sheng-shan, DONG Xiu-jun, FAN Xuan-mei, TANG Ming-gao. *The Jiweishan landslide of June 5, 2009 in Wu Long, Chongqing: Characteristics and Failure Mechanism* [J]. China Institute of Geo-Environment Monitoring, 2009, 17(4):433-444
3. WANG Jia-chen, TAN Wen-hui. *Three dimensional stochastic analysisof slope progressive failure* [J]. Journal of Coal Society. 1997, 22(1):27-31

4. ZHANG Yong-xing, LU Li, ZHANG Si-ping, HU Dai-wen. *Development and Failure Principle of Differential Weathering Overhanging Rock* [J]. Journal of Chongqing Jianzhu University, 2010, 32 (2):1-6
5. Frayssines, M. and Hantz, D. *Modelling and back-analysing failures in steep limestone cliffs. International* [J]. Journal of Rock Mechanics and Mining Sciences, 2009, 46 (7):1115-1123
6. Riccardo Fanti, Giovanni Gigli, Luca Lombardi, Deodato Tapete, and Paolo Canuti. *Terrestrial laser scanning for rockfall stability analysis in the cultural heritage site of Pitigliano (Italy).* Landslides, 2012, DOI 10.1007/s10346-012-0329-5
7. B.Palma, M.Parise, P. Reichenbach, and F. Guzzetti. *Rockfall hazard assessment along a road in the Sorrento Peninsula,Campania, southern Italy*. Nat Hazards, 2012, 61:187-201
8. Claudia Mignelli, Stefano Lo Russo, and Daniele Peila. *Rockfall risk management assessment: the RO.MA.approach* [J]. Nat Hazards, 2012, 62:1109-1123
9. CHEN Hong-kai, XIAN Xue-fu, TANG Hong-mei, WANG Lin-feng. *Stability Analysis Method for Perilous Rock* [J]. Chinese Journal of Applied Mechanics, 2009, 26 (2): 278-282
10. CHEN Hong-kai, TANG Hong-mei, YE Si-qiao. *Damage model of control fissure in perilous rock* [J]. Applied Mathematicsand Mechanics , 2006, 27(7):967-974
11. CHEN Hong-kai, TANG Hong-mei. *Method to calculate fatigue fracture life of control fissure in perilous rock* [J]. Applied Mathematics and Mechanics, 2007, 28 (5): 643-649
12. CHEN Hong-kai, ZHANG Rui-gang, TANG Hong-mei, ZHAO Xian-tao. *Elastic & Impulsive Dynamic Parameters of a Ruptured Compression-shear Perilous Rock* [J]. Journal of Vibration and Shock, 2012, 31(24):33-36.
13. TANG Hong-mei,WANG Zhi, CHEN Hong-kai, XIAN Xue-fu. *Contribution ratio of excitation action triggered by collapse of perilous rock to stability of perilous rock* [J]. Journal of Vibration and Shock, 2012,31(20):32-37
14. CHEN Hong-kai, TANG Hong-mei, WANG Lin-feng, XIAN Xue-fu. *Mechanical mechanism for retreat evolution of steep rock slopes with gentle dip*. Chinese Journal of Geotechnical Engineering 2010, 32(3):468-473
15. CHEN Hong-kai, XIAN Xue-fu, TANG Hong-mei. *Stability analysis method for perilious rock by fracture mechanics* [J]. Journal of Chongqing University, 2009, 32(4):434-437.
16. Chinese Aeronautical Research Institute. *Stress intensity factor handbook* [M]. Beijing: The Science Press, 1981.

Study on the formation mechanism of ground collapse of Yujiawan Reservoir in Chongqing Gele Mountain

Hong-Kai Chen*, Xiao Yin
Institute of Geotechnical Engineering, Chongqing Jiaotong University, Chongqing, 400074, China
**E-mail: chkcxf@163.com*

Ground collapse of soil cover is currently one of the main types of geological hazards in China, with the features of widely developed, randomness, sudden and so on in limestone and modern urban areas. Make the Yujiawan reservoir ground collapse in Chongqing Gele Mountain as an example, established a mechanical model of collapse, proposed collapse criterion,and the distinguishing factor K of collapse pit T1 is 1.31 without the vacuum suction which generated by groundwater quickly landing, it will not collapse. After considering the vacuum suction, the distinguishing factor K reduced to 0.92, collapse inevitably produce, consistent with the facts. Analysis concluded that ground collapse of soil cover can be divided into two categories as Yujiawan reservoir type and Zhazidong type. The prerequisite for Yujiawan's ground collapse is that there exists karst crevice or cave system at the bedrock surface in a certain depth below ground, its cover soil compacted and in saturated conditions, and the key links of Zhazidong's ground collapse is temporary blocking after the burst of underground water pipe, making the pipe water fill quickly in the embankment, forming a saturation area, piping discharge rapidly drained sandy (soil) water, saturated sand (soil) rapid loss, negative pressure cavity appears between the drain and embankment areas, and top soil of the negative pressure cavity cracking down is the main driving mechanism for the formation of this class of d collapse. This research results is positive for further implement on study the formation mechanism of ground collapses, assess the risk of ground collapses effective and control ground collapses.

Keywords: Geotechnical Mechanics; Collapse Mechanisms; Vacuum Suction; Ground Soil Cover; Yujiawan Reservoir.

1. Introduction

Ground collapse is one of typical geological hazards. By the end of 2012, ground collapse have occurred throughout China's 24 provinces,mainly in the provinces of Chongqing, Sichuan, Guizhou, Yunnan, Guangdong, Guangxi, Liaoning, Hebei, Jiangxi, Hubei and Hunan. So far, domestic and foreign scholars have done many researches on the mechanism of karst collapse and propose erosion on the vacuum suction, latent erosion theory, vibration theory,

theory of liquefied, gas explosion theory and other theory to predict karst collapse [1]; Kyriaki Papadopoulou-Vryniotia et al, use multivariate statistical methods and geographic information systems to study the various factors of karst collapse produced, prepareing karst collapse susceptibility map of the study area [2]; Haibo Bai et al, use variable mass mechanics and nonlinear dynamics to establish a mechanical model of groundwater intrusion karst formations roof's damage mechanism, which showing that the greater the groundwater seepage velocity the smaller karst formations roof stability [3]; He Keqiang et al, analyzed groundwater exploitation and other factors that affect the full collapse of Zaozhuang Shili, then they put forward some countermeasures prevention of karst collapse and protection of karst water resources and environment [4]; Haijun Zhao et al, analyzed the problem of karst collapse widespread in southern China. They find that high-intensity rainfall accelerates karst hydrodynamic conditions, changing the physical and mechanical properties of overburden on the karst cave, and trigger or promote the formation of caves and collapse [5]; Mingtang Lei et al, through the large-scale model tests, established a methods of karst collapse risk assessment based on geographic information system [6]; Xu Weiguo et al, think that in a relatively closed network of confined karst groundwater, pumping (rowing) groundwater to lower the groundwater level below the surface of cap layer which can form vacuum suction cups, vacuum cavity suction and vortex suction funnel. This three kinds of effects will damage the soil cover layer in karst voids [7]; He Keqiang et al, established a formula for the critical height of soil hole in inter collapse phase and a mechanical model for infiltration pressure effect in surface collapse stage. The models that He Ke-qiang et al, rasied based on Platts theory and limit equilibrium theory [8]. Visibly, the aforementioned study focused on the problem of karst collapse. Recent years, along with mountain highway, urban shallow tunnels and a large number of underground space construction and development and utilization, ground collapse problem occurs frequently, mainly soil cover collapse [9]. The Yujiawan reservoir of Chongqing Gele Mountain Jingang village occurred six times ground collapses during the construction of Zhongliangshan tunnel of Chongqing Light Rail NO.1 line from October 2010 to April 2013, reservoirs have dried up.Yujiawan reservoir's ground collapse belongs to surface karst fissure system filling soil collapse problems. The formation mechanism of this research has a positive meaning for the effective prevention of surface soil cover collapse hazard.

2. Geological Conditions of Ground Collapse Area

Yujiawan reservoir ground collapse area is located in Chongqing Gele Mountain town Jingang village Longtangkan cooperatives. Before the ground collapse disaster strikes reservoir is normal (Fig.1). As shown in Figure 2 ground collapse have appeared six places in the east side of the reservoir, during October 20 to 2013 April 6 during the construction period of Zhongliangshan tunnel of Chongqing Light Rail NO.1 line, respectively using T1, T2, ..., T6 marked. The largest one of them is T1 collapse pit which is 5.0 metres in length, 3.7metres in width, 6.0 metres in depth, oval-shaped pit, the area of it about 16 square metres(Fig.3). Karst collapse belongs to dissolution cave collapse in this area that the collapse area is located in the southeast wing Gele Mountain anticline, over the Jialing River Fm limestone and Leikoupo Fm marlite, Surface soil's thickness is 3 ~4 metres which was composed by eluvial, talus, alluvium and other quaternary loose soil. NE - SW trending underground river are developed below ground 20 ~ 30m deep and above The underground river is Karst, developing a number of sinkholes, ground fissures, solution dolines, karst depression and other karst landforms, and they were filled by Quaternary loose soil. Before the ground collapse occurred, surface loose soil water table is high and depth is only 1.5 ~ 2 m. But after the ground occurred, groundwater depth increased to 5 ~ 11 m.

Fig. 1. Outline of Yujiawan reservoir before ground collapses.

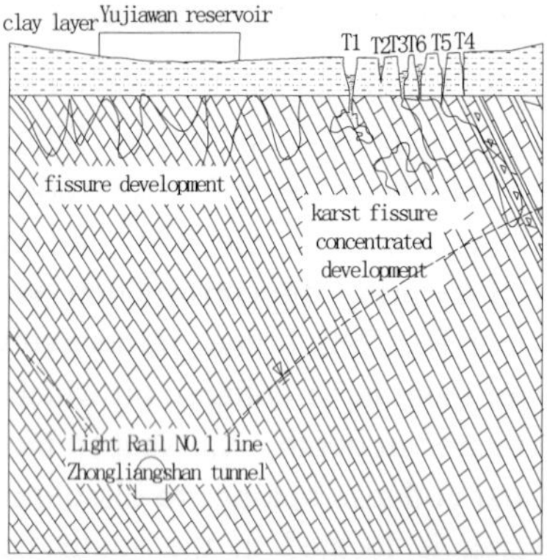

Fig. 2. Geological section at ground collapse area inYujiawan reservoir.

Fig. 3. T1 ground collapse pit in Yujiawan reservoir.

3. Forming Mechanism of T1 Collapse Pit

Yujiawan reservoir ground collapse occurred when Zhongliangshan tunnel of Chongqing Light Rail NO.1 line Boring to K26 + 845 Synchronization. The location is south of the east wing of anticline Gele Mountain aquifer and inflow from 549.89m^3/d cut through before suddenly increase to 1591.64m^3/d site measured. It can be inferred that the aquifer that Zhongliangshan tunnel excavation cut through is key link Yujiawanan reservoir ground collapse occurred. We can generalize Yujiawan reservoir T1 collapse pit into a column (Fig. 4). The mechanical model of which is shown in Figure 5. Focus on cover layer soil column at the top of the soil cave to establish formula of collapse force F_{down} and anti-collapse force F_{up} as follow:

$$F_{down} = G + \Delta P S_{bottom} \tag{1}$$

$$F_{up} = f + c S_{side} \tag{2}$$

In the formula, F_{down} is Karst collapse force (kN); F_{up} is soil cover layer anti-collapse force (kN); G is the weight cover layer soil column (kN); ΔP is cave vacuum suction cavity (kpa); f is the sidewall friction of cover layer soil column (kN); c is soil cohesion (kPa); S_{bottom} is the bottom area of cover layer soil column(m^2); S_{side} is the side area of cover layer soil column(m^2). Related load is calculated as follows:

Weight of soil column:

$$G = S_{bottom} \mathrm{h}\,\gamma \tag{3}$$

In the formula, γ is the bulk density of cover layer soil (kN/$m^{3)}$;h is the height of cover layer soil column (m).

Vacuum suction Δ P: The extreme of vacuum suction is a atmospheric pressure, equal to 10 metres of water pressure; But in the natural environment, it is difficult to have a absolute closed condition in the actual collapse of karst cavities formed According to Chen Guoliang's research, the vacuum negative pressure difference is generally no more than 50kPa, so the proposed value is between 0 ~ 50kPa [10].

Frictional resistance f:

$$f = \frac{1}{2} K_0 \gamma h^2 S_{side} \tan\phi \tag{4}$$

In the formula, φ is the soil friction angle (°) K_0 is lateral pressure coefficient of soil mass; The physical meaning of remaining variables are the same as before.

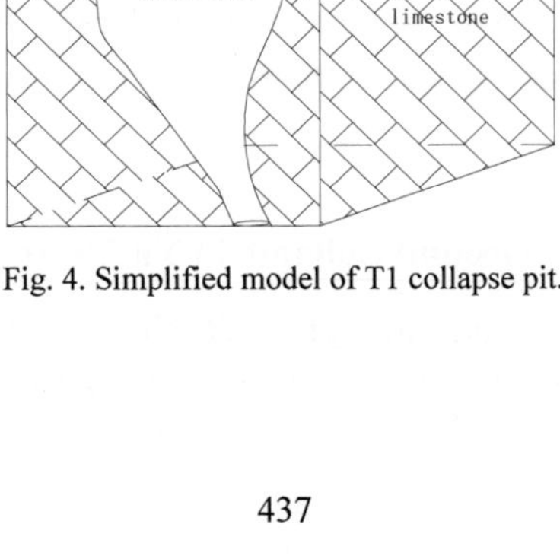

Fig. 4. Simplified model of T1 collapse pit.

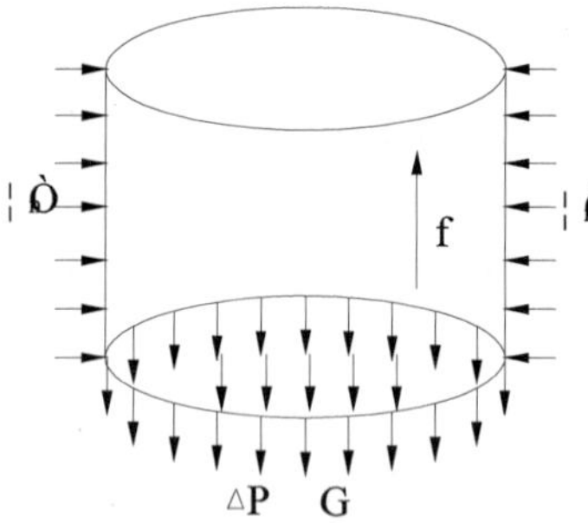

Fig. 5. Mechanical model of T1 collapse pit.

According to the collapse force and anti-collapse force of soil column at the top of the collapse cave to define the collapse distinguishing factor K:

$$K = \frac{F_{up}}{F_{down}} = \frac{\left(\frac{1}{2}K_0\gamma h \tan\phi + c\right)S_{side}}{\left(h\gamma + \Delta P\right)S_{bottom}} \tag{5}$$

When K < 1, it is easy to occur ground collapse;

When K=1, it is in the critical state of the occuring ground collapse;

When K > 1, it is not easy to occur ground collapse. To simplified Yujiawan reservoir T1 collapse pit into an ellipsoid: major axis a=5.0 m, minor axis b=3.7m, about 6.0m in depth, and $S_{bottom} = \pi ab = 58.12\,m^2$, $S_{side} = [2\pi b + 4(a-b)]\ h = 46.50\,m^2$. the cover soil is silty clay and the thickness h is 4 m. The results of site sampling soil test: bulk density γ=22.3 kN/m^3, cohesive force c= 68kPa, internal friction angle φ=23°; K_0=0.71,ΔP=25kPa. Take all the parameters into the formula (5):

$$K = \frac{F_{up}}{F_{down}} = \frac{\left(\frac{1}{2}\times 0.71\times 22.3\times 4^2\times \tan 23^\circ + 68\right)\times 46.50}{\left(4\times 22.3 + 25\right)\times 58.12} = 0.85 \quad \langle 1$$

Indicates that the T1 will occur ground collapse. If the vacuum suction is not considered, that is P=0, the result K=1.09 in (5), it is not easy to occur ground collapse.

4. Evolution Model of Ground Collapse in Yujiawan Reservoir

The soil soaked by groundwater gradually losing water and consolidation settlement under its own weight which was caused by slow descent of groundwater level. It is easy for the top cave soil cover to occur collapse, when the groundwater level suddenly drop. Zhongliangshan tunnel of Chongqing Light Rail NO.1 line is located at the elevation of 70~ 80m height, which is lower than that of Yujiawan reservoir. Construction of tunnel cuts through permeable zone, introducing extra water gushing quantity that sharply increased about 2 times. At the same time, the surface soil cover layer of underground water level dropped quickly which have evoked 6 pieces of cracks, and consequently, collapse. In order to analyze the influence of the construction of Light Rail NO.1 line Zhongliangshan tunnel on the ground surface deformation from the macro scale analysis. Three dimensional numerical simulation of tunnel excavation is carried out for the geological landform structure (Fig. 6) and the physical mechanical parameters (Table 1) in Gele Mountain to get the vertical displacement along the tunnel to the ground (Fig. 7) and the ground collapse area is shown in figure 8. Visible, Yujiawan reservoir is located in the biggest collapse area that tunnel excavation produced. It can be inferred that the excavation and construction of Zhongliangshan tunnel of Chongqing Light Rail NO.1 line is easy to cause the water level in the area of the Yujiawan reservoir to drop significantly because of the large deformation. It has a positive effect to increase the magnitude of ΔP in the formula (5).

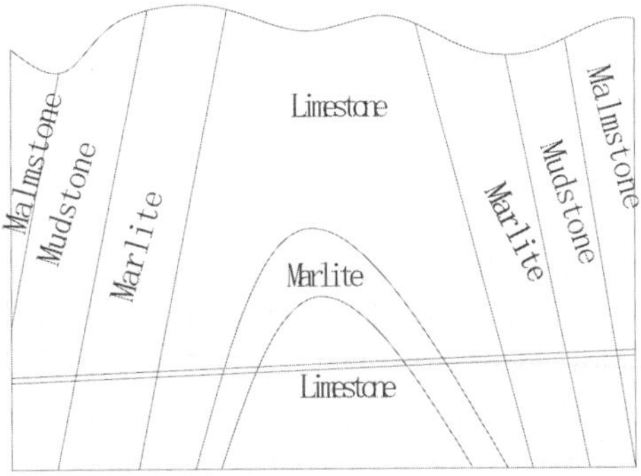

Fig. 6. Geological section along Zhongliangshan tunnel.

Table 1. Physical and mechnanical parameters of strata in Mt. Gele.

Lithology	E kPa	μ	γ kN/m^3	c kPa	φ°
Limestone	7.50×10^7	0.19	25.9	3.00×10^4	48.5
Mudstone	3.4×10^6	0.23	23	5.00×10^3	31
Marlite	5.5×10^6	0.21	24.6	1.00×10^4	38
Malmstone	5.5×10^7	0.2	25.5	2.40×10^4	42.5

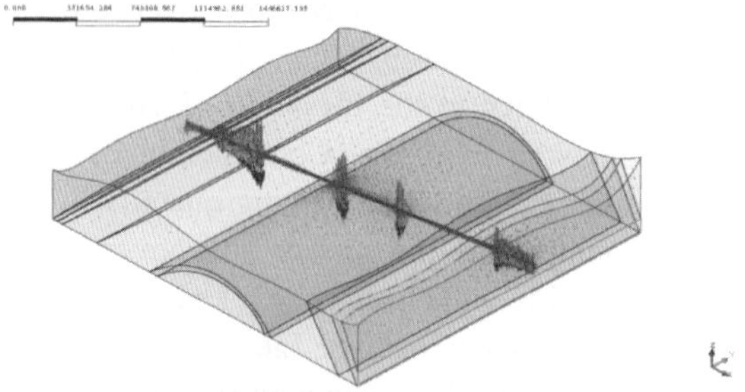

Fig. 7. Vertical ground displacement along Zhongliangshan tunnel.

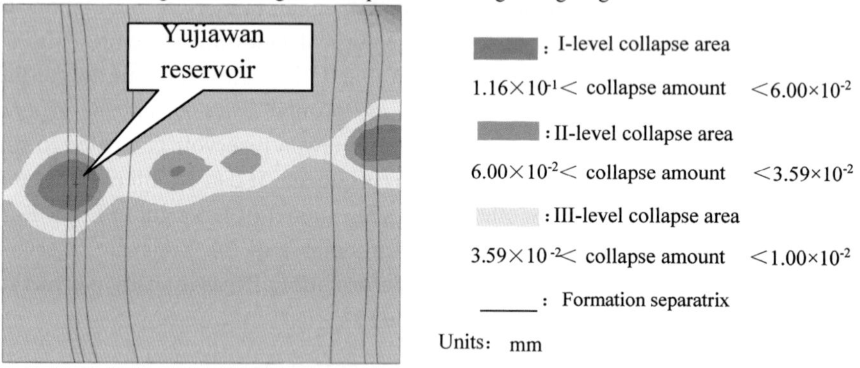

Fig. 8. Zonging of ground collapse along Zhongliangshan tunnel.

The evolution model of ground collapse in Yujiawan reservoir is shown in Figure 9, which can be divided into five stages: stage (a) is the geological conditions before the occurrence of ground collapse; stage (b) is that the groundwater level sudden drops because of permeable layer of mountain body was cut through by tunnel excavation and groundwater seepage caused by the mountains deformation; stage (c) is that a negative pressure cavity appear at the top of the cave;stage (d) is that the process enlargement of the negative pressure cavity in the top of the cave and extensional deformation appears in bottom of soil cover of cavity;stage (e) is that soil cover at the top of the negative pressure cavity occurred collapsed (Figure 3). Visible, prerequisite for Yujiawan reservoir ground collapse is produced by the lower portion of the surface of soil bedrock presence of karst fissures or cave system. The upper cover layer soil is compact and has a certain thickness of saturated layer and the key point is that there is a cavity between the soil cover and the top of the cave during the sudden drop of groundwater level. Negative pressure generated in the cavity. The greater the sudden drop of groundwater level, the greater the value of negative pressure which strongly drawn to the top of the cavity, and the volume of the negative pressure cavity increases quickly. When the volume of the negative pressure cavity is increased to a one certain, the sudden collapse of the soil cover at the top of the cavity occurred.

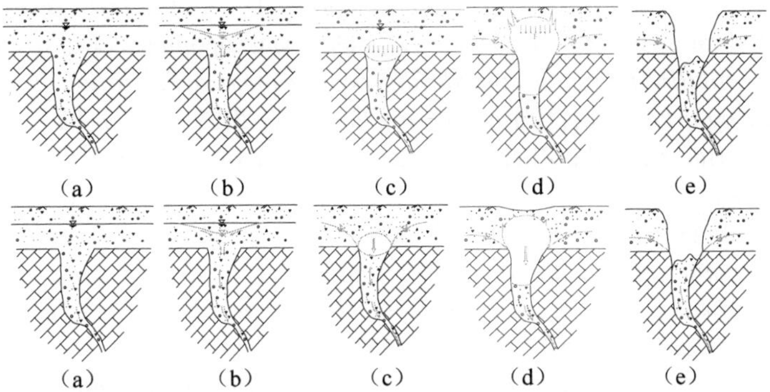

Fig. 9. Evolution model of ground collapse at Yujiawan reservoir.

5. Conclusions

(1) Yujiawan reservoir ground collapses is that the soil cover because of groundwater level fast landing induced ground collapse in the karst area which is one typical hazard. Based on the collapse force and anti-collapse force of soil column at the soil cover of the cave, the ground collapse criterion of the soil cover is established. The rationality of the construction of the ground collapse discriminant is verified by the T1 collapse pit of Yujiawan reservoir.

(2) Numerical simulation of three dimensional displayed that Yujiawan reservoir is located at the top of the maximum vertical deformation area in the top of Zhongliangshan tunnel of Chongqing Light Rail NO.1 line. It can be inferred that areas have the greaterd surface placements the tunneling-induce more likely to occur ground collapse. The tunnel excavation cut off the mountain aquifer then the groundwater level drops quickly, which should be the key factors causing ground collapse..

(3) Proposing ground collapse evolution model of Yujiawan reservoir, a prerequisite for Yujiawan's ground collapse is that there exist karst crevice or cave system at the bedrock surface below ground; The groundwater level sudden drop induced cave at the top of the cavity vacuum suction is the trigger mechanism of ground collapse of the soil cover layer; Top soil cover layer cavity collapse is macroscopic appearance of ground collapse.

References

1. Xing Chen. *Study on prediction and evaluation of karst collapse mechanism*[M], Bei Jing: Geological Publishing House, 2005.
2. Kyriaki Papadopoulou-Vrynioti, George D. Bathrellos, Hariklia D. Skilodimou,George Kaviris, Kostas Makropoulos. *Karst collapse susceptibility mapping considering peak ground acceleration in a rapidly growing urban area*[J]. Engineering Geology, 2013,158: 77–88.
3. Haibo Bai, Dan Ma, Zhanqing Chen. *Mechanical Behavior of Groundwater Seepage in Karst Collapse Pillars*[J]. Engineering Geology, 2013,3:103-124.
4. Keqiang He, Yuyue Jia, Min Zhao, Weigong Chen. *Comprehensive analysis and quantitative evaluationof the influencing factors of karst collapse in groundwater exploitation area of Shiliquan of Zaozhuang, China*[J]. Environ Earth Sci, 2012, 66:2531–2541.
5. Haijun Zhao, Fengshan Ma, Jie Guo. *Regularity and formation mechanism of large-scale abrupt karst collapse in southern China in the first half of 2010*[J]. Nat Hazards, 2012, 60:1037–1054.
6. Mingtang Lei, Xiaozhen Jiang, Li Yu. *New advances in karst collapse research in China*[J]. Environmental Geology, 2002, 42: 462– 465.
7. Weiguo Xu. *Discussion on the vacuum effect of ground collapse in karst mining area*[J]. Geological Review, 1981, 27(2):174-180.
8. Keqiang He, Bin Wang, JiTao Wan. *Research in forming mechanism and collapse model of Zaozhuang karst collapse*[J]. Rock and Soil Mechanics, 2002, 23(5): 564-574.
9. Chengping Zhang,Dingli Zhang, Mengshu Wang, Qianqian Li, Shengchun Liu. *Construction mechanism and control of city tunnel-induced ground collapse*[J]. Rock and Soil Mechanics, 2010, 31(s1): 303- 309.
10. Guoliang Chen. *Causes and prevention of karst ground collapse*[M], Bei Jing: China railway Publishing House, 1994.

Finite element analysis of joints seismic performance of frame structure under strong earthquake

Hai-Xu Yang, Yu Liang[1,a], Xi-Guang Chen[1,a], Tian-Qiong Xue[1, a]
[1]*School of Civil Engineering, Northeast Forestry University. Harbin, China 150040*
E-mail: yhxcumt@163.com, [a] 984104934@qq.com

Abstract: Using an airport expansion project with frame structure as an example, this paper studied the seismic performance of reinforced concrete frame structure's beam-column joints under the action of strong earthquake. With Abaqus software, the response of structure in different seismic magnitude and peak acceleration under EL-Centro wave, Accel artificial wave and Taft wave were calculated, and the seismic performance of local frame node is also analyzed. The analysis results indicate that considering floor, the effect on stiffness of the beam structure is significantly improved and structural lateral stiffness is increased, but the "strong beam weak column" destroy appears easily. If reinforcement ratio and stirrup ratio are improved, it will enhance the strength of column and restrict the concrete at the end of the play, which also contributes to achieve "strong column weak beam" type.

Keywords: RC Frames; Earthquake; Joints; Seismic Performance; Abaqus Analysis; Slab; Reinforcement Ratio; Damage.

1. Introductions

Nodes of the reinforced concrete frame structure includes core area of beam-column intersected and beam-column ends connected to it, plays a transmission pivotal role [1]. Liu Haiqing [2] pointed out in the study, The main reason of frame structure collapsed is the underlying seismic capacity is weak, beam-column joints destruction. Jiaoshuang Jian et al [3] study found that node subjected complex stress under the earthquake, Column end showing the plastic hinge side easily, Structure is not to achieve "strong column weak beam" destruction. So, to research seismic performance of framework nodes is necessary and practical significance.

Based on Abaqus finite element analysis software, Studied by the finite element simulation method, Abaqus analysis software can solve simple linear problems with complex nonlinear problems, it can simulate concrete, mechanical properties of steel accurately, and its analysis results are reliable.

Meanwhile, the use of finite element analysis method can save resources and improve the efficiency of research [4].

Through the study of earthquake affected areas on-site analysis, nodes of frame structure appear partial concrete crushing, end-column plastic hinge, member cracking and et al phenomenon, structure will quickly enter the state from the plastic work flexible work status [5]. Impact on the seismic behavior of frame structure with two large factors, as follows: (1) Floor slab. Since most of the existing framework of the construction methods using whole cast, slabs and beams casting together as a whole, beam turn into a similar T-beam member, thereby increasing the flexural capacity and bending stiffness of the beam to form a "strong beam weak column". (2) Column reinforcement ratio. In order to ensure frame column flexural capacity, we must ensure that the pillars with enough longitudinal reinforcement, the minimum reinforcement ratio of basic constraints play a role in the protection of concrete, to ensure the strength of concrete when the column in large plastic deformation occurs.

This paper establish three analysis model, they are "pure frame" model, the model taken into account floor slab, the model taken into account floor while improving the model stirrup ratio. By calculating the data and observing the form of structural damage to examine the influence of slab and column stirrup ratio on seismic performance of frame structure node.

2. "Pure Framework" Finite Element Analysis Model

2.1. *Finite element modeling*

An airport terminal expansion project as an example, use Abaqus finite element modeling software, which beams and columns in accordance with the actual size of the modeling structure. Length of the beam is 8.5 m, height 0.8 m, width 0.65 m, the first layer and the second layer column height 4.5 m, the third layer column height 5.6 m, the size of the column of 0.65 m × 0.5 m. The strength of concrete beams and columns are C40.Model select solid elements, type of unit is being hexahedral 8 reward points reduced integration (C3D8R). Use an ordinary steel constitutive relationship, and consider the elastic and plastic phase stage of development. Rebar choose to use beam elements B31, because the beam elements can simulate the reaction of rebar in the earthquake field such as stress [6]. Structural finite element model and reinforcement is shown in Figure1, this model is named Model 1.

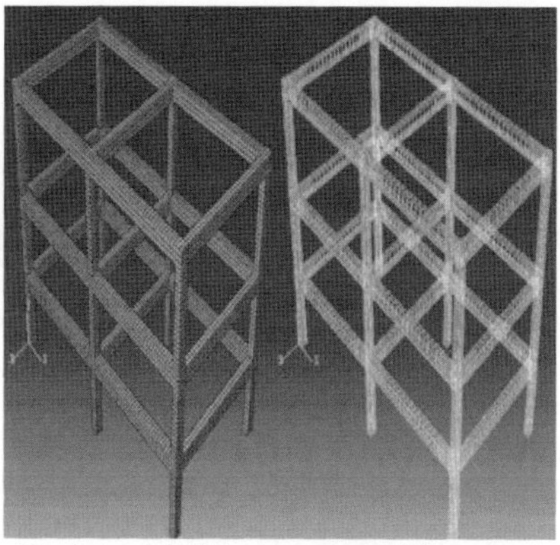

Fig. 1. model of structure and reinforcement

Constitutive relation of concrete in strict accordance with "Design of Concrete Structures" (GB20010-2010) [7] Appendix C formula C.2 provisions plastic stage select materials and damage factor. Mass density of concrete input to the Abaqus is 2 400 kg/m^3, Elastic modulus of 3.25×10^4N/mm^2, Poisson's ratio is 0.3. Density of a rebar input 7800 kg/m^3, elasticity modulus 2×10^5N /mm^2, Plastic modulus 4.4×10^2 N/mm^2, Poisson's ratio is 0.3.

2.2. *Finite element analysis*

Project is located in Shenzhen, fortification intensity is 7 degrees, the basic design earthquake peak acceleration of 0.1g, selection of three seismic waves: EL-Centro wave, Accel artificial wave and Taft wave, develop seismic wave of six different acceleration peak, conditions shown in Table 1, figure 2.

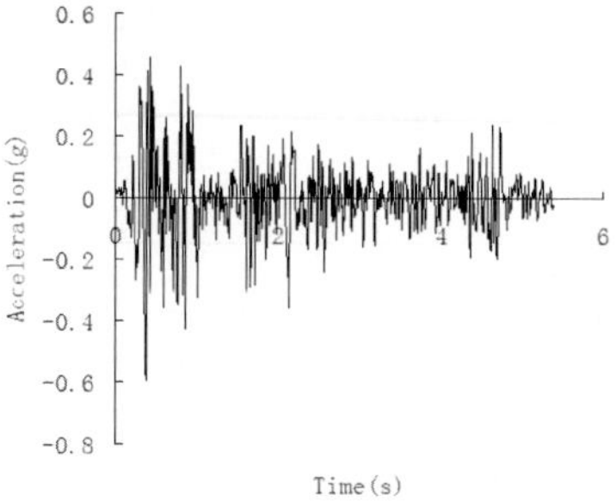

(a) EL Centro wave

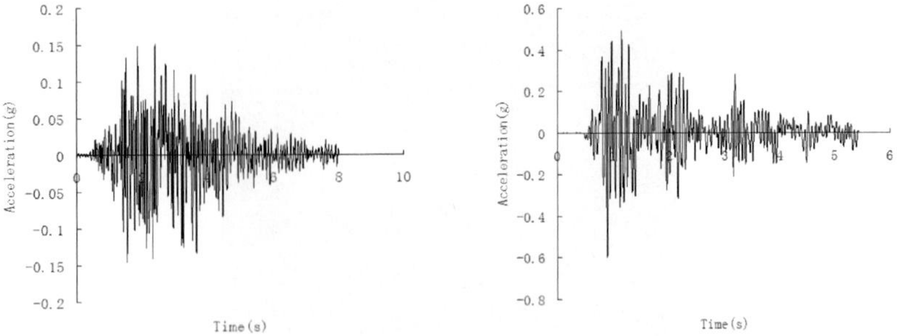

(b) Accel wave (c) Taft wave

Fig. 2. inputting earthquake wave

Table 1. earthquake waves inputting scheme

Test	Dir	Seismic wave	Acceleration (m/s^2)	Excitation direction
	X	El Centro-1	0.9	
1	*Y*		0.8	Three-way
	Z		0.6	
	X	Accel-1	1.1	
2	*Y*		0.8	Three-way
	Z		0.7	
	X	Accel-2	2.8	
3	*Y*		2.0	Three-way
	Z		1.6	
	X	El Centro-2	3.0	
4	*Y*		2.3	Three-way
	Z		2.0	
	X	El Centro-3	6.0	
5	*Y*		5.2	Three-way
	Z		4.6	
6	*X*	Taft	6.2	
	Y		5.3	Three-way
	Z		4.7	

Note: The predetermined short axis direction as the X direction, Y axis direction is the direction vertical to the Z direction.

2.3. *Structural seismic response analysis*

(1) Acceleration response analysis

The amplification factor of three directions on each floor are plotted in Figure 3, Figure 4.We can see the variation of the amplification factor under different seismic wave of different peak acceleration. Because of elevated floors, acceleration response increases rapidly with height and ground motion input intensity, the third layer on top of the seismic response maximum acceleration. The response of structure is different when enter different seismic waves; for example acceleration peak is close when input middle seism and small seism. Contrast Accel wave and El Centro wave, the impact generated by Accel wave on the structure is larger. It shows that the frequency of the Accel wave is closer to the natural frequency of the structure itself; therefore generate the effect of resonance amplification.

(2) X, Y horizontal displacement

Relative displacement of model is small when input the small seism, so ignore the seismic response of the first two groups. Angular displacement between the structural layer case plotted in Figure 5.

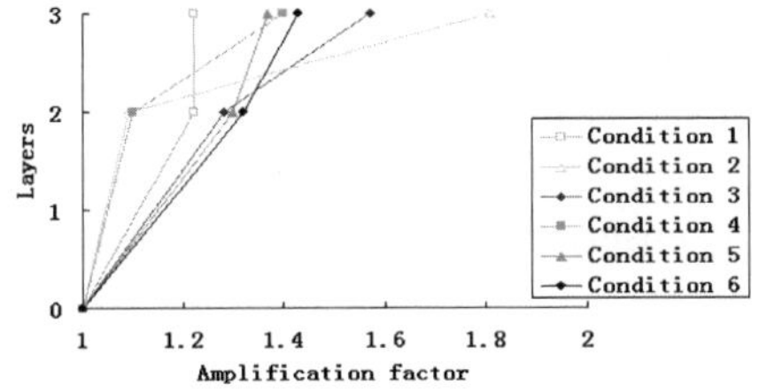

Fig. 3. X-amplification coefficient

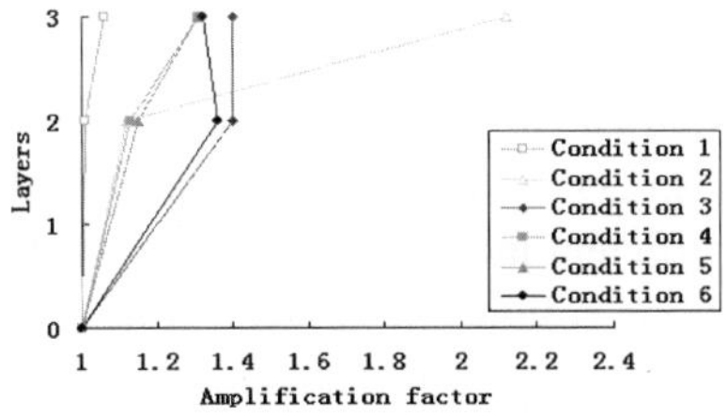

Fig. 4. Y-amplification coefficient

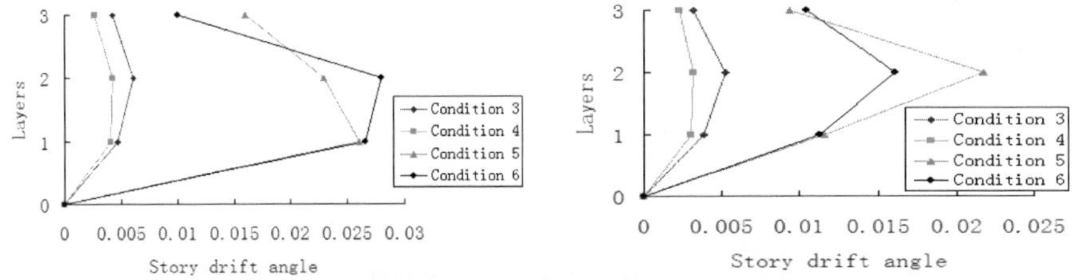

(a) X-Inter-story displacement angel of model 1 (b)Y-Inter-story displacement angel of model 1

Fig. 5. X-Inter-story displacement angel of model 1

In condition 4, the second layer interlayer displacement angle is 1/210rad, the third one is 1/306rad, and story drift angle has been greater than the required specification 1/550rad, This shows that the structure of this case has entered the elastic-plastic stage, but did not enter serious nonlinear stage. When the applied peak is substantially equal, the impact generated by Accel wave on the structure is larger than El Centro wave done. It shows that the frequency of the Accel wave is closer to the natural frequency of the structure itself; Under the effect of the strong shock –Taft wave(α_{max}=6.2m/s^2), the second layer between the displacement angle reached 1/26rad. At this time, there have been structural damage node, structure exhibited severe nonlinear, part of the end beam bending failure has occurred, portion of the end column produce plastic hinge.

3. Impact on the Seismic Performance of Slab

In modern reinforced concrete frame structure, slab, beams and columns always entire cast as a whole, increase the stiffness and flexural capacity of the beam, meanwhile, it is easy to cause structure to appear "strong beam weak column" under earthquake. seismic design specification (GB5011—2011) specified reinforcement area of a beam end including compression reinforcement of beams and the slab reinforcement. Then increase the likelihood of "strong column weak beam". But there is no detailed explanation about the impact of the floor slab on the seismic performance of frame structures. American ACI318 structural design specifications think the floor slab poured on site can improve flexural capacity of beam largely, when conducting seismic design, we must consider the effective flange width strengthened the beam, specification requirements take six times the length of the beam on each side of the plate thickness as an effective flange width, on this basis, the column end moment coefficient increases η_c is still considered, to ensure that the structure can achieve "strong column weak beam" [9].

In order to investigate the impact on the seismic performance of the floor, and based on the established model, increase the thickness of 120 mm reinforced concrete slab, each of the other parameters remain unchanged, calculation results analyzed to compare “pure frame” structure. Model 2 overall appearance shown in Figure 6.

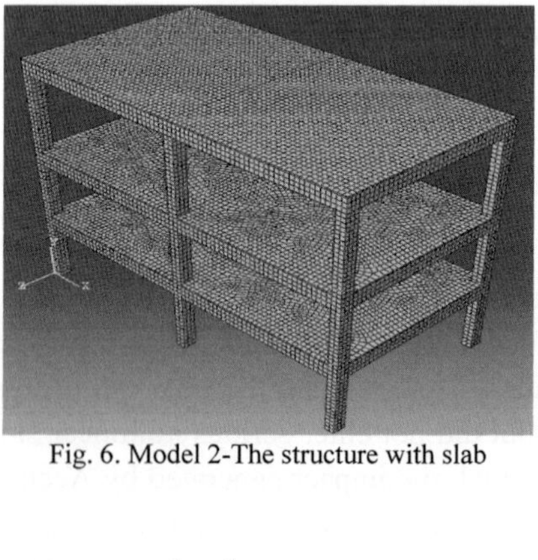

Fig. 6. Model 2-The structure with slab

3.1. *Acceleration response contrast*

Taking into account the effect of the case floor, acceleration response of the structure in general has increased (Figure 7), this is because the inertial force generated by the weight of the floor of its own enlarged the overall structure of the acceleration response, and thus the seismic performance of the structure is very unfavorable.

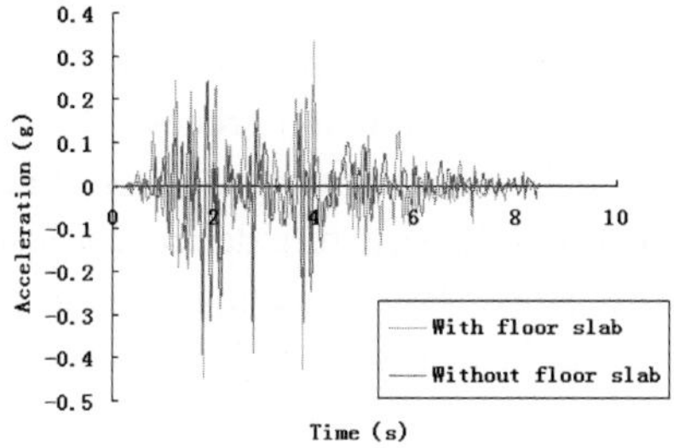

(a) Comparison of acceleration on *X*-direction under work condition 3

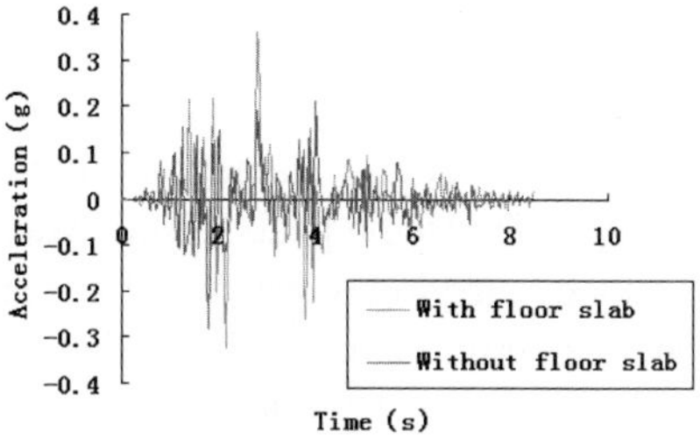

(b) Comparison of acceleration on *Y*-direction under work condition 3

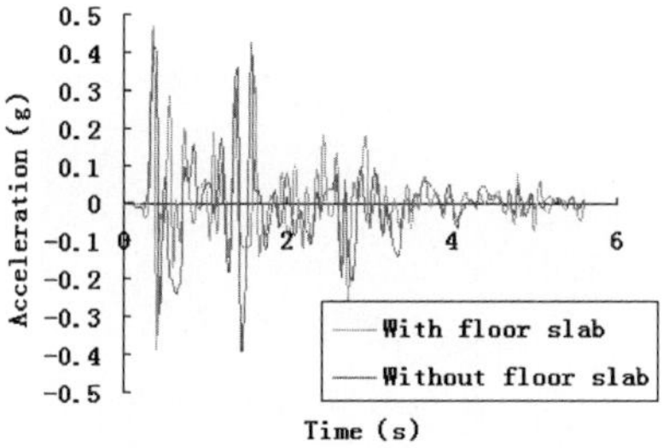

(c) Comparison of acceleration on *X*-direction under work condition 4

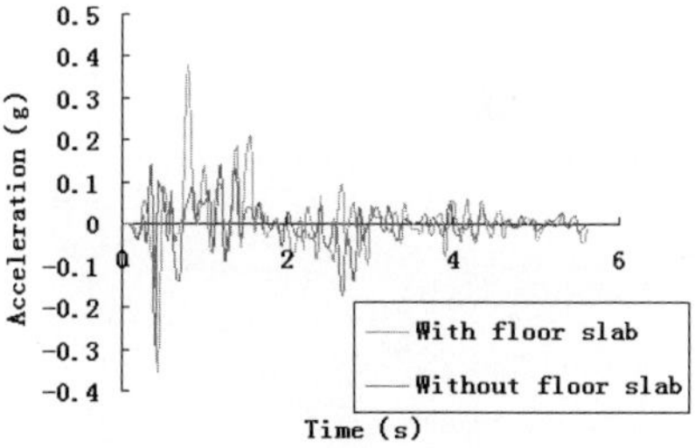

(d) Comparison of acceleration on *Y*-direction under work condition 4.

Fig. 7. Comparison of Model 1 and Model 2 on X-acceleration under Work condition 3, 4

3.2. *Compare the relative displacement response*

By calculating the results (Figure 8) can be found that no structural damage occurred and the structure did not enter fully plastic stage after entering the inelastic state. If we consider the impact of the floor slab, it'll increases the lateral stiffness of the structure of the force, resulting in the displacement response of the structure to reduce. After considering the impact of the structural slab, displacement response and story drift angle decreased greatly.

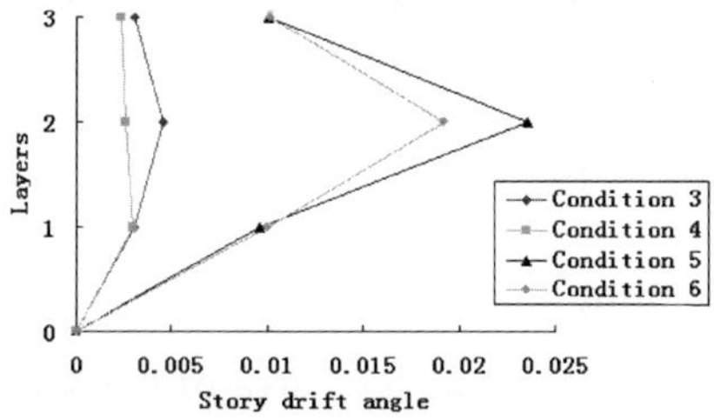

(a)Angular displacement on the X-direction (/ rad)

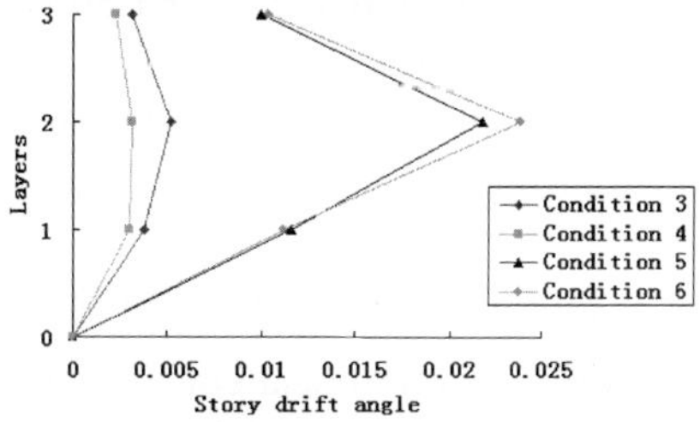

(b) Angular displacement on the Y-direction (/rad)

Fig. 8. inter-story displacement angel of model 2

3.3. *Comparative analysis of structural damage*

Under the action of rare earthquake, finite element modeling software will terminate when model 1 is calculated in 1.01s in the calculation process, this is due to the structural damage during analysis. It has already exceeded the time corresponded to the peak seismic wave. This shows the structure destroyed, lost the load carrying capacity after earthquake peak. By extracting structural concrete strain data, we found that the tensile strain of beam end has exceeded the limit strain; Therefore, the structure of the beam end produced a tensile failure, column end did not appear the yield damage and brittle fracture before beam end damage, achieve specification requirements "strong column weak beam" (Fig 9).

Fig. 9. Damage of Model 1

Fig. 10. damage of Model 2

In the case of considering the role of the floor slab, we can see the governing injury is in the bottom of the column and capitals and the column end appear a serious plastic hinge zone however beam members did not show significant damage (Fig10). Because the frame column had occurred early yield and column reduced ability to withstand the load, led to cause excessive lateral displacement. At this point the displacement angle of the second layer has reached 1/40rad; It has exceeded the elastoplastic range specified in the specification requirements for 1/50rad. At this point the whole structure remains high stiffness. The maximum load structure withstood is in the bottom part of the column, frame column occurred shear failure after the emergence of plastic hinges, led to overturning the whole structure collapses.

4. Column Reinforcement Ratio Effect on the Seismic Performance

The forces of reinforced concrete frame structure of the core area of the node, the world is still not unified analysis. Currently, there are American scholars advocated "pillar model" theory and as the representative of New Zealand "Truss" and "diagonal strut body" [9]. After contrasting different seismic design code from different states, we knew "Pillar model" theory is widely used for structural seismic design. The node of such model will be seen as a piece of pillar structure received a large shear, we need to configure additional stirrups to constraint column, so as to ensure the safety of members of the effect [10].

There is no clear division of seismic rating in the United States ACI318 specification, but take measures to ensure the safety of structure through increasing the length of stirrup encrypted area, reducing the spacing and increasing the strength of rebar for "column model" theory, its basic requirements are as similar as Chinese seismic code highest level. Requirements for non-encrypted area of the column are the smaller value between 6 times the diameter of longitudinal reinforcement and 152mm, the requirements are stricter than Chinese. In China the latest specification, stirrups are HPB300, HRB400, HRB500, etc. Most stirrups are HPB300 in China framework.

In this regard, establish model 3 to contrast to model 2. Stirrups strength increased to HRB500, length of stirrup encrypted area is full length encrypted column, consider floor slab and remain other parameter.

4.1. *Structural acceleration response*

Acceleration response of the structure at the core area of the node decreases slightly under the terms of increasing stirrup ratio of column end, due to acceleration under the conditions of the structure of the core area of the node decreases marginally, so do not consider a detailed comparison here.

4.2. *Displacement response contrast*

As can be seen by the results (Fig. 11), model 3 plays a role in constraining structural lateral movement after increasing stirrup ratio, and the relative displacement of the structure has been reduced.

4.3. *Structural damage analysis*

In model 2, nodes in the core area of the concrete structure cracks have appeared after condition3 and condition4 inputting seismic waves, then structure is in elastic-plastic stage. In condition5 and condition6, column end damage occurs mostly, story drift angle has reached 1/40, structure is into the plastic stage, finally column end yielding occurred first and appeared serious consequences "strong beam weak column" type of damage.

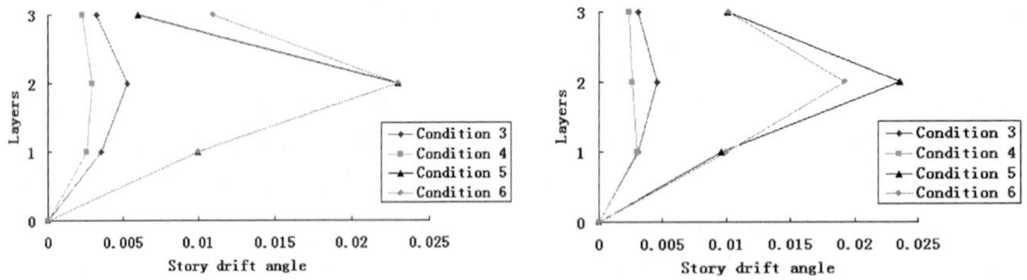

(a) X-Inter-story displacement angel (b) Y-Inter-story displacement angel

Fig. 11. inter-story displacement angel of model 3

In condition 6, damage of the structure shown in Figure 10, part of the node of the column end appeared damage failure. Increasing the stirrup ratio and stirrups strength limit damage of concrete crushed in the node area of column end, improve the bearing capacity of the column ends, amplify structural lateral stiffness, and achieve a "strong column weak beam" type of damage.

5. Conclusions

Based on two specimens of three reinforced concrete frame structure, establish a "pure frame" structure model, a model considered floor slab, and a structure model considered floor slab with increasing stirrup ratio and improving the strength of stirrup. By Abaqus finite element software to calculate, conclude as follows:

(1) In the "pure frame" structural form, beam yields before column does and generates bending failure, then the structure achieves a "strong column weak beam" type of damage.

(2) The situation considered floor slab is better than the situation of "pure frame", because floor slab enhances the stiffness of beam. Structure relative displacement has been reduced, and the damage of the core area of the node is end-column concrete failure, it belongs to a "strong column weak beam" type of damage.

(3) By increasing the column end stirrup ratio an d improving strength grade steel stirrups used, angular displacement between the structures of the layer is reduced , column end concrete compressive deformation is limited effectively as well, crack development is prevented, and the possibility of structure "strong beam weak column" type of damage is reduced.

References

1. Zhang Shaokun, Tu Ming. Study of force transmission mechanism of Reinforced Concrete Frame Joints seismic [J].Shaanxi Architecture, 2007 (141):12-16.

2. Liu Haiqing, Chen Zhenxi, Yuam Jing.3-D simulation on collapse of high-rise structures under strong earthquake[J].Journal of Liaoning TechnicalUniversity:NaturalScience,2010,29(6):1 074-1 077.
3. Jiao Shuangjian, Wei Weifeng, Qimin. A summary on research of RC frame subject to seismic damage [J].World Information on Earthquake Engineering, 2000, 16 (4):42-46.
4. Ding Tao, Yu Shaochun. Frame construction mode numerical simulation analysis based on ANSYS [J].Journal of Liaoning Technical University Natural Science, 2010, 29(S1):43-44.
5. Zhang Yingbo. The analysis of features of earthquake disasters of buildings in Wenchuan earthquake,and suggestions [J]. Shangxi Architecture, 2011, 37(2):38-39.
6. Wang Jinchang, Chen Yekai. Abaqus applications in civil engineering [M]. Zhejiang University Press, 2006.
7. GB50010-2010, Code for design of concrete structures[S].Beijing: China structure industry Press, 2010.
8. Ye Lieping, Qu Zhe, Ma Qianli,et al. Study on ensuring the strong column-weak beam mechanism for RC frames based on the damage analysis in the Wenchuan earthquake[J]. Building structure, 2008, 38(11):52-67.
9. Chen Xi, Wu Jianhua. The load transformation of RC frame joints analysis [J].Sichuan Building Science, 2006, 32(2):15-24.
10. Fu Jianping. Seismic of Reinforced Concrete Frame Joints and Design Method [D]. Chongqing: Graduate school of Chongqing Unicersity, 2002.

Numerical study on air quality in a comprehension transit hub

Yan-Feng Li[1,*], Han-Xiao Feng, Zhao Meng, Hong-Yi Wang, Chen-Liang Xu

Architecture and Civil Engineering, Beijing University of Technology, Beijing, China

**E-mail: liyanfeng@bjut.edu.cn*

Indoor air quality of a public transport interchange station in a comprehensive transit hub has been investigated by experimental studied and numerical simulation. Carbon monoxide was selected as the main pollutant for description of air quality. Ventilation systems, bus traffic and passenger flow have been get by on-site measurement. Large eddy simulation technology has been applied to analyze indoor air quality of public transport interchange station. The boundary conditions are determined according to the measured date. The distributions of carbon monoxide at heights of 0.8m and 1.6m under different time periods and operating conditions of public transport interchange station are calculated. Time periods include the peak hour and trough hour in this study. Results show that carbon monoxide concentration in peak hour is about twice than that in trough hour in most area except the rear area of the vehicle. Air quality would reach the harmful degree in period between 5min to 10min if the ventilation system is not operating. The ventilation system should be operated during the peak hour and intermittent operation mode can be adopted during trough hour in order to meet the requirement of air quality standards.

Keywords: Public Transit Hub; Air Quality; Large Eddy Simulation; Ventilation.

1. Introduction

With the development of the different city traffic system, more and more interchange station becomes have been built. Some public transit hubs are designed as a semi-enclosed or semi-underground space for being closely adjacent to subway stations. The characteristics of hubs include: 1) The air quality of transit hub is directly influenced by vehicular emission because the space is isolated from outside world. 2) Bus traffic and passenger flow varies in relation to rush hour and peak hour. 3) The mechanical ventilation system does not operate appropriately in winter according to economic problem. 4) Due to bus traffic and passenger flow in such interchange hub are much higher than those in a common garage, it has limitation in design ventilation system according the code for common underground garage, etc.

[*] *Correspondent authour*

Comfort and hygiene standard are important factors in measuring air quality of transit hubs. Generally, time period of during which passengers stay transit hub is short. The nitrogen oxide (NO_x) concentration usually fulfills environmental standard requirements. Therefore, Carbon monoxide (CO) concentration is selected as a reference for sanitary standard. The highest allowable concentration is 115.0mg/m^3 in 5 minutes and 30.0mg/m^3 in one hour.

A semi-underground bus terminal transit hub in Beijing has been selected as a research object where the air quality in winter has been investigated by on-site measurement. The main reasons for conducting research in water are described as follows. Firstly, there is a large temperature difference between inside and outside environment and passenger flow is higher than that in summer because of coldness. Therefore, analysis on indoor air quality in winter is more meaningful for investigate the effect operation mode of ventilation system. During the air quality measurement, the bus traffic, passenger flow and concentrations of CO in rush hour and those in trough hour were also recorded. The object of this present study is to simulate CO levels of the different periods and determine operating mode in total year based on the measured data.

2. Numerical Simulation

2.1. *Physical model*

A comprehensive transit hub is 154 meters long in north-south direction and 166 meters wide in east-west direction, and 7.3 meters high. It is a single-floor building structure including subsidiary room and equipment room, etc. There are five entrances in this hub and entrance 1 and 2 connect subway station passageway. Entrance 4 is used as a public transport gateway and entrance 3 is used as passage for both bus and passengers. Entrance 5 is an emergency exit for public transport.

2.2. *Methodology*

Fire dynamics simulator (FDS), which was developed by National Institute of Standards and Technology (NIST) was selected in this paper. Large eddy simulation (LES) performed by FDS, a computational fluid dynamics (CFD) model is used for computing buoyant-driven turbulent equation, which was validated by full scale experiments in building fire. Courant-Friedrichs-Lewy (CFL) number can be as a criterion to justify the convergence and time step.

Because the length and width of the hub are beyond 150m and the square area of each air-outlet is only around $3m^2$, it is better divided into three parts for determining mesh size. The bottom air-outlet and CO producer are placed on the space 0.3m high above the ground and upper air-outlets are placed on the space 6.7m high above the ground. As a result, the grid size of these area is 0.25m×0.25m×0.25m, and the other regions grid size is 0.5m×0.5m×0.5m. The total grid number is 2.1 million.

The parameters in boundary condition of the transit hub including wind velocity, temperature and CO are determined according to the measurement date. Values of temperature and wind velocity are listed in Table 1.

Table 1. On site test results at portals in peak hours.

Test position	Peak hours		Trough hours	
	Temperature /□	Wind velocity/(m/s)*	Temperature/□	Wind velocity/(m/s)
Entrance 1 (underground plaza)	1.1	0.2	1.5	0
Entrance 1(exit)	-0.4	0.48	0.2	0
Entrance 2	-2.5	0.35	-2.5	0.2
Entrance 3	-2.3	0.36	-2.2	0.3
Entrance 4(traffic)	-2	-0.31	-1.8	-0.04
Entrance 5(traffic)	-2.1	-0.2	-1.8	-0.45

Note: when the value is negative is means wind flow from hub to outside.

The surface of the heat producer of burner in FDS is 8.993kW/m2. By test, on the condition of the ambient temperature is 273K, discharge temperature is 350K. As a result, the temperature was set at 350K and the product of combustion was YCO=1.0, YS=0.0. Each car was considered as a module with size of 12m×4.3m×2.5m, with heat source glued on the surface in the rear of car. The releasing rate of CO and temperature of the exhaust pipes can be simulated by FDS software.

Although transport vehicles are non-load running in the transit hub, the number of heat and pollutant producers are arranged based on the number of bus in peak hours and trough hours in order to analysis the running effects of vehicles.

Video data has been adopted to calculate the total number of the bus filled with passengers and the passengers getting in and out of the hub in one hour. It is revealed that there are 22 buses in peak hour from 8:00 to 9:00A.M and 14 buses trough hour from 10:00 to 11:00A.M in average.

Owing to ventilation system was off during the on site measurement, numerical simulation is the only way to analyze the distribution of the CO in the hub with the operation of ventilation system to provide guidance for determining operation mode. Four scenarios have been studied and listed in Table 2. No. 1 scenario is the peak hour condition and ventilation system is not operating. No. 2 scenario is peak hour condition and ventilation system is operating. No. 3 and No. 4 scenarios are trough hour conditions with and without ventilation system respectively.

3. Results and Discussion

According to Indoor Air Quality Standard (IAQS) (GB/T18883-2002), 10 sampling points were selected in the hub. The concentrations of CO at 1.6m and 0.8m high were taken into consideration carefully according to breathe heights of adults and children respectively. TSI8762 air quality tester with the range of 0 to 500×10^{-6} and precision of 3% was applied in measurement. Experiment was taken at 8:00-9:00A.M., 10:00-11:00A.M. Jan.11th.2012. Data was recorded and every 15s and averaged value was give in Table 2.

Table 2. CO concentrations in different scenes in the station.

Scenario	Z=0.8	other areas	Z=1.6	other areas
peak hour, fan off	120.1/116.3*	15.0/14.0*	80.0/76.2*	8.0/7.5*
peak hour, fan on	60	3	25	2
trough hour, fan off	100.0/95.5*	6.0/5.6*	40.0/38.0*	5.0/4.6*
trough hour, fan off	28	3	20	2

Note: value in "*" represents the average value, concentration unit is mg/m^3.

The distributions of CO concentrations at heights of 0.8m and 1.6m respectively at peak hour are shown in Fig. 3. It can be clearly demonstrated that CO concentrations near the exhaust pipes reached 120mg/m^3 in the scenario one. It was higher than the allowed value of 115mg/m^3 in 5 minutes. However, CO concentrations in other zone were also about 10mg/m^3. On the contrary, in scenario 2, CO concentrations of most area met the public health requirements. At the child breath height of 0.8m, CO concentrations were higher than those at 1.6m high. The reason is mainly that the exhaust pipes were at low height. Therefore, it is appropriate to increase the percentage of exhaust air flow volume for vents at low level when designing ventilating system.

Fig. 2 presents the distributions of the CO concentration at the trough hour. It can be seen that CO concentrations at trough hour are apparently lower than that at the peak hour because of the reduction in the number of no-loading vehicles. CO concentration near the exhaust pipes is about 100mg/m^3 without ventilation system. However, CO concentration decreased significantly when ventilation system was operated. Compared with those at the peak hour, CO concentrations in other zones were also decreased to enough low to meet the hygiene standard.

It can be concluded from the results of these four scenarios, the ventilation system must be operated to ensure air quality at the peak hour meet hygiene standard in winter because of poor air exchange condition. Intermittent operation mode was considerable to be applied at the trough hour.

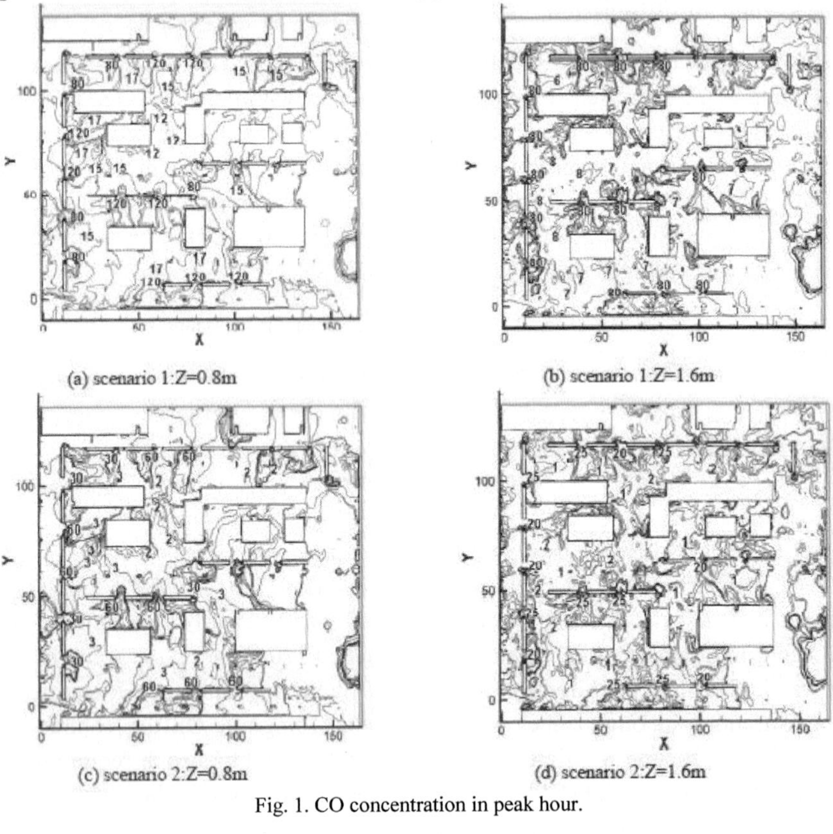

(a) scenario 1:Z=0.8m (b) scenario 1:Z=1.6m

(c) scenario 2:Z=0.8m (d) scenario 2:Z=1.6m

Fig. 1. CO concentration in peak hour.

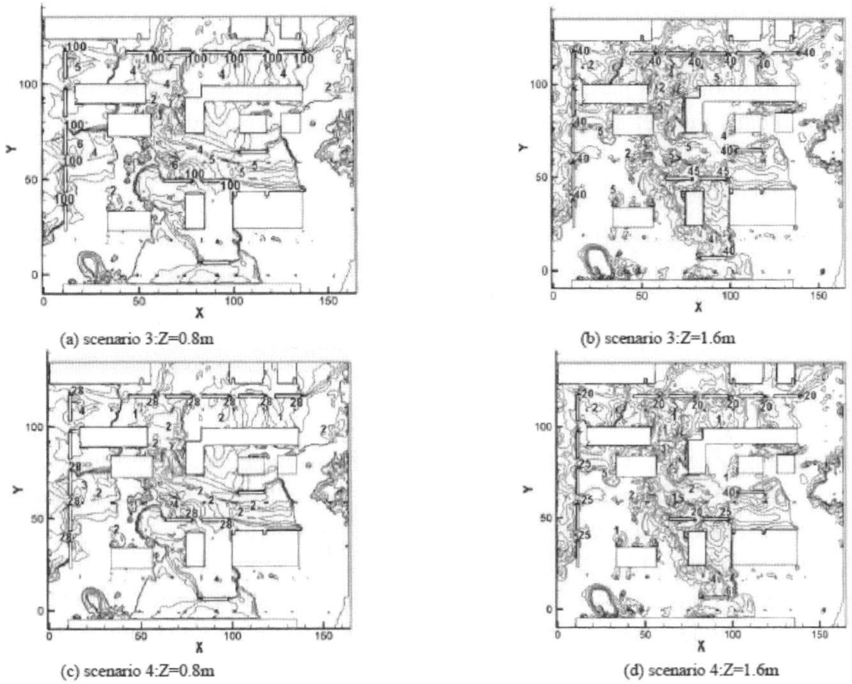

Fig. 2. CO concentrations in trough hour.

4. Conclusion

This paper deals with the simulation of CO concentration inside a transit hub at both peak hour and trough hour in winter. The conclusions are listed as follow:

Comparing with value of on-site testing data, the predicted results are slightly higher and error range is in a reasonable limit. By the validation results, FDS can be suitable for analyzing pollutant concentrations in hub under different conditions.

Air quality varies significantly with traffic volume in the transit hub. CO concentration at the peak hour is twice as that at trough hour even the ventilation system is turn on. CO concentration is obviously higher than that in other zone at 0.8m which is average child breath height level. Much attention should be paid for health protection for children.

When ventilation system is turn off, CO concentration at the rear of the car exceeded the time period hygiene limits of 5 minutes at the peak hour. When ventilation is operated, CO concentration of the rear of the car reduced to value less than limit in 5 minutes, but still higher than the time period hygiene limit of one hour. For considering healthy of workers and operators in the transit hub, their daily activity range should be kept away from the exhausts pipes.

In addition, it is hard to summarize an overall assessment thanks to lack of detailed data. Further studies are therefore should be developed.

Acknowledgments

This work was supported by Beijing Joint Science Foundation (Grant No. 14L00001) and Natural Science Foundation (Grant No. 51378040).

Work partially supported by Beijing Joint Science Foundation (Grant No. 14L00001) and Natural Science Foundation (Grant No. 51378040).

References

1. Huang Zhenyi, Pan Jincheng, Huang Chunying, et al. Investigation on health effect of indoor air pollution in the toll-gates [J]. *Occupational Health and Emergency Rescue*, 4(1) pp: 43-44. (2006).
2. Xue Jiaping, TIAN Weili, Zhang Qingyu. Development of NO_x Emission Inventory from Motor Vehicles in Hangzhou and Study on Its Influence on AirQuality[J]. *Research of Environmental Sciences,* 23(5) pp: 613-618. (2010).
3. T.T. Chow, Zhang Lin, Wei Bai. Assessment of alternative ventilation schemes at public transport interchange [J]. *Transportation Research Part D: Transport and Environment.* 11(6): pp: 447-458. (2006).
4. Zhang Lina, Feng Jiang, T.T. Chow, C.F. Tsanga, W.Z. Lu. CFD analysis of ventilation effectiveness in a public transport interchange [J]. *Building and Environment,* 41(3) pp: 254-261. (2006).
5. TAO Wenquan. *Numerical Heat Transfer (Second Edition)* [M] Xi'an: XI'AN Jiaotong University Press. (2001).

Calculation method of anti-slide piles spacing considering the soil arching effect

Tao Chen, Ke-Xuan Guo, Ming Yang, Si-Qi Chen, Hong-Kai Chen
College of Hydraulic & environmental Engineering, China Three Gorges University, Yichang, Hubei,443002,China
E-mail: 972130341@qq.com

Based on the researches of soil arching effect between anti-slide piles in cutting slope projects, establishing a mechanical calculation model. Firstly, according to mechanic analysis, deriving the reasonable arch axis equation, and calculate the soil arch height based on the state of stress at a point in soil arch interchange. Secondly, on the premise of both sides of the soil arch friction and slide thrust meet the static equilibrium condition, according to the stability of the largest stress of section and the strength requirements of arch springing and mid-span section, the most reasonable calculation formula of piles spacing is derived through comparative analysis.

Keywords: Anti-Slide Pile; Spacing; Calculation Method; Soil Arching Effect.

1. Introduction

China is the high incidence area of landslide disaster, anti-slide pile is the main engineering measures of project management, and the soil arching effect between anti-slide piles is a common phenomenon in slope engineering, the key technique problem that design the anti-slide pile is the rational pile spacing.

From the discovery of the soil arching effect between anti-slide piles up to now, many domestic and international scholars do study the soil arching effect. Roberts discovered the "grain effect" for the first time, the force at the bottom of the barn remains the same when grains accumulate to a certain degree and reach maximum, this is called the soil arching effect [1]. Terzaghi confirmed the existence of the soil arching effect through the famous trap-door test [2]. Zhou Peide et al. and Zhao M H et al. calculated pile spacing based on the static equilibrium condition and the strength requirements of mid-span section, but they didn't consider the influence of the soil arching effect to the pile spacing [3,4]. This article explores the calculation method of anti-slide piles spacing considering the soil arching effect.

2. The Calculation of the Soil Arching Height

The soil arching effect as shown in Figure 1, the formation of the soil arch changes the stress state of the medium that causes the stress redistribution, and transfers the pressure that acting on arch to the arch springing and the surrounding stable medium [5].

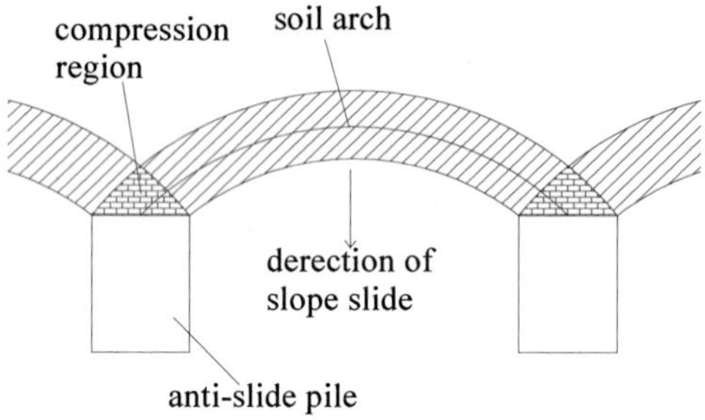

Fig. 1. The soil arching effect diagram.

The soil arching stress as shown in Figure 2, according to the structural mechanics, we can calculate reasonable arch axis equation:

$$y=\frac{4h}{l^2}\left(lx-x^2\right) . \tag{1}$$

The reaction force of vertical and horizontal direction of the anti-slide pile:

$$F_x \quad \frac{ql^2}{8h}, F_y \quad \frac{ql}{2}. \tag{2}$$

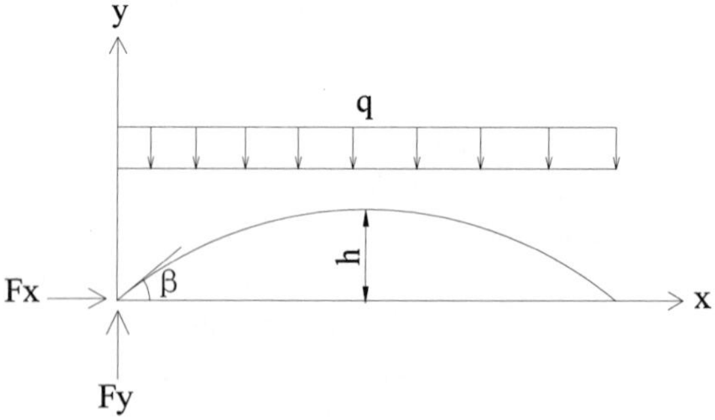

Fig. 2. The soil arch stress diagram.

As shown in Figure 3, suppose when the soil arch arrive the limits balance condition, the forces of both sides of the arch springing reach the maximum, and the points of this section arrive the ultimate stress state at the same times.

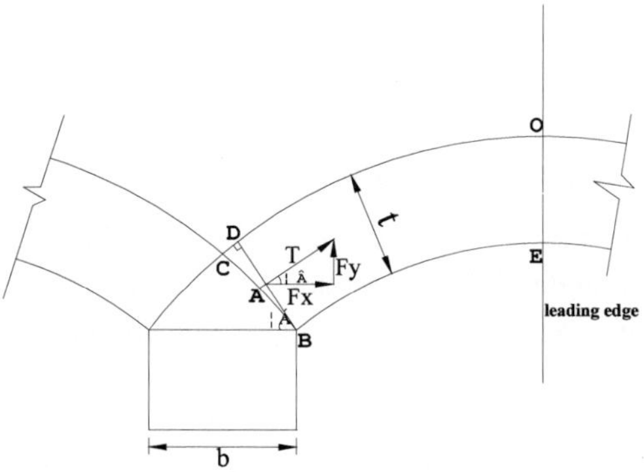

Fig. 3. The mechanical model of the compressive zone of soil arch interchange.

Where A is the arch axis starting point of arch springing, its stress state is shown in figure 3, and its direction of maximum major principal stress σ_1 is along the tangent direction of the arch axis starting point that angle with the horizontal direction is β=45°-$\varphi/2$, where φ is internal friction angle of soil, the minor principal stress is perpendicular to the maximum major principal stress. Therefore, based on formula (2) :

$$\tan\beta = \tan\left(45^\circ - \frac{\varphi}{2}\right) = \frac{F_y}{F_x} = \frac{4h}{l} . \tag{3}$$

Based on formula (3), we can calculate the expression of the arch height:

$$h = \frac{l\tan\left(45^\circ - \frac{\varphi}{2}\right)}{4} . \tag{4}$$

3. The Control Conditions of Soil Arch Terminal

3.1. *Equilibrium condition of soil arch*

The soil arch should meet the static equilibrium condition, that is to say the sum of both sides of anti-slide pile side friction resistance should not be less than downslide thrust of soil arch by following formula:

$$2\left(F_x \tan\varphi + c\left|BC\right|\right) \geq ql . \tag{5}$$

Where c is cohesive force of soil mass, φ is internal friction of angle soil mass, b is the width of anti-slide pile. As shown in Figure 4, the triangle BCD can approximate as a right triangle, the calculation formula of $|BC|$ is:

$$|BC| = \frac{t}{\sin(\alpha+\beta)} = \frac{b}{2\cos\alpha} . \tag{6}$$

3.2. *Equilibrium condition of cross section BC*

The shape that in the compressive zone on one side of the soil arch can approximate the triangle, the cross section BC is under maximum pressure, so it should meet the requirements of the corresponding equilibrium condition:

$$T\cos(\alpha+\beta) = c|BC| + T\sin(\alpha+\beta)\tan\varphi . \tag{7}$$

Where α is the angle between horizontal and BC; T is the resultant force of horizontal and vertical direction:

$$T = \sqrt{F_x^2 + F_y^2}$$

Put formula (4) into (2), we can calculate F_x :

$$F_x = \frac{ql}{2\tan\beta} = \frac{ql}{2\tan\left(45^\circ - \varphi/2\right)} . \tag{8}$$

Based on formula (8), we have following result:

$$T = \frac{F_x}{\cos\beta} = \frac{ql}{2\tan\beta\cos\beta} = \frac{ql}{2\sin\beta} . \tag{9}$$

Through formula (5), (7), (8) and (9), we can calculate a relation by following formula:

$$\frac{\cos(\alpha+\beta)}{\sin\beta} = 1 - \frac{\tan\varphi}{\tan\beta} + \frac{\sin(\alpha+\beta)\tan\varphi}{\sin\beta} . \tag{10}$$

Finally, put formula (6) and (8) into (5), we can calculate clear distance of pile l_1 :

$$l_1 = \frac{bc\tan\left(45^\circ - \frac{\varphi}{2}\right)}{q\cos\alpha\left(\tan\left(45^\circ - \frac{\varphi}{2}\right) - \tan\varphi\right)} . \tag{11}$$

4. The Stress Intensity Control Conditions of Soil Arch Terminal

When soil arch meets the condition of static equilibrium, the ultimate stress strength condition should also be ensured. This article analyzes centroid point of arch springing A and leading edge point of vault section.

4.1. *Stress intensity of point A*

The stress of both sides of the arch springing is maximum under the effect of slip thrust, when A reaches the stress limit state, in order to ensure the strength conditions of the arch springing, we can use Mohr- Coulomb strength criterion [6] by following formula:

$$\sigma_3 = \sigma_1 \tan^2\left(45^\circ - \frac{\varphi}{2}\right) - 2c\tan\left(45^\circ - \frac{\varphi}{2}\right). \tag{12}$$

Trough formula (8), we have following result:

$$\sigma_1 = \frac{F_x}{|BC|} = \frac{ql\cos\alpha}{b\tan\left(45^\circ - \frac{\varphi}{2}\right)}, \sigma_3 = \frac{F_y}{|BC|} = \frac{ql\cos\alpha}{b} \tag{13}$$

Put formula (13) into (12), we can calculate clear distance of pile l_2 :

$$l_2 = \frac{2bc\tan\left(45^\circ - \frac{\varphi}{2}\right)}{q\cos\alpha\left|\tan\left(45^\circ - \frac{\varphi}{2}\right)\text{-}1\right|}. \tag{14}$$

4.2. *Stress intensity of point E*

Point E is the most unfavorable point in the soil arch section, in the context of soil mass that in front of pile is dug, the direction of maximum principal stress σ_1 is along the X axis, the minimum principal stress σ_3 is 0.When point E reaches stress limit state, according to Mohr- Coulomb strength criterion:

$$\sigma_1 = \frac{F_x}{t} = \frac{ql\cos\alpha}{b\tan\beta\sin(\alpha+\beta)}, \sigma_3 = 0. \tag{15}$$

Put formula (15) into (12), we can calculate clear distance of pile l_3 :

$$l_3 = \frac{2bc\sin\left(\alpha + 45^\circ - \frac{\varphi}{2}\right)}{q\cos\alpha}. \tag{16}$$

5. Calculation Method of Reasonable Pile Spacing

Under the condition that the soil arch meets stability and strength requirement, through formula (13), we can calculate the result of α, and put it into formula (14), (17) and (19), the three clear distance of pile can be calculated. The most reasonable clear distance of pile is the minimum value of the formula (11), (14) and (16), that is to say $l=\min(l_1, l_2, l_3)$, so the expression of pile spacing is:

$$S = b + l . \tag{17}$$

6. Conclusions

(1) The paper analyzes the reasonable arch axis equation through mechanic analysis of soil arch, and Calculate the soil arch height based on the state of stress at a point in soil arch interchange.

(2) Under the condition that the soil arch meets static equilibrium, according to the stability and strength requirement of arch springing, and strength requirement of soil arch mid-span section, the paper calculate three pile spacing, the smallest one is the most reasonable pile spacing.

References

1. Hou Mei-ying, Lu Kun-quan. Exotic particles [J]. New Material Industry, 2001, (2):26-28.
2. TERZAGHI K. Theoretical soil mechanics [M]. New York: John Wiley&Sons, 1943.
3. Zhou De-pei, Xiao Shi-guo, Xia Xiong. Discussion on rational spacing between adjacent anti-slide piles in some cutting slope projects [J]. Chinese Journal of Geotechnical Engineering, 2004, 26(1) :132-135.
4. Zhao M H, Chen Y H, Yang C W. Methods for determining rational spacing between anti-slide piles considering soil arching effects [J].Yantu Gongcheng Xuebao/Chinese Journal of Geotechnical Engineering, 2015, 37:16-21.
5. Sun J S, Chen M, Jiang Q H. Numerical Simulation of the Forming and Failure Process of Soil Arching Between Anti-slide Pile [J].Advances in Building Materials, 2011, 261-263:1614-1618.
6. Liu Ying-yu, Li Hong. Application of Mohr-Coulomb yield criterion in geotechnical engineering [J]. Global Geology, 2010, 29(4):633-639.

The renovation and planning design of road system in beautiful country — Taking the Dongbu Village's T Junction as an example

Yun-Yang Zheng[1,*], Yuan-Yuan Tian[1,†]
School of Civil Engineering and Architecture, Nanchang Hangkong University, No. 696, south, Fenghe Avenue, Nanchang, Jiangxi, 330063, China
*E-mail: *zyunyang@163.com, †1264488039@qq.com*

This paper mainly deals with the T-junction rectification which is located in intersection of County Road 087 and Township Road 069 in Dongbu, China. We designed the rural road and proposed reasonable schemes by analyzing the problems in green, facade and sidewalk based on the principles of village planning and requirements of eco-village construction. After planning and renovation, the traffic environment is greatly improved, and a delightful view and better residential environment are achieved at the same time. On the other hand, the Hui-style buildings are well kept as a core architectural style of the village, which is also a good representation of cultural inheritance. In a sense, this study, made under the framework of beautiful countryside construction, carries both practical and academic values, and it can act as a reference and basis for similar cases.

Keywords: Road Planning; Street Renovation; Green Construction; Humanization Design.

1. Introduction

In recent years, sustainable development of cities are more and more noted. The need to protect the environment from the wanton ecological destruction of unfettered economic growth and conspicuous consumption is unquestionable [1], so optimal design of rural road is a vital way to protect ecological environment and preserve resources. Meanwhile, because local government responded actively to the country's appeals for beautiful countryside, a large number rural-urban construction has been taken on the agenda. Although the national macro-control determines the direction of development of construction, better coordination of government functions is required [2]. To sum up, the construction and reconstruction of road is important to sustainable development, so the design of rural road is of great significance.

A large number of researches focus on urban road design. Urban roads are critical to the efficient use of urban areas, to their growth and development and to their environmental protection. Many local authorities in cities in the developing world are grappling with increasingly complex challenges regarding urban road provision [3]. Li et al. also argued that traffic congestion and ground space scarcity are increasing the need to find more space in the urban underground [4]. But rare research is concerned with rural road design. The research made by Junbin Xu et al. is comparatively relevant. It covers campus road from the road network layout, the road classification, the road cross section design, and the appropriate length and width of the road [5], but these studies are far from being applicable. According to Raymond Asomani-Boateng et al., they only analyzed how rural road could be effected without discussing plan and design of rural road [6]. In general, the research of rural road of reconstruction has not been paid enough attention and explored deeply, so it is quite necessary for us to further the study in this area.

2. Project overview

2.1. Project location

Dong Bu village, located in northeast Jingdezhen in Jiangxi Province of China, was known for kaolin which is an important raw material of Jingdezhen ceramics production. Meanwhile, kaolin is transported to Jingdezhen via the East River of Dong Bu village, so Dong Bu Village's traffic condition is of vital importance to development of Jingdezhen. The focus in this project is the connection of County Road 087 which straightly leads to Yao Li with Township Road 069 to kaolin mine park. We managed to establish an intersection of the two roads (See Figure 1).

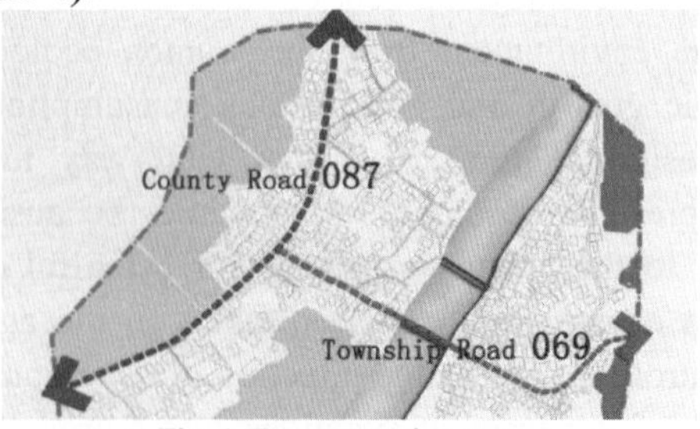

Fig. 1. Reconstruction zone.

2.2. Situation project

The old road design simply includes the factors like motorcycles, bicycles and pedestrians, without fully thinking of possibility for the rapid increase of cars, infrastructure, green environment and humanity, which fails to meet the growing demand. Due to the use over the years, the road bears serious wear and tear, dirty and dusty, which produces an undesirable environment. To solve these problems, we redesigned and reconstructed the road. The present situation is shown in Figure 2 and Figure 3.

Fig. 2. County Road 087 Situation of road 1.

Fig. 3. County Road 087 Situation of road 2.

3. Buildings Renovation Along the Roadside

3.1. *Current problems*

The buildings on both sides of County Road 087 lacked integrity and continuity, and the architectural style was very messy. Some were low-rises with only 1-3 floors, in old fashion and with poor qualities, and some were temporary buildings. Within the planned road, residential, commercial, public buildings are distributed in an improper order, and one major problem in renovation is that some buildings stand in the way of the planned road which need to be displaced for the planned road width of 12m.

3.2. *Facade renovation along the street*

By investigation, some buildings which greatly affect the whole sight of the street need to be removed, and some can be improved by roof renovation or wall decoration.

(1) Roof reform

In Dong BU village, the architectural styles are varied with a combination of the predominant Anhui Style and some modern ones. In order to achieve uniform and harmony, the Anhui architectural style should be kept with some slight changes in modern buildings for both economical and cultural values. We reformed the roofs by increasing the wharf walls, tiles and some other small decorations characteristic of Anhui architectural style.

(2) Exterior walls decoration

Most of the buildings along the roadsides inherited Huizhou architectural style. Some walls of the buildings owing to the age and weathering have been seriously damaged that need to be large-scale rectification. Some measures could be taken such as: repainting walls, removing outbuildings. Suitable light-colored were selected to paint external walls and other measures to improve aesthetics. Impression effect is shown in Figure 4 .

Fig. 4. After renovation of Township Road 087.

4. Road Renovation

4.1. *Principles of the landscape planning along the road*

Currently, there are only a few disorganized green layouts along the planned road. In order to increase green and improve the landscapes along the road, we concentrate on rejuvenating the existing vegetation, like grass and shrubs without using lawns, and then focus on coordinating landscape with village style, through a reasonable collocation of vegetation, water and buildings, creating the effect of green in all year round but with distinctive colors and features in every season.

(1) Native species selection

According to the local conditions, native trees are given priority and some new varieties are appropriately introduced.

(2) Principles of planting

The distance of trees should be not less than 4m. When the sidewalk is wider than 4m, chamfer before planting should be adopted. When the sidewalk is wider than 6m, we need to plant mainly deciduous trees at least in the second row. Trees nearby Bus Stations can be planted at the back of the waiting structures in order to increase the effect of shade.

(3) Principles of planting arbor and shrubs

When the width of dividing greenbelts is larger than or equal to 1.5m, arbor should be planted as the major species combined with shrubs, ground cover. And the crowns on both sides should not be overlapped over the lanes. When the width of dividing greenbelts is smaller than 1.5m, shrubs should be taken over, combined with shrubs, ground cover plants. If dividing greenbelts were broken by crosswalk or road entrance, transparent configuration should be adopted in the termination.

Generally distance between arbors ranges from 4m to 8m, depending on species and crown when planting. We have two planting schemes. One is that single plant is planted in one or two lines, and the other is that pattern or piece is made from 20 to 30 per square meter. The effect of the reconstruction is shown in Figure 5 and 6.

Fig. 5. Before renovation of Township Road 069.

Fig. 6. After renovation of Township Road 069.

4.2. *The scheme of reform along with the road*

The road planting should be based on economy and practice, convenient maintenance, diversity. A diversified green landscape could be formed by combining arbor, shrubs, flowers with other species,.

(1) Road planning

Along County Road 087, less than 1.5m green belt is planned, where privet and holly are planted in 4m distance interval. The rest of green belt could be holly. A gap should be broken per 25m, and the width of gap was 2.5m equal to that of sidewalk. Meanwhile, for Township Road 069, less than 1.5m green belt where shrubs or mix of shrubs and ground cover are planted, and a gap should be broken per 10m to 25m, and the width of gap is 2.5m equal to that of sidewalk. Transparent configuration should be adopted in the termination. The impression drawing is shown in Figure 7 and 8.

Fig. 7. Before renovation of County Road 087.

Fig. 8. After renovation of County Road 087.

(2) Landscape nodes

The section of designing landscape nodes within planning of road improvement is also a very important part. The attention to the design of key nodes and elaborate distinctive landscape nodes will enhance aesthetic value for the whole road. Crape myrtle, plum, chrysanthemum, peach, cherry and other trees and shrubs can be applied in this project to create desirable color combination, so beautiful sceneries in spring and summer are presented, and the scale of landscape can be achieved both from the perspectives of quality and quantity.

(3) Green belts

The shops which along the roadsides occupied large space, which leaves the bad impression of being dirty, chaotic, and poor, and the condition gets worsened with the time. In order to improve our environment, reduce pollution, reduce road noise, regulate effectively the behaviors of the villagers, we must set the green belts along the roadsides.

4.3. *Road section*

For the design of the County Road 087, we adopted dual carriageway with width of 4.5m, and width of the green belt and sidewalk was 1.5m and 2.5m, respectively. The detail is shown in figure 9. For the design of the Township Road 069, we also adopted dual carriageway with width of 4m, width of the green belt and sidewalk was 1m and 2m, respectively.

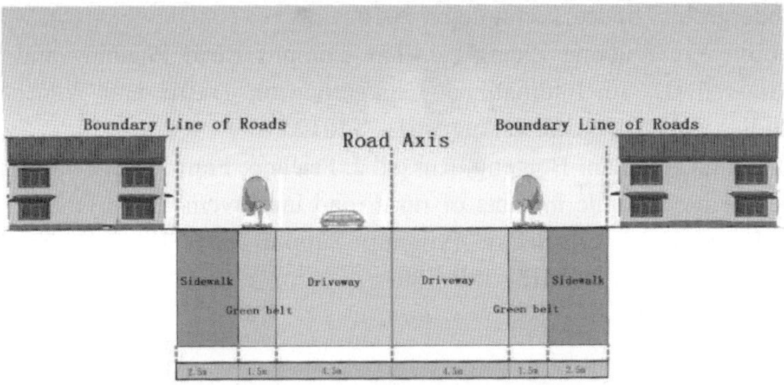

Fig. 9. Impression drawing of road planning of County Road 087.

5. Conclusion

This paper analyzes the problems in green, facade and sidewalk in village road planning in a real case. Based on the principles of village planning and requirements of eco-village construction, we made on-the-spot investigations and proposed reasonable schemes. This is a case typical of eco-village planning. After planning and renovation, the traffic environment is greatly improved and a delightful view is achieved at the same time. A better environment in local residence is accordingly created. On the other hand, the Hui-style buildings are well kept as a core architectural style of the village, which is also a good representation of cultural inheritance. In a sense, this study, made under the framework of beautiful countryside construction, carries both practical and academic values, and it can act as a reference and basis for similar cases.

References

1. Mark Deakin, Alasdair Reid. Sustainable urban development: Use of the environmental assessment methods [J]. Sustainable Cities and Society 10 (2014) 39–48.
2. Ying Tang, Robert J. Mason etc. Governments' functions in the process of integrated consolidation and allocation of rural–urban construction land in China [J]. Journal of Rural Studies, pages 43-51.
3. Mauro Peter Henrique Chissano, John Minnery. Roads, rates and development: Urban roads and growth in Xai-Xai, Mozambique [J]. Habitat International 42 (2014) 48-57.
4. XiaoZhao Li, Congcong Li ect. Multiple resources and their sustainable development in Urban Underground Space [J]. Tunnelling and Underground Space Technology, 2016.
5. Junbin Xu, Zhiyong Zhang etc. The Campus Road Planning and Design Research [J]. 8th International Conference on Traffic and Transportation Studies Changsha, China, August 1–3, 2012.
6. Raymond Asomani-Boateng, Russell J. Fricano, Frank Adarkwa. Assessing the socio-economic impacts of rural road improvements in Ghana: A case study of Transport Sector Program Support (II) [J]. Case Studies on Transport Policy 3 (2015) 355–366.

Research of the review of building automation

Ying-Jie Bai[1,2], Zhi-Yong Mo[1,2,*], Jin Chen[1,2], Ji-Quan Yang[1,2]
[1]*School of Electrical and Automation Engineering ,Nanjing Normal University,*
[2]*Jiangsu Key Laboratory of 3D Printing Equipment and Manufacturing ,Nanjing Normal University,*
Nanjing, Jiangsu Province,210042,China
**E-mail: mozhiyong@njnu.edu.cn*
www.njnu.edu.cn

3D Architectural printing is produced by combining 3D printing technology with architectural design and construction technology. In this paper, its supporting technologies, including building information modeling, materials modeling, digital manufacturing and virtual manufacturing technology are reviewed and summaried, and its research status and development trends are analysised. Building information model BIM uses IFC interface as its information exchange standard, integrates building geometry, physical, structure, economic information and so on, provides services for building design, construction and management stage, and builds full life cycle model; Contour Crafting uses concrete for printing materials, builds the entire house automatically under control of computer. D-Shape printing technology uses fine aggregate and cement materials as printing materials, forms a sturdy stone buildings with specific shape. Multiple Material Modeling combines building structure with a variety of building materials according to its geometric distribution, and it is the important basis of 3D architectural printing. Besides, CNC manufactures building components automatically. Virtual manufacturing stimulates construction process using visualization technique, tests design flaws, finds clash of construction, so as to guarantee implement of minds of the designer. In the final part of the paper, 3D Architectural Printing is regarded as an ideal numeric control way for building construction, which does great help to raise the level of automation, energy saving and environmental protection in the construction industry.

Keywords: Building Information Model; 3D Printing Building; Digital Constructing, Virtual Constructing.

1. Building Information Model

1.1. *Development background*

With emergence of new architectural forms and the development of new building materials, two-dimension Architecture CAD cannot meet the requirements. Traditional way of building costs a lot of resources, and it is neither energy-efficient nor environmentally friendly. Intelligent manufacturing and advanced manufacturing have been leading to profound changes in the manufacturing industry already, and now, these technologies or concepts are introduced into construction projects. Through a large number of documents, key technologies in the field of building automation are introduced and discussed as follows.

At present, there are two key problems to be solved urgently in the global construction industry: (1)Lack of coordination in the production process which causes low production efficiency and severe waste of resources.

(2) Repeat work, especially in a early period, and structural design revision in architectural project which push up the cost of construction.

As a example, on one hand, social productivity is increased greatly in the United States, on the other hand, the productivity in the construction industry has not risen steadily, showing a downward trend. The status is due to the inferior level of automation in construction industry while a lot of work is done by hand. Other industries in the same period, due to introduction of automation technology and information technology, productivity levels increase obviously.

1.2. *Feature of BIM*

In order to improve the exchange of information in the construction industry, to improve productivity and quality, and to reduce production costs, INDUSTRY ALLIANCE FOR INTEROPERABILITY (IAI) develops a building information-sharing standard IFC(INDUSTRY FOUNDATION CLASSES), which is an international standard for data exchange in the building industry now. Based on IFC, many kinds of BIM software are developed. These softwares are architecture engineering model which include three-dimension structure information as well as a variety of relevant engineering information. In BIM , models are formed by digital elements, and all informations are stored in the consolidated database in digital form, which including geometry information, physical information, construction information, technical information, and economic information.

Li Jiancheng believes that BIM is a mainstream technology in building digital technology, [1] digital design and BIM should work closely, and provide service whole construction life cycle; Liu Zhaoqiu holds that BIM visualization techniques could be applied in design of some large, complex structural system to construct a comprehensive model with topology, node information, stiffness data, material properties, boundary supporting conditions and load distribution of finite element structural analysis model, which can enhance the processing power of calculation software; [2] Yu Qiang analysised 9 classes necessary relationship during developing CAD data platform based on BIM. [3] He described that BIM platform should achieve the Division of objects. Extensibility of the platform, topology, geometric modeling strategies and data-related display technology are also discussed. BIM software is so complex that various objects and relationships should be handled correctly in optimized configuration. Man Qingpeng sets up construction information space based on BIM which includes construction process information model, resource information model, and cost information. [4,5] He also applies ubiquitous computing in information management of construction. Then a unified information model based on BIM is established, it solves the problem of information-sharing during construction. A seamless fusion of physical space and cyberspace is achieved through different types of dynamic information model and information transfer network model. Zhang Jianping establishes a building property information model based on the IFC standards and a IFC data exchange interface , implements an intellectual property management system which integrates building automation equipment management subsystem and property management subsystem.[6] The system makes use of building information modeling BIM, middleware, and Web Service technologies to realize information exchange and sharing between property management stage, design stage and construction stage. She also presented a new method used to semiautomatically acquire Model View Definition(MVD) complying with IFC Standard. [7] In some cases, MVD is superior to BIM because the later has a great amount of data, and low efficiency of parsing. Besides, data of MVD is easy to be modified, and data confidentiality is good.

1.3. *Differences between BIM and CAD*

There are essential differences between BIM software and traditional CAD:

(1) From point of view of Architectural CAD, building comprises geometric elements, such as points, lines, and surfaces. In BIM, there are many basic building elements, such as walls, Windows, beams, columns. They can be described as relevance, and it means that any change in any element will be immediately reflected to the others,

(2) Component information in BIM Model is rich, it is easy to communicate with CNC equipment, and to transfer data. Then, CNC can produce building elements according to instructions, while the traditional two-dimensional CAD software is unable to do so,

(3) Relying on new kind of BIM model, building constructing process can be stimulated by visualization technology and virtual constructing. Some issues is obvious, such as pipes conflicts, which is difficult to be found during design phase. Such functions in BIM not only ensure smooth progress of the construction, but also save investment and time.

(4) There is not a unified technology platform for collaborating among different specialties in traditional architecture CAD. So a lot of data is entered in repeatedly by different specialties, and it is difficult to ensure these data consistency in BIM, specialties views and design information are drawn directly from the same model, thus to share these information is not only easier but also safer.

2. 3D Architecture Printing

Research on building automation technology in Europe and other developed countries is from different directions. For example, information technology is frequently used in building design and construction by for now, such as building information model BIM described earlier. Besides, from point of view of advanced manufacturing, research is being carried out on building directly automatically manufactured, and 3D building Printing technology is just the case which the aim is to build house automatically without any man's power.

2.1. *Contour crafting*

Contour technology was proposed by Professor Khoshnevis, university of Southern California, to build houses for the first time in the world, the entire process is completely automated, and it has successfully applied in lab environment.[8] According to media reports, its goal is that a 200-square-meter house will be built in 24 hours.[9] Diagram of the whole device is displayed in Figure 1(a), and the nozzle, the core component is shown in Figure 1(b).

The whole sliding device has a gantry structure, and it can run on two parallel rails. The beam has a special structure on which there are two printing heads moving freely. Among these two printing heads, one is for jetting concrete, used in building walls, the other is actually a robot, and is responsible for capturing building elements, used in the construction of roof. Extrusion nozzle has an extremely complex structure with a multi-axis feature, which is determined by its tasks. Its basic moving forms include:

(1) Concrete extrusion operation. If necessary, it can also mix a variety of components, such as concrete, accelerators, and reducing agent,

(2) Head rotating operation. Head always faces the center line of wall, spinning necessarily according to the shape of the wall with a rotating angle generally less than 360°,

(3) Pose control of side mud knife. Its inclining angle is controlled according to wall's situation, vertical or curved. The aim is to make the outer wall surface always smooth.

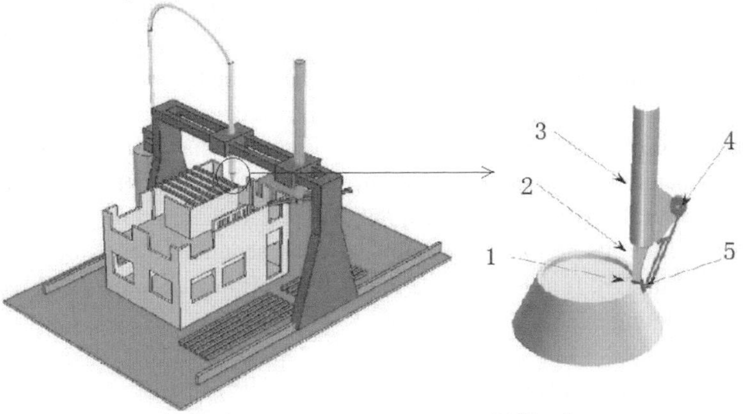

(a) Equipment panorama (b) Nozzle

1 : The level of mud knife; 2 The nozzle; 3 : Material conveying pipe;
4 : Lateral mud control mechanism; 5 : Lateral mud knife

Fig. 1. Contour crafting.

According to contour crafting technology, the constructing process begins with designing building model. CNC machine instructions is generated in terms of model data. The printer is driven by a servo motion system to move and jet concrete on-demand to form building's wall. The operations start from the base up until the roof. After steel building components have been laid out on wall, concrete will be cast by nozzle, then a house is finished. The entire process is characterized by automatic computer control. During the process, a very little artificial work is needed to monitor in case of exceptional circumstances.

2.2. *D-Modeling technology (D-Shape)*

Engineer Enrico invented D-Molding technique for constructing building with maximum sizes of six meters' long, six meters' wide and three meters' high, and it is currently the largest size of building 3D Printer. [10,11] Its characteristics lies in several hundreds of minor nozzles divided into several groups, each nozzle can eject magnesium adhesive materials on a layer of very fine particles of sand of which thickness is among 5mm and 10mm. After another layer of sands is laid on top of previous one, the nozzles eject again according to building model. Hundreds of layers are accumulated, the building is ready.

Building texture made by this technology is similar to marble and its process is similar to three dimensional printing 3DP which is developed by 3D System Company, but the difference is different kinds of ink or glue. This process has the feature of self support, which means that lower layer of material is the support of the upper one. When the whole constructing process is complete, remaining material can be removed for next usage. Another important future of the technology is that roof can be manufactured directly without any steel lines as long as the structure made of magnesium quality adhesive and sands is firm enough. Generally, the roof is not flat-topped but arched.

Compared with the contour technology, its constructing material is not concrete mixed well in advance. The key is the special nozzles which spray adhesive materials, and are also great in number. Operations are controlled by the architectural CAD software. Digital imaging technology is applied in the manner that the image of each printing layer is acquired, then is compared by the original design. If a deviation is found, a modifying operation will be taken to ensure printing quality of each layer.

2.3. *Multiple materials technology*

Plain concrete is usually not suitable for three dimensional printing architecture mainly due to the requirement of different solidification time, the late strength performance of concrete, and the shrinkage factor of concrete. In order to achieve rapid solidification, accelerating agent and other additives are needed to be added into cement. Therefore, used multiple materials or composite materials are necessary for 3D architectural printing. Some kinds of materials are prepared in advance, whereas others may be prepared during the printing process when different material is cast by different nozzle proportionally. In order to construct building with help of Computer Aided Manufacture (CAM), a new kind of CAD model is essential which not only building structure but also materials distribution is displayed. Traditional CAD is only able to reflect structure information of building. So multiple-material modeling theories and methods are current research focus around the world.

Because the traditional three dimensional CAD models only include the building's geometry or structure information, the heterogeneous or multiple materials distribution information of building cannot be reflected which is important for 3D pringting. Therefore, materials information expression in heterogeneous materials CAD models is a key problem. Yang and Qian presented a modeling method based on B spline.[12] Kou and Tan developed B-Rep modeling.[13] Wang described structure information of multiple continuous phase parts by the concept of heat conduction.[14] Patil, who presented R function modeling method for describing the structure of materials, he and his group used RM destination model in describing a heterogeneous solid model.[15] Biswas proposed a field modeling method based on geometric domain.[16] Wu and Liu proposed a volumetric-data-sets modeling method.[17] Zhou and Liu developed made a multiple-color-distance-field modeling methods.[18] Wang studied modeling method of finite element for heterogeneous materials parts.[19] Xu Anping has studied equidistant migration FGM modeling and heterogeneous modeling.[20] There are other scholars who have suggested dynamic modeling theory or cellular unit construction model [21] .

3DP (three dimension print) technology cannot only spray ordinary adhesive but also fine ceramic particles, metal particles, special solutions and medicine liquid on each layer of powder as filler materials, with which a multiple-material component is produced. The component has so-called local special property for special purpose.[22] Mott made multiple-material components with internal structure by directly spraying formation technology.[23] Lappo has developed a multiple-material selective laser sintering (SLS) equipment based on discrete formation, and it can be used to make discontinuous material prototypes.[24] Cho Reported a new kind of forming equipment developed in terms of 3DP craft created by MIT, the device produced three dimensional models directly by way of numerically controlled nozzles spraying materials.[25] Yang developed a set of multiple-material powder spraying equipment employing SLS craft, and has successfully made three-dimension functional gradient material components.[26] Bremnan developed a laminated manufacturing equipment for commercialization to produce electric ceramic components.[27] Choi studied a multiple-material laminated manufacturing process based on route planning algorithm on topology structure [28] .

3. Digital Constructing and Virtual Manufacturing

3.1. *Digital constructing*

Industrial robot applications in automotive manufacturing have been a long time, but in the field of architecture, usage of CNC machinery just begins. Representative works of digitally constructing building is Guggenheim Museum completed by renowned architect Frank. Gehry. The whole process can be divided into several steps. Firstly, an ideal model of the museum is made by hand. Then, the model is scanned with a scanner, therefore three dimensional structure information is acquired. Secondly, bearing structure and building envelope are designed in computers, and real building components are produced by NC machine. Finally, all components are assembled on site controlled by computers.[1] It is so perfect that it has been drawing tremendous attention from all over the world. Since then, a era of digital constructing begins.

At present, study on digital construction is mainly divided into three categories:

(1) How to produce architectural components with more and more complex surface by CNC machines;

(2) How to produce different kinds of concrete component template by CNC machines, with which novel shapes and different patterns then can be gotten by casting concrete, and these components are ready for digital construction;

(3) Develop new kinds of method for building design, which makes three-dimensional positioning and digital constructing easier by CNC machines.

Through simulating human behavior patterns and individual motion characteristics the ability to, building can be formed from bottom to up. Individual units of building become digitally manufactured objects, whole building is formed according to structure logic. This kind of algorithm mainly includes Cellular Automata, Fibrous Aggregation Algorithm and Morphgenetic Algorithm. Building forming technology can be realized by robot intelligent NC processing method; In addition, Environmental Responsive Algorithm turns ecological effectiveness such as sound field, light environment, airflow environment and structure performance into adjustably regulated parameter. Genetic Algorithm is used to differentially calculate epidermal cells of the building, epidermal cells of micro variation and evolution can be accurately processed by robot CNC machine technology [29].

Digitally constructing tools are CNC machines, new design methods different to traditional ones are essential. In the field of machinery, CNC machine tools receive CAD data of parts, generate a series of processing instruction, and production starts immediately because there are tools and softwares developed for purpose of coordinative work, such as CAD/CAM/CAE/PLM. However, the situation in building industry is different from machinery one. There is not integrated software and hardware system suitable for digital design and construction. Computer aided design CAD is essential as a computer-aided design tool, computer itself can not generate a formation yet. The process from building modeling to digital construction is not seamless. Therefore, the digital modeling theory and constructing technology of buildings should require study in-depth.

3.2. *Virtual manufacturing*

In the course of construction, there often need to know some details or what the result looks like at the end. The purpose is to ensure the implementation of architect's design intent. Virtual manufacturing is the key to above problems, which stimulates the entire construction process and builds relationships between architectural models and building entities. Building Information Model or BIM, supports to implement virtual manufacturing.

One of the basic function of virtual manufacturing is three-dimensional visualization of object by which a large amount of data produced by modeling, calculation and result can be turned into graph or image. They can be displayed on computer screen clearly, and the correctness and reasonableness are easy to be analyzed. In addition to above function, dynamic presentation of local modification is very useful after finite structures analysis model has been set up. The model should include unit topology, node information, stiffness data, material properties, boundary support conditions, load distribution information. Structure's change along with different constructing stage can be presented dynamically [2].

4. Conclusion

Advanced manufacturing not only changed the traditional manufacturing industries, but also has a major influence on the construction industry. Usage of Building Information Modeling on large scale and practice of digital constructing in building industry have important meaning, and they are breakthrough to architectural CAD. Building Information Model is a platform including all information of architectural design and construction information, and BIM software is an essential tool for modern large buildings especially for skyscrapers. The aims of digital constructing is to upgrade from manual operation to CNC machinery constructing, there is a considerable distance from the object because of the complexity of building structures and huge volume of building compared to machine parts. Currently, construction formworks and building components are automatically manufactured by CNC machines, and three dimensional printing is a promising technology for true digital manufacture of buildings.

Acknowledgment

This work is supported by grant 61273243 of the National Nature Science Foundation.

References

1. Li Jiancheng, *BIM and Digital Fabrication.* Time Architecture. 2012, (5): 64-67.
2. LIU Zhao-qiu, LI Yun-gui. *The Development of BIM and Its Application in Design of Project.* BUILDING SCIENCE. 2009, Vol. 25, No.1:96-99,108.
3. YU Qiang. *Design of CAD Data Kernel Platform based on BIM.*New Technology & New Process. 2011, (9):90-93.
4. Man Qingpeng Sun Chengshuang. *Construction information model based on IFC.* CHINA CIVIL ENGINEERING JOURNAL. 2011, Vol. 44: 239-243.
5. Man Qingpeng Li Xiaodong. *Collaborative construction based on ubiquitous computing and BIM.* CHINA CIVIL ENGINEERING JOURNAL. 2012, Vol. 45.:311-315.
6. ZHANG Jianping, GUO Jie, WANG Shengwei, XU Zhengyuan. *Intelligent facilities management system based on IFC* standard and building equipment integrat ion. J Tsinghua Univ (Sci & Tech). 2008, Vol. 48, No. 6:940-942,946.
7. Yu Fangqiang, Zhang Jianping, Liu Qiang. *Semi-automatic generation of the BIM model view definition based on the IFC standard.* J Tsinghua Univ (Sci & Technol), 2014, vol. 54, No 8:987-992.
8. BehrokhKhoshnevis. *Automated construction by contour crafting—related robotics and information technologies*[J]. Automation in Construction. 2004, (13): 5-19.
9. Solar Valley Advanced Equipment Research Center. United States development of 3D printing: Creating 232 square meters of buildings in 24 hours[J] Advanced Equipment, 2014, (44): 44-46.
10. Enrique Canessa, Carlo Fonda and Marco Zennaro. *Low-cost 3D Printing for Science*, Education & Sustainable Development[M], ICTP—The Abdus Salam International Centre for Theoretical Physics 2013 ICTP Science Dissemination Unit.
11. Li Fuping, Deng Chunlin, Wan Jing. 3D printing building technology and outlook of concrete industry [J]. Concrete World, 2013, (3): 28-29.
12. Yang Pinghai, Qian Xiaoping. *A B-spline-based approach to heterogeneous objects design and analysis.* Computer-Aided Design. 2007, 39: 95-111
13. Kou X Y, Tan S T, Sze W S. *Modeling complex heterogeneous objects with non-manifold heterogeneous cells.* Computer-Aided Design. 2006, 38: 457-474

14. Jianfeng Wang, James K. Carson, Mike F. North, Donald J. Cleland. *A new structural model of effective thermal conductivity for heterogeneous materials with co-continuous phases.* International Journal of Heat and Mass Transfer. 2008, (51) :2389-2397.
15. Patil L, Dutta D, Bhatt AD, et al. *A proposed standard-based approach for representing heterogeneous objects for layered manufacturing.* Rapid Prototyping Journal. 2002, 8(3):134-146.
16. Biswas A, Shapiro V, Tsukanov I. *Heterogeneous material modeling with distance fields.* Computer Aided Geometric Design. 2004, 21(3):215-242.
17. Xiaojun Wu, Weijun Liu, Michael Yu Wang. *A CAD modeling system for heterogeneous object.* Advances in Engineering Software. 2008, (39):444-453.
18. Zhou Hongmei, Liu Zhigang, Lu Bingheng. *Heterogeneous object modeling based on multi-color distance field.* Materials and Design. 2009, (30):939-946.
19. Su Wang, Nanfei Chen, Chin-Sheng Chen, Xinxiong Zhu. *Finite element-based approach to modeling heterogeneous objects.* Finite Elements in Analysis and Design. 2009, 45(8-9): 592-596.
20. AnpingXu, Zhihua Liu, Yunxia Qu. *Heterogeneous primitive modeling method based on material feature classification.* In the Proceedings of the 7th WSEAS International Conference on Applied Computer & Applied Computational Science (ACACOS '08), Hangzhou, China, April 6-8, 2008: 471-476.
21. Kou X Y, Tan S T. *Heterogeneous object modeling: A review.* Computer-Aided Design. 2007, 39:284-301.
22. Chol W, Sachsl E M, Patrikalakisl N M et al. *Methods for Distributed Design and Fabrication of Parts with Local Composition Control.* Proceedings of the 2001 NSF Design and Manufacturing Grantees Conference, Tampa, Florida, 2001.
23. Mott M, Evans J R. *Zirconia/alumina functionally graded material made by ceramic ink jet printing.* Materials Science and Engineering, 1999, A271:344-352.
24. Lappo K, Jackson B, Wood K, et al. Discrete multiplematerial selective laser sintering (M2SLS) : experimental study of part processing[C]/Solid freeform fabrication symposium. Austin, TX: The University of Texas, 2003: 109 -119.
25. Cho W J, Sachs E M, Patrikalakis N M, et al. *A ditheringalgorithm for local composition control with three-dimensionalprinting*[J]. Computer-Aided Design, 2003, 35(9): 851-867.

26. Yang S F, Evans J R G. *A multi-component powder dispensing system for three dimensional functional gradients*[J]. Material Science Engineering, 2004, 379 (1/2): 351-359.
27. Bremnan R E, Turcu S, Hall A, et al. *Fabrication of electroceramic components by layered manufacturing (LM)*[J]. Ferro-electrics, 2003, 293(1) : 3 -17.
28. Choi S H, Cheung H H. *A topological hierarchy-based approachtotoolpath planning for multi-material layered manufacturing*[J]. Computer-Aided Design, 2006, 38: 143 -156.
29. Philip F. Yuan, *From Scripting the future to Fabricating the Future.* Time Architecture. 2012, (5): 10-21.

A new method based on improved Newmark method of solving nonlinear static-dynamic coupling analysis

Gui-Xuan Wang, Xun-Qiang Yin, Hao-Ran Li

Civil Engineering Technology R&D Center, Dalian University, Dalian, 116622, China

E-mail: lc-tm2008@163.com

www.dlu.edu.cn

Static-Dynamic Coupling Analysis is the key technology issue for analysis of structure response in nonlinear condition. Combined with the implicit solving character of Newmark integral algorithm, a nonlinear Static-Dynamic Coupling algorithm based on improved Newmark method is raised, by studying the inner relation between nonlinear finite element dynamic and static equilibrium equations, to obtain rational nonlinear earthquake response of the structure under static and dynamic coupling excitation. The method is verified by analysis of Static-Dynamic coupling response of Fengman gravity dam in linear and nonlinear condition. The result of analysis shows that the method is effective and rational. Meanwhile, combined with abundant structure element model and nonlinear analysis of general finite element software, the method can be used in implicit solving general finite element program conveniently to carry out Static-Dynamic Coupling Analysis of various structure.

Keywords: Static-Dynamic Coupling Analysis; Improved Newmark Method; Earthquake Response; Nonlinear.

1. Foreword

Significant buildings, like dam, high-rise building and nuclear power station etc., are affected by kinds of static and dynamic loads in operation period. Kinds of static and dynamic loads should be combined when checking the ultimate bearing capacity of significant buildings. Generally, soil pressure, gravity of high-rise building, prestress and permanent loads that are applied on the structure are belonged to static load. Seismic load, wind load, explosive power and impact force are dynamic loads [1]. In many complicate loads, the seismic responses of structure are most concerned by engineers. The seismic responses of structure are related to static loads, such as structure gravity, soil pressure or uplift pressure, and dynamic loads influenced by seismic excitation. Therefore, for seismic response of structure under effects of many loads, static-dynamic coupling analysis of structure should be performed.

Static and dynamic loads being considered separately is the general way of static-dynamic coupling analysis. Firstly, static loads are calculated independently and saved as corresponding conditions. Then calculate the dynamic loads. At last, linear superposition of static and dynamic results is performed so that seismic response of structure under static-dynamic coupling excitation is obtained. This method is not only complex in operation and needs be calculated many times, but also is just applicable to linear elastic analysis only [2]. In nonlinear condition, the method cannot perform the superposition of static and dynamic response, and has no practical meaning in initial condition. So, static-dynamic coupling effect in nonlinear condition is one of the important issues for nonlinear seismic response analysis of structure [3]. Static-dynamic unified issue is widely studied by scholar domestic and international. A unified viscous-spring artificial boundary for 3-D static and dynamic applications was established by Liu Jingbo [4] which is applicable to static and dynamic problem in the meantime, and was verified by linear elastic material. A static-dynamic coupling infinite element was proposed by Wang Zhengzheng [5] and by elastic half-space. Jiang Xinxin [6] proposed a method that could directly solve structure response under static and dynamic loads, and verified the applicability of method under linear condition. However, the studies above were verified only under linear condition, the applicability of corresponding issues wasn't verified under nonlinear condition.

For solving problems above, combined with the implicit solving character of Newmark integral algorithm, a nonlinear static-dynamic coupling algorithm based on improved Newmark method is raised, by studying the inner relation between nonlinear finite element static and dynamic equilibrium equations for solving this issue. The algorithm could obtain rational nonlinear earthquake response of the structure under kinds of loads. Meanwhile, the algorithm can be used in kinds of implicit solving general finite element program conveniently easily. At last, this method was verified by static-dynamic coupling analysis of Fengman dam under linear and nonlinear condition.

2. A Nonlinear Static-Dynamic Coupling Algorithm Based On Improved Newmark Method

2.1. *Nonlinear static equilibrium equation and n-r method*

Nonlinear static equilibrium equation is showed as below:

$$[K(u)]\{u\}=[F^u] \tag{1}$$

where: $[K(u)]$ is a function of the unknown DOF values, and if $[K(u)]$ is constant, equation (1) is linear equilibrium equation; $\{u\}$ is nodal displacement vector; $[F^a]$ is vector of applied loads. For nonlinear problem, the response of structure can't be solved by linear equilibrium equation directly. Generally, nonlinear problem is solved with the Newton-Raphson iteration method (N-R method) [7]. Iterative solutions of N-R method are shown as below:

$$\left[K_i^T\right]\{\Delta u_i\} = \{F^a\} - \{F_i^{nr}\} \tag{2}$$

$$\{u_{i+1}\} = \{u_i\} + \{\Delta u_i\} \tag{3}$$

where: $[K_i^T]$ is Jacobian matrix; i is subscript representing the current equilibrium iteration; $[F_i^{nr}]$ is vector of restoring loads corresponding to the element internal loads. For dynamic problem, $[K_i^T]$ is the effective coefficient matrix; $[F_i^{nr}]$ is the effective applied load vector which includes the inertia and damping effects

2.2. *Nonlinear dynamic equation in incremental form*

Dynamic equilibrium equation of linear structure is shown as below:

$$[M]\{\ddot{u}\} + [C]\{\dot{u}\} + [K]\{u\} = [F] \tag{4}$$

By simultaneous equation(4), equation(2) could be modified as:

$$[K_i^T]\{\Delta u_i\} = [F^a] - [M]\{\Delta\ddot{u}_i\} - [C]\{\Delta\dot{u}_i\} \tag{5}$$

Dynamic equilibrium equation is solved by Newmark integration [8] generally. Newmark integration is implicit integration and is a common analysis method of structure response under dynamic loads.

Now, in time increment of N-R method which has incremental form, is used to solve dynamic nonlinear problem. From time increment i to i+1, consider Newmark assumption as below:

$$\{\dot{u}_{i+1}\} = \{\dot{u}_i\} + [(1-\delta)\{\ddot{u}_i\} + \delta\{\ddot{u}_{i+1}\}]\Delta t \tag{6}$$

$$\{u_{i+1}\} = \{u_i\} + \{\dot{u}_i\}\Delta t + [(\frac{1}{2} - \alpha)\{\ddot{u}_i\} + \alpha\{\ddot{u}_{i+1}\}]\Delta t^2 \tag{7}$$

where: α, δ is Newmark integration parameters.

Equations below could be obtained by rearranging equation (6) and (7):

$$\{\ddot{u}_{i+1}\} = a_0(\{u_{i+1}\} - \{u_i\}) - a_2\{\dot{u}_i\} - a_3\{\ddot{u}_i\} \tag{8}$$

$$\{\dot{u}_{i+1}\} = a_1(\{u_{i+1}\} - \{u_i\}) - a_4\{\dot{u}_i\} - a_5\{\ddot{u}_i\} \tag{9}$$

where:

$$a_0 = \frac{1}{\alpha \Delta t^2}; a_1 = \frac{\delta}{\alpha \Delta t}; a_2 = \frac{1}{\alpha \Delta t}; a_3 = \frac{1}{2\alpha} - 1;$$
$$a_4 = \frac{\delta}{\alpha} - 1; a_5 = \frac{\Delta t}{2}(\frac{\delta}{\alpha} - 2); a_6 = \Delta t(1-\delta); a_7 = \delta \Delta t$$

equations below could be obtained by simultaneous equation (8) and (9):

$$\{\Delta\ddot{u}_i\} = \{\ddot{u}_{i+1}\} - \{\ddot{u}_i\} = a_0(\{u_{i+1}\} - \{u_i\}) - a_2\{\dot{u}_i\} - (a_3+1)\{\ddot{u}_i\} \tag{10}$$

$$\{\Delta\dot{u}_i\} = \{\dot{u}_{i+1}\} - \{\dot{u}_i\} = a_1(\{u_{i+1}\} - \{u_i\}) - (1+a_4)\{\dot{u}_i\} - a_5\{\ddot{u}_i\} \tag{11}$$

$$\{\Delta u_i\} = \{u_{i+1}\} - \{u_i\} \tag{12}$$

equation (10), (11) and (12) can be substituted into equation (5) and rearranged. Then obtain:

$$[K_i^e]\{u_{i+1}\} = \{F_i^e\} \tag{13}$$

where:

$$[K_i^e] = [K_i^T] + a_0[M] + a_1[C] \tag{14}$$

$$\begin{aligned}\{F_i^e\} = \{F^a\} &+ [M][a_0\{u_i\} + a_2\{\dot{u}_i\} + (a_3+1)\{\ddot{u}_i\}] \\ &+ [C][a_1\{u_i\} + (1+a_4)\{\dot{u}_i\} + a_5\{\ddot{u}_i\}] + [K_i^T]\{u_i\}\end{aligned} \tag{15}$$

Note that equation (13) is similar to equation (1). Equation (13) has time-effect for being introduced Newmark integration, meanwhile $[K_i^e]$ is equivalent stiffness matrix with time-effect. Therefore, influence of time-effect can be cancelled, by setting time increment Δt in Newmark integration. Then, combining with initial condition, the inner relation between nonlinear static and dynamic equilibrium equations can be obtained and transmuted into each other.

2.3. *New method based on improved newmark method of static-dynamic coupling analysis*

Under initial condition, acceleration, velocity and displacement are all zero, namely $\{u_i\} = \{\dot{u}_i\} = \{\ddot{u}_i\} = \{0\}$. While, except of $\{F^a\}$ in the right of equation (15), the rest of it are all zero. Equation (14) and (15) can be substituted into equation (13) and rearranged. Then obtain:

$$([K_i^T] + a_0[M] + a_1[C])\{u_{i+1}\} = \{F^a\} \tag{16}$$

meanwhile, a_0 and a_1 tend to zero, when Δt tends to infinite, namely $\lim\limits_{\Delta t \to \infty} a_0 = \lim\limits_{\Delta t \to \infty} \frac{1}{\alpha \Delta t^2} = 0$, $\lim\limits_{\Delta t \to \infty} a_1 = \lim\limits_{\Delta t \to \infty} \frac{\delta}{\alpha \Delta t} = 0$. Then equation (16) can be transformed:

$$[K_i^T]\{u_{i+1}\} = \{F^a\} \tag{17}$$

Therefore, under initial condition, namely $\{\ddot{u}_{i+1}\} = 0$, $\{\dot{u}_{i+1}\} = 0$, and if Δt tends to infinite, *N-R* equilibrium iterative equation can be transformed to the form of nonlinear static equation, while ensuring velocity and acceleration as zero, which is consistent with static solution in theory. Thus, the feasibility of this algorithm is proved under nonlinear condition. So, reasonable nonlinear seismic response of structure under kinds of loads can be obtained by this algorithm. Moreover, the algorithm can be used in kinds of implicit solving general finite element program conveniently.

3. Numerical Verification

The method will be verified by linear and nonlinear static-dynamic coupling analysis of Fengman dam, on general finite element analysis software ANSYS.

The size of geometric model is shown in figure 1. Finite model is simulated by element Plane182 with plane strain, as shown in figure 2. Parameter of concrete is shown in table 1 [9].

In calculation of verification, applied loads compose of static and dynamic loads. Static loads are constituted by gravity of dam and upstream hydrostatic pressure; seismic load is considered as dynamic load. And hydrodynamic pressure is calculated by formula Wastergaard [10] and simulated by mass element Mass21. Rigid foundation is used in model. Hydrostatic and hydrodynamic pressure is applied on normal pool level, as shown in figure 1. Damping ratio is 0.05. Horizontal and vertical record of Fengman earthquake is used as seismic wave [11], as shown in figure 3, and applied in horizontal and vertical direction. The duration of earthquake is 10.54s, and the peak of horizontal direction is 0.474g, peak of vertical direction is 0.312g.

Table 1. Material parameter of dam concrete.

Density (kg/m^3)	Poisson Ratio	Elastic Modulus (GPa)	Uniaxial Tensile Strength (MPa)	Uniaxial Compressive Strength (MPa)
2648	0.2	3.1027	24.1	2.9

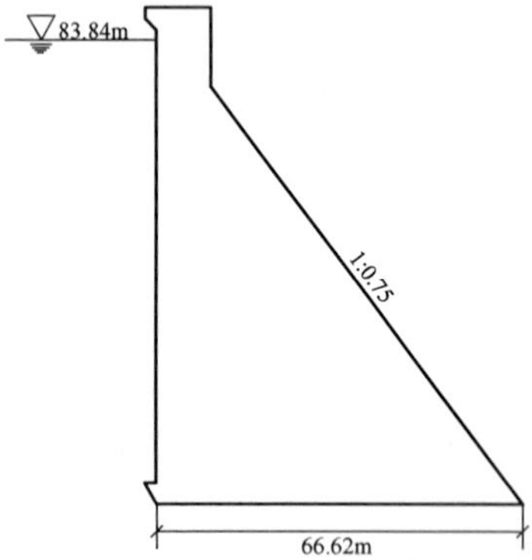

Fig. 1. Geometry model.

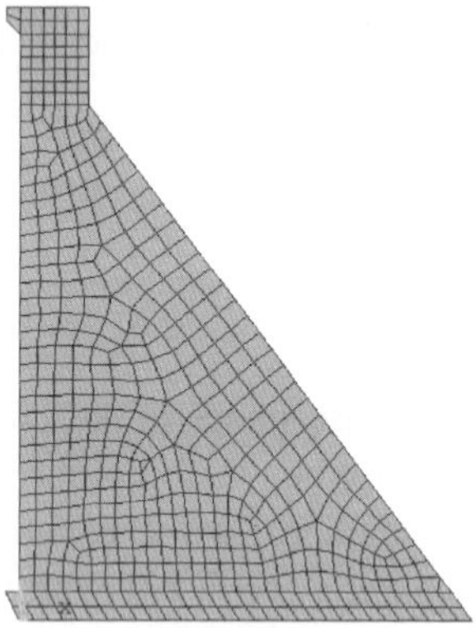

Fig. 2. Finite element model.

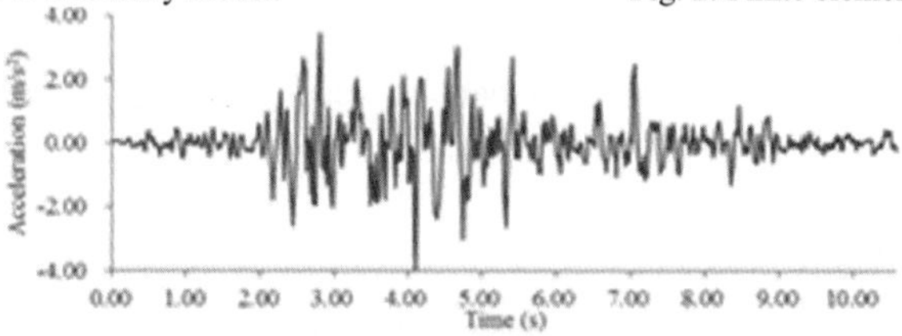

(a). Horizontal

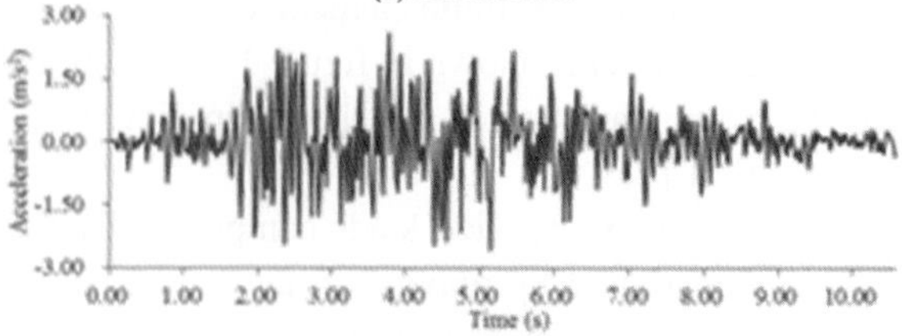

(b). Vertical

Fig. 3. Earthquake record.

3.1. *Numerical verification without considering nonlinear characteristic of concrete*

Under condition without considering nonlinear characteristic of concrete, namely linear condition, initial time step of transient analysis is set as 100000s. Static-dynamic coupling analysis is performed by using the new method, which could use dynamic analysis to solve static-dynamic coupling analysis directly.

According to calculation, the value and layout of dam displacement contour solved by initial time step of the new method is totally the same as the result solved by normal static analysis, as shown in figure 4, which means that the result of the new method is equivalent to the result of normal static analysis, and consistent with the conclusion of math derivation. It proves that static analysis can be performed in dynamic analysis by setting initial time step, and precision meets requirement.

In dynamic analysis,two response displacement time-histories of node 277 in dam crest, obtained respectively by the new method and condition superposition, are extracted and compared with each other, as shown in figure 5.

Under the initial condition, node displacement response in figure 5 is not start from zero, for the hydrostatic pressure and gravity of dam. Figure 5 shows that result of the new method is well consistent with result of condition superposition. Meanwhile, this method can perform dynamic analysis directly after equivalent static analysis, and result of it is consistent with result of condition superposition. However, operation of this method is more convenient compared with condition superposition. It shows that the method is effective, precise and feasible under linear condition.

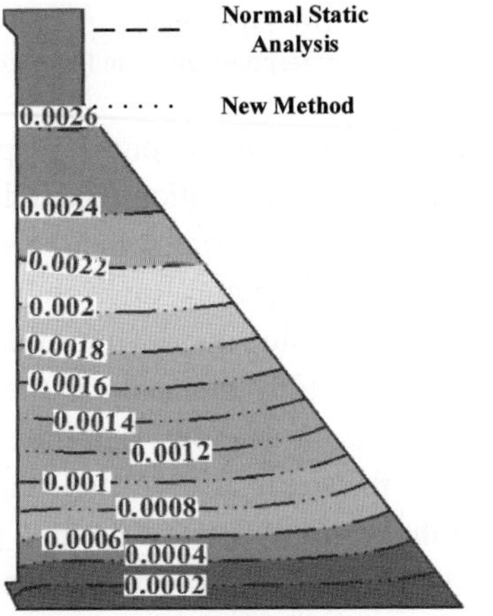

Fig. 4. Comparison of displacement contour plot in linear condition.

3.2. *Numerical Verification Considering Nonlinear Characteristic of Concrete*

Considering nonlinear characteristic of concrete, constitutive relation of concrete is simulated by multilinear isotropic hardening model MISO [12]. Stress-strain curve of concrete is obtained according to code for design of concrete structures [13], as shown in figure 6.

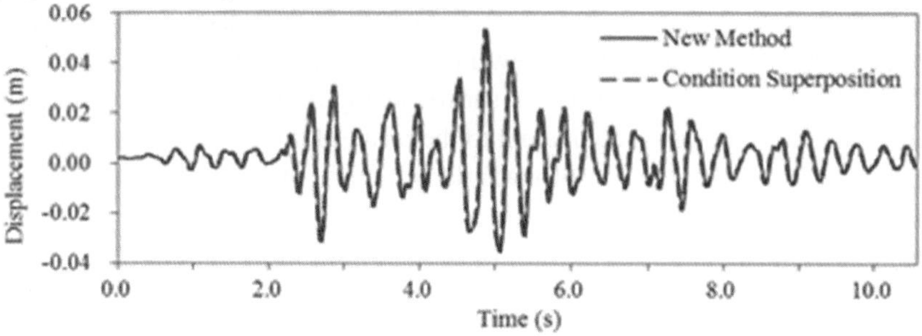

(a). X-Direction

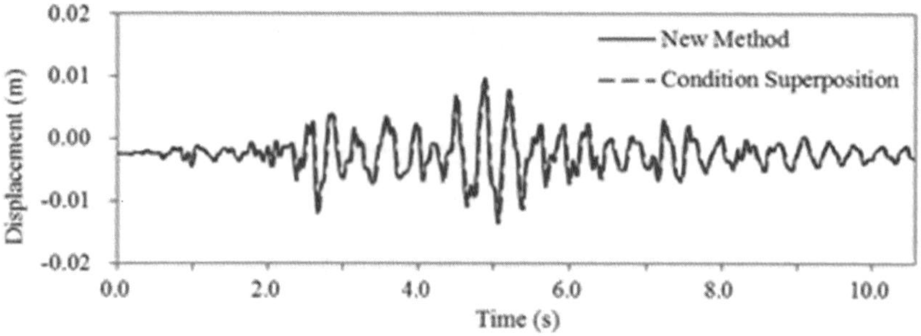

(b). Y-Direction

Fig. 5. Displacement response curve in linear condition.

Considering nonlinear characteristic of concrete, applied static and dynamic loads are the same as loads under the condition which did not consider nonlinear characteristic of concrete. The initial time step is also setting 100000s. This nonlinear problem is solved by full Newton-Raphson [14].

According to calculation, the value and layout of dam displacement contour solved by initial time step of is totally the same as the result solved by normal static analysis, as shown in figure 7. It shows that the nonlinear result is similar with the linear one, because the loads are so small that the concrete is still in linear status. It shows that the result of the new method is equivalent to the result of normal static analysis, under nonlinear condition, which is consistent with the conclusion of math derivation.

In dynamic analysis, same as section 2.1, two response displacement time-histories of node 277 in dam crest, obtained respectively by the new method and condition superposition, are extracted and compared with each other as well, as shown in figure 8.

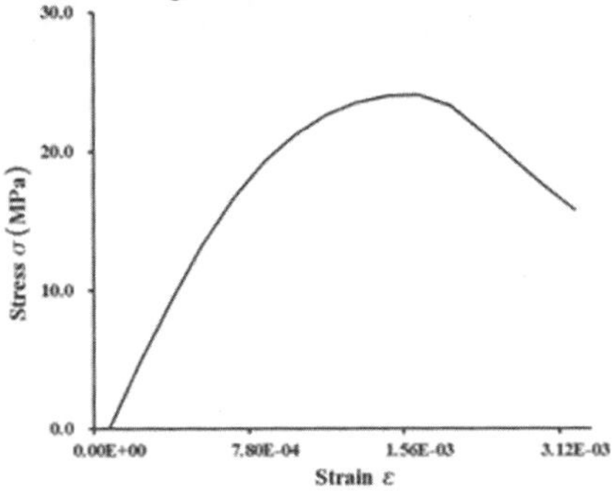

Fig. 6. Stress-strain curve of concrete.

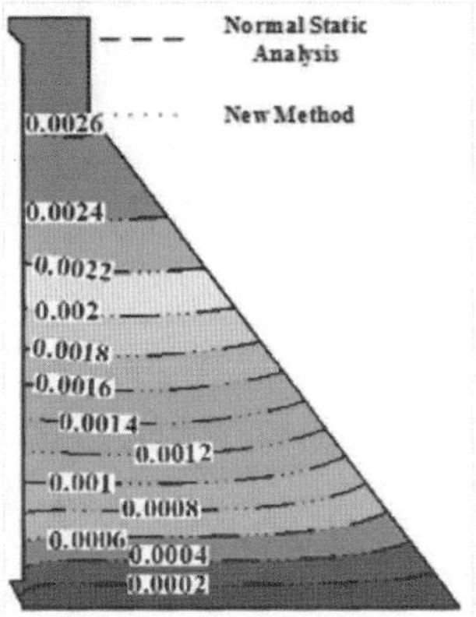

Fig. 7. Comparison of displacement contour plot in nonlinear condition.

The trend of curve in figure 8 shows that the result of the new method is apparently different from the result of condition superposition, for nonlinear characteristic of concrete. The reason of phenomenon above is that condition superposition is applicable only to linear result, not to nonlinear. Under nonlinear condition, condition superposition cannot obtain comprehensive response of static and dynamic loads. In initial condition of dynamic analysis, namely before seismic load and hydrodynamic pressure being applied, the dam is deformed by gravity and hydrostatic pressure. Therefore, displacement response curve of node is not started from zero. With the passage of time, the displacement response curve of the new method shows a clear deviation from condition superposition and is apparently greater than condition superposition, because condition superposition is meaningless in nonlinear condition. At last, for nonlinear characteristic of concrete, plastic deformation is appeared in some part of material, and the curve did not return to starting value.

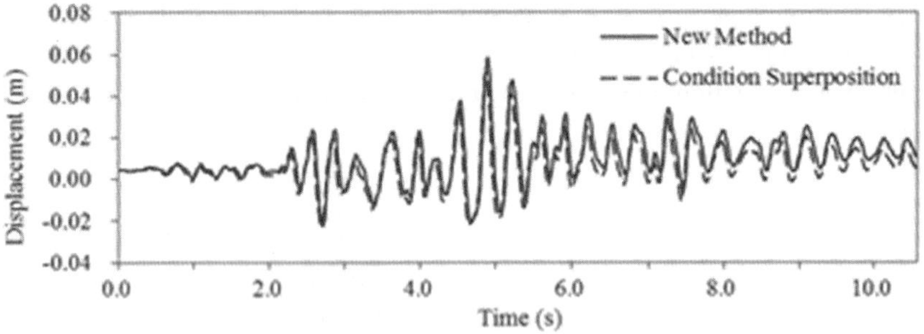

(a). X-Direction

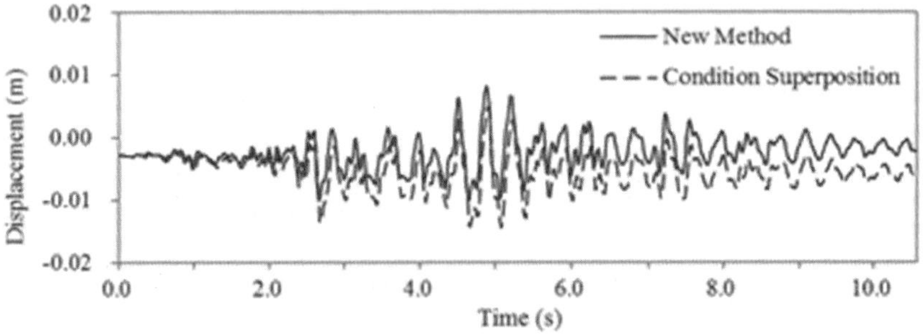

(b). Y-Direction

Fig. 8. Displacement response curve in nonlinear condition.

4. Conclusion

Combined with the implicit solving character of Newmark integral algorithm, a nonlinear static-dynamic coupling algorithm based on improved Newmark method is raised, by studying the inner relation between nonlinear finite element static and dynamic equilibrium equations for solving this issue. The algorithm could obtain rational nonlinear earthquake response of the structure under static and dynamic coupling excitation. Meanwhile, the algorithm was proved by mathematical derivation and can be used in kinds of implicit solving general finite element program conveniently easily. Then static-dynamic coupling analysis of Fengman dam in linear and nonlinear condition is performed respectively to testify the applicability of algorithm under corresponding condition. Numerical analysis shows that:

(1) In condition without considering nonlinear characteristic of concrete, namely linear condition, nodal displacement of all dam nodes solved by initial time step of the new method is totally the same as the result solved by normal static analysis, which shows that the method is consistent with the conclusion of math derivation. response curve of the new method is well consistent with condition superposition. It shows that result of the method is the same as

condition superposition, and the method makes static-dynamic coupling analysis much easier without switching between some analysis types and calculating a large amount of conditions.

(2) In condition considering nonlinear characteristic of concrete, nodal displacement of all dam nodes solved by initial time step of the new method is totally the same as the result solved by normal static analysis, which shows that the method is consistent with the conclusion of math derivation. In dynamic analysis, condition superposition cannot obtain reasonable static-dynamic coupling result, for it is meaningless in nonlinear condition. Therefore a greater difference appeared between two methods. The above shows that the complex operation and the disadvantage that cannot obtain reasonable seismic response of structure under kinds of loads of condition superposition were settled by new method.

Acknowledgments

This work is supported by: National Grand Science and Technology Special Project of China under Grant Number 2011ZX06002-10.

General Project of Department of Education of Liaoning Province (Grant No.l2014498).

References

1. DL5077-1997 *Specification for load design of hydraulic structures*[S]. Beijing: China Electric Power Press, 1998. (in Chinese).
2. LIU Jingbo, LV Yandong. A direct method for analysis of dynamic soil-structure interaction[J]. *China Civil Engineering Journal*, 1998, 31(3): 55-64. (in Chinese).
3. TU Jin. Static-dynamic combined analysis method of nonlinear dam-foundation system[J]. *Journal of E Zhou University*, 2006, 12(6): 8-11. (in Chinese).
4. LIU Jingbo, LI Bin. A unified viscous-spring artificial boundary for 3-D static and dynamic applications[J]. *Science in China (Ser. E)*, 2005,48(5):570-584. (in Chinese).
5. WANG Zhengzheng, GAO Bo. Static-dynamic coupling analysis model of structure-soil system[J]. *Chinese Journal of Computational Mechanics*, 2012, 29(2): 165-170. (in Chinese).
6. JIANG Xinxin, LI Jianbo, LIN Gao, et al. Direct method for static-dynamic analysis if structure based on transient analysis in ANSYS[J]. *Journal of Computer Applications*, 2014, 34(S1): 338-340. (in Chinese).

7. K. J. Bathe. *Finite Element Procedures*[M], Prentice-Hall, Englewood Cliffs (1996).
8. R. W. Clough, J. Penzien. *Structural Dynamics*[M]. Translated by WANG Guangyuan. Beijing: Higher education Press, 2010. (in Chinese).
9. YIN Xunqiang. Improvement and engineering application of structure-foundation dynamic interaction model[D]. *Dalian: Dalian University of Technology*, 2013. (in Chinese).
10. H. M. Westergaard. Water pressures on dams during earthquakes[J]. *Trans. ASCE*, 1933, 98: 418-433.
11. J. Lee, G. L. Fenves. Plastic-damage Model for Cyclic Loading of Concrete Structures[J]. *Journal of Engineering Mechanics*. 1998, 142(8): 892-900.
12. CHEN Huifa. *Constitutive Equations for Materials of Concrete and Soil*[M]. Beijing: China Architecture & Building Press, 2004: 25-29. (in Chinese).
13. GB50010-2010 Code for design of concrete structures[S]. *Beijing: China Architecture & Building Press*, 2011. (in Chinese).
14. ANSYS Incorporated. Programmer's Manual for ANSYS.

Study on the mechanical performance of height adjustable friction pendulum system

Hai-Xu Yang[1,a], Tian-Qiong Xue[1,b], Xi-Guang Chen[1], Liang Yu[1]
[1]School of Civil Engineering, Northeast Forestry University, Harbin, China, 150040
E-mail: [a]yhxcumt@163.com, [b]984104934@qq.com

A new type of height adjustable friction pendulum system boasting circular sliding surface has been developed in order to resist severe vibration and the differential settlement of the base. This paper has probed into the mechanical performance and parameters of the friction pendulum system and has concluded its unidirectional restoring force model and equivalent stiffness. The coefficient of friction of the adjustable friction pendulum system has been set through the sliding test. The equivalent natural period of vibration of friction pendulum system can be obtained on this basis. The result of the analysis shows that the curve shape of the restoring force of friction pendulum system is determined by the curvature radius of its convex surface, the weight of its superstructure and its coefficient of friction. The natural period of vibration of friction pendulum system is only related to the radius of cambered surface R, coefficient of frictionμ and the horizontal displacement of friction pendulum Dd. It is not related to other parameters, especially unrelated to the mass of its superstructure, thus possessing good stability. The height adjustable friction pendulum system can resist strong vibration and the differential settlement of the base. It also boasts self-limiting, self-restoring capacity as well as good stability against vibration. There is no need to attach centripetal damping structure. Furthermore, it is not necessary to take complicated seismic measures with regard to the superstructure, facilitating convenient application and enhancing the reliability of seismic isolation devices.

Keywords: Friction Pendulum; Seismic Isolation System; Mechanical Performance.

1. Guidelines

The seismic isolation device is used to obstruct the energy input of building structure so as to achieve the objective of damping control [1,2]. Seismic isolation device layer are installed between the infrastructure and the superstructure of buildings, dividing the structure into superstructure, seismic isolation layer and infrastructure. The seismic response is not magnified, only equivalent to 1/12-1/4 of the seismic influence that is sustained by the original structure. Even when the seismic isolation structure is subjected to strong earthquakes, the function of its superstructure can be maintained, the building is not seriously demolished or toppled down and life security is guaranteed.

In this paper, a new type of height adjustable friction pendulum system(AFPS) has been designed, which can provide great deformability and self-storing capacity. The key component of the seismic isolation layer of the seismic isolation structure, namely, the mechanical performance and parameters of friction pendulum system has been studied and the unidirectional restoring force model and equivalent stiffness of the friction pendulum system has been concluded. The coefficient of friction of the adjustable friction pendulum system has been set through the sliding test. The equivalent natural period of vibration of friction pendulum system is obtained on this basis.

2. Fundamental Principles of Height Adjustable FPS

FPS (Friction Pendulum System) can separate the superstructure and the base fixed in the foundation and prolong the natural period of vibration of the structure. If the damping of the building is constant, the acceleration response will greatly weaken so that the seismic response of the structure will attenuate and the seismic wave will be hindered to spread to the superstructure. The friction pendulum device boasts unique circular sliding surface with self-limiting, self-restoring capacity and good stability against vibration. There is no need to attach centripetal damping structure, facilitating convenient application and enhancing the reliability of the seismic isolation device[3,4].

Compared with the construction site of ordinary buildings, the relative motion of both sides of tectonic faults generally presents in two special forms, one is that when an earthquake occurs, the amplitude and waveform of seismic vibration between the upper side and the lower side of faults have big differences; the other is that there is slow relative motion between both sides of faults. It is a common phenomenon that both natural active faults and underground coal mining might give rise to differential settlement of ground surface, tilting of building structure and cracking of wall. It is proved by both theory and practice that seismic isolation can resist strong seismic vibration and the height adjustable system can adapt to differential settlement. If the height adjustable friction pendulum system is applied in basic seismic isolation projects, it can resist not only strong vibration, but also differential deformation of active faults. Moreover, the height adjustable system is an effective dry friction sliding seismic isolation device, which can also resist differential settlement via the adjustment of geometric parameters of the seismic isolation layer.

3. Fundamental Structure of the Height Adjustable FPS

The sectional dimension of the height adjustable friction pendulum system is illustrated in Fig.1-Fig. 2 (the unit of length in the Fig. 1-2 is mm).

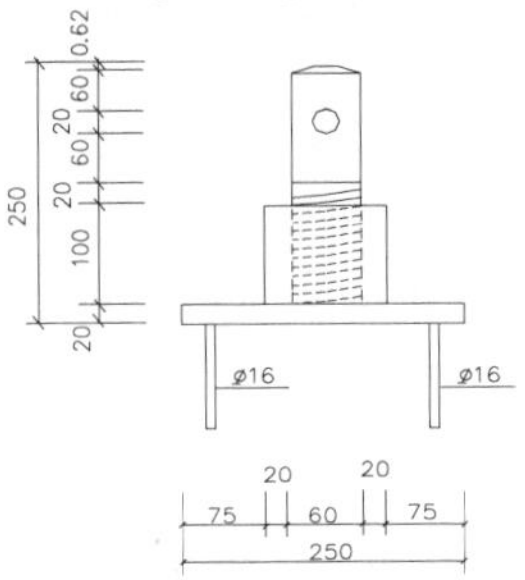

Fig. 1. Elevation Drawing of the Lower Device of AFPS

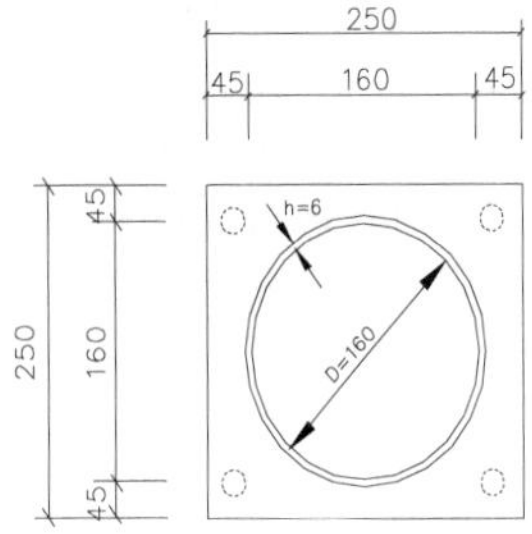

Fig. 2. Plan of the Upper Support Plate

3.1 *Mechanical Performance Study and Parameter Analysis of FPS*

The Unidirectional Restoring Force Model and Equivalent Stiffness of Friction Pendulum System. Owing to the fact that the friction on the sliding surface of friction pendulum system possesses high nonlinearity, the seismic response of its basic vibration isolation structure presents high nonlinearity. Therefore, there are still a lot of problems to solve. Currently, Coulomb Friction Model and Continuous Function Model [5,6] are often adopted to cope with the difficult friction model of friction pendulum system. These models are unidirectional so each direction is subjected to force independently.

The computational model of AFPS can be simplified into a structure that conducts small amplitude motion with a sliding block whose mass is m sliding along the convex slide rail of the circular arc on the friction pendulum system, as can see in Fig. 3. The sliding block is subjected to gravity W, bearing force of the sliding block N, friction f and inertia force. The highest point of the convex surface is set as the origin in the horizontal direction. θ indicates the intersection angle, which is formed when the sliding block moves relatively to the vertical axis of the slide rail, taking the counterclockwise as the positive direction. The contact between the sliding surface and the sliding block is surface contact, which facilitates energy dissipation via friction in the sliding process. When an earthquake occurs, the horizontal motion of the ground surface will drive the sliding block to slide horizontally within its circular arc surface, thus the superstructure will be slightly elevated. Then the sliding block will get back to its original equilibrium position under the influence of its own gravity. The function of friction pendulum system is: on one hand, it supports the total weight of the structure; on the other hand, it prolongs the natural period of vibration of the structure, making the fundamental frequency of the structure go beyond the frequency range of high energy earthquakes, thus effectively mitigates the seismic response of the structure. When the friction pendulum system supports the structure, it will not lose its bearing capacity. What is more, it can sustain the displacement between the base and the superstructure.

The horizontal displacement of friction pendulum of AFPS $X = R\sin\theta$; X and X signify velocity and acceleration of motion respectively; the bearing force of the slide block $N = WCOS\ \theta$, W is the vertical load of the superstructure, W=mg; friction f can be calculated according to the following Eq.1:

$$f = \mu N sign\ (X) \tag{1}$$

In Eq.1, μ is the coefficient of sliding friction of the slide block[7]; from the experimental study of AFPS, the coefficient of sliding friction is related to sliding velocity and stress. $sign(X)$ is sign function,

$$sign(X) = \begin{cases} 1 X > 0 \\ -1 X < 0 \end{cases} \tag{2}$$

The equilibrium equation of the horizontal motion of the structure:

$$mX + N \sin \theta + sign\ (X) f \cos \vartheta = 0 \tag{3}$$

Compared with the amplitude of structural motion X, the curvature radius of the convex surface R (namely, the curvature radius of the sliding surface) is big. Nevertheless, the coefficient of friction of AFPS is tiny. Thus, it can be approximately concluded that $N = mg$, $\cos\theta = 1$.

If the vertical vibration is negligible, the Eq.4 can be simplified into

$$mX + mg\frac{X}{R} + (signX)\mu mg \approx 0 \tag{4}$$

The horizontal restoring force of AFPS F is:

$$F = mg\frac{X}{R} + (signX)\mu mg \tag{5}$$

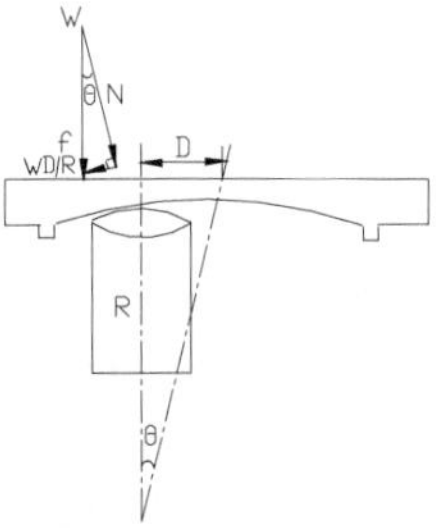

Fig. 3. Computational Model of AFPS

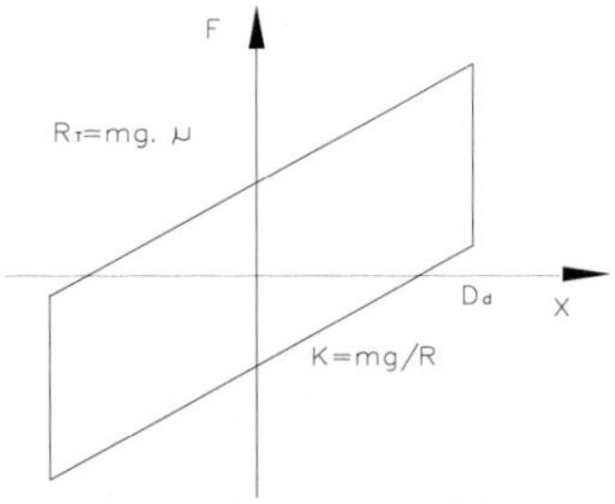

Fig. 4. Restoring Force Model of FPS

3.2 *significant parameters of AFPS*

The horizontal force F includes lateral restoring force caused by gravity and friction. The restoring force model of AFPS is illustrated in Fig. 4. The friction of AFPS is much less than the restoring force of gravity. The horizontal stiffness generated by the gravity of the superstructure is:

$$K = \frac{mg}{R} \tag{6}$$

The vibration period can be obtained on the basis of the Eq.6

$$T = 2\pi\sqrt{\frac{m}{K}} = 2\pi\sqrt{\frac{R}{g}} \tag{7}$$

From Fig. 4 the equivalent linear stiffness of AFPS is:

$$K_E = \frac{\mu mg}{D_d} + \frac{mg}{R} \tag{8}$$

The equivalent viscous damping coefficient[8,9] is

$$\varsigma_E = \frac{E}{2\pi K_E D_d^{\ 2}} \tag{9}$$

In Eq.9, E is hysteresis dissipation energy. Substituting Eq.9 into Eq.10, the following Eq.11 can be obtained.

$$\zeta_E = \frac{2}{\pi}\left[\frac{\mu}{\mu + (D_d / R)}\right] \tag{10}$$

Suppose Ks is the equivalent stiffness of the vibration isolation system of AFPS, Ku is the equivalent stiffness corresponding to the basic vibration mode of the superstructure,

$$\frac{1}{Ks} = \frac{1}{K_U} + \frac{1}{K_E} \tag{11}$$

$$Ks = \frac{K_U K_E}{K_U + K_E} \tag{12}$$

When the load that is exerted in the horizontal direction of AFPS overcomes friction, the sliding device will start sliding on the convex surface so that the superstructure of AFPS will begin simple pendulum motion of small amplitude. The AFPS plays a part in changing the period of motion and dissipating the kinetic energy. The curve shape of the AFPS restoring force hinges on the curvature radius of the convex surface of AFPS, the weight of the superstructure and the coefficient of friction of AFPS. The bigger the dimension of hysteresis loop is, the stronger the energy absorption capacity it is.

AFPS has three significant parameters, namely, the curvature radius of the convex surface related to period and stiffness, coefficient of friction connected with friction and horizontal displacement of friction pendulum D_d. When the horizontal load that is exerted on the AFPS is smaller than friction, the motion of the whole structure will be the same as there is no AFPS. Once the horizontal load exceeds friction, AFPS will play a role in changing the period of motion and dissipating the energy of motion.

3.3 *Coefficient of Friction of AFPS*

When the coefficient of friction of AFPS is set, the maximum base shear of the earthquake input structure is the limiting friction Fmax:

$$F_{\max} = \mu W = \mu Mg \tag{13}$$

Theoretically, the smaller F is, the better it is, for in this way the seismic influence on the structure is less. When μ=0, the structure is subjected to zero seismic influence. According to the actual situation of the materials, the above condition can not be attained in practice.

3.4 *Period of AFPS*

When the superstructure presents stiffness, the stiffness $K_u = \infty, K_s = K_E$. Therefore, the equivalent natural period of vibration of FPS is about

$$T = 2\pi\sqrt{\frac{M}{K_S}} = 2\pi\sqrt{\frac{D_d R}{(\mu R + D_d)g}} \tag{14}$$

The Eq.14 is only related to the radius of cambered surface R, coefficient of friction μ and horizontal displacement of friction pendulum μ, not related to other parameters.

4. Conclusions

In conclusion, AFPS boasts simple structure and superior performance and can be used for seismic isolation of new and existing buildings in seismic region. The utilization of this seismic isolation device can enhance the seismic resistance capacity of structure and prevent the structure and the internal facilities from seismic damage. A research on the mechanical performance and parameters of AFPS has been conducted and the unidirectional restoring force model and equivalent stiffness of AFPS have been concluded. The main conclusions are as follows:

(1) The natural period of vibration of AFPS is only related to the radius of cambered surface R, coefficient of friction μ, horizontal displacement of friction pendulum μ, not related to other parameters.

(2)The curve shape of the restoring force of AFPS hinges on the curvature radius of the convex surface of AFPS, the weight of the superstructure and the coefficient of friction of AFPS.

(3)Compared with ordinary base sliding, the convex surface of AFPS can generate centripetal restoring force when the structure stops moving. Thus, after the earthquake, the residual deformation of the structure can be mitigated.

(4)The stiffness of AFPS is directly proportional to the weight. The center of rigidity coincides with the center of mass, which can diminish the twisting response of the structure.

Acknowledgments

This work is supported the Heilongjiang Province Nature and Science Foundation (E2015057).

References

1. Jinping Ou: Structural Vibration Control (Science Press Publications, Beijing 2003)
2. R.I.Skinner, W.H.Robinson and G.H.Mcverry: Introduction of Engineering Isolation, translated by Lili Xie and Xingquan Zhao(Earthquake Press Publications, Beijing1996)(in Chinese)
3. Yicheng Jing, Jiaxiang Tang and Shishu Xiong, in: Geotechnical Mechanics (English), Vol. 22 (2001), p. 28
4. Bashshi, A.,Arkai,H. and Shimazu, T., in: Earthquake Engineering & Structure Dynamics. 1998. Vol.27(1998), p. 29
5. MostaghelN and DavisT. In: Earthquake Engineering and Structural Dynamics, Vol.5(1997), p. 541

6. Jian Fan and Jiaxiang Tang, in: China Civil Engineering Journal, Vol. 33 (2000), p. 11(in Chinese)
7. MokhaA, ConstantinouM C and ReihornAM. In: Testing. Journal of Structural Engineering, ASCE, Vol.116 (1990), p. 438(in Chinese)
8. Lin Yang, Xiyuan Zhou and Youpo Su, in: Special Structures, Vol. 22 (2005), p. 43(in Chinese)
9. Chiyu Jiao, Shide Hu and Zhongguo Guan, in: Vibration and Shock, Vol. 26(2007), p.113(in Chinese)

Numerical simulation analysis of metal plate impacted by rigid ball at high-velocity

Ze-Tian Kang*, Bo Zhou, Shi-Feng Xue
Department of Engineering Mechanics, College of Pipeline and Civil Engineering,
China University of Petroleum,
Qingdao, Shandong 266555, China
**E-mail: zetiank@163.com*
www.upc.edu.cn

The penetration of a plate under a high-velocity impact loading exhibits a very complex dynamic response process with highly nonlinear characteristics. A failure criterion was presented for the cracking model of homogeneous isotropic material. Based on this failure criterion, numerical simulation was conducted on the process of homogeneous isotropic metal plate impacted by a rigid ball at high-velocity by using LS-DYNA program. The relationship between effective stress and time at the impact point was expressed and analyzed. The curves of energy versus time were also plot to illustrate the impact process obviously.

Keywords: Damage Model; High-Velocity Impact; LS-DYNA; Energy.

1. Introduction

Object impacted at high speed is easy to damage or failure, and it has brought great harm to our work and life. For example, the windshield of a high-speed car impacted by a stone and the aircraft impacted by a bird, they all bring the hidden trouble to our life and property. Many scholars have studied on impact damage failure criterion. Considering the influence of stress component on different failure modes, Hou et al.[1] put forward the failure criterion of four kinds of impact damage models. On the basis of research of predecessors, according to the damage mechanism of composite laminated plate delamination impacted at low-velocity, Weidong Wen [2] developed a layered failure criteria to distinguish the fiber extrusion damage of positive impact and the fiber fracture damage on the back, considering the influence of interlaminar tensile stress, interlaminar shear stress between the layers and matrix cracking on delamination. Combined with the interlayer connection unit, Zhuo Zhuang et al. [3] simulated the damage of laminated plates impacted at low-velocity and compression failure problem of laminated plates containing prefabricated layers and got the ultimate tensile strength, damage evolution and stiffness reduction respectively by introducing damage models using ABAQUS software. They also put forward the equivalent mechanics model applied to static test of structure level and expand to the structural impact simulation.

In order to investigate the destructiveness under high-speed impact loading, referring to the work of predecessors, we simulate the dynamic process of metal plate containing holes impacted by a high-speed rigid ball and analyze the energy transformation in the failure process using the finite element software LS-DYNA.

2. Failure Criterion

There are four kinds of damage models connected with the composite laminated plate impacted at low-velocity, including extrusion and cracking of the matrix, fiber breakage and stratification. The failure criterions for the above damage models are respectively expressed as [4-7] Eq. (1), Eq. (2), Eq. (3) and Eq. (4).

$$\frac{1}{4}\left[\frac{-\sigma_{22}}{Q_{12}}\right]^2+\frac{C_{22}^2\sigma_{22}}{4Q_{12}^2C_{22}}-\frac{\sigma_{22}}{C_{22}}+\left[\frac{\sigma_{12}}{Q_{12}}\right]^2\geq 1, \tag{1}$$

$$\left[\frac{\sigma_{22}}{T_{22}}\right]^2+\left[\frac{\sigma_{12}}{Q_{12}}\right]^2+\left[\frac{\sigma_{32}}{Q_{32}}\right]^2\geq 1, \tag{2}$$

$$\left[\frac{\sigma_{11}}{T_{11}}\right]^2+\left[\frac{\sigma_{21}^2}{Q_{12}^2}\right]+\left[\frac{\sigma_{31}^2}{Q_{12}^2}\right]\geq 1, \tag{3}$$

$$\left[\frac{\sigma_{33}}{T_{33}}\right]^2+\left[\frac{\sigma_{13}}{Q_{13}}\right]^2+\left[\frac{\sigma_{23}}{Q_{23}}\right]^2\geq 1, \tag{4}$$

where, T_{ii} denotes the tensile strength and C_{ii} denotes the compression strength in the direction of i, Q_{ij} represents the limiting shear stress in the plane of ij.

Referring to the above damage models of composite laminated plate, the homogeneous isotropic material failure criterion in the case of cracking degrades into Eq. (5).

$$\left[\frac{\sigma_{1i}}{Q_{1i}}\right]^2+\left[\frac{\sigma_{2i}}{Q_{2i}}\right]^2+\left[\frac{\sigma_{3i}}{Q_{3i}}\right]^2\geq 1, \tag{5}$$

where, Q_{ij} ($i \neq j$) represents the limiting shear stress in the plane of ij, Q_{ij} ($i = j$) denotes the tensile strength in the direction of i or j.

3. Simulation and Analysis of Impact

The homogeneous piecewise linear plastic material metal plate with four holes is impacted by a rigid ball at high-speed. The radius of the rigid ball is 25cm, so is the hole. It is 100cm from the ball center to the plate, and the initial velocity of the ball is 100cm/ms and the solve time is 25ms. Both the length and width of the plate are 400cm, the thickness is 1cm, and it is clamped around, $\rho = 0.00783\text{kg/cm}^3$, $E = 200000\text{MPa}$, μ=0.3, yield stress 207 MPa, failure plastic strain 0.75. The Table 1 list the plastic characteristics of the plate.

Table 1. The plastic characteristics of material metal plate.

effective plastic strain	0.00	0.08	0.16	0.40	1.00
Stress / Mpa	207	250	275	290	300

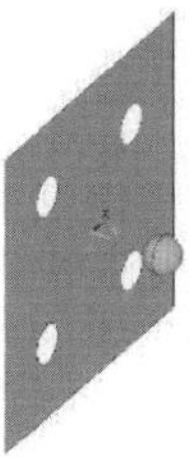

Fig. 1. Geometric model.

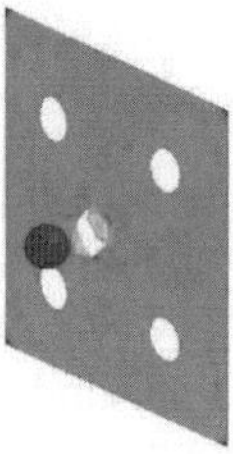

Fig. 2. Disruptive metal plate.

Figure 1 and Figure 2 show the geometric model and the damage process of the metal plate.

Four elements in the plate corner are selected, and the max effe-ctive stress is 516.47MPa as show in Figure 3, the initial damage time is 0.75ms.

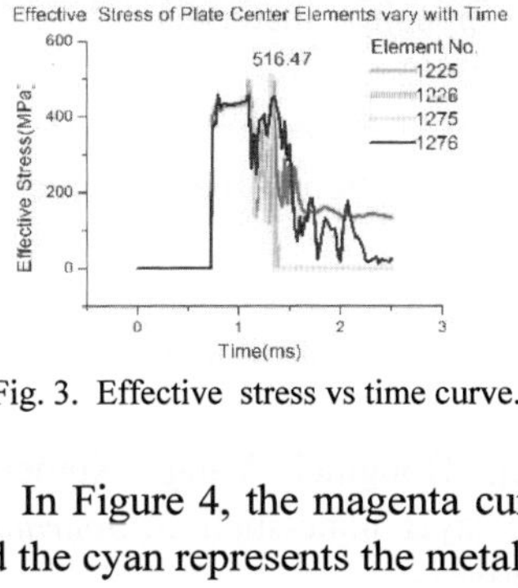

Fig. 3. Effective stress vs time curve.

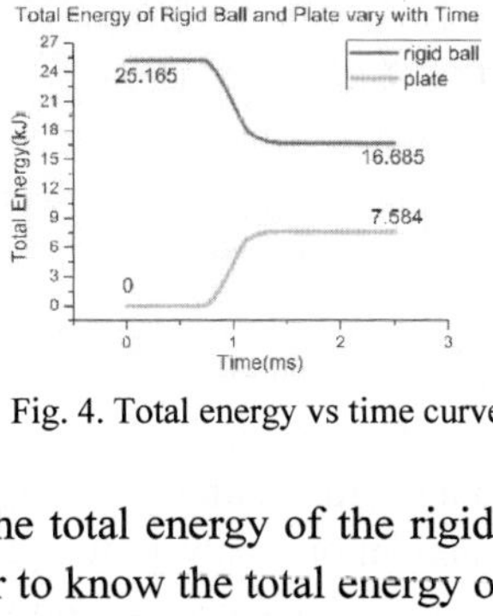

Fig. 4. Total energy vs time curve.

In Figure 4, the magenta curve represents the total energy of the rigid ball and the cyan represents the metal plate. It is clear to know the total energy of thc system reduces about 0.90kJ and the kinetic energy of the rigid ball reduces from 25.165kJ to 16.685kJ.

4. Summary

A failure criterion was presented for the cracking model of homogeneous isotropic material. We got the Effective Stress-Time relationship of the impact point on the plate by describing the Effective Stress-Time curves of four elements around the point. The Energy-Time curves of the plate and the ball were also described in this paper.

Acknowledgment

Authors wish to acknowledge the support of the National Natural Science Foundation of China (11472309).

This work is supported by the National Natural Science Foundation of China (11472309).

References

1. Hou J P, PETRINIC N, RUIZ C, et al. Prediction of impact damage in composite plates. *J. Composites Science and Technology*. 60 (2000).
2. Weidong Wen, Ying Xu, Haipo Cui. Damage Analysis of Laminated Composites Under Low Velocity Impact Loading. *J. Materials Engineering*. 21, 7 (2007).
3. Zhuo Zhuang, Hengxu Song, Tao Peng, Liyuan Wang. DAMAGE OF COMPOSITE UNDER LOW-VELOCITY IMPACT AND APPLICATION OF MECHANICAL EQUIVALENT MODEL. *J. Engineering Mechanics. A* 02 (2012).
4. Yinghua Zhao, Jingzhong Wang. Interface damage process and the elastoplastic constitutive relation of composite materials. *J. ACTA MATERIAE COMPOSITAE SINICA*, 16, 4 (1999).
5. Weishen Zhu, Qiangyong Zhang, Shucai Li. Three-dimensional crisp elastic-plastic fracture damage model in the application of the fractured rock mass engineering. *J. ACTA MECHANICA SOLIDA SINICA*, 20, 2 (1999).
6. Guiping Zhao, Tianjian Lu. Dynamic response of cellular metallic sandwich plates under impact loading. *J. Chinese Journal of Theoretical and Applied Mechanics*, 40, 2 (2008).
7. Jian He, Yufeng Xiao, Shiquan Yang, Dongmei Yang. Numerical simulation analysis for penetration of one layer plate-shell. *J. Journal of Harbin Engineering University*, 27, 3 (2006).

Research on key technology of construction monitoring of long-span curved bridge with high piers

Long Zhang

China Communication North Road & Bridge Co., LID,
Beijing, 100024, China
E-mail: 190826230@qq.com

For construction monitoring of long-span curved bridge with high piers, a construction monitoring system based on modern control theory is proposed, take "construction-measuring-recognition-fix-warning-construction" cycle control as the core, construction process was overall controlled. The WuJiaWan Bay Bridge is taken as an example, the finite element model of the full-scale bridge is established by using Midas/civil 2013, the construction process is simulated by calculation. The theoretical values and measured values of alignment, stress and other indicators was made comparative analysis. The result show that this method can effectively control the construction quality, alignment and stress of bridge can meet the design requirement, and bridge can closure smoothly.

Keywords: Construction Monitoring; Long-Span Curved Bridge; High Piers.

1. Introduction

Long-span bridge with high piers is favorable with advantage of large spanning ability and smooth traffic. Uncertain factors during construction progress make a big different between ideal state and actual state after bridge completed. Thus, research on construction monitoring play an important role in alignment and stress of structure. So far, many scholars have done a lot of research on construction monitoring of bridge [1-9]. But, the construction monitoring research for curved bridge was not much. Especially, for large-scale complex projects, it was difficult to control alignment and stress accurately [2].

Aiming at the above problems, a new construction monitoring system based on modern control theory was put forward. The bridge construction process was taken as a complicated dynamic process; error analysis was take basis for ideal state of structure, tests in-situ, error information. And, adjustment operations of the construction site was guided, making the actual construction state tends to the predetermined ideal state. Then, according to the current state of structure, forward-calculation was made, possible state of stress and deformation in the future was forecasted, and the construction control parameters of next phase was provided. Because a new error was introduced in each step during construction, prediction and adjustment is constant circulation process. Process of circulation control " construction- measuring- recognition- fix- warning- construction " was introduced. Construction control process is shown in figure 1.

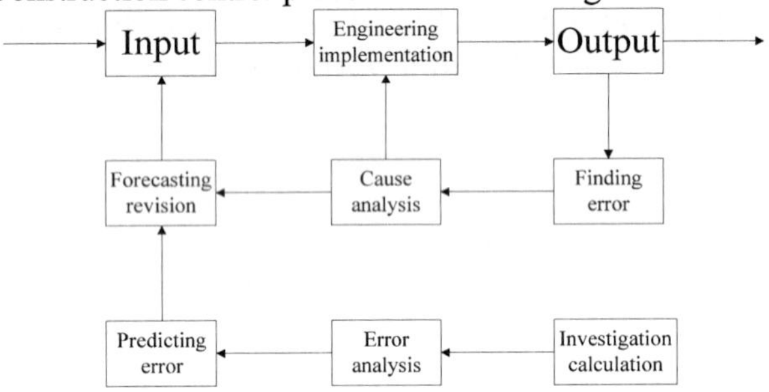

Fig. 1. Construction control process.

2. The Meaning of Construction Monitoring

For long-span curved bridge with high piers, the particularity of material and the complexity of construction technology, combined with many unpredictable factors in the construction process, and may result in artifacts in some parts of the stress reserve is insufficient or excessive, alignment deviation, thus causing hidden danger. By measuring stress, strain, deformation, temperature and other physical quantities in critical control, the actual stress state and deformation of each component during each construction stage was known, and some technical measures was taken. Make the bridge structure is reasonable, the alignment meet with the design requirements, and ensure the safety of the bridge construction, construction quality, beautiful, reliable and durable for a long time [7].

3. The Main Content of Construction Monitoring

The main content of construction monitoring including monitoring adjustment calculation, monitoring, testing and error adjustment. Monitoring adjustment calculation is the calculation of construction control parameters such as stress and displacement of each segment during construction based on the simulation technology; monitoring test is the measurement of alignment, stress and temperature of each component; error adjustment is the dynamic adjustment of construction process based on actual test data combined with the theoretical calculation result. Construction monitoring process is shown in figure 2.

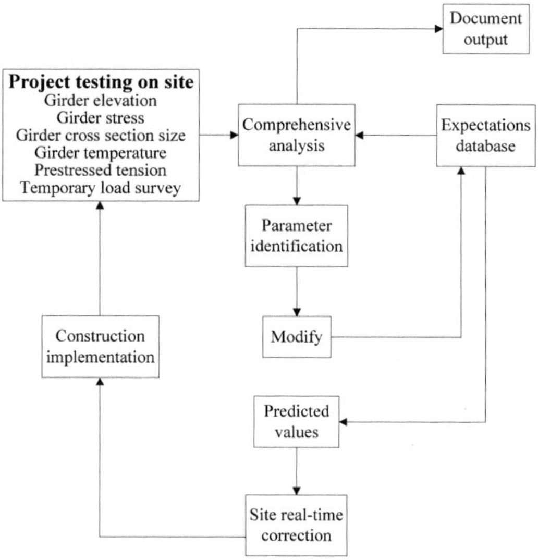

Fig. 2. Construction monitoring process.

4. Engineering Survey

WuJiaWan Bay Bridge is a long-span continuous rigid frame curved bridge with high pier.The length of the highest piers reaches 58.645m with a combination of the main bridge span which is 58 + 100 + 58m. The main pier and transition pier of Girder of main bridge is thin-wall hollow pier. The radius of curvature of the bridge is 750m. Main bridge arrangement is shown in figure 3.

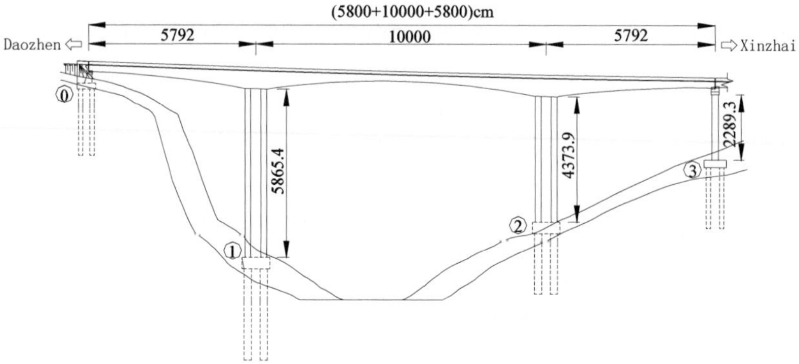

Fig. 3. Main span of the bridge lay out.

5. Construction Monitoring Calculation

The finite element model of the full-scale bridge is established by using Midas/civil 2013 with 152 elements and 159 nodes. The boundary condition of model: the bottom of the pile caps and the foundation is consolidation, double limb end of the pier and pile caps is master-slave node of rigid connection, thin-walled pier and girder adopt stiffness in elastic connection, abutments in both sides adopt general bearing with two degrees. The finite element model of WuJiaWan Bay Bridge is shown in figure 4.

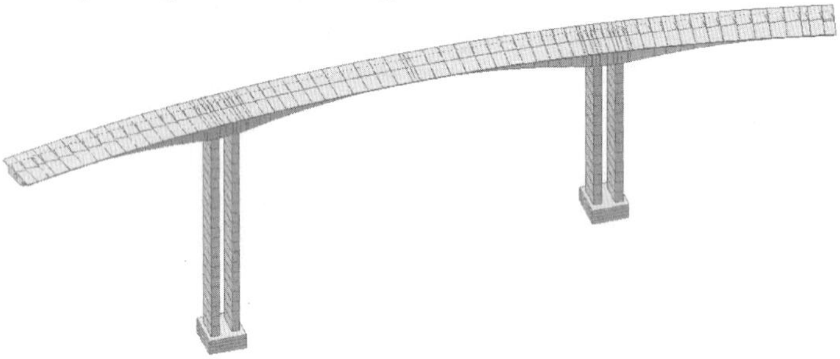

Fig. 4. The finite element model of WuJiaWan Bay Bridge.

6. Construction Monitoring Testing

As the most direct way for calculation parameters, construction monitoring testing is the foundation of construction monitoring for long-span bridge. In order to ensure construction quality and provide quantitative control parameters, alignment, stress, temperature was set as main test content. The alignment test include girder height and formwork placing elevation of girder. The stress, temperature test include stress measurement of girder and temperature measurement of girder. Measuring points of alignment layout is shown in figure 5. Measuring points of stress and temperature layout is shown in figure 6.

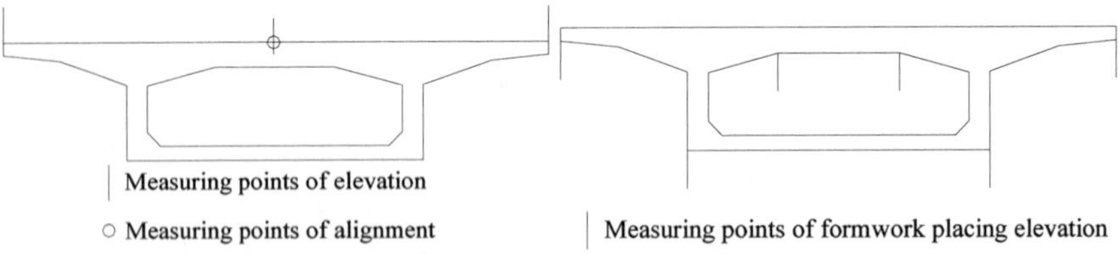

a)Measuring points of box girder cross-section b)Measuring points of box girder formwork elevation

Fig. 5. Measuring points of alignment layout.

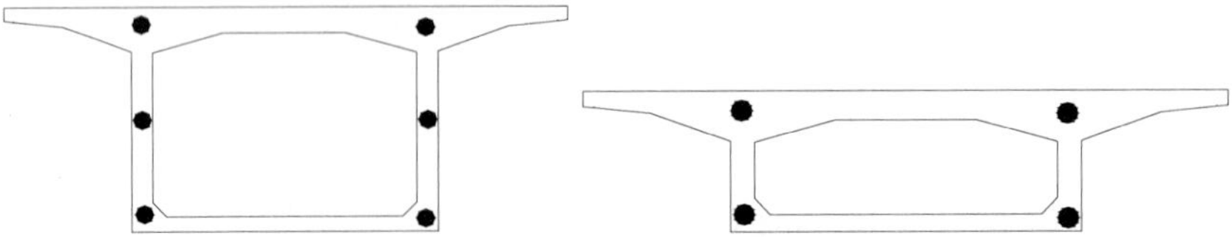

a) Measuring points of box girder root b) Measuring points of side span and closure segment

Fig. 6. Measuring points of stress and temperature layout.

7. Error Adjustments

Along with the promotion of the segment cantilever construction, the corresponding test values for section in each segment was collected, then compared with the relevant numerical in simulation model for the error analysis. The difference between measure value and theoretical value of elevation viewed from midspan and side span is shown in figure 7. The difference between measure value and theoretical value of alignment viewed from midspan and side span is shown in figure 8. The measure value of maximum offset of pier upstream and downstream is shown in figure 9. The measure value of maximum and minimum stress is shown in figure 10.

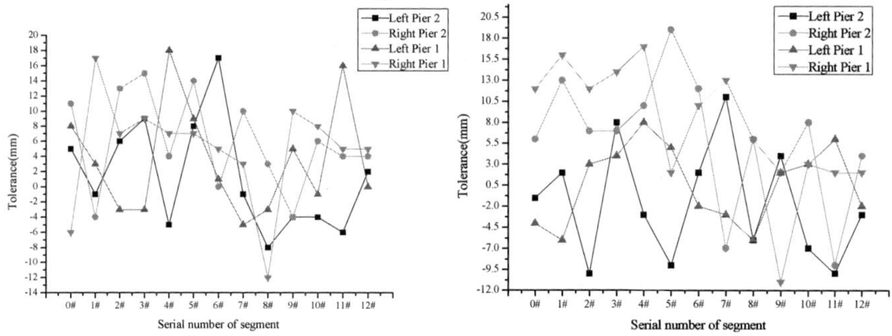

a) section in midspan b) section in side span

Fig. 7. Elevation difference between measure value and theoretical value

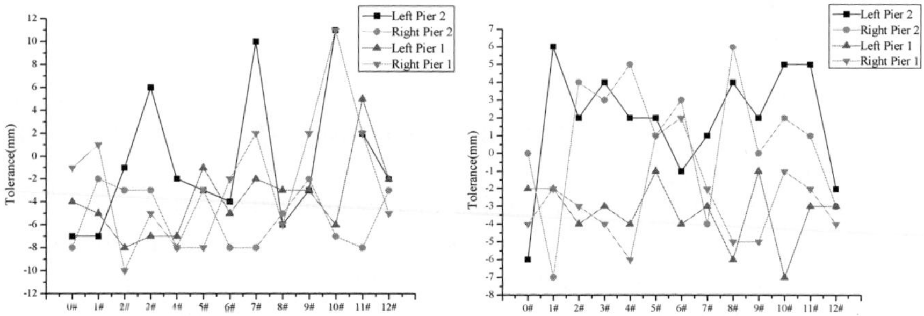

a) section in midspan b) section in side span

Fig. 8. Alignment difference between measure value and theoretical value.

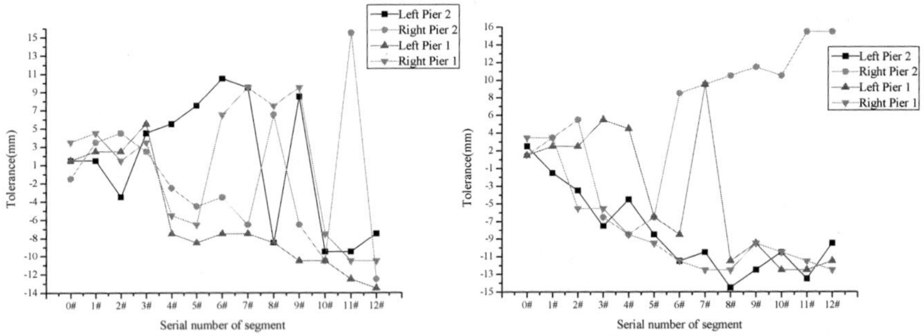

a) upstream b) downstream

Fig. 9. The measure value of maximum offset of pier.

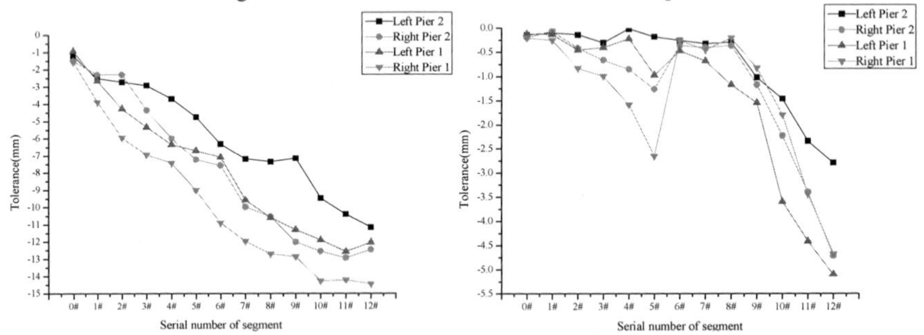

a) maximum stress b) minimum stress

Fig. 10. The measure value of maximum and minimum stress.

8. Conclusions

(1) During the whole construction process, the state of girder stress meet with design requirement, and normal tensile stress in cross-section was not exceed the value of code for design.

(2) In view of the measure value of elevation after completion, girder elevation meet design requirement, the alignment was smooth, the whole process of construction monitoring was successful.

(3) This theory method system provide effective decision support for construction monitoring of long-span curved bridge with high piers, and provide a good guidance for control of structure stress and alignment during construction process.

References

1. Chun Wang, Lu Gan, Mingyu Fang, Yizhou Ye, *Control analysis of cantilever construction of super high pier rigid frame bridge, J. Technology of Highway and Transport*, 1, 62 (2009) (In Chinese).

2. Dali Shi, Xiujun Li, Zhixiao Zhang, *Construction monitoring of cantilever construction for continuous rigid frame bridge, J. East china highway*, 6, 32 (2013) (In Chinese).
3. Qi Yang, Wuming Leng, Rusong Nie, Wei Wei, *Construction control research and analysis of the large span prestressed concrete continuous rigid frame bridge, J. Journal of Railway Science and Engneering*, 7, 11 (2010) (In Chinese).
4. Feng Zhou, Huidong Wang, *Study and Practice on construction monitoring of cantilever construction for long span continuous box-section beam, J. Journal of ShiJiaZhuang Railway Institute (Natural Science)*, 22, 15 (2009) (In Chinese).
5. Tianliang Xu, *Study on construction monitoring technology of long-span continuous rigid bridge with high piers, J. Highways & Automotive Applications*, 150, 169 (2012) (In Chinese).
6. Chunhui Huang, Juntao Shi, *Continuous rigid frame bridge construction monitoring, J. Journal of Guizhou University(Natural Science)*, 30, 99 (2013) (In Chinese).
7. Renpei Qin, Dairong Qiu, *Construction Supervision and analysis of large span high pier continuous rigid bridge, J. Technology of Highway and Transport*, 4, 71 (2007) (In Chinese).
8. Liangjiu Guo, Xiaohui Liu, Rong Wang, *Research on construction control technology for curved continuous rigid frame bridge, J. Technology of Highway and Transport*, 2, 99 (2009) (In Chinese).
9. Weigan Yu, *Research and application on key technology for design and construction of mountain area highway, J. East china highway*, 6, 12 (2005) (In Chinese).

Prediction of foamed concrete compression strength and thermal conductivity based on bp neural network

Guan-Sheng Yin, Chen Fu, Yan-Fei He
School of Science, Chang'an University,
Xi'an, Shaanxi, China
E-mail: yings@chd.edu.cn

In this paper, BP neural network model is used to predict the compression strength and thermal conductivity of the foamed concrete. The experimental data were divided into training dataset and control dataset. On the training dataset, the proposed BP neural network model was applied. The fitted model was obtained by tuning the parameters of mixing proportion with error rate controlled at pre-defined level. The prediction accuracy of the model was verified by comparing the results of the fitted model on the control dataset with true values. The results show that the predicted error rate is less than 8%, indicating that BP neural network is capable of predicting the experimental data accurately.

Keywords: BP Neural Network; Foamed Concrete; Compression Strength; Thermal Conductivity; Error Rate.

1. Introduction

With the development of wall innovation and energy saving in China, a growing number of problems emerge in the field of low-energy and passive building materials and structures. The foamed concrete, as a new type of lightweight and environmentally friendly material, is widely used in external wall thermal insulation and energy saving area for its excellent performance of heat insulation and sound insulation. However, its main defects including the weak strength, large value of dry shrinkage and high water absorption, hamper its application.

The strength of foamed concrete is influenced by many factors, such as dry density, porosity, raw materials and curing methods, among which dry density is the most influential factor linearly affecting the strength of foamed concrete [1]. Zhu et al. (2013) studied the relationship between pore structure and thermal conductivity of foamed concrete, and found that thermal conductivity increases as dry density increases, resulting a decrease of thermal insulation. They concluded that the relationship between strength and thermal conductivity of foamed concrete is nonlinear. The higher the strength of foamed concrete is, the worse the performance of heat insulation and sound insulation becomes.

The experimental methods and mathematical models have been developed to predicting the strength and other properties of foamed concrete [4]. These mathematical models include the strength-mixing proportion model, the strength-porosity model, the gel-space model, etc [3]. Many methods have been proposed to determining the mixing proportion of foamed concrete, one of the most critical factors affecting its strength, but there is not a generally accepted standard [5]. A traditional approach to calculate the strength of the ordinary concrete uses the Bowromi equation, based on which the mixing proportion can be determined with absolute volume method and bulk density method [6]. However, this approach cannot be applied to foamed concrete for certain factors such as foaming agent. Zhu et al. (2013) conducted orthogonal experimental design with either a single or multiple factor(s) for the mixing proportion of foamed concrete and established the linear regression model for it [7].

In this paper, BP neural network is used to predict the performance of foamed concrete through its autonomous training model. It is then retrained to meet the requirements of the mixing proportions under multiple working conditions.

2. BP Neural Network Theory

2.1. *Summary of BP neural network*

BP neural network uses the steepest descent algorithm to simulate nonlinear mapping relationship through several times of training, as long as the number of the neurons in the hidden layer is appropriate.

The illustration of BP neural network method is shown in Figure 1.

As shown in Figure 1, $P_1, P_2,..., P_i$ are input values $a2_1, a2_2,..., a2_k$ are predictive values, ω_{ij} and ω_{jk} are weights, and $b1_j$ and $b2_k$ are thresholds. BP neural network is like a function system, where the input and output parameters can be treated as independent and dependent variables. Suppose the node number in the input layer and output layer is i and k, respectively, then BP neural network model simulates a nonlinear function mapping from these i independent variables to k dependent variables.

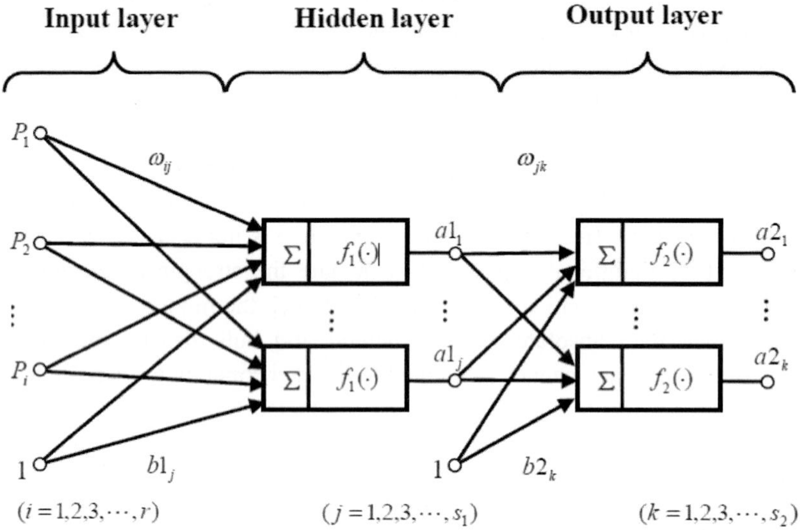

Fig. 1. The workflow of BP neural network method.

2.2. *Activation function*

The activation function $f(\cdot)$ is the core of a neural network, the functions of which include (1) control the effects of input variables on the output variables; (2) create a functional mapping from input variables to the output variables; and (3) transform the infinite domain of input variables into the finite domain of output variables.

In this paper, $f_1(\cdot)$ takes the S type function (logsig function)

$$A = f(n) = \frac{1}{1+e^{-n}} \tag{1}$$

$f_2(\cdot)$ takes the linear function (purelin function)

$$A = f(n) = n \tag{2}$$

and another type of activation function takes the form

$$A = f(n) = \begin{cases} 1, & n \geq 0 \\ 0, & n < 0 \end{cases} \tag{3}$$

3. Case Study

3.1. *Analysis of experimental data*

We use the experimental data from Zhu et al. (2013), who conducted the experiments to investigate the relationship between pore structure and thermal conductivity of foamed concrete, to test the effectiveness of the BP neural network model. The method of testing compressive strength and thermal conductivity are as follows.

In the test, two groups were taken out randomly as control samples from twelve groups of experimental data, say No.3 and No.9, and the remaining ten groups are used as training samples. The data contains six parameters. They are four proportioning parameters (cement, fly ash, water and foaming agent usage) and two property parameters (compressive strength and thermal conductivity coefficient). The BP neural network model is used to associate the compressive strength and thermal conductivity coefficient, the dependent variables, with the mixing proportion values, the independent variables. Comparing the output values of the model with the control data, the weights of the neural network can be adjusted adaptively until reaching the preset accuracy. The results are shown in table 1.

Table 1. Experimental data of different mix proportions, compression strength and thermal conductivity coefficient of foamed concrete

Sample number	**Design of mix proportion**				**Compression strength/** MPa	**Thermal conductivity** / $[W\cdot(m\cdot K)^{-1}]$
	cement/g	**flyash/g**	**water/g**	**foaming agent/g**		
1	800	200	360	0	49.9	0.213
2	800	200	360	2.7	3.82	0.098
3	800	200	360	5.4	1.24	0.066
4	800	200	360	8.1	0.41	0.054
5	650	350	380	0	48.8	0.201
6	650	350	380	2.7	3.39	0.095
7	650	350	380	5.4	0.97	0.061
8	650	350	380	8.1	0.39	0.046
9	500	500	400	0	47.3	0.196
10	500	500	400	2.7	2.58	0.093
11	500	500	400	5.4	0.70	0.057
12	500	500	400	8.1	0.34	0.043

3.2. *Network design*

Ten training groups and two control groups were randomly selected using rand function in Matlab, Data are normalized by mapminmax function.

$$y = f(x) = (y_{\max} - y_{\min}) \times \frac{x - x_{\min}}{x_{\max} - x_{\min}} + y_{\min} \tag{4}$$

where $y_{\min} = -1$, $y_{\max} = 1$, $x_{\min}$ and $x_{\max}$ are the minimum and maximum values of the original data. After the normalization, data is in the domain of $[-1,1]$. This avoids the network prediction error caused by the difference in the magnitude of each dimension data.

The BP neural network is implemented by using the function newff in the neural network toolbox in Matlab. The parameters are initialized as follows. The number of hidden layer is set to 5. The activation function of the hidden layer is set to 'logsig'. The activation function of the output layer is set to 'purelin'. The learning function is set to 'trainlm'. The expected error is set to 0.0001. The learning efficiency index is set to 0.001, and the maximum iteration number is set to 10000.

After normalizing data, the train function is used to train the BP neural network, and the predicted output is obtained using the function sim. Finally, the output is transformed to the original scale and the prediction error is calculated.

3.3. *Analysis of experimental results*

(1) Results of BP Neural Network model on training data

The analysis of BP neural network training error and BP neural network regression are shown in Figure 2 and Figure 3.

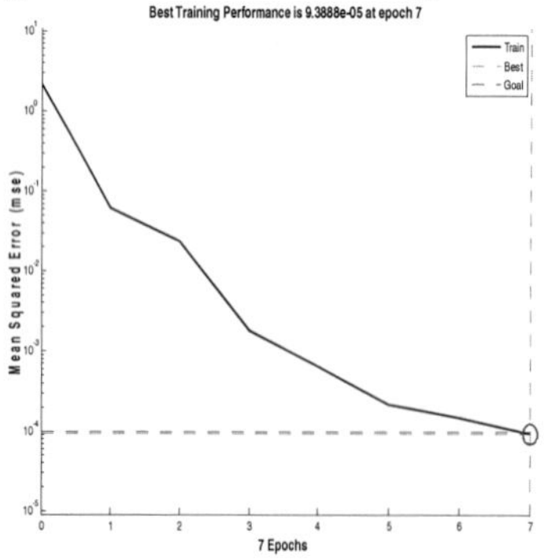

Fig. 2. The training error of BP neural network model

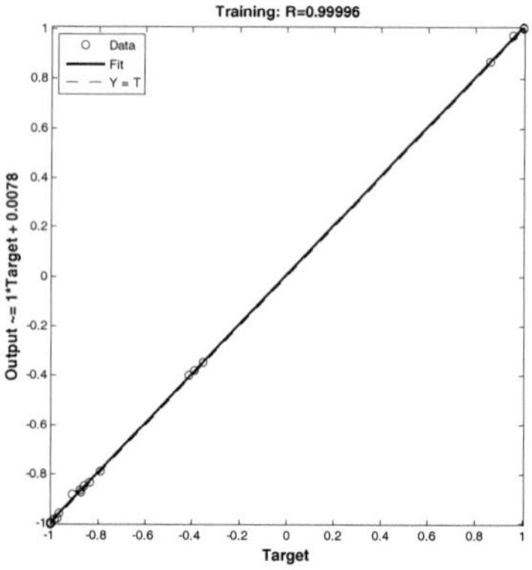

Fig. 3. BP neural network regression

The training error curve of BP neural network model on the experimental data shows that the prediction error meets the criteria in the 7th step, indicating that the Levenberg-Marquardt algorithm converges fast. The fitted line of BP neural network regression shows its goodness-of-fit for the experimental data. It in return validates the appropriateness of parameters settings. For instance, setting 5 hidden layers, choosing logsig and purelin functions as activation function for the hidden layer and output layer, etc. Table 2 presents the results of BP neural network model on training data.

Table 2. Results of BP neural network model on training data

Group number	**Fitted values**		**actual value**		**Error value**	
	Compression strength/ MPa	**Thermal conductivity /** [W·(m·K)$^{-1}$]	**Compression strength/** MPa	**Thermal conductivity /** [W·(m·K)$^{-1}$]	**Compression strength/** MPa	**Thermal conductivity /** [W·(m·K)$^{-1}$]
1	50.0499	0.2132	49.9000	0.2130	0.1499	0.0010
2	4.1215	0.0985	3.8200	0.0980	0.3015	0.0005
4	0.3620	0.0538	0.4100	0.0540	-0.0480	-0.0002
5	49.2726	0.2021	48.8000	0.2010	0.4726	0.0011
6	3.6528	0.0958	3.3900	0.0950	0.2628	0.0008
7	0.8926	0.0612	0.9700	0.0610	-0.0774	0.0002
8	0.3425	0.0464	0.3900	0.0460	-0.0475	0.0004
10	3.2273	0.0940	2.5800	0.0930	0.6473	0.0010
11	0.7593	0.0573	0.7000	0.0570	0.0593	0.0003
12	0.3130	0.0435	0.3400	0.0430	-0.0270	0.0005

(2) Analysis of BP neural network model for prediction

The prediction of the compressive strength and thermal conductivity of the control samples can be obtained using the mix proportion values obtained from the training samples. The results are shown in table 3.

Table 3. Analysis of prediction results of BP neural network

		actual value/MPa	**BP network output value**/MPa	**relative error /%**
NO.3	**Compression strength**/MPa	1.2400	1.1531	7.0081
	Thermal conductivity/[W·(m·K)$^{-1}$]	0.0660	0.0615	6.8182
NO.9	**Compression strength**/MPa	47.3000	48.3382	2.1949
	Thermal conductivity/[W·(m·K)$^{-1}$]	0.1960	0.2002	2.1428

It can be seen from Table 3 that the prediction errors of the two predicting samples are less than 8%. One may notice that the prediction error of sample 3 is significantly larger than that of sample 9. This can be explained by the following reasons: (1) the pre-treatment process of data brings in the error; (2) the pre-set training accuracy is not good enough; (3) the BP neural network model brings in certain error.

In allusion to the problems mentioned above, we offer three possible solutions: (1) collect accurate experimental data; (2) set high accuracy precision threshold for the training process; (3) optimize the training process of BP neural network by other algorithms.

4. Conclusions

(1) The BP neural network model was successfully implemented by taking advantage of the neural network toolbox in Matlab. It was able to predicting of the compressive strength and thermal conductivity of foamed concrete.

(2) The analysis of prediction error of BP neural network and regression analysis verifies the effectiveness of selected activation functions and other input parameters. Furthermore, the normalization of training data increases the prediction accuracy.

(3) The results show the high predicting accuracy of BP neural network model. Control group No.9 and No.3 have relative error less than 3% and 8%, respectively, resulting an average relative error of 4.541%. Therefore, BP neural network is an effective method for the prediction of compressive strength and thermal conductivity of foamed concrete. It also has potential applications in other engineering practices.

(4) In the future research, other optimization algorithms can be attempted for the training process and normalization techniques can be adopted to both training and predicting samples, to improve the accuracy and generalization of BP neural network.

References

1. Meng Zhang,Qing-Bo Tian. Research status and Prospect of foamed cement[J].Silicate Bulletin,2014,10:2547-2551.(In Chinese).
2. Ming Zhu,Fang-Gang Wang. Study on the relationship between pore structure and thermal conductivity of foamed concrete[J]. Journal of Wuhan University of Technology,2013,03:20-25.(In Chinese).
3. E. K. Kunhanandan Nambiar,K. Ramamurthy. Models for strength prediction of foam concrete[J]. Materials and Structures,2008,412.

4. Xiang-Yong Guo,Jian-Li Guo.Research Progress on properties of foamed concrete[J]. construction quality,2013,04:17-23.(In Chinese).
5. Xian-Yan Ren,Yu-Rong Zhang. Research status and Prospect of foamed concrete[J]. concrete,2011,02:139-141+144.(In Chinese).
6. Jun-Yu He.Design method of mix proportion of fly ash concrete [J]. Subgrade engineering,2006,02:101-103.(In Chinese).
7. Hong-Yin Zhu.Study on design and performance of foam concrete mix ratio[D].Northwest Agriculture and Forestry University,2013.(In Chinese).
8. Shuang Cong.Neural network theory and application oriented to MATLAB toolbox[M].Anhui:University of Science & Technology China press,2009:8-68.(In Chinese).
9. Xiao-Chuan Wang,Feng Shi. 43 case analysis of MATLAB neural network[M].Beijing: Beihang University press,2013:1-2.(In Chinese).
10. GB/T 10294-2008,Determination of thermal resistance and related properties of thermal insulation materials by thermal barrier method[S].Beijing: China Standard Press,2009.(In Chinese).
11. Jong,In Kim,Doo Kie Kim. Application of neural networks for estimation of concrete strength[J]. KSCE Journal of Civil Engineering,2002,64.
12. Yeh, I. Modeling Concrete Strength with Augment-Neuron Networks[J].Journal of Materials in Civil Engineering , November 1998, Vol.10, No.4:pp.263-268.
13. Jie You,Yi Che. Prediction of existing building concrete strength based on BP neural network[J]. Journal of Architectural Science and Engineering,2011,01:70-75.(In Chinese).
14. Wen-Shan Pi,Hong-Biao Zhou. Prediction of compressive strength of concrete based on BP neural network[J]. Low temperature construction technology,2011,04:14-16.(In Chinese).
15. Hong Li,Tao Peng. Prediction of compressive strength of concrete based on BP and RBF neural network[J]. Journal of Wuhan University of Technology,2009,08:33-36.(In Chinese).
16. Cheng Yu.Research on the relationship between concrete pore structure and strength based on artificial neural network[D].Zhejiang University,2006.(In Chinese).
17. Xin-Ping Hu,Xiang-Yu Li. Study on thermal conductivity and strength of foamed concrete[J]. Silicate Bulletin,2014,11:2940-2945.(In Chinese).

Analysis of flow induced noise sources in a horizontal Francis turbine

Hong-Ming Zhang*, Li-Xiang Zhang, Ji Yao

Department of Engineering Mechanics, Kunming University of Science and Technology, Kunming, Yunnan, China

**E-mail: hmzhang8888@126.com*

www.kmust.edu.cn

The paper presents numerical study of flow induced noise sources in a horizontal Francis turbine. The spiral case, the runner and the draft tube are used as noise sources. The sound pressure, power spectrum density and sound pressure level caused by the various noise sources at the receiver are numerically simulated. Analysis was performed by OpenFOAM code. FW-H acoustic model is used to calculate the sound pressure signal at the receiver. The numerical simulation results show that the spiral case is the main noise source of the horizontal Francis turbine.

Keywords: Flow Induced Noise Sources; Horizontal Francis Turbine; OpenFOAM.

1. Introduction

The noise of fluid machinery can cause adverse effects on the working environment. The horizontal Francis turbine is mainly used in small hydropower station. Its main noise is the structure vibration noise and the noise induced by internal flow. Structural vibration noise can be effectively reduced by improved manufacturing and installation processes. But the noise induced by internal flow due to the internal flow mechanism is very complicated, it is attracting more and more scholars to carry out research In recent years. Wen et al.[1] analyzed the noise source of a centrifugal turbocharger compressor using CFD techniques. Their research results show that the shedding of thick boundary layer and the vortex motion result in the high-level noise. The main noise sources are the regions of the separate flows, the secondary flows and the wake flows. Liu et al.[2] investigated the performance, pressure pulsations, hydraulically generated vibration, and noise level in a centrifugal pump. Their research results show that the influences of the sound pressure of centrifugal pumps on the structure appear insignificant. Wang et al. [3]calculated the flow noise caused by dipole source from pressure fluctuation on the wall of guide vane and analyzed the relationship between the pressure fluctuation and flow noise. Their results show that the end of the guide vane and the back of the blade inlet are main dipole noise source.

In this paper, the flow induced noise sources in a horizontal Francis turbine is simulated using OpenFOAM code. The spiral case, the runner and the draft tube will be used as noise sources. A formulation of the Ffowcs-Williams–Hawkings analogy will be employed in OpenFOAM code.

2. Numerical Method

2.1. *Governing equations and acoustics model*

The analysis of flow induced noise source of horizontal Francis turbine is carried out by using CFD and Lighthill sound analogy theory. The governing equations include the three-dimensional unsteady Reynolds-average Navier-Stokes equations and FW-H equations.

Navier-Stokes equations:

Continuity equations:

$$\frac{\partial}{\partial x_i}(\rho u_i) = 0 \tag{1}$$

Where ρ is the density of the liquid, u_i is the Reynolds averaged relative velocity.

Momentum equations:

$$\frac{\partial}{\partial t}(\rho u_i) + \frac{\partial}{\partial x_j}(\rho u_i u_j) = -\frac{\partial P}{\partial x_i} + \frac{\partial}{\partial x_j}(\mu \frac{\partial u_i}{\partial x_j}) + \frac{\partial \tau_{ij}}{\partial x_j} \tag{2}$$

where, P is the static pressure; μ is the viscosity of the liquid; τ_{ij} is Reynolds stress.

FW-H equations[4]:

$$\frac{1}{c_0^2}\frac{\partial^2 p}{\partial t^2} - \nabla^2 p = \frac{\partial^2}{\partial x_i \partial y_j}\{T_{ij} H(f)\}$$

$$-\frac{\partial}{\partial x_i}\{[P_{ij} n_j + \rho u_i (u_n - v_n)]\delta(f)\} \tag{3}$$

$$+\frac{\partial}{\partial t}\{[\rho_0 v_n + \rho(u_n - v_n)]\delta(f)\}$$

where c_0 is the far-field sound speed, p' is the sound pressure at the far field. T_{ij} is the Lighthill stress tensor, $H(f)$ is the Heaviside function, P_{ij} is the compressive stress tensor, n_j is the unit normal vector, u_n fluid velocity component normal to the surface f=0, u_i is the fluid velocity component in the X_i direction, V_n is the surface velocity component normal to the surface, $\delta(f)$ is the Dirac delta function.

2.2 *Numerical algorithm*

In this paper, the analysis of flow induced noise sources contains two steps. First, a time-accurate flow solution is generated from which time histories of the relevant variables on the selected source surfaces are obtained. Then sound pressure signals at the user-specified receiver locations are computed using the source data collected during the first step. In the OpenFOAM, the LES turbulence model is adopted and the pressure-velocity coupling is handled via a Pressure Implicit with Splitting of Operators (PISO) procedure. When the flow field calculation is completed, we can use acoustics model FW-H to calculate the sound pressure signal of each receiver. Analysis of power spectrum density and sound pressure level can be carried out using the obtained sound pressure signal.

3. The Calculation Domain, Boundary and Initial Conditions

Fig. 1 shows the calculation domain of a horizontal Francis turbine. The main parameters are listed in Table 1. In our study, the spiral case, the runner and the draft tube are defined as the noise source zones. A acoustic pressure signal receiver is placed on the right side of the turbine, which is located 10 meters from the center of the runner.

All computational grids are generated by CFD ICEM, including four domain of the spiral case, the guide vane cascade, the runner and the draft tube. The total grid number is 20700000. The runner is rotating domain and others are stationary domains. The grid number of runner domain is 12,300,000.

In the algorithm, the rotation speed is 500 rpm, the time step is the time that the runner turns round the 1/360 ring. Thus, the time step was 0.0003333 s. The inlet bulk mass flow rate and outlet pressure boundary conditions are applied. In the present study, the inlet bulk mass flow rate is 10550kg/s, the outlet pressure is 0Pa. All calculations are done on the SUGON high performance computer, using the 224CPU core and it takes 36 hours.

Table 1. Parameter of horizontal Francis turbine.

Diameter of runner(cm)	110
Number of runner blade	16
Number of stay vane	16
Number of guide vane	16
Rated head (m)	44
Rated power (kw)	4189
Rotational speed (rpm)	500
Rated flow rate (m^3/s)	10.55

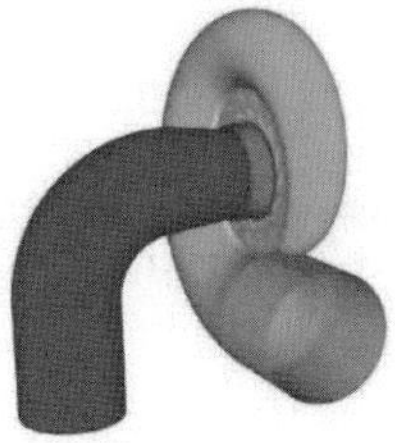

Fig. 1. Horizontal Francis turbine.

4. Results and Discussion

The acoustic pressure, sound power spectrum density and sound pressure level are calculated at the receiver due to different noise sources. Fig. 2 shows the acoustic pressure distribution caused by different noise sources at the receiver. The peak value of acoustic pressure caused by the spiral case noise source is larger. It is similar to the sound pressure distribution caused by all noise sources. The acoustic pressure caused by the runner and draft tube noise source is relatively small. Fig. 2d has a certain periodicity. Fig. 3 shows the acoustic power spectral density distribution caused by different noise sources at the receiver. It can be seen the acoustic power spectral density distribution caused by the spiral case and runner noise source over a wide frequency band. But the amplitude of power spectrum density distribution caused by spiral case noise source is large. The power spectrum density caused by the draft tube noise source is only a very low frequency range and the amplitude is very small. Fig. 4 shows the 1/3 octave band sound pressure level distribution caused by different noise sources at the receiver. It can be seen that the sound pressure level caused by the spiral case noise source is larger and reached 145dB. The sound pressure level caused by the runner noise source is also up to 117dB. The sound pressure level caused by the draft tube noise source is relatively small. Compared with the sound pressure level spectrum curve caused by all the noise sources, the spiral case is the main noise source of the horizontal Francis turbine.

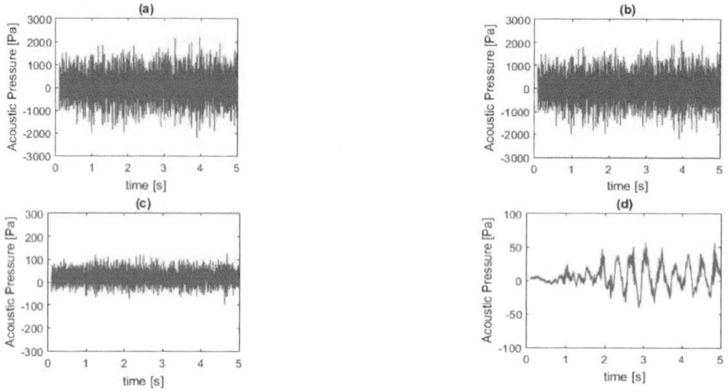

Fig. 2. The acoustic pressure distribution caused by different noise sources at the receiver. (a) All noise sources, (b)Spiral case noise source, (c) Runner noise source, (d) Draft tube noise source.

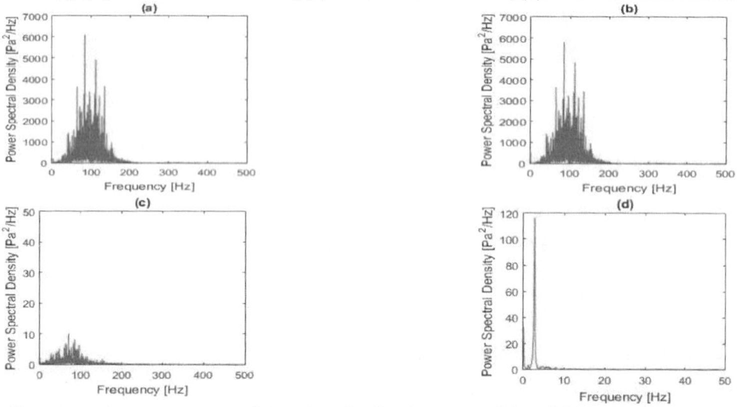

Fig. 3. The acoustic power spectral density distribution caused by different noise sources at the receiver. (a) All noise sources, (b) Spiral case noise source, (c) Runner noise source, (d) Draft tube noise source.

5. Conclusion

In this paper, the problems of flow induced noise sources in a horizontal Francis turbine is analyzed using numerical methods. The spiral case, the runner and the draft tube are used as noise sources. The acoustic pressure, power spectrum density and sound pressure level caused by the various noise sources at the receiver are numerically simulated. The numerical simulation results show that the spiral case is the main noise source of the horizontal Francis turbine.

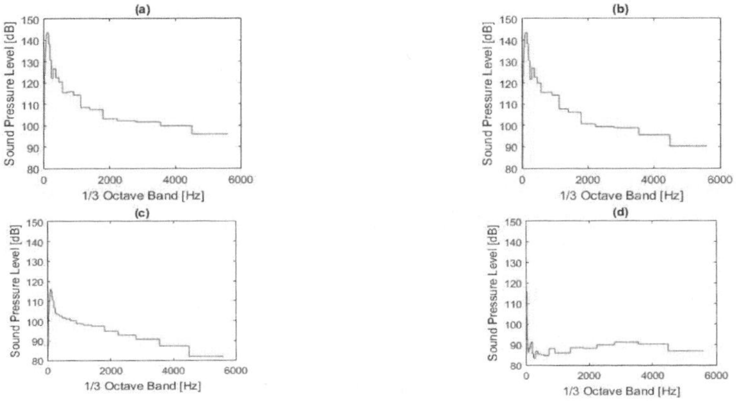

Fig. 4. 1/3 octave band sound pressure level distribution caused by different noise sources at the receiver. (a)All noise sources, (b)Spiral case noise source, (c) Runner noise source, (d)Draft tube noise source.

Acknowledgments

This work was financially supported by the National Natural Science Foundation of china (No. 51279071) and the Natural Science Foundation of Kunming University of Science and Technology (No. 20149x16).

References

1. Wen H, Sang J, Zhong Q, et al. Optimization and Analysis of Aerodynamic Characteristics and Noise Source in a Centrifugal Compressor with Vaned Diffuser[J]. U.P.B. Sci. Bull., Series D, Vol. 77, Iss. 4, 2015.
2. Liu H, Ding J, Dai H, et al. Numerical Research on Hydraulically Generated Vibration and Noise of a Centrifugal Pump Volute with Impeller Outlet Width Variation[J]. Mathematical Problems in Engineering, 2014.
3. Wang huizhi, Zhou daqing, Zhang languo. Numerical simulation of flow noise in Francis turbine with low specific speed[J]. Chinese Journal of Water resources and Power, 2014, 32(8): 153-156. (In Chinese).
4. Casalino D, Jacob M, Roger M. Prediction of rod-airfoil interaction noise using the Ffowcs-Williams-Hawkings analogy[J]. AIAA journal, 2003, 41(2): 182-191.

An evaluation index system for cable-stayed bridge from multi-information sources

Xiao-Ling Liu*, Qiao Huang, Yuan Ren
School of Transportation, Southeast University,
Nanjing, 210096, China
**E-mail: liuxiaolingseu@163.com*
www.seu.edu.cn

Evaluation data sources for modern large-scale bridges include manual inspection information and long-term monitoring data and nondestructive testing results. It is generally difficult to reflect the overall condition of structure from single source. Therefore a comprehensive evaluation index system based on multi-data sources was conducted for making better maintenance strategy. A cable-stayed bridge in Nanjing city was chosen as an example in this paper to illustrate the operation condition, evaluation result based on annual inspection report and the current problems of long-term monitoring system in detail. The system mainly contained the defects information from manual inspection in assistance of long-term monitoring data, and the evaluation result can be used as the basis for bridge maintenance and the data renewal of the base index.

Keywords: Cable-Stayed Bridge; Evaluation System; Maintenance; Condition Evaluation; Long-Term Monitoring.

1. Introduction

Long-span cable-stayed bridges are mostly built at the key places of traffic to guarantee the transportation safety. However, they face great challenges from several aspects, including rising traffic volume, fatigue damage of the component, material aging and emergency incidents in daily operation. The demand of the safety, feasibility and durability of bridges calls for a scientific evaluation system [1] for them. Currently, many cable-stayed bridges in the world are installed with management and long-term monitoring systems [2], which are used to monitor the operation condition of the bridge and collect data for maintenance strategy. The evaluation data sources include manual inspection information, long-term inspection data and nondestructive testing results. The data obtained from visual manual inspection or measurement of portable instrument are generally subjective [3]. The reliability of long-term inspection data depends on the quality and accuracy of the sensor. However, the identification of damage based on monitored data still remains a technically difficult problem due to the complexity and uncertainty of the bridges [4]. Therefore, condition evaluation based on single data source is generally unable to reflect the whole technical condition of the structure. Instead, the structural long-term monitoring system and the inspection maintenance management system should be combined to investigate the bridge evaluation system.

The Nanjing No. 3 Yangtze River Bridge was built in 2005, which is a cable-stayed bridge with steel tower and steel box girder. The inspection system and long-term monitoring system of the bridge have been in service since 2005. The existing problems in manual inspection system and long-term monitoring system of the bridge were illustrated. Furthermore, a comprehensive evaluation system for long-span cable-stayed bridge was established by considering multi-information sources.

2. Conditions of Manual Inspection System

According to the summary of regular inspection reports from 2009 to 2013, some problems exist in the manual inspection.

(1) The manual inspection system is commercial software of independent operation, and its data has not been shared with long-term inspection system yet;

(2) Because of the limitation in both inspection methods and professional knowledge of the workers, some hidden and hard-to-be-reached parts are not completely inspected. Moreover, missing damage number and area add difficulty to the bridge evaluation. Besides, the inspection content and emphasis on bridge by different inspection organizations are different.

(3) Cable-stayed bridge consists of 17 components according to the "Standards for Technical Condition Evaluation of Highway Bridges (JTG/T H21—2011)" (SCE2011). However, cable-stayed bridge is a complicated system with the components more than that of general bridge. Some components, such as damping device, tuyere and dehumidification system, are not listed in SCE2011.

3. Operation Condition of Long-Term Monitoring System

The long-term monitoring system mainly includes sensor subsystem, data collection subsystem, data processing and control subsystem and data visualization subsystem. There are more than 600 sensors for static property, dynamic property and environmental property of the structure. Some problems were found in the system as follow:

(1) Increasing data distortion

As quality problem and improper protection during construction, the damage rate of the inspection sensor rises, increasing the difficulty of extracting the damage index. Data distortion occurs in all the parts. According to the operation condition of the bridge, some representative data distortion features are summarized and shown in Table 1.

Table 1. Data distortion features.

Distortion type	Detailed features	Feature descriptions
Single data distortion	Single data missing	Common problem; Maximum proportion can reach 20%.
	Outliers	Happens in several sensors
Continuous data distortion	Data bias	Mostly happens in cable force and stress sensors
	Data missing	Happens for system sudden failure
	Data disorders	Sensor failure

Two kinds of distortion are given in Table 1, which are single-point data distortion and continuous-point data distortion. In more detail, data missing and outliers are common. While the continuous data distortion is complex and It requires calibration of the sensors and inspection data. In the actual operation, with accumulation of massive data, most data is lack of check and renewal.

(2) Immature evaluation method

Currently, it is still a technical problem to realize the damage diagnosis from monitoring data, and the individual monitoring evaluation [5] cannot completely reflect the comprehensive property of the whole bridge.

4. Establishment of Evaluation System with Multi-source Integration

The following conclusions can be obtained from the above analysis: although the subjectivity of manned inspection is strong, its result is of great value for the determination of subsequent maintenance scheme; the long-term monitoring inspection data has great unreliability and inconsistency, but the beneficial safety information can also be obtained with the verification of finite element model calculation.

A comprehensive evaluation index system with consideration of multi-information source is established, which is mainly based on the manual inspection information with the assistance of long-term monitoring data.

Based on the description of the comprehensive evaluation system in the last section, the following evaluation framework is established, which is shown in Fig. 1.

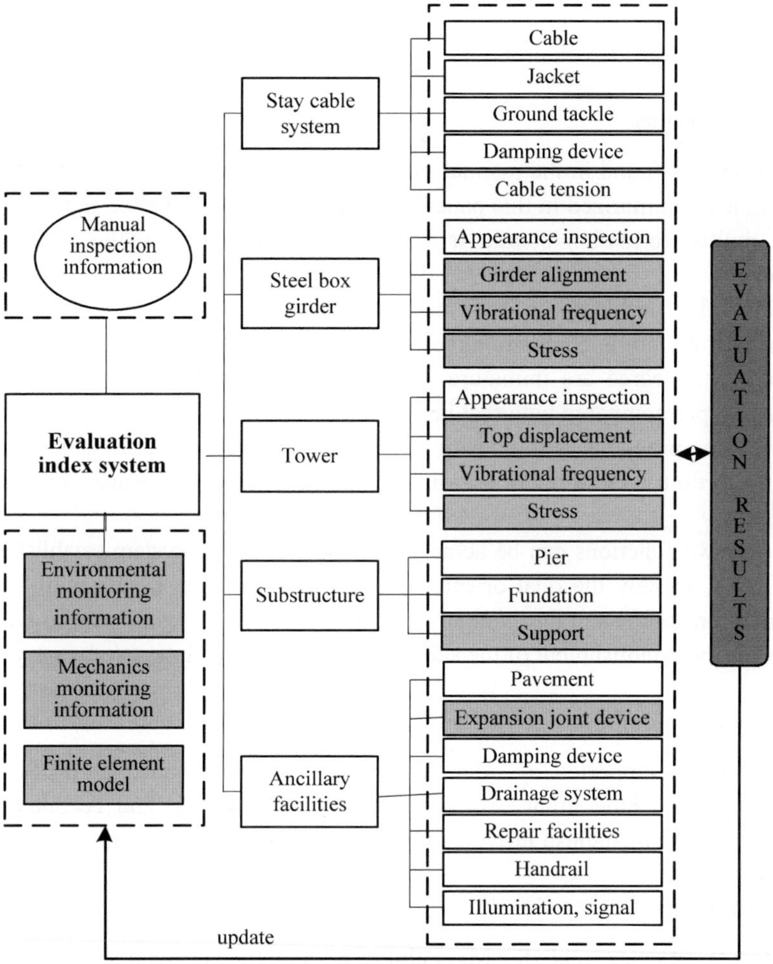

Fig. 1. The comprehensive evaluation index system.

Two functions are achieved by this evaluation framework.

(1) The index system is established with consideration of multi-information source. As seen in Figure 1, the hierarchical index system established in this paper can be divided into three layers. The 1st layer consists of some kinds of large components of cable-stayed bridge. The 2nd layer includes some sub-components and long-term monitoring data. And the 3rd layer has detailed damage indexes.

(2) Evaluation result has dual orientations. The evaluation result can be divided into total score and individual score. This can be used to provide the

basis for bridge maintenance scheme determination and to assist the data renewal of the ground-floor index.

5. Conclusions

The problems found in the manual inspection system and long-term monitoring system are summarized in this paper. A comprehensive evaluation index system with multi-information source is established. The main conclusions are as follows:

(1) The problems in the manual inspection system include: manual inspection system is commercial software of independent operation; Inspection content and degree are immature; there are large amount of component and damage types compared with that in SCE2011

(2) The problems of the health monitoring system include: large amount of data distortion, lack of relation to bridge maintenance and immature evaluation method.

(3) Two functions can be achieved by the evaluation system established in this paper, namely the dual oriented property of the index system with multi-information source and the evaluation result. The research result can provide the basis for the establishment of the subsequent multi-source integrated system.

Acknowledgment

This work is supported by grant 51208096 of the National Natural Science Foundation of China, grant 2014Y02 of the Major Scientific and Technological Special Project of Jiangsu Provincial Communications Department.

References

1. G. E. Tsiatas and S. M. Palmquist, Fatigue evaluation of bridges, *Probabilistic Eng. Mech.*, 14, 189 (2014).
2. H. Wenzel, *Health Monitoring of Bridges*, (John Wiley & Sons, NY, 2008).
3. S. Lee and N. Kalos. Bridge inspection practices using non-destructive testing methods, *J. Civil Eng. Man.*, 21, 654 (2015).
4. J. Collins, G. Mullins, C. Lewis, et al, State of the practice and art for structural health monitoring of bridge substructures, *Phil. Trans.*, 365, 515 (2014).
5. M. Xu, Q. Shen, Y. Liu, et al, Large-scale bridge structural health monitoring system based on smart sensor networks, in *Proc. Int. on Transportation Information and Safety* (*ICTIS 2011*), (Wuhan, China, 2011).

Effect of water to binder ratio on C30 high titanium blast furnace slag self-compacting concrete workability

Chen Wei, Lin-Ze Li, Wei Wang, Song Zhong
College of Civil Engineering and Architecture, Panzhihua University, Panzhihua 617000, Sichuan, China
E-mail: 514806690@qq.com
www.pzhu. Edu.cn

The biggest difference between Self-Compacting Concrete(SCC) and ordinary concrete is mainly reflected in the special requirements of Self-Compacting Concrete(SCC) for workability, while the water to binder ratio is one of the main factors that affects workability. In this paper, industrial solid waste of high titanium blast furnace slag in Panxi region is used to compound C30 SCC to study the effect of water to binder ratio on its workability. The test results show that: water to binder ratio has a great impact on the workability of high titanium slag SCC. When water to binder ratio is between 0.44 and 0.46, the compounded high titanium slag SCC has a better workability.

Keywords: Self-Compacting Concrete (SCC); High Titanium Blast Furnace Slag; Workability; Water to Binder Ratio.

1. Introduction

Self-compacting concrete (Self-Compacting Concrete (SCC) in Japanese English, and Self-Consolidation concrete (SCC) in American English) refers to one of concrete that has the characteristics of high fluidity, uniformity and stability, and can flow and fill the template space under its own weight without external vibration when pouring [1]. After hardening, the workability of SCC is similar to that of ordinary concrete, and the biggest difference between SCC and ordinary concrete is mainly reflected in the special requirements of SCC for workability. SCC's workabilities mainly include three aspects: filling ability, passing ability and segregation resistance.

Filling ability: it is one of the workability that SCC mixtures can only rely on its own gravity to completely fill the template molded. Slump expansion degree test can be used to study it.

Passing ability: it is one of the workability that SCC mixtures can evenly go through the narrow gap. J-Ring and L-instrument tests can be used for verifying it.

Segregation resistance: it is one of the workability that the various components maintain a uniform dispersion within SCC mixtures. It can be tested by using the segregation ratio tube test to make the experiments.

High titanium blast furnace slag (referred to as high titanium slag) is the waste residue generated in the smelting process of vanadium-titanium magnetite, which is produced by Panggang Group. The study on high titanium slag composition found that high titanium slag does not contain dicalcium silicate, so silicate does not decompose; there is no free calcium oxide and samples have good structural stability with no powder and no bursting; sulfur is less than 1% and high titanium slag does not contain FeS and MnS. Therefore, high titanium slag has the features of structural stability, good workability and is not easy to decompose, so it can replace ordinary natural sand as concrete aggregate to prepare concrete [2]. In this paper, the high titanium slag is used to compound C30 SCC to study the effect of W/B on SCC's workability.

2. Raw materials

2.1. *Cement*

Adopt the P.O.42.5R produced by Panzhihua Ruifeng Building Materials Company as the experimental cement.

Table 1. Physical and chemical properties of cement

Fineness(80μm residue on square hole sieve)%	Water demand for normal consistency %	Setting time h:min		Compressive strength Mpa		Flexural strength Mpa		Stability (Boiling)
		Initial setting	Final setting	3d	28d	3d	28d	
0	29.6	167	240	26.1	49.5	4.6	6.9	Qualified

2.2. *Chemical admixture*

Take polycarboxylate superplasticizer prepared by Panzhihua Jiyuan Chemical Admixture Company as the experimental superplasticizer. Solid content is 40% and water-reducing rate is over 30%.

2.3. *Coarse and fine aggregates*

Use high titanium blast furnace slag produced by Panzhihua Huanye Company as the coarse and fine aggregates, and its physical properties are shown in Table 2 and Table 3.

Table 2. Physical properties of coarse aggregate

Apparent density Kg/m3	Bulk density Kg/m3	Crush index /%	24h Water absorption	Saturated surface-dry water absorption
2830	1390	17.8	12%	4%

Table 3. Physical properties of fine aggregate

Apparent density Kg/m3	Bulk density Kg/m3	Crush index /%	Fineness modulus	Powder content	24h Water absorption	Saturated surface-dry water absorption
3130	1770	21.3	2.6	10.8%	12%	11.5%

2.4. *Fly ash*

Take the Grade I Class F fly ash producted by Panzhihua Gangcheng Group Comprehensive Industrial Filiale as the experimental fly ash. The technical index is shown in Table 4.

Table 4. Technical Index of Fly Ash

Fineness(45μm residue on square hole sieve, method: negative pressure sieve method) (%)	Water demand ratio (%)	Loss on ignition (%)	Water content (%)	Stability
9.8	102	3.7	0.6	Test samples after boiling: no cracks, no curved bottom. Rating: Qualified

3. Mix Proportion of C30 High Titanium Slag SCC

W/B is a major factor in determining the strength of concrete in mix design, which identifies key indicators of binder content and amount of the slurry. The greater of the W/B indicates more water and less binder content, which will lead to the bleeding of concrete. In other words, because of lack of bonding material, concrete is not dense and filling ability is poor, thus affecting the flowability of concrete. The smaller of W/B indicates less water and more binder content, which will lead to too little free water, insufficient paste, and too dry concrete, so the flowability of concrete is poor, and concrete even does not flow and so on. Or from the theoretical explanation of flow resistance of concrete, if W/B is smaller, cement plays a larger role in bonding, and concrete aggregate can not resist self-compacting flow [3].Thus, when the W/B is in a certain range, high titanium blast furnace slag SCC achieves optimal performance.

In this experiment, W/B is respectively 0.50,0.48,0.46,0.44 and 0.42 to test the workability of high titanium slag SCC to evaluate the impact of W/B on the workability of SCC and determine the optimal W/B. The mixing ratio is shown in Table 5.

Table 5. The mix proportion of water to binder ratio effect on High Titanium Slag SCC workability

Experimental No.	High titanium slag gravel kg/m3	High titanium slag sand kg/m3	Cement kg/m3	Fly ash kg/m3	Chemical admixture g/m3	Water kg/m3	W/B
1	730	1095	309	104	76.1	273	0.50
2	730	1095	313	105	76.1	267	0.48
3	730	1095	317	106	76.1	262	0.46
4	730	1095	322	108	76.1	258	0.44
5	730	1095	327	109	76.1	249	0.42

4. Experimental Results and Analysis

Make the slump extension degree, T500, L instrument flow, orimet flow rate and segregation resistance experiments on SCC of different mix ratios. The experimental results are shown in Table 6.

Table 6. Experimental result

Number	W/B	Slump degree/ expansion degree mm	T500 /s	L instrument /s	L instrument H2/H1	Orimet /s	The ratio of separation /%
1	0.42	250/500	5.8	11.6	4/14	137	11
2	0.44	255/650	5.2	8.6	6/12	99.4	8.6
3	0.46	265/700	4.2	4.4	7/10	55.5	8.6
4	0.48	270/715	3.9	3.4	6/11	61.3	9.3
5	0.50	260/720	3.3	5.6	3/15	74.1	12

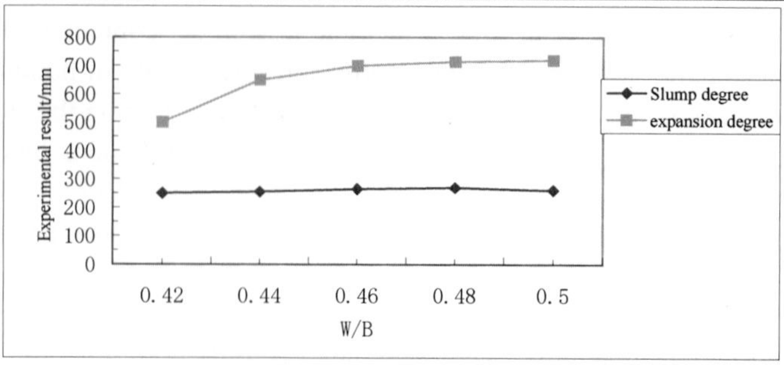

Fig. 1. Slump extension degree

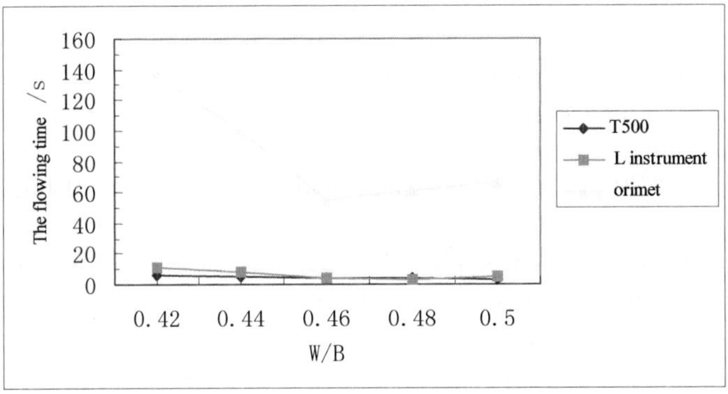

Fig. 2. The flowing time of T500, L instrument and orimet instrument

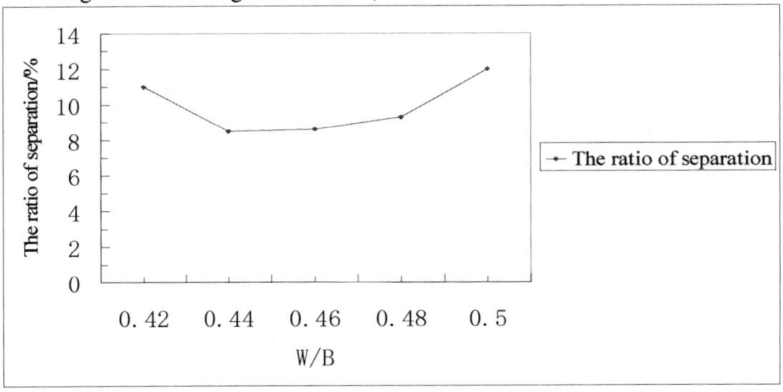

Fig. 3. The ratio of separation

From Figure 1,it can be drawn that with the increase of W/B, flowability of high titanium slag SCC is not reinforced. When W/B is between 0.42 and 0.44, the flowability increases the maximum. Between 0.44 and 0.5, flowability changes little. So it can be concluded that when W/B is between 0.44 and 0.46, the flowability of high titanium slag SCC is better. From Figure 2,it can be drawn that with the increase of W/B, the flowability, filling ability and passing ability of high titanium slag SCC reinforce, and when W/B is 0.46, it is optimal. Analysis showed that when W/B is 0.46-0.50, since the W/B increases, the binder content reduces and water becomes more, so concrete aggregate is not well wrapped by slurry, thus it can not carry concrete to flow, and flowing time in L instrument and orimet instrument becomes longer. From Figure 3, it can be drawn that too large or too small W/B will cause a certain degree of segregation and bleeding of high titanium slag SCC. When W/B is between 0.44 and 0.48, segregation resistance is optimum.

5. Conclusion

(1)Through making experiments on C30 high titanium slag SCC of different water to binder ratios, it is obtained that W/B has a greater impact on the workability of high titanium slag SCC.

(2)When W/B is between 0.44 and 0.46, the workability of high titanium slag SCC is better

(3)As the high titanium slag itself is rough, its flowability is poor and W/B is too small. So the adhesive force between each aggregate increases and concrete can not flow automatically. Therefore, the W/B of high titanium slag SCC at the same level is higher than the W/B of average SCC.

References

1. Liu N. Study on preparation and shear properties of self-compacting concrete [D]. Hunan: Hunan University, 2015.
2. Huang SH, Chen W. Application of high titanium slag on concrete materials [J]. New Building Materials. 2006.
3. Luo Y. Optimization research on mix parameters of medium-low grade SCC [D]. Hunan University, 2008.

Experimental study on the feasibility of hydraulic prestressed concrete filled steel tube

Jin-Lan Wang[1,*], An-Guo Tian[2], Xiao-Ning Cai[2], Xin-Xin Gu[1], Zheng Zhao[1]

[1]*School of Mechanics & Civil Engineering, China University of Mining and Technology, Xuzhou, Jiangsu, China*

[2]*School of Civil Engineering, Huaihai Institute of Technology, Lianyungang, Jiangsu, China*

[*]*E-mail: wjl1490770311@163.com*

A new production technique for hydraulic prestressed concrete filled steel tube is proposed. First of all, the experimental research on the loss of prestress is carried on, showing the feasibility of the method of applying hydraulic prestress. Then experimental study on hydraulic prestressed concrete filled steel tubular short column and concrete filled steel tubular short column under axial loads is conducted. The results show that the method can realize exerting active quantitative prestress of concrete filled steel tube, at the same time, applying prestress value is stable and reliable. Applying hydraulic prestress to the concrete filled steel tube, it can take full advantages of steel tube and concrete, improving the mechanical properties of the concrete filled steel tubular column, enhancing compressive bearing capacity. The results can provide experience in the subsequent construction and design.

Keywords: Prestress; Concrete Filled Steel Tube; Hydraulic Pressure; Feasibility Guidelines.

1. Introduction

Engineering structure toward heavy-duty, large cross, towering, using conventional reinforced concrete columns make a larger cross-sectional dimension. To improve this problem, using concrete filled steel tube column can reduce about 50% in the case of approximately the same amount of steel. The advantage of concrete filled steel tube column is high bearing capacity, convenient construction, good plastic toughness and preventing the shear failure of short column [1-3]. However, both the steel and concrete materials have different deformation properties, resulting in deformation of steel and concrete is not synchronized, so that core concrete cannot be in a three-compressed state from the outset, cannot give full play to the advantages of the compressive strength of concrete, the tensile strength of steel. The prestressed concrete filled steel tube can solve the shortage of the ordinary steel for lake of restrictions of concret. The key to the prestressed concrete filled steel tube is reasonable and effective applying prestress, there are several ways: (1) Traditional post-tensioned. Due to limited strength and deformation of concrete results in very limited prestressing. (2) Self-stressing concrete. This method is influenced by many factors, which is unable to determine the size of prestressed worth [4-5]. (3) Expansion pipe. Prestressed value is too small. Based on the current problems on applying prestress, we propose a novel method of applying hydraulic prestress. The method can realize exerting active quantitative prestress of concrete filled steel tube, at the same time, applying prestress value is stable and reliable.

2. The Implementation Methods of Hydraulic Prestressed Concrete Filled Steel Tube

The steel pipe is using circular section, both ends of it are sealing plate by welded. At 20mm from the sealing plate set two 20mm diameter holes, one is a vent, and the other for the infused and pressurized interface (Figure 1). Experiment applying prestress the pipe through the interface 5, until concrete emerged in No. 2 hole, then perfusion can be completed, can be closed the vent. And then pouring mortar by the pressurized equipment, until monitoring pressure reaching the goal, stoping injection, stabilizing after half an hour to complete applying prestress [6].

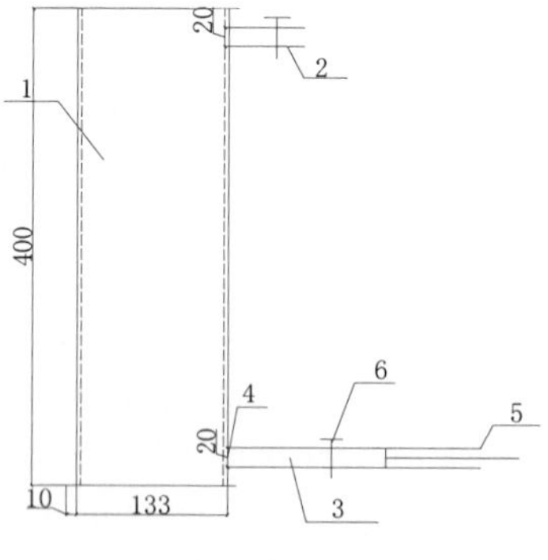

1 Steel pipe; 2 Vent; 3. Infusion tube; 4 Perfused and pressurized interface; 5 Perfusion mouth 6 Valve;
Fig. 1. A model for the specimen.

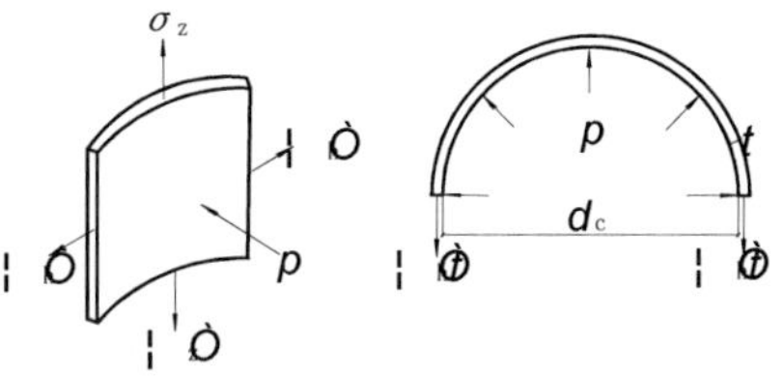

Fig. 2. Steel pipe stress diagram.

According to the diagram of steel pipe under preloading pressure as shown in figure 2, the equilibrium equation of hoop stress and longitudinal stress is established as in Eq. (1).

$$\begin{cases} \sigma_h = \dfrac{E_s}{1-\mu_s^2}(\varepsilon_h + \mu_s \varepsilon_z) \\ \sigma_z = \dfrac{E_s}{1-\mu_s^2}(\varepsilon_z + \mu_s \varepsilon_h) \end{cases} . \tag{1}$$

(Note: E_s, μ_s, ε_z, and ε_h respectively represent the elastic modulus, poisson's ratio, axial strain and hoop strain.)

Confining force that is preloading stress is established by the figure 2, as in Eq. (2).

$$p = \frac{2t}{d_c}\sigma_h . \tag{2}$$

3. Experimental Study on the Prestress Loss

The size of the prestressing value directly affects the mechanical properties of the prestressed concrete filled steel tube. The inadequate sealing of flange sealing and the shrinkage and creep of concrete can lead to the loss of prestress. In order to study the size of the prestress loss, three specimens were tested with different prestress level, loaded prestress is 3Mpa, 5Mpa, 7Mpa. The process of applying prestress shows in Figure 3. Figure 4 shows the strain-time curve. Figure 4 illustrates the loss of prestress which is controlled within 20%, achieving the expected result, verifying the feasibility of the method of applying hydraulic prestress.

Fig. 3. Pressing process.

Fig. 5. Failure mode of G2.

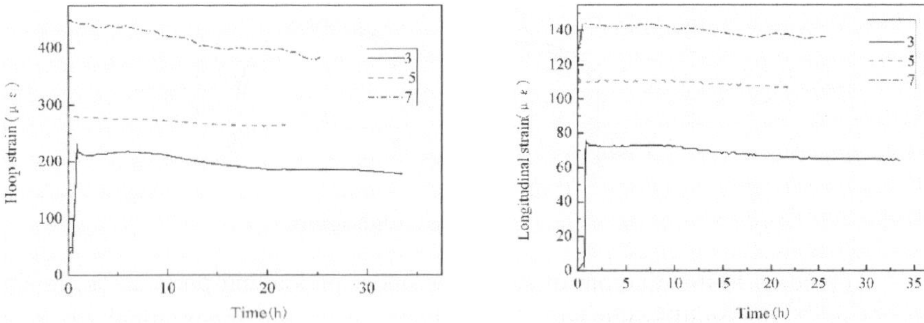

(a)Hoop Strain-time curve (b)Longitudinal Strain-time curve

Fig. 4. Strain-time curve.

4. Experimental Study on Axial Strength and Deformational Behavior

To study the effect of prestressed concrete steel tube performance by applying prestress, a hydraulic prestressed concrete filled steel tube and a concrete filled steel tube are tested. The design parameters are shown in Table 1. Two displacement meters are arranged symmetrically to monitor the longitudinal displacement of the specimen. Pasting four pairs of strain gauges to monitor the strain in the middle of the specimen.

Table 1. Specimen size and material strength.

number	D/mm	t/mm	F_y	F_{cu}	P/MPa	N_u/kN
G1	133	4.5	301.8	33.17	0	1203
G2	133	4.5	301.8	34.33	5	1364

Note: D, t, Fy, Fcu, P and Nu respectively represent the pipe diameter, wall thickness, tensile strength, ultimate compressive strength of concrete, prestress and ultimate bearing capacity of the specimen.

4.1. *Failure modes*

In the process of applying hydraulic prestress, the damage form of prestressed concrete filled steel tubular specimen and concrete filled steel tubular specimen is similar. As the load increases, the prestressed specimen continues to thicken, rust peeling after reaching yield strength of steel, then load can also continue to increase. The local buckling can be observed when the strain is about 0.01 to 0.03, and finally shear slippage is appearing (Figure 5). Throughout the course of the experiment, prestressed specimen shows good ductility and high carrying capacity in the later stage.

4.2. *Analysis of the load-deformation curve*

Load-deformation curve of the specimens is obtained according to the result (Figure 6). The figure shows that the loading process includes the elastic stage, elastic-plastic stage and reinforcement service stage. The elastic stage of hydraulic prestressed concrete filled steel tube short columns can reach more than 85% of the ultimate bearing capacity, while the concrete filled steel tube at about 75%. The elastic stage of prestressed specimen is obviously promoted, ultimate bearing capacity is also improved. The core concrete of hydraulic prestressed concrete filled steel tube is in the three-state pressure from the outset, compared with concrete filled steel tube, improving the stress state of the structure, slowing the crack width and the generation of longitudinal cracks, improving the compressive bearing capacity and deformation capacity of the structure.

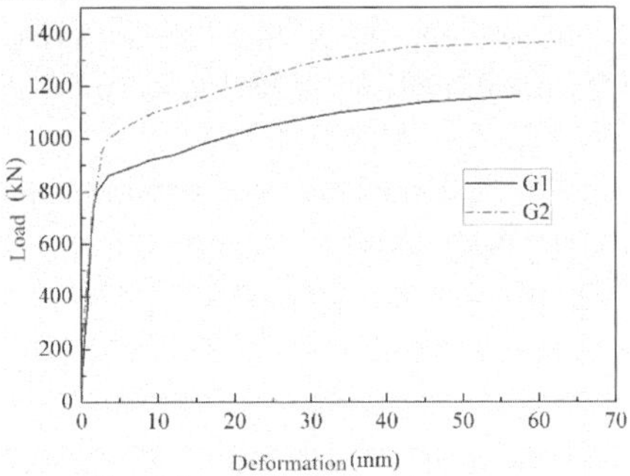

Fig. 6. Load-displacement curve.

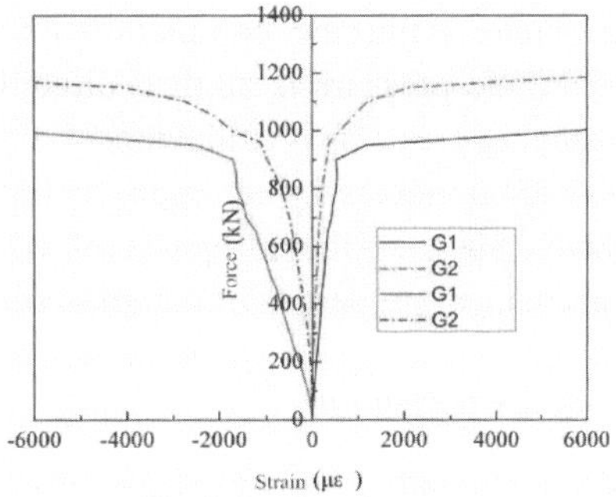

Fig. 7. Load-strain curve.

4.3. *Analysis of the load-strain curve*

Load-strain curve of the specimens is obtained according to the result (Figure 7). As can be seen from Figure 7, longitudinal strain and hoop strain of the elastic phase gradual increase in steel cross, but the longitudinal stress is greater than the hoop stress. After the peak load, hoop stress will increase rapidly because of the lateral deformation of core concrete. Compared with concrete filled steel tube, the strain of prestressed specimen grows slowly and the elastic phase improves significantly due to the effect of the presence of prestress under the same load. Prestressed specimen has obvious elastic stage, the elastic limit is about 0.12%. The stage strain is about 0.2% from elastic-plastic stage to reinforcement service stage, which is also a turning point strain.

5. Summary

This paper presents a new method of prestressing, through the analysis of test results, the following conclusions can be obtained.

(1) The study verify the feasibility of the method of applying hydraulic prestress. The loss of prestress is controlled within 20%, achieving the expected effect.

(2) The elastic stage of hydraulic prestressed concrete filled steel tubular short columns can reach more than 85% of the ultimate bearing capacity. At the same time, the ultimate bearing capacity is improved.

(3) The method can realize exerting active quantitative prestress of concrete filled steel tube, compared with the self-stressing steel tube, fully considering the effect of prestress on the structure.

(4) This method can be easily combined with the process of pumping concrete. After the completion of the concrete pumping, closing the stock, hydraulic pressure is applied directly in the mouth filling, then the purpose of applying prestress is achieving.

(5) It necessary to improve further theoretical analysis and experimental study, such as the influence of concrete strength, steel ratio as well as the mechanical properties of complex stress.

Acknowledgment

Work partially supported by grant 15KJB560001 of the Natural Science Foundation of the Colleges and Universities in Jiangsu Province and BK20130408 of Natural Science Fund Project in Jiangsu Province.

References

1. L. H. Han, *Theory and practice of concrete filled steel tube structure. (The second edition)*, (Science press, 2007). (In Chinese).
2. S. Aghdamy, D. P. Thambiratnam and M. Dhanasekar, Computer analysis of impact behavior of concrete filled steel tube columns. *Advances in Engineering Software.* 89, 52-63 (2015).
3. J. Liu, J. A. Abdullah, S. Zhang, Hysteretic behavior and design of square tubed reinforced and steel reinforced concrete (STRC and/or STSRC) short columns. Thin-Walled Structures. 49(7), 874–888 (2011).
4. C. K. Huang, X. Chang and D. C. Jiang, Constitutive relations for concrete core of self-stressing concrete-filled steel tube under axial load. *Journal of Dalian University of Technology,* 50(1), 81-85 (2010).

5. C. K. Huang, X. U. Lei and Y. Liu, Research on mechanical performance of self-compacting and self-stressing concrete-filled steel tube short column under axial loading. J*ournal of Dalian University of Technology,* **46**(5), 696-701 (2006).
6. A. G. Tian, M. D. Li, Z. L. Zong, Method for producing a prestressed concrete steel, China, 201310179503.5 (2013). (In Chinese).

Comparison of stoker feed boiler and circulating fluidized bedboiler

Ming-Hao Li, Jun-Feng Qi

School of municipal and environmental engineering ,Jilin Jianzhu University , Changchun, Nanguan District ,China

E-mail: liminghao754742472@163.com

This paper describes in detail the quantitative comparison of energy saving in the three aspects, such as chain grate boiler and circulating fluidity bed furnace in the boiler heating efficiency, unit in accordance with the power consumption ratio and excess air coefficient in the same boiler room.We choose Stoker-feed Boilers type is:DHL64-1.6/150/90-AII,Circulating Fluidity Boilermakers type is:DHX45-1.6-H.

Keywords: Stoker Feed Boiler; Circulating Fluidity Bed-Boiler; Boiler Efficiency; Unit Load Power Consumption Ratio; Excess Air Coefficient; Quantitative Comparison.

1. Preface

Energy is the lifeblood of the development of the national economy, also is an important material foundation to improve the living standards and conditions of the people. In the system of energy consumption, the boiler and heating system has a large proportion. According to an uncompleted statistic, The annual coal consumption of heating system accounts for about 1/3 of the total coal production in China. Therefore, energy conservation is a long-term strategic policy for China's social and economic development. The Communist Party of China in the Fifth Plenary Session of the Sixteenth Central Committee passed the "national economy and social development in the Eleventh Five Year Plan", and puts forward the implementation unit GDP energy consumption reduced by 20% of the binding targets, the deployment of the emphasis of the ten energy saving projects and energy saving evaluation analysis and transformation of coal-fired industrial boilers are listed as the first key energy saving projects, among these projects.

In coal-fired industrial boilers, more varieties of different specifications, the type of furnace each are not identical, combustion is the main way of layer combustion, accounting for about 95% of the total capacity of coal-fired boiler (which Stoker-feed Boiler accounted for about 65 percent ,Reciprocating boiler accounted for 20% of reciprocating grate, a fixed grate accounted for about

10%), the Circulating Fluidity Bed-boiler is about 3% ~ 5%.We made a comparison and analysis of energy saving on quantitative Stoker feed Boiler and the Circulating Fluidity Bed-boiler, promote the use of furnace can save more energy in the same working condition,according to the proportion of the type of boiler used in our life.

2. Economic Comparison of Operating Efficiency under Boiler Unit Load

The main economic indicators of boiler thermal efficiency of the boiler (including the gross efficiency of the boiler, boiler efficiency), boiler load rate, combustion efficiency, production costs, etc. Boiler thermal efficiency is a long used technical and economic indicators, The combustion efficiency of the boiler total heat input in the boiler system in a part not transformed into heat, it reflects the degree of complete combustion of fuel. We use the thermal efficiency of the boiler (is boiler gross efficiency) (hereinafter referred to as the boiler efficiency) and boiler load rate (hereinafter referred to as the loading rate) as the basis of chain boiler and circulating fluidity bed boiler operation economic efficiency analysis and contrast.

In order to make a comprehensive analysis and comparison of the economic and combustion conditions of the above two kinds of boiler, and then find out the main factors that affect the efficiency and load rate of the boiler, We need to calculate the boiler efficiency, We got a formula(1), which is made by the ratio of the heating capacity of the boiler and the heat generated by the combustion of coal within a certain period of time. The expression of load rate is formula(2), The running efficiency expression under the unit load is formula(3).

$$\eta = 4.2 \times \frac{G \times (t_g - t_h)}{B \times Q_{net,\ ar}} \times 100\% \tag{1}$$

$$\phi = \frac{Q_{function}}{3.6 \times 10^6 \times Q_{rated}} \tag{2}$$

$$Q_{function} = 4.2 \times G \times (t_g - t_h)$$

$$\eta_{ratio} = 100\% \times \frac{\eta}{\phi} \tag{3}$$

- η is Boiler efficiency,
- G is Average water supply flow,
- t_g is Outflow water temperature,

- t_h is Return water temperature,
- $Q_{net,ar}$ is Low calorific value of coal,
- ϕ is Load rate,
- $Q_{function}$ is Actual evaporation of boiler,

Due to the need of quantitative analysis of the two types of boiler operating efficiency, So use formula (1), (2), (3), Stoker feed Boiler and Circulating Fluidity Bed-boiler in a certain period of time the average water flow, water temperature, water temperature, etc. in order to meet the relevant heating specifications, according to the actual situation of the same boiler room. And burn the same batch of coal, which have the same low calorific value,So we only need to calculate the corresponding operating efficiency according to the formula. According to the actual observation, we take 4 hours for the observation time period, respectively, measured two boiler related data, and calculate something as shown in Table 1 below:

Table 1.Table of quantitative measurement data of operational efficiency under boiler unit load

Type of boiler	Outflow water temperature (celsius degree)	Return water temperature (celsius degree)	Rated evaporation (MW)	Coal consumption (kg/h)	Low calorific value of coal(kJ/kg)
Stokerfeed Boiler	86	39	64	1.6×10^4	1.8×10^4
Circulating Fluidized Bedboiler	140	80	58	9.2×10^3	1.8×10^4

According to the data in Table 1 we can get the running efficiency of Stokerfeed Boiler is 79.9%, Circulating Fluidized Bedboiler operating efficiency is 97.8%. According to the calculation, we can see that the efficiency of Circulating Fluidized Bedboiler is 16 percentage points more than that of Stokerfeed Boiler .

According to the principle we know that Circulating Fluidity Bed-boiler to coal quality requirements is not high, wide adaptability. In Circulating Fluidity Bed-boiler by weight, fuel accounted for only 1% to 3% of bed material, the other is non combustible solid particles, such as pasteurization agent, ash, etc. Therefore, fresh coal particles to the bed is surrounded by hot ash particles. Due to intense mixing in the bed, these hot ash particles actually played the infinite"ideal arch" effect, the coal is heated to the ignition temperature and it starts to burn. The combustion of coal particles, the release of heat, which can make the bed temperature to maintain a certain level, it is generally not difficult ignition of Fluidity bed, cause wide adaptability to coal. The chain boiler can not burn the inferior coal and the slag of the chain furnace can be stable combustion in a Fluidity bed boiler. Circulating Fluidity Bed-boiler combustion efficiency is high because of the following characteristics: air (oxygen) and small particles (diameter 0 13llnn) of the coal particles in the Fluidity state of good mixing; combustion rate is high, the second is the fly ash recalculation combustion. The Stoker feed Boiler fuel adaptability, because the water is too high or volatile matter is too low volatile will slow fire, so that at the end of the combustion of carbon particles is not completely burned; carbon particles too large will lead to incomplete combustion. Coal decentralization, formation of porous coke, coke after combustion, ash formed coke, to put a thin layer of "gray". Then the "gray" will be thickened and finally will become soft because of high temperature, or melt will be difficult to enter the coke tightly wrapped in air, resulting in a very slow progress of the combustion process is very slow, resulting in a larger solid incomplete combustion loss. So the running efficiency of the chain furnace is more than 16 percentage points higher than that of the circulating Fluidity bed furnace.

Through the analysis we draw the conclusion: the circulating fluidized bed boiler is higher than the ordinary chain furnace in the operation efficiency, and the fuel combustion efficiency is higher.

3. Economic Comparison of Power Consumption Per Unit Load Ratio

In the boiler room operation, excluding some repairs, kindling, labor costs may be ignored operating costs, of which most of the expenses including part of coal consumption, another part is for fan, coal processing, and other electrical power consumption for chain furnace power part includes a blower fan, induced draft fan, grate speed reducer, coal equipment, coal equipment, desulfurization denitration and slag removal equipment. CFB boiler has no grate speed reducer but has a material return fan. According to the power consumption ratio, coal feeding equipment and coal crushing equipment are two common power boiler and desulfurization and denitrification in addition to slag device power consumption as an approximate equivalent. As for CFB boiler, in order to overcome the resistance of the cloth wind plate and the material layer, high pressure air blower is needed. In order to compare the power consumption of the above parts, we select the two boiler at the same time to work for 4 hours, respectively, measuring the voltage and current of each machine at the same time, taking the average value,using the formula (4), to calculate the total electricity consumption in the time period. Formula (5) for calculating the power consumption ratio of unit load.

$$P_{\text{electric}} = 1.4722 \times 10^{-6} \times \sum UI \quad (4)$$

(Sum of the product of voltage and current at both ends of the machine)

$$K = \frac{3600 \times p_{\text{electric}}}{Q_{function}} \quad (5)$$

- $Q_{function}$: The actual evaporation of the boiler has been obtained
- K: Power consumption per unit load ratio

Due to the quantitative analysis of the power consumption of the boiler, using formula (4) and (5), using related instruments were within four hours of measurement chain boiler andcirculating fluidized bed furnace the power machine of actual voltage and currentseveralgroups,and the values were averaged and fill in the table 2.

Table 2. Boiler unit load power consumption ratio quantitative measurement data record form

Type of boiler	Primary air fan		Return fan		Induced draft fan		Grate speed reducer	
	U(V)	I(A)	U(V)	I(A)	U(V)	I(A)	U(V)	I(A)
Stokerfeed Boiler	380	135	\	\	380	390	380	8
Circulating Fluidized Bedboiler	380	315	380	158	380	526	\	\

According to the data in Table 2 we can get the Stokerfeed Boiler unit load power consumption ratio of 6, the Circulating Fluidized Bedboiler unit load power consumption ratio of 26.2.

Through the analysis, the chain furnace technology is mature, the operation is simple, the boiler auxiliary power supply air draught fan's wind quantity, the wind pressure is small, the power consumption is little, the load adaptability is strong. And circulating fluidized bed furnace hearth tall, separation recycling system is complicated in structure, high resistance of cloth wind plate resistance and system, the auxiliary power consumption high.

4. Quantitative analysis of excess air coefficient in exhaust gas

Excess air coefficient of industrial boiler directly influence the economy of boiler operation, accurate determination of it is necessary to supervise the economic operation of the boiler, the excess air coefficient reflects the boiler air distribution modes. It has a direct impact on the burning in the furnace condition and the quality of the boiler efficiency to the level of. It is defined as:

$$\text{Excess air coefficient}\ (\partial) = \frac{\text{Actual supply air volume}\ (V)\ Nm^3/Kg}{\text{Theoretical air volume}\ (V_0)\ Nm^3/Kg}$$

At present, the basic use of flue gas composition analysis to determine the excess oxygen content in flue gas, so as to determine the excess air coefficient under the operating conditions. It is related to the fuel types, combustion mode and the improvement of combustion equipment.Excess air coefficient is too large, not only reduce the furnace temperature, combustion effect, but also increases the flue gas emissions, causing boiler exhaust gas heat loss and citation, fan power consumption increases.Similarly, the excess air coefficient is too small, cannot guarantee the sufficient combustion of the coal, will cause the increase of mechanical and chemical incomplete combustion loss. Chemical incomplete combustion, mechanical incomplete combustion heat loss sum for the most hours of excess air coefficient is optimum excess air coefficient. Usually need to be determined by experiment. In this paper, we have measured the oxygen content at the outlet of the tail flue to determine the excess air coefficient, and the final result is filled in the following table 3.

Table 3. Table of excess air coefficient of boiler smoke

project	Stokerfeed Boiler	Circulating Fluidized Bedboiler
Excess air coefficient	1.29	1.14

According to the state of smoke at the excess air coefficient specification (standard following table) comparison of the measurement results we know that the excess air coefficient is higher than that of chain boiler circulating fluidized bed furnace

5. According to The Quantitative Analysis

Through the above three points of quantitative analysis, we know circulating fluidized bed furnace operating efficiency was higher than that of the chain furnace, high combustion efficiency, fuel adjustment, can make full use of the fuel calorific value calorific, reduce unnecessary waste of fuel; But in the work at the same time, the furnace of circulating fluidized bed boilers power consumption is high, more energy consuming than the chain furnace, but the power is part of the cost in the dust, harmful gases, reduce air pollution, environmental protection. At the same time, the excess air coefficient of the smoke exhaust is reflected from the side of the boiler fuel combustion and the combustion efficiency. The circulating fluidized bed furnace is superior to the chain furnace in the determination of the excess air coefficient at the tail end. Generally speaking, the circulating fluidized bed furnace has a better performance in the process of heat supply.

6. Concluding Remarks

Generally speaking, the circulating fluidized bed furnace has a better performance in the process of heat supply, which has wide fuel adaptability, full combustion and less pollution, In real life, electricity consumption compared to coal consumption consumption can be said that the proportion is very small, manufacturers tend to pay more attention to the process of fuel combustion. Therefore, the technology is more and more mature Circulating Fluidized Bedboiler may be widely used.But because the technology is not mature and the circulating fluidized bed furnace maintenance cost is high, the operation period is short, the total cost is higher, we also have to consider.

Acknowledgement

Work partially Supported by Jilin Architecture University Students Innovation and Entrepreneurship Training Program, project number 201510191010.

Project building energy saving and heating system evaluation index system of Jilin University of Architecture.

Reference

1. Lijun Wei . Measures and ways to realize energy saving and consumption reduction for coal fired industrial boilers [Z], *Technology frontier* .
2. Yanqiu Wang, Yujing Li, Guomin Zhang. Analysis and Discussion on energy saving ways of industrial boilers [A] , *Applied energy technology* 2008(11).
3. Yongliang Wang, Yantong Xiao, Qi Cui, Shengli Zhang . Population, Qiu Dong. Chain grate boiler with low load operation optimization combustion economic analysis [b] .*Industrial Boiler* 2015(4).
4. Jing Yan. Analysis of circulating fluidized bed boilers [Z], *Beer technology* 2006(11).
5. Daming Shi. Determination of excess air coefficient and its error analysis under the operating condition of boiler [Z], *Chinese Journal of Electrical Engineering* 1991.

Discussion on train of thought about defense safety design on building's outside environment

Ke-Qi Wang
School of architecture and urban planning, Jilin JIANZHU University, Changchun, Jilin province, China
E-mail: wkq0431@126.com

Building's outside environment ab.:BOE) are the spatial carriers and condition factors of urban crimes. Valid design of BOE is powerful mean of preventing crimes as well as important guarantee for constructing safety human settlements. This paper takes CPTED (crime prevention through environmental design) theories as foundation, takes the carrier and condition of crime as research objects, and introduces the necessities and significance of implementing defense safety design, and comes up with the basic train of thought about defense safety design of BOE.

Keywords: Building's Outside Environment; Defense Safety; Train of Thought.

1. Introduction

Urban crime is one of the most important issues affecting people's life. The relationship between physical environment and urban crime inspires the interests of multi-disciplinary researchers such as criminology, urban and rural planning as well as architecture. Oscar Newman put forward the concept of "defense safety design" in the book named "Defensible Space" [1] in 1972.

Crime prevention and control is a complex system engineering. Crime prevention through environmental design (ab.:CPTED) is a branch of criminal prevention and control system, the basic idea is to take physical spatial environment as the research objects and to raise the cost of committing crime, augment the difficulty of crime, increase the possibility of crime behavior being found and prevented in the process of crime implementation, thus achieve the effect of crime prevention.

Crime prevention and control measures should not rely on non-spatial factors such as monitoring technology and management and ignore the functions of situational factors those have close relationship with crime. It should consider the function of BOE in terms of crime prevention and control in urban planning, architectural design, landscape design and construction project management. We need to carry through defense safety design for BOE based on local conditions. It help to create a more harmonious living environment.

2. BOE is the Spatial Carrier and Condition Elements of Crime

BOE refers to the space building's surroundings or inter-buildings. It is the spatial carrier and condition elements of criminal behaviors. Urban crime has strong spatial character, BOE are both the elements of the crime and the way of crime prevention. The location, distance, territory, visibility, trafficability, attraction, cohesive force, and control force have significant impacts on crime. Criminal psychology considers that crime behaviors have three elements which are criminal tendency, criminal environment and the resistance of crime. Among them, the character of environment are factors that stimulating, induction and conceal criminal behavior. This bridges the relationship between urban crime and BOE, the urban crime has close relation with spatial location, spatial form and spatial atmosphere [2] of BOE, BOE is one of the most important factors affecting urban crime, it's the spatial carrier of urban crime.

3. Necessity of the Implementation of Crime Prevention Through Environmental Design

3.1. *Carrying out CPTED improves the quality of human habitat environment effectively*

With the emergence of high-rise buildings in the city, the traditional blocks and approachable scale architectural form are gradually disappearing, because of the interaction between people is less, and lack of necessary communication, it is hard to form a harmonious neighborhood relationship, this lead to interpersonal indifference and public security blind spots increases. To some extent, good design of BOE design can eliminate some induced factors of crime, reduce the occurrence of violence, criminal behavior, so, to reduce the negative space in the BOE, and to create a positive humanistic space and defensible space, it can promote the communication and mutual cooperation between people, improve the neighborhood relationship, increase the natural supervision to strangers, build safe living environment, enhance people's sense of security, it can effectively improve the quality of the residents' living environment.

3.2. *Carrying out CPTED provides new ideas and strategies for urban public security administration*

In the face of various intra-regional unsafe factors, the security measures people preferred are mostly depends on the property management and high-tech security technology, or rely on the residents' committees, regional police force, etc. But, although some areas set the burglar meshes, security doors, theft affairs are still happen everywhere. Investigate the reasons, Mainly due to ignore the functions that BOE acts on prevention and control to crime. CPTED helps urban security departments setting more specific goal in the processes of management, regulation and maintenance work, it helps thinking and analyzing crime issues from a new perspective. Meanwhile, it enables administrators dealing with the crime wave caused by urbanization in a more practical way, and protects people' life and property safety more effectively.

3.3. *Carrying out CPTED provides new perspectives and methods on studies of urban crime and urban space*

The issue of crime prevention and control is huge system problem involves wide ranges, the studies and applications about urban crime prevention and control from the perspective of urban planning, architectural design and landscape design are relative lag and scarce, especially the studies systematically from the perspective of BOE is very few, this illustrates that architecture and its correlative disciplines did not engagement to solve the crime problems adequately. Therefore, from the perspective of architecture, the research train of thoughts that take the BOE as direct object, provide new perspectives and methods for the studies of architecture, criminology and planning.

4. Basic Ideas about BOE Defense Safety Design

To make the physical environment into behavior environment through defense safety design are the prime aim, the main ideas and contents are descript as the follow Figure 1.

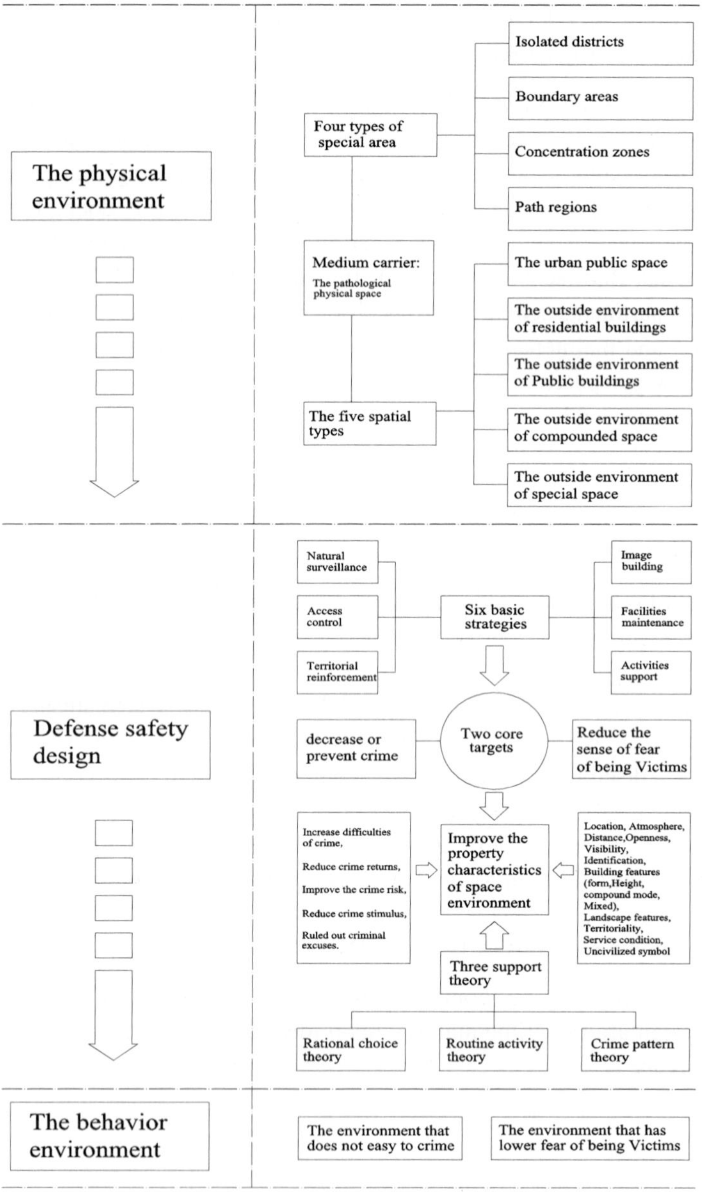

Fig. 1. The main ideas and contents of crime prevention through building's outside environmental design.

BOE defense safety design, focuses on digging the relationship between building's outside physical environment and criminal behaviors, as well as the precautionary measures of planning, architecture and landscape design. Therefore, BOE defense safety design should take the on-the-spot investigation as foundation, reference the experiences of domestic and foreign researches, based on the specific local customs, climate condition, geographical features, and the characteristics of space environment, select the place of Vulnerable to Crime of urban physical environment to study deeply, such as (1) urban villages, colonies, low income districts, etc, the locations vulnerable to be rejected; (2) the junction regions of different types of communities, the boundaries of different social classes, urban-rural fringe locations cause by rapid urbanization; (3) rich regions, the places where all levels, specific functional areas, gathered large numbers of people, properties, goods, information; (4) regions of urban transportation system that connect and circulation various elements , and regions along the transportation system, such as the public space that citizen's daily life occupied, and so on .

The basic links about BOE defense safety design are to research and analyze the above locations, to explore the mechanisms and environmental elements that influence the crime behaviors in BOE, to comb the existing problems of BOE defense safety design, to summarize the relationships between BOE and crime behaviors, and then put forward crime prevention strategies and methods in specific outside environments, to construct defense safety theory, principles, countermeasures and methods of urban planning, architectural design and urban management.

For the crime behaviors of rational criminal (potential), below are the basic ideas of BOE defense safety design: Through the reasonable design of spatial location, spatial form, spatial atmosphere of BOE, using design strategies such as reducing crime motive, decreasing crime opportunity, strengthening crime control, and with the aids of domains controlling of crime targets, the trafficability of potential crime subjects, the visibility of criminal process, and then enhancing the security of BOE.

For example, in order to strengthen the control of visibility of criminal process, namely strengthen the importance of natural surveillance, the windows of buildings can be designed to watch larger outside range, restrict the use of building elements that blocks or reflects view of the window; Gateway's form and location are designed to help more people watching, and maximize the sight range; Use fewer buildings construction and environmental landscape that blocks vision, such as extra higher concrete balustrades, plants, signboards, etc. In order to restrict trafficability of potential crime subject, to stop irrelevant crowds going in, to use physical methods prohibit or delay potential crime subject entering or access, we can control the form and the number of intersections of road, control the width of roads, set up reasonable road hierarchy and road network density, set speed obstacles that affect trafficability, clear spatial mark character, strengthen the road illumination and neighborhood monitoring. In order to strengthen the control of domain of crime target, mainly control the boundary, entrance, scope of perceive, we can set up plants, fences or walls around particular areas to stop pedestrians crossing, using special logo to impede trespassers moochers or alcoholics; Use of public art, sculpture, flags and banners to do tag; Control the space size in order to strengthen contact among neighbors; Strengthen territoriality using landscape, decorative columns or fences to define territory or property ownership, and so on.

Certainly, whether to strengthen the control of visibility of criminal process, to restrict trafficability of potential crime subject, or to strengthen the control of domain of crime target, they should be incorporate with the emergency alarm facilities, public security sentry boxes, and environment maintenance, lights assist, barriers setting, access restriction, and to apply tactics and strategies of gathered popularity such as strengthening recreation and interesting of space. In addition, to strengthen the control of visibility of the criminal process, sometimes conflict with the requirements for residential health horizon of current code, I think that from view of human's hierarchies needs, health horizon requirement should give way to security requirements, after all, the security requirements are human's basic needs. And sometimes in order to strengthen the control of domain of crime target, it will weaken the control of visibility of criminal process. In addition, plant size, density, genre and form are also the important contents of BOE defense safety design. Public housing residents with nearby trees and natural landscapes reported 25% fewer acts of domestic aggression and violence [3].

5. Epilogue

Urban crime is a social problem that any city can't avoid. There are close relationship between crime and BOE, they influence each other, induce mutually. To some extent, the design of BOE can eliminate some factors that induce crime, reduce the occurrence of violence and other criminal behaviors cause by adverse BOE, thus affecting the occurrence of crime, it can help to build a safe living environment and strengthen people's sense of security.

Acknowledgment

This paper is Supported by science and technology plan projects of Ministry of Housing and Urban-Rural Development, 2013 (2013 - R2-16); The Research Fund for the Doctoral Program of Jilin JIANZHU university.

References

1. Oscar Newman. Defensible Space: *Creating Prevention Through Urban Design*. The Macmillan Publishing, 1972:264.
2. MA Rui. *Research on the Urban Physical Environment Vulnerable to Crime*. Dissertation Submitted to Tsinghua University in partial fulfillment of the requirement for the degree of Doctor of Engineering, 2010: 42-58 (In Chinese).
3. Kuo, F.E., and W.C. Sullivan. 2001. *Aggression and Violence in the Inner City: Effects of Environment Via Mental Fatigue*. Environment and Behavior 33, 4:543-571.

Study for piled raft foundations stress calculation and design

Kui Dai

School of Civil Engineering and Architecture, Wuhan Polytechnic University, Wuhan, Hubei, China, 430023

E-mail:Daikui888@163.com

Based on the specific design conditions and material analyses, the paper analyses the distinguishing features of piled raft foundations, calculates stress for piled raft foundations, presents the design priciples of piled raft foundations, sets up calculation formulas for piled raft foundations, and sums up piled raft foundations in theory.

Keywords: Piled Raft Foundations; Whole Function; Potential Energy Principle; Pile-Soil Foundations.

1. Introduction

There is no specification of existing about piled raft foundations design, however, in practical work, because of the geological conditions, with the natural foundation as a basis, although it can meet the needs of multi-layer or more of high-rise buildings, but they cannot meet the needs of high-rise building design; as the same if the pile capacity is not enough, but building load is greater and the spacing of shear wall is too intensive, that will made desktop pile to hard to implement. The greater column loads need more piles, but if these piles are been put down, there will conflict with the surrounding piles, and the distance of piles cannot meet the requirements of specifications. Individual cases can be combined with a double or more column of the united cap, but the ability pass more complex, and sometimes also between the united caps have conflicts and it does not meet the specification requirements of the pile spacing. Pile bearing capacity is not been full used, but they cannot play the role of a surplus, while piled raft foundations can solve this contradiction.

2. The Characteristics of Piled Raft Foundation

2.1. *The concept of force as a whole*

The entire load of the building material through the raft of this conversion occurred in relation with the pile. It has a more definite force, and it can make full use of the potential of the pile, while it also can make full use of the potential of soil. Under the case of without increasing the stress of concrete and steel , it can use the end of the raft to take 5% to 10% of the building load [1]. We can see the building, raft, pile and soil , that it gets together for a stable whole. At the same time it is easy for us to analyze about a single pile and play their respective advantages in together. At the same time it can help to reduce local damage to the entire building damage, and ease the local buckling on the overall stability of the whole building damage.

2.2. *Adjustment of piled raft foundation adaptable*

If the load between the piles or large wall, is too difference we can primarily by adjusting the number of piles and pile spacing to a reasonable arrangement, by increasing the strengthening of the stiffness of the load roft can solve the problem of increasing stress, such as thickening of the local board, and additional dark beams and increasing the reinforcement and so on.

2.3. *Economic rationality*

As the overall load of the building beared by all the piles, it is essential to adjust the spacing of the piles of columns or shear walls. Encrypted under the appropriate pile wall, pile arrangement based around a more dense, more sparse within the principles of pile layout can make the best use of the bearing capacity of piles to achieve the purpose of saving a few piles, we can make the full potential of the pile. If the local pile can is not enough, we can adjust the raft and conver the load to other pile to bear; to ensure the power transmission, can strengthen the local raft stiffness, improve reinforcementratio, set the dark beams to achieve. Also by adjusting the upper structure, when the overall load in the same situation, we can change the load of the local column and shear wall to meet requirement of the local force. Relatively with pile, the suspended desk on pile can not take full use of its potential, such as the post count to be 5.3, the design is to use 6. Cap quantity is more, the more piles can not take full use of their potential, which is being wasted.

3. Theoretical Design Basis of Piled Raft Foundations

3.1. *Foundation model*

The key about piled foundation is to build a model, this model should be simulated as accurately as possible the interaction between the foundation and basis of the performance of the main mechanical properties, but also to facilitate the use of existing mathematical methods of analysis, especially in the contemporary and it should facilitate the full use of computer, allowing to connect with computer easily. From the existing model view three foundations, are not applicable to pile raft foundation.

Boussinesq solution and Winkler foundation model proposed ground intensity of the pressure at any point P and the settlement of the point S is proportional to the assumption[2], that P = KS, K as the coefficient of subgrade, in units of MN/m^3, in essence, the foundation as the number of days separating the composition of the small soil column system, column representing the spring. But the piles of piled raft foundation clearly can not see as the spring.

Half-space soil model to a homogeneous elastic half-space model as the foundation, soil settlement at any point in S (X, Y) and the substrate reaction force, and close to the distribution of load, whereas the stiffness of the pile soil stiffness ratio much larger pile of soil can not constitute a homogeneous elastic half-space.

Confined compression layer ground model is limited in the side of each layer under the conditions of compressive strain and additional stress is proportional to the linear deformation of the foundation is seen as. However, the land base can not be regarded as linear pile deformed.

Piles of land-based stiffness [K] is extremely important to determine, according to the type of pile, mainly based compression, the pile can be divided into compressible rigid piles and pile, and so land-based species can be divided into two categories. Due to space limitations, detailed in this issue for later.

3.2. *Working properties of piled raft foundations*

It can be approximately regarded as the nature of the rigid foundation with elastic foundation, pile-soil system may be regarded as an ideal linear elastic or elastic-plastic body for consideration.

3.3. *About the pile and raft shared building loads*

It can be considered to afford the vaule of 5-10 raft building load.

3.4. *The anti-force distribution size of the top of the pile and the pile arrangement*

For high-rise buildings on piled raft foundations, the anti-force distribution of the top of pile is the angle of pile big, the side of pile second, the inter of pile small. Under the pile wall or column the pile spacing should be more dense, and the same as the around the foundation of the pile spacing, but at the inter the pile spacing should be less.

3.5. *How to determine the raft thickness of piled raft foundations*

The thickness of the raftcan be determined by a single pile punching load of 1.5 times the average pile load.

3.6. *The internal stress of pile raft foundation*

The internal stress is about to schedule of the building construction that is the formation of the stiffness of buildings which is result of the upper structure and piled raft foundations, but also with thetemperature.

4. Interaction of piled raft foundation of the whole

Comparative analysis of the current prevalence of upper structure and foundation interaction method is based on the finite element method, based on the upper part of the structure were obtained stiffness matrix [3], based on land-based stiffness and the pile stiffness matrix by matrix composition, the balance of power established by the equation and the deformation compatibility conditions through the establishment of the displacement equation to arrive at the internal force and displacement we need. But we know that a building of the pile, foundation, the upper structure is closely linked to the lower part of the upper part is also affected by the shock, the impact of the upper part of foundation settlement. The upper part of stiffness, the overall effect is strong. Box foundation stiffness is better than the whole board basis [4], the floor, roof, wall can work together.

Now this paper is based on the characteristics of piled raft foundations, think: we should be a whole raft foundation as board shape from the deformation of raft foundation and boundary conditions, the paper developed a displacement model, based on structural features of the upper soil and pile nodes acting on the raft the base of the force, according to the overall potential energy principle to establish the role of the equation to find unknowns. These unknowns include the raft foundation at any point in the moment, torque, normal stress, shear stress, given numbers for the columns and walls of the reaction force, raft, soil, pile the basic parameters, as well as the basic parameters of the upper.

4.1. *The displacement formula of piled raft foundation formula*

If the raft foundation is thick enough, the stiffness is large (such as foundation piles box), even a whole raft settlement will occur or differential settlement, forming lean. The settlement and displacement is

$$W_i = \theta_y X_i + \theta_x Y_i + W_0 \qquad (1)$$

(i=1,2,3……,n)
in which x and y are raft foundation which uniform sedimentation value in the direction of X and Y axis.

If raft is not too thick, stiffness is not significant, it can be seen as flexible, the settlement consists of two parts: One part of the overall settlement for the raft foundation and the tilt, the other part of the raft foundation of the bending deformation, Deformation which can be used, said the convergence of double trigonometric series: [1].

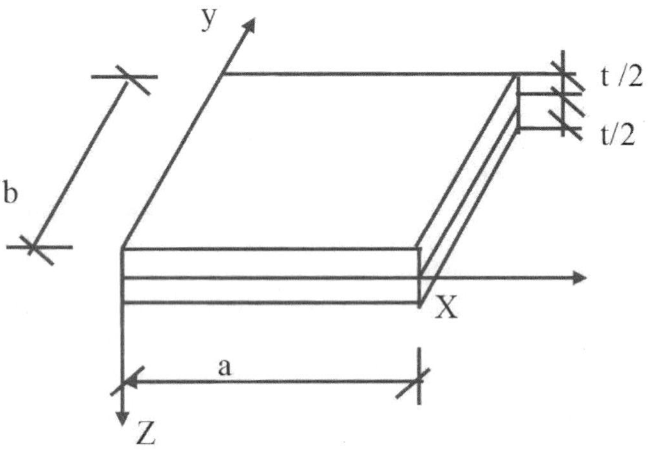

Fig. 1.

$$W_W(x,y) = \sum_{m=1}^{\infty}\sum_{n=1}^{\infty} Amn \sin\frac{m\pi x}{a}\sin\frac{n\pi y}{b} \qquad (2)$$

Merging the above, we can see the total displacement

$$W_Z = W_i + W_w = S_0 + xX + yY \qquad (3)$$

4.2. *The establishment of the overall interaction model*

Displacement from the raft-based model based on structural columns from the upper edge node, Node applied to pile-soil force on the raft and the raft base weight, the potential energy expression of raftfoundation can be obtained, potential energy principle can be established according to the upper structure of a raft a joint action of a basic equation of the foundation.

4.2.1. *The deformation energy*

The strain energy density of raft foundation established is *e*, the deformation energy of the raft foundation can be written as

$$W_{w} = \int\int\int e dx dy dz \tag{4}$$

In which $0 \le x \le a, 0 \le y \le b, -\frac{t}{2} \le z \le \frac{t}{2}$

4.2.2. *The work done by external force*

(1) Weight of raft foundation work done Wr. $W_r = GW$, G is raft foundation for the weight, W is the average settlement for the raft foundation.

(2) Pile of land-based reaction force by the hard Wps.

Let the land-based pile stiffness matrix [K], the number of nodes for the M, pile-soil column node reaction force and displacement vectors are column vectors {Pi} and {Ui}

When

$$\{P_{\mathrm{i}}\} = [K]\{Ui\} \tag{5}$$

$$W_{ps} = -\{P_i\}^T\{U_i\} = -\{U_i\}[K]\{U_i\} \tag{6}$$

(3) Column node forces the upper structure of the work done Ws.

Divided into two parts: the work done by the column node reaction force when the foundation settlement is uniform ,and the work done by the column node reaction force when the foundation settlement is not uniform.

4.2.3. *The raft foundation of the potential energy E*

The potential energy of a system of external forces for change can be made to work with the poor.

So

$$E = W_b - W_r - W_{ps} - W_s \tag{7}$$

4.2.4. *Solving the equation*

According to potential energy principle, a system in equilibrium, its potential energy minimum.

By (7) differentiate the different variables, so that it is zero.

It is possible to acquire a series of equations to obtain the unknown parameters.

5. Conclusions

More than we made all kinds of analysis of piled raft foundation, and thus come to a conclusion that the formation of a preliminary theoretical model of the design of piled raft foundation makes a certain basis, and shows a clear outline. Due to space limitations, details of piled raft foundation of theoretical analysis and computer paper for later explanation.

References

1. DONG Jian-guo, ZHAO Xi-hong. Foundation for structures of tall building[M]. Tongji University Press, 1997. 9:270-271 (in Chinese).
2. HAN Xiao-lei. Soil mechanics and foundation. Beijing: Metallurgy industryr Press, 2004. (in Chinese).
3. BAO Shi-hua, FANG E-hua. Design for structures of tall building[M].Tsinghua University Press, 1990.10 (in Chinese).
4. JGJ3-2002 Technical specification for concrete structures of tall building[S]. Beijing: China Architecture Building Press, 2002.7. (in Chinese).

Demolition Accounting in Residential Building Project; Case Study of the Wuhan Tiandi Mixed-Use Development (WHTD) *
(April)2016

A.S. Damaneh† and S. Darbalaei and X. Fengju

Management School, Wuhan University of Technology,
Wuhan, Wuchang 430070 ,Hubei, China
†E-mail: Santa_m1984@yahoo.com.com
http://www.whut.edu.cn/

L. Shui

Wuhan Shui On Tiandi Property Development Co Ltd
Wuhan City, Hubei Province, China
http://www.arup.com/projects/wuhan_tiandi

Abstract: In this Decade, First visitors to China are impressed by two things; the skyscrapers and Demolition sites, this paper strives to settle down the demolition dust in Residential Building Projects (RBP) by Accounting's help. Between 2008 and 2010, around 1.8 billion square meters(SQMT) of floor space was demolished [1] .And only in 2011, 1.9 billion SQMT of residential floor space was built in China. The Ministry of Housing and Urban-Rural Development has not released a figure for housing demolitions for nearly a decade [2]. All profit and loss of Demolition process should add to the new construction process to reach a better and accurate Environmental Accounting result for the best decision-making. Calculates the Total Demolition Cost (TDC), by finding the 5 main demolition costs are available in this paper. Five main costs consists; Workers wage(Ww), Rental Equipment cost(REc), Permit Fee(Pf), Sorting Cost (Sc) and Removal cost of hazardous material (Rhmc).Transportation cost from demolition site to the landfill site and landfill cost are not calculated in this paper because , these figures shall be calculated in Waste process. The major questions that this paper strives to answer is: what is the TDC for 1 SQMT of RBP of Wuhan Tiandi? And , what is the contribution of 5 waste material groups of total demolition cost? Refere to the table 2 and from the results of this study , The average cost for demolition process in Wuhan Tiandi is about 115 Yuan (19.04 Dollar) per square meter, provided this figure is much bigger in US (about 50 to 80 Yuan per SQMT) this is a rough estimate for RBP in China.

Keywords: Residential Building, Demolition, Demolition Accounting,

---

* This work is supported by Wuhan Tiandi Construction Project, ARUP Company
† This paper and Work partially supported by Wuhan university of Technology

1. Introduction

The question that I always find myself asking is; why is it so difficult to build a sustainable economy for a country or to develop a desirable culture for a society, this may take years or even decades, but the destruction or demolition will be done in a much shorter period of time?. The research is carried out through two categories, namely that of Construction and Demolition(C&D), which has a deep root in me. However, the destruction of the environment is the hidden intellectual concern in this study. The total real estate investment in china is Yuan 6,441.3 bn, Annual construction of residential real estate by real estate developers is about 1,306.95 million m^2 and the average real estate sale price in china is about $5791/m^2$,[5].

2. Methodology

This paper used a quantitive methodology to find the answer for the two main paper's question which is; What is the TDC for 1 SQMT in RB? what is the contribution of 5 waste material groups of total demolition cost? This methology is going to find out the seven figures such as: duration of demolition process, permit fee , renal equipment cost, total demolition area, workes wages, removal hazardous material's cost and sorting and seperation cost of building waste. as in Eq. (1) shows the formulation way for TDC calculation.

Where;

Worker's wages =Ww, Rhmc= Removal cost for hazardous materials

Ec= equipment cost, Sc= Sorting cost, Pf= permits fee , DA= Demolition Area

TDC = (1)

$$\frac{\sum Ww + Rhmc + Ec + Sc + Pf}{Demolition\ Area}$$

This paper is dependent upon a number of assumptions, and we consider the sensitivity of paper to these assumptions.

-Demolition process will be happened after 40 years of construction processes; thus the cost of demolition should be added to the cost of new building project.

-we assumed that the old building that suppose to be demolished is exactly the same as Wuhan Tiandi project in terms of area, materials used and so on.

-Demolitions area assumed to be same as construction area of Wuhan Tiandi exactly (592,000 SQMT).

- Transportation cost and landfill cost are not add to the Total Demolition Cost(TDC) becuase these costs should be compute in Waste process.

3. Types of Materials in Residential construction and Demolition Projects:

Before arriving to the demolition discussion, this paper displays the 5 main building waste material categories that during the demolition process should be separated. In this study, waste should transfer to the 5 big bins or sites, the separation and sorting of demolition materials plays a crucial role in this paper. it is the one of the main demolition expenses. this paper makes 5 bulk group of demolition waste to calculate a little bit more accurate result for demolition costs in RBP. This paper works on a residential project because Residential Buildings (RB) normally have a long life cycle with only limited interaction among stakeholders involved in different phases of a building's lifetime. According to Worldwatch Institute, the construction of buildings consumes 40% of the stone, sand and gravel, 25% of the timber and 16% of the water used annually in the world. [4] This section of demolition process (or accounting) has an essential effect on Environment cost in RBP.Calculation of demolition cost is approximate and uncertain for accountants but to achieve a better result, we split the demolition materials into 5 main categories and make a distribution percentage for each group. The cost of demolition is not equal for all 5 groups of materials and material wastes, But due to lack of adequate information, the total cost of demolition will be allocated into 5 group of building materials equally.Although, the author thinks that there is a better classification for building materials as well, but due to lack of information, can't classify the material, this classification splits the building materials into 3 groups which is very useful for computing the waste process. These 3 groups of building materials are; recyclable materials, reusable materials and materials which are more transfer to landfill site. Five Main building material groups are:

3.1 Used Structural Metal

First group is Used Structural Metal that construction requires steel, which in turn requires iron ore. There is an estimation shows that residential construction uses about 14 per cent of China's crude steel output[5]. Metal is used as structural framework for larger buildings such as skyscrapers, or as an external surface covering. There are many types of metals used for building such as Steel (The estimated steel use in residential construction grew by over 70 per cent from 2004 to 2011)[6]. While most modern apartments are constructed with reinforced concrete in China. Estimates by Walsh in 2011, suggest low-rise Building construction requires, on average, around 40-50 kg of steel per SQMT of floor space, whereas around 60 to 100 kg of steel per SQMT is required for a

30-storey building, hence Wuhan Tiandi project used about 50×592000Sq = 296000000 kg steel for its RBP and demolition process produces 29600000 kg steel as well . There is an assumption that the average steel intensity for newly constructed and demolished Residential Buildings in China was 60 kg per SQMT in 2010. [6]. Given that approximately 1.54 billion SQMT of residential floor space was constructed with reinforced concrete in 2010, this implies steel use of about 93 million tonnes[7].

3.2. Existing Structure

Existing structural Materials contains, concrete or partition walls, clay, stone and cement are the most important material for this group.

3.3. Body Structure or Wall and ceiling structure

This category covers the framework of buildings. We assume Bricks, Ceramics and Plaster can be in same category because, in the process of demolition and waste management can be re-used an equal percentage of these materials.

3.4. Used Timber and plastic and glass

Wooden houses are more common in South America; therefore, in China for concrete structure, plastic, timber and glass are in same tomb.

3.5. Other Materials and light weight Aggregates:

Each substance or materials of building materials that do not include the 4 above groups, will be part of this group .(Only 5%)

4. Demolition Process:

This paper focuses on a big building project in Wuhan-China, Wuhan Tiandi Riverview Plaza site is located in the central area of Wuhan, adjacent to Tangtze River and the Bridge across Tangtze River. Wuhan Tiandi With a site area of 48ha, the total gross floor area of approximately 1,500,000m² is planned to accommodate a population of 20,000 people. The total amount of investment is around 10 billion Yuan. Wuhan XinTiandi completed its construction in November, 2009 with a gross floor area of 57,000 SQMT above ground, And it includes a residential building area of 529,000 SQMT(6,372,234.9 Sfeet)[8]. Demolition process is a part of new construction process hence we have 3 different accounting methods in building sector; Demolition Accounting, Waste Accounting and Construction Accounting. Between 2008 and 2010, around 1.8 billion SQMT of floor space was demolished to reduce the number of dilapidated buildings in China [9]. These demolitions imply an annual demolition rate of at least 22.2%, based on our estimates of the stock of floor Space. It is unlikely; however, that China will sustain such a high building demolition rate, because the stock of dilapidated residential buildings projects (RBP) will gradually decline. As these buildings are replaced with higher quality construction, the average life-span of residential buildings will increase for new Chinese RB. Also, the legal system has moved to enforce property rights for citizens, which will add cost to the expropriation process and reduce the rate of building demolition. Hence, we assume the average building life span of 40-50 years (Lardy 2012). These include governmental regulations on the use of property, and obligations that are created in the process of demolition. This process will be happened after the building's or project's life but Accountants must find this costs and add it to the building cost for make a green life for current and future generations, [9]. Generally, costs of demolition are expenses in the period in which they are incurred except for the purchase of land and building with the intention to demolish the building.

5. Calculation Model

To better understanding the prospects for urban residential demolition, this section attempts to present a calculation model for RB demolition. Some of these figures are accurate and extractdirectly from Wuahn Tiandi RBP project but some of them are approximately used, because there is no data available. Demolition process meet seven needs that includes A , B, C, D E , F and G.

A: Strives to find out the accurate measurement of demolition area of RBP. The residential Building area is 529,000 SQMT (assumed that demolition area is same as constructed area)

B: It's going to find the time for demolition process, how long will take demolition process.the amount of demolition time varies signifinatly depending on many factors including proximity of building to other buildings, etc. To setup the building for demolition process many tasks need to be completed; Obtain all necessary permits, All utilities need to be disconnected , Perform asbestos abatement, Remove all other hazardous materials, etc.remove all RB materials from inside RB including all drywall, furniture, loose metals - basically perform interior demolition to expose all metal beams, Have a good implosion demolition firm perform study to determine proper placement of explosives,Precut beams and set charges, implode and cleanup the demolition site. According to some projects managers perspective and some projects, the time is for this project is about 240 days(8months)

C: The cost for equipments rental (an hour) is an important figure for calculation model, there are different demolition machine available such as: Compaction Equipment, Crawler, Wheel and Backhoe Loader , Dozer - Crawler, Dumper, Excavator - Crawler and Wheel and Mini, Light Equipment, Off-Highway Truck, Skid Steer, Attachments. Refere to the table number 2. We asked the cost of Rental equipment directly from several demolition site that average cost is 50 $ per hour in china for different machine. Rates of equipments for Single Shift = 160 Hours per month [10] .(It is the general practice of the industry and Highway Equipment Company to charge overtime for any hours used in excess of 160 hours in any given month at the rate of 1/160 of the monthly rate for each hour in excess of 160 in any one 28 consecutive day period.) shown in Table 1.

Table 1. working time and rental fee for demolition equipments

No	Name of equipments	Rental rate for an hour	Working time	Total fee
1	Compaction Equipment	50	150 days-800 hours	$40000
2	Crawler Loader, Wheel and Loader Backhoe	50	140 days- 750 hours	$37500
3	Dozer - Crawler,	50	50 days- 250 hours	$12500
4	Excavator - Crawler-mini and wheel	50	180 days-965 hours	$482500
5	Dumper,	50	200 days-800 hours	$40000
6	Light Equipment	20	200 days-830 hours	$16600
7	Off-Highway Truck	20	should add to waste process 1200 hours	
--	Total			$184850

D: A demolition Plan should be submitted to the Buildings Department for approval. demolition needs some certificate such as: Hoarding Permit and excavation permit and some more permits from Environmental protection department, Leon Berkelmans and Hao Wang(2014). This step going to find out the Cost of any permit or certificate for demolition process. The cost of the demolition permit is based on the size of the building to be demolished and depending on local rules and the size and nature of the project(in USA). Permit costs vary considerably, for example, The permit fee for a garage or shed is $25.. Permit cost starting around $25-$50 but going up to several hundred dollars or more in China. According to the section 3.2 building permit fees in US the permit fee for Residential Structure is $13.00 per $1,000 of estimated cost of demolition. If we do the same in china , (592×13)7696 dollar it's the cost of demolition permit for this project. [11]

E: It‘s the most important part of calculation model that going to find the cost of asbestos removal. Hire an inspector to do it properly. when hazardous materials e.g., asbestos containing materials, or petroleum, are present, they should be removed and cleaned/disposed of according to the statutory requirements Administered by the EPD, Fire Services Department, Labour Department(LD). The asbestos waste should be handled, stored and disposed of as chemical waste in accordance with the Waste Disposal Ordinance & Waste Disposal Regulati on Budgeting and Financial Management Manual (2014).If you need asbestos removal to accompany normal demolition that will cost an extra $3 to $5 per square foot. [12] Asbestos removal requires your demolition company to take extra care in disposing of waste. It also requires them to send their employees through an asbestos certification class also, 108800SQMT is an approximate kitchen area for Wuhan Tindi RBP 89200 SQMT is an approximate Bathroom and area of hazardous material for Wuhan Tindi.

1 SQMT = 10.7639 Square Feet, hence 592000 SQMT = 6,372,234.9 Sq Feet

Rhmc=108800 +89200 =198000SQMT×10.7639 = 2131252 × $ 5(30Yuan) (asbesteos removal fee for 1 QSFT = $10656261 rate of removal asbestos

F: for calculation of worker's wages for an hour .Hiring some workers to remove and dispose of the old materials. 25 permanaents workers who are worked for 240 days and 8 hours per day , according to the worker's rate in china , each worker should get paid 3.25 dollar for an hour , therefore;

25 workers × 8 hours × 240 days × 3.25 wage = $360,568

G: This step going to measure the cost of separation and sorting waste into 5 big groups.10 workers work for 100 days (8 hours a day= 5 dollar) to sort the waste material into 5 different bins or sites that they are ready for transport to the Recycling,Reuse and landfill sites.

10 workers× 8 hours a day × 5 dollar wages per an hour × 100 days = 40,000

2 quipment work for almost 450 hours × 50 dollar for an hour rental fee = 22500

Hence: 40000 + 22500 = 62500 dollar is a sorting cost for Wuhan Tinadi

Table No 2 . Calculation Table

A C& D Area	B Time of Demolition Process	C Rental equipments cost	D Permit fee for D process	E Asbestos rate for 1 SQFT	F Workers Wages(Hour)	G Sorting & seperation cost	Number of workers
592000 m²	no sources in current documents - 240 days approximately	$50 And, $ 20	1000 SQ is about $ 13 , hence; 592 * 13= 7696	$ 5.2 Totall area made up of hazardous material is 198000 Sq	$ 3.25 for Full-time workers $ 5 for Part-time workers	10 workers for 100 days	25 25permnent workers 10temperory workers

Hence; Ww= $ 360568 , Rhmc= $ 10656261, Ec– $ 184850, Sc= $ 62500

Pf= 13×592= $ 7696

$$\frac{\sum 360568+10656261+184850+62500+7696}{592000}=\frac{11271965}{592000}=$$ 19.04dollar is ademolition DCost per SQMT , 19.04× 6 (rate of Dollar) = 114Yuan is a total Dc 1 SQMT.

Table 3 . TDC contribution to the each group of materials in Wuahn Tiandi RBP

No	Materials Group	Total Waste in Demolition	TDC contribution of 5 GBM for 1Sq
G-1	Used Structure metal: Metal, steel, bar...	30%-29600000kg	$5.7
G-2	Existing Structure: Concrete, cement, clay, stone,	35%	$6.664
G-3	Body Structure or Wall and Ceiling Structure: Bricks, , ceramics, Plasteraw	20%	$3.808
G-4	Used Glass, plastics, Timber...	10%	$1.904
G-5	Other Materials light weight aggregates (chemical and hazardous materials...)	5%	$0.95
--	Total	100%	19.04

6. Conclusion:

Demolition Accounting is a new financial accounting term for Environmental Accounting that requires extensive and articulate studies, this is an ideal area for future research, but if we look at the statistics released through Books and articles on internet about demolition process, you can find, all are approximate. From the study results, we can suggest that builders should mix three different accounting methods to get the best green result. The conclusion of this article is two-pronged; theory and practical. Theory aspect complains about 3 problems: firs problem and biggest problem is the lack of sufficient information in the field of green accounting. The second problem; Governments do not have enough volition to make green decisions in this sector. The third problem is that lack of cooperation between companies and researchers, for various reasons such as: the language barrier and lack of trust. China is very advanced in the process of demolition this decade. The average cost for residential demolition is different for kitchen and bath room and living rooms; therefore the total cost for demolition is about 115 Yuan per SQMT. This is a rough estimate. It's difficult to pinpoint an exact cost for TDC, because each property is different and can present its own set of unique challenges depending on the age of the property.

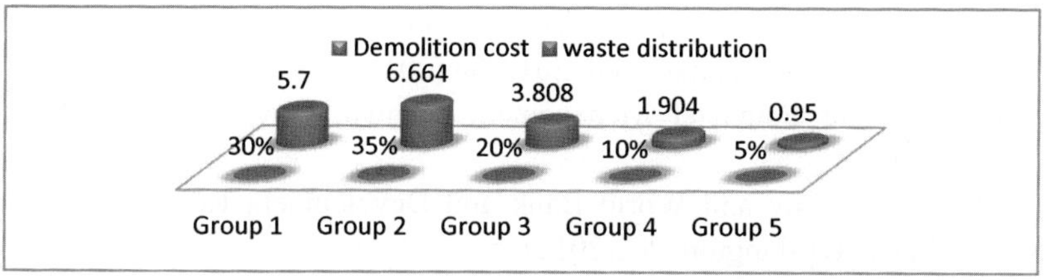

Fig 3. TDC Distribution for 5 RB Materials Group

Acknowledgments

I would like to express my gratitude to Prof. Xu Fengju at Wuhan University of Technology who guided me in the completion of this inquiry through discussion and feedback. I would also like to thank Dr. Sanaz Darbalaei and Mr.Mohammadreza Soleimani for their kindness in providing me with valuable information and support which allowed me to carry out this research.

References

1. State-Owned Assets Supervision and Administration Commission of the State Council, 2011
2. Rosealea.Y, Thomas.G of GK Dragonomics, (State-Owned Assets Supervision and Administration Commission of the State Council 2011)., Bloomberg (2011).
3. Information on, (http://www.statista.com/topics/1500/real-estate-in-china/) (2015). NBS China Statistical Yearbooks(2012-2014).
4. Woeztel, J, L Mendonca, J Devan, S Negri, Y Hu, L Jordan, X Li, A Maasry, G Tsen and F Yu, ' Preparing for China's Urban Billion', report,(2009).
5. Wu J, 'The Recent Developments of the Steel Industry in China', Document No DSTI/SU/SC(2009) 15 presented to the OECD Steel Committee Meeting, Paris, 8–9 June (2009).
6. Ashely Smith, 'How Much Does Demolition Cost? ',USA(2012) and Walsh S, 'New Opportunities for Industry Growth', Presentation to Metal Bulletin Conference 'China Iron Ore 2011', Beijing, 23–24 March(2011).
7. Woetzel et al (2009) and Walsh in 2011. Given that 1.54 billion Sq of residential floor space was constructed with reinforced concrete in 2011
8. ARUP website & information on, http://www.arup.com/Global_locations/Greater_China/Wuhan.aspx, (2015)

9. Leon.B and Hao.W, 'Chinese Urban Residential Construction to 2040' Research Discussion Paper, RDP 2012-04 (2012)
10. Budgeting and Financial Management Manual, Government of Nova Scotia(2014)
11. Highway company and World Bank and Development Research Center World Bank, Washington DC,(2012).
12. Budgeting and Financial Management Manual - Government of Nova Scotia(2014)and Au C-C and JV Henderson, 'Are Chinese Cities too Small?', The Review of Economic Studies, 73(3), pp 549–576,(2006).

Analysis of brick structure tower rectification technology — Protection of Kaifeng iron tower

Li Li[1], Li-Sha Wang[2]
[1]*College of Civil and Architecture Engineering, Henan University, Kaifeng 475004, Henan , China*
[2]*Zhengzhou Normal University, Zhengzhou 450000, Henan , China*
E-mail: lylylee@henu.edu.cn

Through to the state-level key cultural relic Kaifeng iron tower's damage origin analysis, according to its brick masonry structure stress characteristic, the iron tower incline key aspect "the water" "the earth" the nature, the tower body and the tower footing foundation coordinated work and so on, proposed that the Kaifeng iron tower incline corrects an error the technical processing plan, provides the essential reference for the brick structure tower's protection.

Keywords: Brick Structure; Kaifeng Iron Tower; Tilt; Rectification.

1. Introduction

Chinese retained many historical civilization and ancient buildings. There are over 500 important ancient towers, and nearly 200 of them have a history of more than 700 years, most of which are of brick masonry structure. So, brick ancient tower is an important part of ancient architectures and it is the representative of the ancient high-rise building. Because of the influence of natural factors and destruction of human activities, such as wind and rain, sun and dew, season to change, freezing and thawing changes, earthquake, thunderbolt [1-2], and so on, brick ancient towers inevitably affected by different degrees of injury and tilt, even face the danger of collapsed [3-4]. Ancient tower has very high value of the buildings, the art, the history and culture, and the archaeological, therefore, the protection of ancient tower is imminent, it has very important practical significance and it's also a hot topic in the protection of cultural relics.

2. General Situation of the Iron Tower in Kaifeng

"Kaifeng iron tower" formerly known as Kai bao temple tower, located in iron tower park which is in the northeast corner of Kaifeng, Henan. In 1961, iron tower was first time approved as a national key Protect the Cultural Relics, and enjoy the laudatory name of "The first tower in the world". The tower's predecessor is a traditional wooden tower, octagonal and thirteen storeys, 360 feet high. The main function of this tower is worship the Buddha Sarira. Supervision by the famous architect Yu Hao responsible for. It was built in the Song Dynasty of Taiping Xingguo seven years (982 years), and completed in Duangong two years (989 years). Named "Fusheng tower", In 1013, Due to the phase wheel of the tower has shined, there is a new name is "Linggan tower". In Qingli four years (1013 years)in June, it destroyed by the thunder fire, only endured 56 years. The first year of Huangyou (1049 years), the emperor Renzong ordered to reconstruction. Experienced previous lessons, In order to prevent the thunder fire and improve the performance of resistance to wind and rain, thus to chooses the Coloured glaze component. Because of the tower whole body was all inlaid with brown Coloured glaze tile, it looked from a distance approximate iron color, so from the Yuan Dynasty up to now has been known as the "iron tower". The tower is thirteen storeys octagonal pavilion style, nowadays, it higher than that of ground 55.08m (the original height is 55.88m), This is the highest one of existing domestic Coloured glaze towers, and It is fully used the form of China's timber frame structure, the slender tower is tall and majestic. The basic parameters of the iron tower as shown in the table 1. Basic structure of the tower's wall: the Keisuke shape door and eight side corner supported by detached column, the headers of the plane surface with the grooved bricks of the elevation component frame, set the elevation stretcher in the middle. The bracket sets between columns distribute very intensively, Ni dao dougong and Gua zi dougong are connected. Architrave above the tile slope were all laying flat brick. The eave purlinat the corner of the tower eaves were intersect into splayed, At the confluence of the place setting the cicada shape cantilevered corner beam, and there are 104 LingDuo under the eaves. each direction opened one room in the tower's base, make the door to the keisuke shaped spire by corbel technique, the brick staircase of north gate can around the core column arrived at the tower's top(brick staircase has 168 steps), The remaining three are all grottos. The tower of each layer has eight sides with window, The first layer's window is facing south, second is facing west, third is facing north, fourth is facing east, These are bright windows and the rest are blind windows, the rest may be deduced by analogy [5]. The core column of tower is the core component which supports the tower, and other external brick components of different uses are connected to the core column through climbing flights, which makes all of them as a whole, and which make the tower has a good seismic capacity.

Table 1. The relevant parameters of Kaifeng iron tower.

Tower layer munber	*Tower layer height*	*Diameter*	*Thickness of the wall*	*Window's area*	*Tower layer munber*	*Tower layer height*	*Diameter*	*Thickness of the wall*	*Window's area*
1	5.44	10.25	1.50	0.79	8	4.65	7.76	0.87	0.66
2	5.05	9.82	1.40	0.90	9	4.46	7.06	1.10	0.45
3	4.59	9.43	1.35	0.83	10	4.44	6.60	1.25	0.54
4	4.08	9.44	0.95	0.75	11	4.44	6.15	1.28	0.45
5	4.08	8.77	1.08	0.77	12	4.32	5.80	1.11	0.40
6	4.58	8.40	1.02	0.74	13	3.09	5.55	1.02	0.43
7	4.86	7.96	0.95	0.67					

3. The Research of Kaifeng Iron Tower's Damage Reason

Kaifeng iron tower rely on its reasonable design, firm structure and other advantages, it has been 962 years is still standing. Because of the characteristics of the iron tower material, it inevitably affected by natural and artificial factors influence and the tower has suffered varying degrees of damage. For example, parts of the surface of Coloured glazed tiles have been shedding, Individual pendant had severe corrosion, canopy tower overgrown with weeds, The whole tower inclined seriously(the spire's displacement deviation in the plane is 0.764m)Etc. The urgent need to analysis the damage reasons of the tower, so as to take corresponding and effective protection measures.

3.1. *Natural factors*

Affected by the Kaifeng local climate: The winter monsoon is northwest wind, the wind velocity is 2~4m/s, the gust is 6~7 level, It has a great effect for the tower to southeast tilted; the summer monsoon is southeast wind, the wind velocity is 1~2m/s, this is beneficial for the tower tilted to the southeast direction. Extreme weather coupled with the solarization and sometimes lightning stroke etc, it had experienced the storm wind 19 times, rainstorm 17 times, hail 11 times. All of these factors caused the tower's local damaged, and lead to the tower's components decay, and the tower body broken and tilt.

Kaifeng is located in the plains, there is little seismic belt, seismic fortification intensity for 7 degrees, the basic seismic acceleration is 0.10g, once experienced earthquake 44 times, and the earthquake also is an important factor which leading to the tower tilt.

Kaifeng is bordered the Yellow River, so the groundwater resource is quite rich. Since the rise of ground water level in flood season, the iron tower foundation soil changes, so the foundation bearing capacity decreased and produce the phenomenon that subsidence, eventually lead to the tower tilt.

The northwest side of the Kaifeng tower, it original has a moat (Now there are site) and Kaifeng water system engineering (the water flow has always been very large), there are a number of lotus pond and lotus ditch on the west side of the iron tower. When the soil below the tower is not uniform or the tower was in the place of alluvial fan mouth, fine grain soil and soluble salt will go along with the groundwater flow away, long term effects will cause the tower foundation uneven settlement and then resulting in the tower tilted.

3.2. *Human factors*

When the Iron tower at the early stage of construction there existed the artificial tilt problem, this caused the stress of foundation has uneven distribution, then,the tower foundation uneven settlement, and exacerbated the uneven distribution of stress, eventually lead to the tower's body is continue to tilt.

Sometimes, human disasters are fiercer than natural disasters. Kaifeng has had more than 6 times of the river trouble, the iron tower is also no escape from fate. Such as the Zicheng Li uprising army to attack Kaifeng, The governor of the Ming dynasty dug the Yellow River levee, try to flood the insurrectionary army, the results, Kaifeng flooded turn into ruins, and the tower base was also flooded into the Yellow River water sediment. Confirmed by exploration, near the tower under about 10 meters there is a layer containing a large amount of residual rotten brick tile cultural layer. When the burst of the Yellow River, the flood is just right from the northwest flow to the Southeast, which is another important factor leading to the tower tilted to the Southeast direction.

In 1938, when the Japanese army invaded the Central Plains, they had once artillery attacks the tower from the Northwest, from fourth to thirteenth layers ,each layer of the eaves and bases have damaged, among them, the eighth and ninth layer are quite serious(So far, still can see the repaired places). In addition, the aircraft had attacked the tee more than 60 bullets, these results all will lead to the tower tilted to the southeast direction.

Kaifeng iron tower has not escaped the “Play”, “smashing”, “robbing” of the persecutions during the cultural revolution ,Tower‘s partial components was damaged or lost, and produce eccentric compression, so that it prompted the tower to tilted..

In a word, the root cause of tower tilt is not single but is complicated, due to the role of a variety of factors and lead to the serious consequences of tower tilted.

4. Corrective Actions of Kaifeng Iron Tower

At present, China is enlarging the protection to the historical building relics, so every year is consuming a large quantity of manpower and material resources. In Henan province, the main measures of pagoda building protection has the following three: The rescue protection; The general protection; The environmental governance protection etc. The tilt correction problem of tower belongs to the rescue protection. The following will explore to the tilt correction problems of Kaifeng iron tower.

4.1. *Process of tower's monitoring*

Select tower's key points as the monitoring points, such as the tower spire, each layer of certain eaves or window top etc, cooperate with the more obviously control points which is around the tower, and monitoring tower's location coordinates and azimuth (this work requires the special team to complete). in the process of Tower’s corrective governance, the work of construction and monitoring should be carried out at the same time, in order to grasp the deformation of the tower, and do know the score.

4.2. *Treatment of groundwater and surface water around the tower*

Use the water drainage or water retaining methods to control water. As far as possible to make the surface water away from the iron tower or use the surface waterproof layer to isolate surface water; the treatment of underground water is to cut off water by building continuous walls. Eventually, so as to guarantee the tower stress is uniform, uniform settlement, do not produce eccentric compression, does not have the phenomenon of stress concentration etc, and can’t create the opportunity for the iron tower inclined once again.

4.3. *Tower base treatment*

Kaifeng tower is tilted to the southeast direction, In the northwest side place of the tower foundation, where is distance the tower foundation 1.2m away, depth is 5m.drilling and took out the excess soil, to make the tower leaning back, achieve the error correcting purpose. Or preload 500t at the northwestern edge of the tower, to make the northwest tower sinking quickly, finally making the tower leaning back. Leaning Tower of Pisa correction had tried this method, but the effect is not obviously.

4.4. *The process of change and shift the tower base*

Because of the Kaifeng tower had a long history, so it tilted seriously. We can try excavating holes around the tower, set the reinforcing cage or steel pipe etc, using the pressure grouting way to let it filling density, and it will re-form a new foundation which has the high strength and strong ability, to make the tower to become the new unity. Avoid the risk of" Without a solid foundation, the strong earthquake will be happen". At this time there are two options to choose:

① we can make use of the new foundation settlement chance, dig the soil in northwest side of the tower foundation to make it subsided or use a hydraulic jack In the southeast side of the tower to lift it, In order to successfully complete the work of tower tilt correction;

② we can use the building displacement principle, choose a favorable location, and then using the special hydraulic jack hoisting the tower, move the tower to the selected location, and then complete the final positioning fixed work. Ancient building relocation is a more and more widespread way of protect the cultural and natural heritage, As long as we move the tower to the safer place, we can make the buried in the underground part of the iron tower recreate, and can avoid the damaged of flood or earthquakes and other natural disasters etc, and It also can provide conveniences for our economic development and city planning. Finally, really realize the protection for iron tower cultural relic.

4.5. *Combinational correction*

Like a big tree transplanting, with a rope to tie up all the dirt and roots of the tree. Use the timbering, steel reinforcement hoops, steel wire hoops and hemp ropes tie up whole of Kaifeng iron tower, to make it as a whole and use cable ropes to control the direction. Then use the steel raft to underpin the iron tower's foundation, then for earth excavation in the northwest side of tower, the southeast side pressure, should be depends, when it reaches the target, using the pressure grouting to make the foundation stability.

4.6. *Sinking well water-jet method rectification*

Still use the hoop to fixed the tower, and with cables controlling direction. In order to dig soil role, near the iron tower's northwest side to water-jet open caisson, In addition, perforating and pumping water. The northwest side accelerated subsidence, finally achieve the purpose of tower rectification. But this method has some risks, especially is crucially for the control of water, this need to be cautious.

5. Conclusions

In Kaifeng iron tower rectification problems, we should be first consider the problem is tower body base's expanding and strengthening, to ensure the tower permanent stability after the tilt correction.

Through the research of tower rectification, we can found the key of brick structure tower inclination governance is the "water" and "soil" processing.

Many factors lead to the tower tilted, however, use the repressive measures of dig soil or top pressure is the relatively effective measures for brick structure towers tilt correction, this need to enough attention.

How to realize the tower body and the tower base coordination is the core of the brick structure tower inclination rectifying works, it's also the hot spot on which professional persons of construction industry are studying at present.

Acknowledgments

Project funding: Basic and frontier project of Henan province department (132300410059).

References

1. Dai Yisu, Cao Shuangyin, etc, Safety and stability analysis for structure of the inclined pagoda [J], Special Structures, 2002.19(4):23-25.
2. CECS 225:2007, Technical code for moving, incline-rectifying, story-increasing and reconstruction of buildings [S].
3. CHEN Ping, ZHAO Dong, etc, A study on the rectification of Jingguang Temple Pagoda in Meixian County[J], Journal of Xi'an University of Architecture and technology, 2003, 35(1):44-47.
4. Jiang yuhua, Liu ling, Method of inclination observation in ancient pagoda rectifying process[D], The Eighth National rock mechanics and engineering academic conference proceedings, 2004.10,1016-1018.
5. The record of Kaifeng[M], Publishing house of Beijing Yanshan, 2001.10.

Return to place spirit in the architectural design in universities — A case study of Xiangshan Campus of China Academy of Art and Caotang campus of Xi'an University of Architecture and Technology

Yun-Yang Zheng[1,*], Hong-Li Guo[1,†]
School of Civil Engineering and Architecture, Nanchang HangkongUniversity, No.696, south
Fenghe Avenue, Nanchang, Jiangxi, 330063, China
*E-mail: *zyunyang@163.com, †474857401@qq.com*

The architectural design in universities is not a kind of technology, but a kind of emotional expression which may bring back memories about traditional culture as well as regional culture. This paper makes a comparison between the Xiangshan campus and Caotang campus on the site selection and the traditional design concept of harmony between buildings and environment, and elaborates the regional features of layout, design, color and material selection of university architectures. To show inheritance to traditional culture and add specific regional cultural characteristics, this paper may provide some insight into the construction of university buildings.

Keywords: Architectural Design in Universities; Traditional Culture; Regional Culture; Place Spirit.

1. Introduction

Norberg-Schulz pointed out that place spirit, obtained in the participatory processes, is the sense of human identity about subjective awareness and action [1]. Place spirit is not the reflection of the surface style of the architecture, but the spiritual resonance with human obtained through the harmony between the natural and artificial environment and through emotional relation between architectural form and traditional cultural concepts.

Liang Sicheng once said, architecture bears the reminiscences and feelingsof human beings [2]. However, currently, material comforts outweigh spiritual pursuit, so the national value and materialistic consciousness reflected by the architecture conflict with each other. Likewise, universities inevitably pursue high and noble architectures and prone to imitate western architectures, which is entirely contradictory to the intrinsic properties that should be borne by the universities. In this way, place spirit is very important in the architectural design. We find some excellent design works in universities, including

Xiangshan campus of China Academy of Art and Caotang campus of Xi'an University of Architecture and Technology. Located respectively in the southern part of China and the northern part, the two universities seem to be not comparable with each other. But actually, they both reflect the traditional and regional culture. The style of these two universities pole apart, representing respectively the gracefulness of the south and the steadiness of the north.

2. Inheritance to the Design Concept of the Academy

The Academy is considered to be traditional Chinese university, which promote the development of educational system and is of immeasurable importance to the development of modern university. Both campuses draw on the traditional design concept of the Academy in terms of site selection and the management of relation between architecture and environment.

Fig. 1. Caotang campus of Xi'an University of Architecture and Technology.

2.1. *Site selection*

Under the influence of Confucianism, the site selection of the Academy pays much attention to the natural environment, so secluded and beautiful mountains are often preferred. This concept is consistent with the norm of site selection of both Xiangshan and Caotang campus. The Caotang campus is surrounded on three sides by magnificent mountains like Qinling and is adjacent to Caotang Temple, with streams winding through it, demonstrating good ecological environment and strong cultural atmosphere (Fig.1). Likewise, Xiangshan campus is built around Xiangshan mountain. It gave up the planned area, namely, the campus city, set by the government, and chose Xiangshan mountain, a place with forest-clad mountains and limpid streams. They believed that natural environment is far more important than architectures and that harmonious environment and nice cultural atmosphere can cultivate students' humanistic spirit.

2.2. *Integration of architecture and environment*

The Academy is always built among the mountains and strives for integration of architecture and environment as well as harmony between man and nature. The overall architectural form of Caotang campus is in line with Qinling mountain. For example, the brick gray surface blends and fits into the whole scenery and becomes part of it. Similarly, Xiangshan campus is shared by half natural scene and half architectures, with small-sized buildings scattered on the foot of the mountain, making the coexistence of nature and architecture scene possible (Fig. 2). Wang Shu once said, the relation between the mountain and architectures in China is never about figure and ground relation, but about coexistence [3]. So it is reasonable to respect the natural environment and build ecological campus environment.

Fig. 2. Xiangshan campus of China Academy of Art.

3. Inheritance to Regional Culture

Regional culture is mainly composed of humanistic factors and natural factors. And architectures of different regions enjoy distinctive particularity in their connotations. So the cultural differences play an important part in the formation of these two styles.

3.1. *The design plan of the campus*

Caotang campus is located in the ancient capital of Xi'an, one of China's oldest cities. The design plan of it imitates the spatial pattern of Lifang unit system of Chang'an city of the Tang dynasty (Fig. 3). Square grid is widely adopted to form a regular layout and to connect different sectors of the campus by the traditional "nine-rectangle-grid" pattern. Unlike the regularity of Caotang campus, the design plan of Xiangshan campus is delicately beautiful, which is typical of Jiangnan landscape. According to specific land conditions, architectures in Xiangshan campus are built well-proportioned around the landscape, creating a secluded and elegant atmosphere.

Fig. 3. The gate of Xuefu city of Caotang campus of Xi'an University of Architecture and Technology.

3.2. *The style design of architecture*

As the ancient capital of thirteen dynasties Qin, Han and Tang, Xi'an has well preserved their traditional cultures. So the style of architecture in Caotang campus mainly inherits the "simple and bold" style of Qin and Han dynasty. The architect makes full use of traditional Hanque model and adds such adaptations as hip roof and exquisite brackets which form a blend of delicacy and force with the regular line of the roof. The gate of Xuefu City of Caotang respects the existing architecture form, on both sides of which the same Hanque model is adopted, making it in line with overall style of the architecture. To some extent, this reproduces the artistic style of the architecture of Qin and Han dynasty. With the aid of traditional Jiangnan gardening techniques, Xiangshan campus adopts the concept of change of the actual situation, extension and separation of space to achieve integration within campus. The rough surface of the architecture and the large number of verandas lead to the complexity of space. At the same time, the concept of frame is widely adopted in the design of windows. Through these windows of different sizes and irregular shapes on the walls, beautiful natural and human sceneries are presented from different angles, typical of the elegance and delicacy of Jiangnan landscape (Fig. 4).

3.3. *The color and material of architecture*

Regional culture is thoroughly demonstrated through the main color and material of architectures of these two universities. Under the influence of the history and culture of Xi'an, Caotang campus uses the typical "brick gray" as the main color to show the moderate and steady cultural atmosphere. Brackets are always reddish brown in color, adding some bright color as well as some interests to the overall gray tone. The use of common rocks as material for the surface of architecture and decoration saved the cost at that time and makes the architecture blended into the overall landscape of Qinling mountain. Xiangshan campus mainly choose black, white and gray to form a clean and elegant image while white walls cast beautiful reflections with gray tiles. And the material is chosen from common southern folk house materials, namely, tiles, bricks, mud, bamboo and timber, filling the campus with strong regional atmosphere. The rough walls and centuries-old bricks may share more intimacy with people.

Fig. 4. The teaching building of Xiangshan campus of China Academy of Art.

4. Inheritance to Energy Conserving Design

Limited resources make the traditional energy conserving mode equally important as the pursuit of traditional and regional culture. University should adopt the energy conserving mode and promote the design concept of sustainable development of architecture.

Xiangshan campus choose the local cedar and bamboo as the main material, which reduces the resource consumption during transportation and also reflects local culture. A large number of local recycled bricks, tiles and stones can be easily found in the campus, demonstrating a return to simplicity and the cyclic utilization of material. This helps to increase the recycling rate of architectural material and to keep a reasonable cost control. Since the southern part of China is hot and humid, the windows on the wall is conductive to ventilation and to reduction of the usage of air conditioners. So the power resources may be saved in this way. The Caotang campus employs the "nine-rectangle-grid" layout, connecting the adjacent institutes in the corresponding grid which is similar to that in the Academy. This is conductive to effective management and energy conserving. Transparent glass roof is widely adopted in the library to make it well lighted. And the university's way of managing natural wind and water reflects the design concept of ecological campus.

5. Conclusion

The designs of both Caotang campus and Xiangshan campus reflect the essence of place spirit, in-depth thinking about historical heritage and integration of conciseness and steadiness of history. The inheritance to traditional culture goes hand in hand with that to regional culture, which provided some insight into the architectural design for other universities. From the ancient Academy to the Chinese and Western combined style in the republic period of China, from the large-size architectures in the early years of People's Republic of China to the architectures which follow the western mode, the design of university architectures finally returns to traditional and regional culture. Through sedimentary accretion, university architectural forms are gradually taking shape. After tortuous development, university architectures will eventually come to the right path of development.

References

1. Christian NorbergSchulz, *Genius Loci: Towards a Phenomenology of Architecture (Chinese translation by Shi Zhiming),* Taiwan: Garden City Publishing, 1986.
2. Wang Yan, Impression of Architectural Style of Singapore, Art and Design, 9 (2015) in Chinese.
3. Wang Liqun, Idyllic Natural Surroundings--A Brief Analysis of Environment Construction of the Xiangshan campus of China Academy of Fine Arts, Beauty & Times, 1 (2015) in Chinese.

Analysis of beam-column joints destruction form about lightweight energy saving concrete for frame structure system

Yan-Min Yang, Ya-Nan Nie, Yan Zhen, Wen-Long Dong

School of Civil Engineering, Jilin Jianzhu University
Changchun, China, 130118
E-mail:1727057066@qq.com

According to the pseudo static test method to observe the lightweight aggregate concrete frame structure mode in anti-symmetry loads .To study the axial compression ratio for 0.1 and 0.3 full lightweight aggregate concrete beam column nodes and edges between nodes failure mechanism, analysis of different axial compression ratio and the same axial pressure than the failure performance of the beam column joint. The results show that the seismic performance of the side joint is better than that of the intermediate node and the bearing capacity decreases with the increase of the axial compression ratio.

Keywords: All-Lightweight Aggregate Concrete; Exterior Beam-Column Joint; Interior Beam-Column Joint; Destruction Form.

1. The Summary of Test

1.1. *The design of test piece*

Table 1. design parameters of each specimen.

node type	test piece number	beam reinforcement (mm)		column reinforcement (mm)		stirrup joints (mm)	compression ratio
		longitudinal	stirrup	longitudinal	stirrup		
intermediate node	ZJD-1	up and down 3Φ18	Φ8@100	3Φ25	Φ8@100	5Φ8	0.1
	ZJD-2	up and down 3Φ18	Φ8@100	3Φ25	Φ8@100	5Φ8	0.3
edge node	BJD-1	up and down 3Φ18	Φ8@100	3Φ25	Φ8@100	5Φ8	0.1
	BJD-2	up and down 3Φ18	Φ8@100	3Φ25	Φ8@100	5Φ8	0.3

According to the design principle of "strong column and weak beam, strong node weak component" [1], the intermediate nodes and side nodes of the middle layer of the intermediate layer under horizontal earthquake action are studied, two intermediate joint specimens and two side joint specimens are designed respectively and the lightweight concrete strength is LC35. The beams and columns and joints by HPB300 grade steel stirrups, beam and column longitudinal reinforcement are using HRB400 grade reinforcement. The design parameters of each specimen are shown in Table 1.

1.2. *Test apparatus*

Used for testing hydraulic jack to the top of the column axial force is applied, in the top of the column using the MTS electro-hydraulic servo actuator under low cyclic loading, the loading device as shown in figure 1.

1.3. *Loading design*

1-counter force frame;2-MTS actuator;3-out of plane protective side support;4-ball hinge; 5-fixed end connector;6-specimen; 7-vertical horizontal bearing

Fig. 1. Test loading device.

The test in 2mm as the loading increment, each load is two times, when the ultimate load decreased to failure load bearing capacity of the specimen (the ultimate bearing capacity of 85%), the specimen failure and this experiment is end. Low cyclic loading test is currently on the seismic performance of the wide range of test methods [2], the so-called low cycle refers to during the test of loading frequency of cyclic load is low [3].

2. Damage Phenomenon and Result Analysis

2.1. *Specimen BJD-1*

Applying the axial compression ratio of axial pressure of 0.1 to the top of the column, the beam end is loaded into the 60kN, the crack in the width of 0.2mm.When the beam is loaded to 22mm, the main crack width of the beam end is 1.4mm, the main crack of the joint core area develops along the diagonal direction, the crack of the beam end develops rapidly and the plastic hinge is produced. Loaded to 40mm, the core zone of the joint concrete off a large area, stirrup exposed, the load at the beam end is 61kN, and less than 85% the ultimate load, so this experiment is terminate. The destructive phenomenon of the test is shown in figure 2.

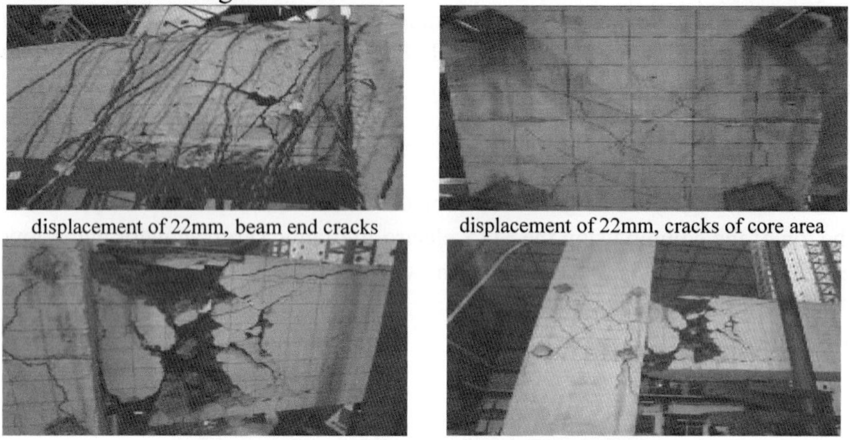

displacement of 22mm, beam end cracks displacement of 22mm, cracks of core area

displacement of 40mm, beam end cracks displacement of 40mm, cracks of core area

Fig. 2. the destructive phenomenon of BJD-1.

2.2. *Specimen BJD-2*

Applying the axial compression ratio of axial pressure of 0.3 to the top of the column, the specimen surface without cracks and the specimen in elastic phase. When the beam is loaded to 24mm, the main crack width of the beam end is greater than 2mm, the main crack of the joint core area develops along the diagonal direction, the crack of the beam end develops rapidly and the plastic hinge is produced. Loaded to 40mm, the core zone of the joint concrete off a large area, stirrup exposed, the load at the beam end is 65kN, and less than 85% the ultimate load, so this experiment is terminate. The destructive phenomenon of the test is shown in figure 3.

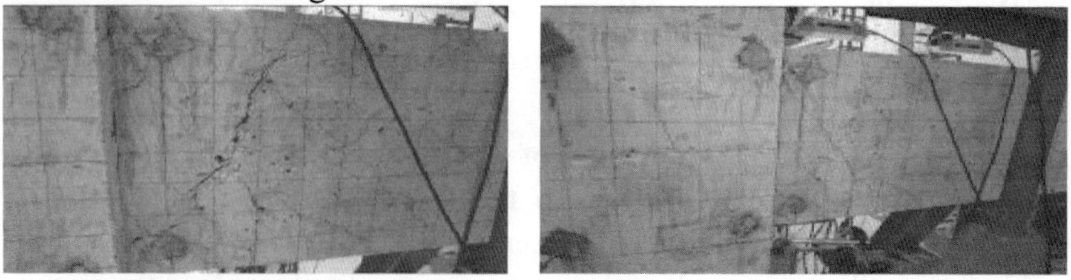

displacement of 24mm, beam end cracks displacement of 24mm, cracks of core area

Fig. 3. The destructive phenomenon of BJD-2.

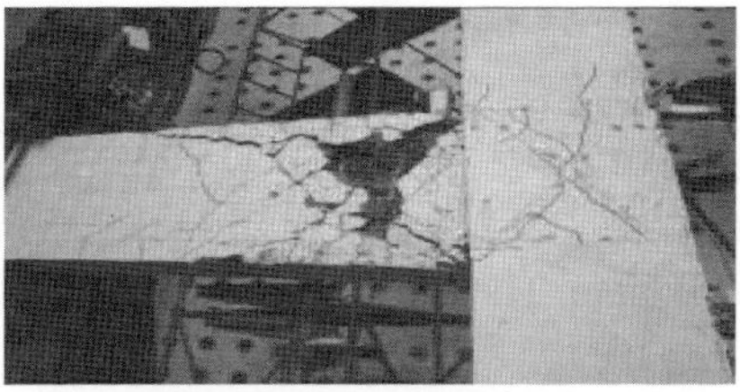

displacement of 40mm, beam end cracks displacement of 40mm, cracks of core area

Fig. 3. (*Continued*)

2.3. *Specimen ZJD-1*

Applying the axial compression ratio of axial pressure of 0.1 to the top of the column, the beam end is loaded into the 37kN, the crack in the width of 0.2mm, the cracks are smaller and less, under cyclic loading, the crack close quickly, is not easy to observe. Loaded to 22mm, the core area appear two 0.4mm wide along the diagonal direction of main crack. Loaded to 38mm, the core zone of the joint concrete off a large area, stirrup exposed, the load at the beam end is 50kN, and less than 85% the ultimate load, so this experiment is terminate. The destructive phenomenon of the test is shown in figure 4.

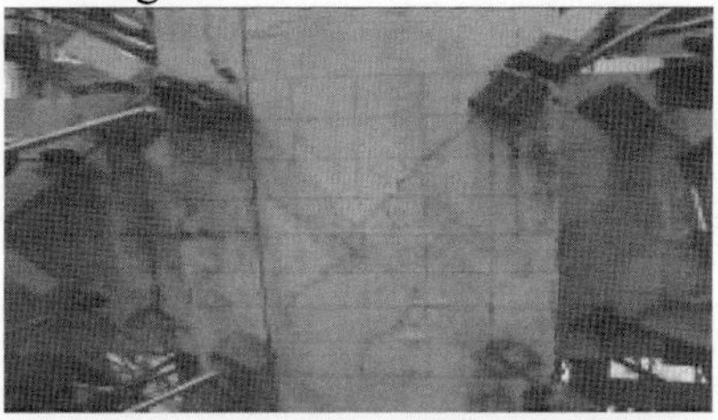

displacement of 22mm, beam end cracks displacement of 22mm, cracks of core area

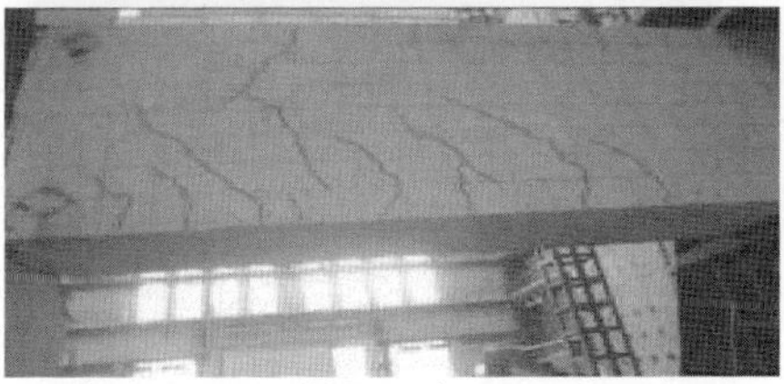

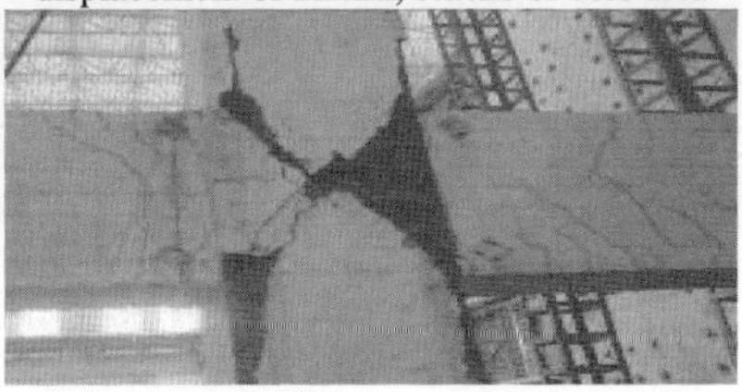

displacement of 38mm, beam end cracks displacement of 38mm, cracks of core area

Fig. 4. The destructive phenomenon of ZJD-1.

2.4. *Specimen ZJD-2*

Applying the axial compression ratio of axial pressure of 0.3 to the top of the column, the beam end is loaded into the 33kN, the crack in the width of 0.2mm, the cracks are smaller and less, under cyclic loading, the crack close quickly, is not easy to observe. When loading to 22mm, the main crack width of the joint core area is 1mm, and extends to the column end. Loaded to 38mm, the core zone of the joint concrete off a large area, stirrup exposed, the load at the beam end is 58kN, and less than 85% the ultimate load, so this experiment is terminate. The destructive phenomenon of the test is shown in figure 5.

3. Conclusion

(1) Shear failure occurs in the core area of the joint and bending shear failure occurs in the beam end.

(2) The seismic behavior of edge joints is better than that of intermediate nodes.

(3) With the increase of the axial compression ratio, the yield load and ultimate load decrease, in high axial compression ratio, the damage is more obvious. so, as the increase of axial compression ratio, the bearing capacity of the component is reduced.

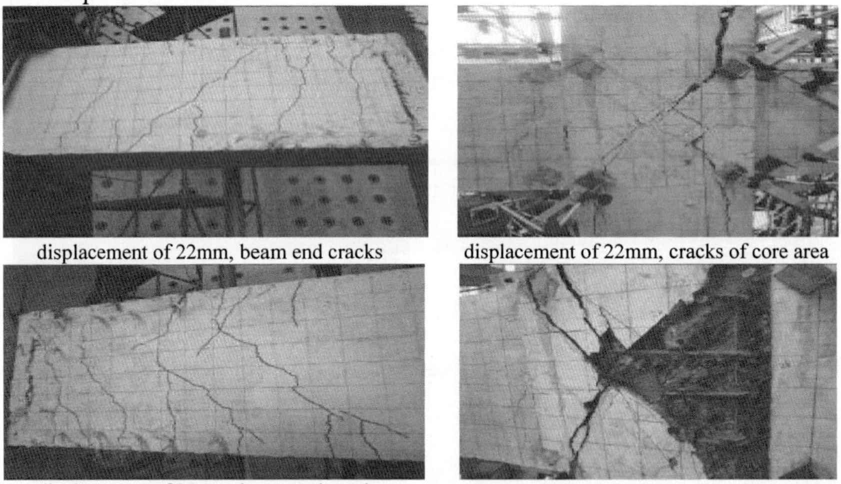

displacement of 22mm, beam end cracks | displacement of 22mm, cracks of core area

displacement of 38mm, beam end cracks | displacement of 38mm, cracks of core area

Fig. 5. The destructive phenomenon of ZJD-2.

Acknowledgments

This work was financially supported by the Jilin Science and Technology Foundation (20150203014SF, 2012200, 2015267).

The authors gratefully acknowledge the Natural Science Foundation of China (No. 51378238).

References

1. Jin-Ping Xu, Jian-Hua Chen, Shao-Hao Zhou. Application research and prospect of lightweight aggregate concrete [J]. foreign building materials science and technology, 2007, 28 (5): 11-13.
2. Wei Huang, Guo-Chen Ma, Cheng-Hua Zhang. Study on seismic performance of ecological composite wall under low cyclic loading [J]. earthquake engineering and engineering vibration, 2013, 33 (3): 155-161.
3. Da Zheng, Li-Dan Liang, Neng-Pan Ju. Under low cyclic loading of prestressed anchor damage characteristics [J]. Rock mechanics and Engineering Journal, 2015, 34:1354-1360.

Study on the strain behavior of steel bar of new lightweight concrete beam column edge joint under quasi-static loading

Yan-Min Yang, Yan Zhen, Wen-Long Dong, Ya-Nan Nie
School of Civil Engineering, Jilin Jianzhu University
Changchun, China, 130118
E-mail:707165968@qq.com
www.jliae.edu

Pseudo static load test of 2 full light concrete frame beam column edge joints under 0.1 and 0.3 axial compression ratio, study on the seismic performance of the lightweight concrete frame beam column edge joints, the strain of steel bar in beam end and the joint of the concrete frame beam column edge joint with different axial compression ratio is analyzed. Test shows, axial compression ratio increases, the constraint effect of stirrups in the core area of nodes increases, but the stress of the node in the core area of the column longitudinal reinforcement no obvious influence, At the same time, once concrete cracking and stirrup strain will appear the obvious mutation, the beam longitudinal reinforcement strain increases, and of all light concrete frame beam column joints seismic performance was evaluated.

Keywords: Lightweight Energy Saving Concrete; Beam Column Edge Joint; Axial Compression Ratio; Strain of Steel Bar.

1. Introduction

Lightweight concrete as a kind of light energy saving, green environmental protection building materials all light concrete, has good structural performance, good seismic performance, thermal insulation, heat insulation and good durability etc. lightweight concrete is lightweight aggregate concrete with light coarse aggregate, lightweight aggregate, cementations material, cement and water [1] The quasi-static load test is carried out on the full light concrete beam column edge joint specimen, study on the strain of the joint core area and the beam end of all light concrete frame beam column joints under different axial compression ratio, analysis of the seismic behavior of the frame beam column joints of the full light concrete frame and the feasibility of its application in the multi story building.

2. Test Scheme

2.1. *Test piece design*

Based on the theoretical research of the existing frame work, according to the design principle of "strong column and weak beam, strong node weak component" [2], design of 2 lightweight concrete frame beam column side joint test piece, lightweight concrete strength LC35. Beam and column longitudinal reinforcement are using HRB400 grade reinforcement, the beam and column, nodes are used HPB300 grade steel stirrups, test piece size, reinforcing bars, as shown in table 1.

Table 1. Design parameters of each specimen.

Node type	Specimen number	Beam end reinforcing bars (mm)		Column end reinforcing bars (mm)		Joints core stirrup (mm)	Design axial compression ratio
		Longitudinal reinforcement	Axial compression ratio	One side longitudinal reinforcement	Stirrup		
Edge node	BJD-1	Up and down each 3Φ18	Φ8@100	3Φ25	Φ8@100	5Φ8	0.1
	BJD-2	Up and down each 3Φ18	Φ8@100	3Φ25	Φ8@100	5Φ8	0.3

2.2. *Test apparatus and arrangement of measuring points*

Due to the test does not consider the effect of P-Δ, so the beam end loading method is adopted, the MTS actuator is symmetrically arranged on the beam end of the test piece, to the test specimen beam end to exert the anti symmetric pseudo static load, the test apparatus is shown in Figure 1. Displacement control loading system is adopted in the loading system, experiment with 2mm as loading increment, each load cycle two times, when the bearing capacity of the specimen is reduced from the ultimate load to the failure load (85% of the ultimate bearing capacity), firmly believe that the test piece is invalid, and the test is terminated. Data acquisition of strain of steel bar using British IMP data acquisition system, the measuring points are arranged as shown in Figure 2.

Fig. 1. The loading device.

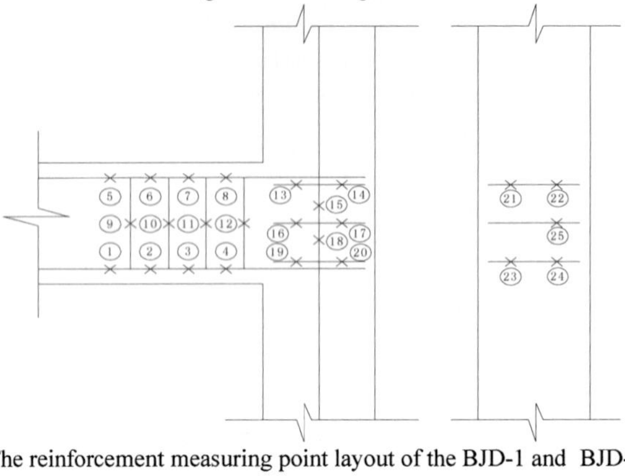

The reinforcement measuring point layout of the BJD-1 and BJD-2

Fig. 2. The reinforcement measuring point layout.

3. Analysis on the Variation Law of the Strain of Steel Bar

3.1. *Analysis of beam end longitudinal reinforcement strain*

Analysis of beam end longitudinal reinforcement strain No. 8 measuring point, as shown in Fig. 3, the figure shows:

(1) Under quasi-static loading, beam end longitudinal reinforcement strain appears positive and negative changes, that beam end longitudinal reinforcement tension, pressure alternation. During the initial loading period, the load - strain relation of the measuring point is approximately linear, the hysteresis loop is not obvious; with the increase of load, cracking of concrete,

the residual strain of reinforcement is getting larger and larger, the load strain relationship of the measuring point is nonlinear, the hysteresis loop area becomes larger.

(2) The specimen of BJD-1 strain between -900με~1100με, specimen BJD-2 strain in -2500με~2000με. At the same time that steel residual stress increases with increasing axial compression ratio; because the axial compression ratio has a limited effect on the rotation of the beam column joints, so with the increase of the axial compression ratio, the plastic hinge region becomes larger, and the strain of the steel bar increases.

(3) The load - strain curve of the edge joint specimen is relatively large jump, it shows that the damage degree of concrete beam end concrete is greater, and the plastic hinge can be more fully formed.

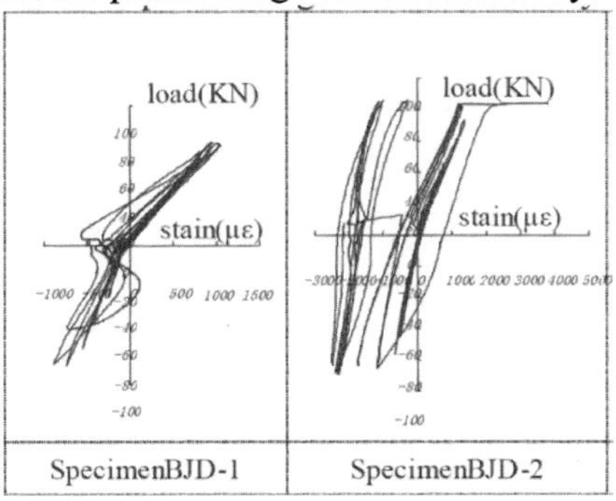

Fig. 3. The load-stain curve of the longitudinal reinforcement in the beam.

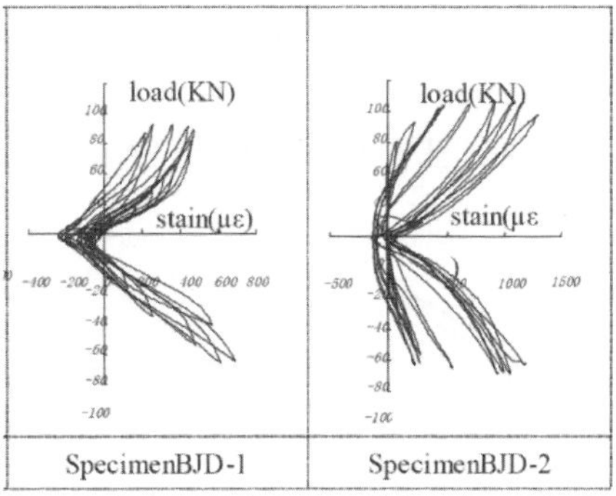

Fig. 4. The load-stain curve of the stirrups in the beam.

3.2. *Analysis of measuring beam end stirrup strain*

Analysis of beam end stirrup strain No. 12 measuring point, as shown in Fig. 4, the figure shows:

(1) Due to the quasi-static loading is loaded on the beam end, beam end cracking earlier, the residual strain of beam end stirrup has also appeared earlier, the stirrup is basically in a state of tension.

(2) When the beam end load is less than 40kN, beam end stirrup strain is negative, stirrups compression, with the increase of load, strain from negative value to positive value, stirrups from compression to tension. when the load is in 40kN ~ 50kN, the residual strain and the hysteresis loop becomes larger, with the increasing of loading, beam end concrete cracking, between the concrete and stirrup bonding failure, then the stirrups on the shear capacity plays a major role.

3.3. *Analysis on the strain of the column longitudinal reinforcing bar in node core area*

Analysis of No. 15 steel strain measuring point, as shown in Fig. 5, the figure shows:

(1) During the initial loading, the load - strain relation of the measuring point is approximately linear; with the increase of load, column longitudinal reinforcing bar in node core area produce residual strain, the load strain relationship of the measuring point is nonlinear, the hysteresis loop increases gradually.

(2) The core zone of the joint column longitudinal reinforcement and stirrup load strain curves are roughly the same, with the increase of axial compression ratio, the strain has a more obvious jump. The specimen of BJD-1 strain between 0~300με, specimen BJD-2 strain in 0~650με, Axial compression ratio has little effect on the stress of the column longitudinal reinforcing bar in node core area of the structure.

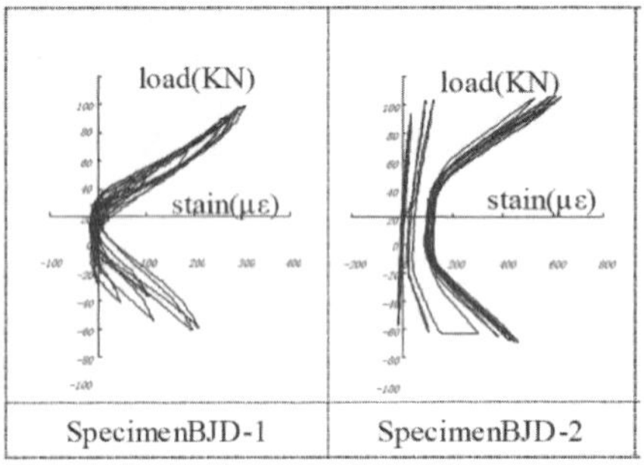

Fig. 5. The load-strain curve of the longitudinal reinforcement in the beam-column interface.

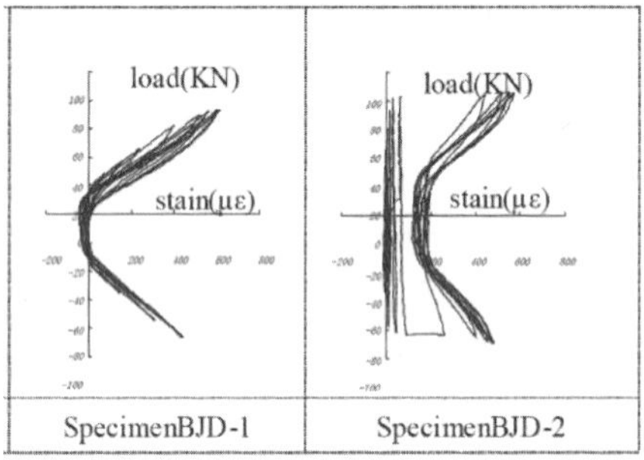

Fig. 6. The load-stain curve of the stirrups in the beam-column interface.

3.4. *Analysis of measuring the strain of the stirrup joint core area*

Analysis of No. 16 steel strain measuring point, as shown in Fig. 6, the figure shows:

(1) Two specimens of stirrup strain are greater than zero, the stirrups are always in tension.

(2) The specimen of BJD-1 strain between 0~300με, specimen BJD-2 strain in 0~650με, joint stirrups strain BJD-2 has a greater leap, but once the concrete cracking, stirrup strain will appear obvious mutation [3].

4. Conclusions

(1) Axial compression ratio has a great influence on beam end longitudinal reinforcement, with the increase of axial compression ratio, The rotation

capacity of beam column edge joint is limited. The plastic hinge is transferred to the beam end, and the plastic hinge zone at the end of the beam is increased, the beam longitudinal reinforcement strain increases.

(2) Edge node is Beam end bending shear failure, load strain curve of joint core stirrup no jumps or mutation value and the hysteresis loop area is smaller.

(3) Test results show that, all light concrete frame beam column joints have good seismic performance, According to the existing specification design [4], Development law of load -strain of reinforcing bar meet the requirements of structural stress.

(4) Test results show that, The application of all light concrete as load-bearing component in the multi story building is feasible, conforms to the development direction of concrete technology, also conforms light weight and high-strength, green environmental protection, energy saving and emission reduction policy concept.

Acknowledgments

This work was financially supported by the Jilin Science and Technology Foundation (20150203014SF, 2012200, 2015267).

The authors gratefully acknowledge the Natural Science Foundation of China (No. 51378238).

References

1. Yan-Min Yang, Study on bearing capacity of all light concrete beams[J], Concrete, 2011, 3:48-50.
2. Jin-Ping Xu, Jian-Hua Chen, Shao-Hao Zhao, Application research and Prospect of lightweight aggregate concrete[J], Foreign building materials science and technology, 2007, 28(5):11-13.
3. Wen-Xiang Chen, Zhong-Xue Gao, Experimental study on frame joints with steel reinforced concrete column[J], Journal of architectural structure, 2002. 23(4):36-39.
4. The Industry Standard Of The People's Republic of China, Technical specification for lightweight aggregate concrete structures (JGJ12-2006) [S], Beijing: Ministry Of Construction, (2006).

Technical application of independent steel tubular support to post-cast strip template construction

De-Feng An
Department of Architectural Engineering, Xuzhou Technician College
Xuzhou, Jiangsu Province, China
E-mail: 1016409445@qq.com

Post-cast strip is common in engineering practice. Though the introduction to presetting independent steel tubular support in post-cast strip and combined with engineering practice, summarization of design and construction is introduced in this paper. The method of cast-in-situ template and post-cast strip template - supporting in independent steel tubular support construction is come into being. The traditional construction technology of post-cast strip template is improved and upgraded. The post-cast strip and cast-in-situ template - supporting can be braced and disconnected simultaneously. It has given consideration to the benefits in safety, quality and economy.

Keywords: Independent Steel Tubular Support; Post-Cast Strip Template; Supporting.

1. Introduction

Post-cast strip, which acts as a temporary settlement joint and shrinkage joint during construction period, is generally existed in engineering practice. By setting post-cast strip, we not only solve the construction problem in pouring concrete, solve the coordination problem of sedimentation deformation in the over-and-under configuration, but also speed up the construction schedule and shorten the construction period. Therefore, post-cast strip is widely used in cast-in-place reinforced concrete structure, but cast-in-place post-cast strip is more common. Because its template support directly affects the entity molding of reinforced concrete structure, how to balance security, economy, quality, efficiency, etc, in template support construction of cast-in-place post-cast strip is a practical problem faced in the the engineering practice.

2. Brief Introduction to Method

Method of steel tube scaffold independently, namely, when pouring concrete, steel tubular support is used on either side of the post-cast strip, differs from construction technology of common template support of cast-in-place post-cast strip support. By means of the method, the template of post-cast strip is no longer independently set up, and after concrete pouring of both sides is completed, all templates are dismantled and keep the steel tube top bracing of post-cast strip. When needed, concrete is poured into post-cast strip template support. After the templates of post-cast strip being allowed to remove, remove them.

3. Project Profile and Solution Selection

The project, which measures construction area of 13000 square meters, is frame-shear wall structure with a basement and 3.8-meter storey height. The top ceiling of basement, whose span of roof board is mostly 8.1 m, is 400-mm-thick cast-in-place reinforced concrete structure. The width of post-cast strip is 800mm; its overall length is 1100m, and edge distance column is 5.43m.

There are more cast-in-place post-cast strip floor plates in this project, and they interconnect up to 1100m. The design requires that concrete construction of post-cast strip must be performed after concrete pouring of either side of floor panel comes up to more than 60 days, and construction of this floor and its upper and lower layers' building wall has been completed, thus a longer interval is required. Besides, in order to avoid the deformation and crack of the beam-slab without stress, post-cast strip must be supported and strengthened temporarily by load bearing body. Independent steel support construction is planned to adopt. Steel support body bears weight by Φ200mmδ=4mm steel pipe, which is set up overall along the longitudinal direction of post-cast strip.

4. Analysis and Calculation on Support Structure

The axis nets of the project are relatively simple, which measure 8.1*8.1m. Support spacing is designed to be 2m. Because a support must be placed at the bottom of the beam, that is, a support is placed at the 550 x 900 mm beam bottom. Cast-in-place slab divides three supports on the basis of 2m spacing, and when calculating, the maximum stress position of beam bottom is only calculated. If the part meets the requirements, cast-in-place support can also meet the requirements. Calculation is done in two ways. One is taking all supports of the cast-in-place floor plate into consideration and the other one is not taking all supports of the cast-in-place floor plate into consideration [1], but this project only consider the latter, that is, if support of the floor plate isn't considered, the support meets the requirements when beam support is only considered, and then the calculation can be passed.

First, the part with the widest span is selected according to the drawings, e.g. see Figure 1

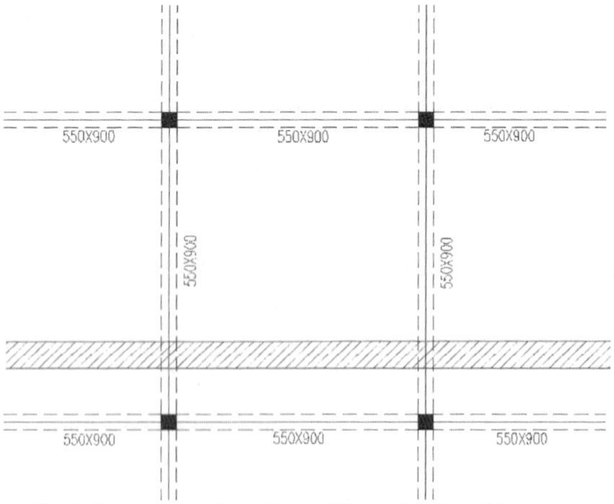

Fig. 1. Post-cast strip at the position with the widest span.

Secondly, the calculation related to independent steel support frame has not been found in PKPM series software, but the calculation related to log support can be found. So, the calculation can be done according to Φ200mm-log strength [2], after the completion of the calculation, the Φ200mmδ=4mm steel pipe is used to conduct substitution checking. ① Strength checking computation: make the strength of the substitution not less than that of no substitution ②Stability checking calculation: because diameter is not varied after the substitution, stability is unnecessary to check.

4.1. *Calculation of wood column below support frame without considering floor plate [3]*

The edge of post-cast strip at the position is 5.43m away from the frame column. The calculation result is (5.43*0.55*0.9+0.4*5.43*7.55)*0.5*2.6 (the weight of reinforced concrete) *1000*9.8=243161(N), calculated according to 243.00KN. The result meets 5.2.5 requirement of *Technical Code for Safety of Forms in Construction* (JGJ 162-2008) [4], therefore the calculation of wood column is passed.

4.2. *Checking computation of strength substitution [5]*

The strength of the steel tube: (100*100*3.14-96*96*3.14)*205 (Strength is 235, and calculation is done by choosing 205 after reserving coefficient of insurance) =504660.8(N); the strength of the log: 100*100*3.14*12 (strength of the log) =37680(N) Because 504.66KN is more than 376.80 KN, the substitution is feasible.

In conclusion, the construction scheme that makes use of independent steel tubular support is feasible.

5. Construction Technology, Method and Points for Attention

5.1. *Construction technology process*

Positioning and setting-out ——→Steel tubular support installation——→ Concrete pouring of each side of post-cast strip——→Normal removal of other template——→Cleaning of post-cast strip template installation——→Concrete pouring of post-cast strip——→Removal of template after concrete hardening

5.2. *Construction technology and method*

The two rows of steel tubular supports are respectively used to sustain every 2 m at 300mm spot on each side of post-cast strip. Post-cast strip isn't considered while template is being put up, and make sure that supporting system of post-cast strip disconnect for independent supporting system. A 5-mm-thick base plate is directly buried at the bottom of the floor plate during the construction, which downward connects Φ200mmδ=4mm steel tube support and can withstand floor plate after dismantling post-cast strip template. When meeting the design requirements of time, concrete is poured into post-cast strip template for the second time. After concrete of post-cast strip meets the design strength requirement, remove the support, e.g. see Figure 2.

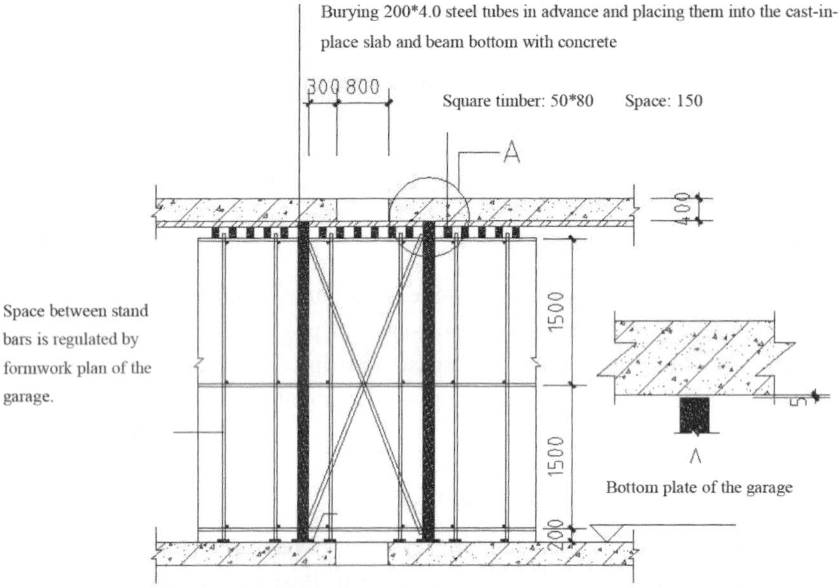

Fig. 2. Steel tube supporting of cast-in-situ template post-cast strip in a garage.
Note: 5-mm-thick steel plate should be used to weld and seal the top and bottom of Φ200mm steel tube.

5.3. *Construction points for attention*

5.3.1. *Sustaining post-cast strip*

When putting up, the location and height of steel tubular support and pre-embedded steel plate must be strictly controlled to make the pressure equilibrate evenly on independent steel tubular support after framed bent is dismantled. Before post-cast strip is sealed, support shall not be dismantled. Other loads must not be applied within the span where the post-cast strip is located, such as construction equipment, material of construction, etc., it is forbidden to remove any component within the support system of post-cast strip before reaching demolition conditions.

5.3.2. *Maintenance of post-cast strip concrete*

According to design requirements, the pouring time of post-cast strip must be confirmed, and can't be brought forward or delayed. Before pouring concrete, the two sides of concrete must be cleaned and chiseled, and watered in advance. Concrete strength must be confirmed according to design requirements. When being poured, concrete must be vibrated densely to make sure of forming quality of concrete, and in the meanwhile, making and maintaining the concrete briquette must be in accordance with the regulations. Maintenance requirements of post-cast strip are the same as general maintenance.

5.3.3. Removal of supports

Because of the steel tubular support on both sides of post-cast strip structure bear loads, removal of independent steel tubular support should be carried out in the end, and when concrete strength can comes up to 100% of the design strength. In addition, removal should be carried out based on intensity of the remained concrete curing briquette with the same conditions [6]. Before dismantling, must submit to the approval to demolition; make corresponding security measures, and compliance with operation code for safety.

6. Conclusion

Method of independent steel tubular support differs from that of ordinary template support, which is mainly used in construction of cast-in-situ post-cast strip. Its main advantages may be concluded as follows:

6.1. *Saving construction cost, shortening construction period, and being safe and applicable*

Compared with ordinary methods, if this method is adopted, steel plates are needed to be pre- embedded, and then steel tubular supports are put up when being put up. Post-cast strip templates and supports are in no need of separate cutting and putting up. When being dismantled, supports can be normally dismantled according to the rule that the support put up first is dismantled first; the support put up later is dismantled later. So the method subverts removal order of the traditional form, and does not affect the removal of other templates. Based on the checking computation, the result is gained that post-cast strip of this project is 1100 meters long, and 0.8meter wide. Because templates are theirs and steel pipes are rented, turnover days of post-cast strip templates can speed up to 42 days, and 30 days of rent of post-cast strip supports can be minimized so that it can bring considerable economic benefits. The support system with solid structure is safe and reliable enough to allow vehicles to pass.

6.2. *Improve construction quality of "post-cast strip"*

"Post-cast strip" is used into the construction in traditional way; the laitance in two sides of post-cast strip concrete must be cleaned, and formed into rough surface. That can completely clean away the waste and adulterant in "post-cast strip". Because of the template at the bottom, the work mentioned above is hard to complete accurately. But, that method can solve the problem completely and greatly improve the construction quality of "post-cast strip".

References

1. MOHURD *Load Code for the Design of Building Structure* GB50009-2012[S] (China Architecture & Building Press, Beijing 2012) (In Chinese).
2. MOHURD *Technical Code for Safety of Timbered Scaffold with Couplers in Construction* JGJ 164-2008[S] (China Architecture & Building Press, Beijing 2008) (In Chinese).
3. China Southwest Architectural Design and Research Institute *Code for design of timber structures* GB 50005-2003[S] (China Architecture & Building Press, Beijing 2003) (In Chinese).
4. Ministry of Housing and Urban-Rural Development of the People's Republic of China(MOHURD) *Technical Code for Safety of Forms in Construction* JGJ 162-2008[S] (China Architecture & Building Press, Beijing 2008) (In Chinese).
5. MOHURD *Code for Construction of Steel Structures* GB 50205-2001[S](China Planning Press, Beijing 2002) (In Chinese).
6. MOHURD *Technical Code for Safety of Steel Tubular Scaffold with Couplers in Construction* JGJ130-2011[S] (China Architecture & Building Press, Beijing 2011) (In Chinese).

Somewhere in northern shaanxi tourism scenic spot design scheme comparison

Xiao-Qiang Wu[1,*], Jian-Zhong Gao[2,†]
[1]*Yan 'an university construction engineering school*
Shaanxi yan 'an, China
[2]*Yan 'an architectural design institute, China*
Shaanxi yan 'an, China
*E-mail: *825318170@qq.com, †nwwolf197724@163.com*

In view of the northern shaanxi somewhere scenic spots surrounding environment situation, through comparing two kinds of design of slope protection, which focuses on a more affordable slope protection project, planning and decision-making for similar areas to provide the related technical indicators, provide reasonable prevention measures and reference.

Keywords: Mountain Slope Protection; Scheme Comparison; Prevention and Control Measures.

1. The Introduction

Qiankun bay, according to the Yellow River scenic areaoverall planning "and" the meander of the Yellow River "master plan of national geological park, an oil company in the Yellow River bay tourism development, realize the regional tourism foreign reception capacity, local folk culture, with the Yellow River Qin Jin grand canyon bare rock as the carrier, enrich tourism products, promote the development of tourism products, according to the planning in the orderlydevelopment, and constantly improve the infrastructure of the scenic spot, gradually expand the content of sightseeing, to extend the time of tourists stranded, the scenic spot in the four seasons long green, three quarters of flowering, facilities, set sightseeing, leisure vacation, folk custom, flavor food as one of the largest tourism, leading role to achieve tourism development areas, increasing the income of the local national economy, solve part of the local people of employment, to improve their living standards, expand qiankun bay in shaanxi province and the country.Degrees, attract more people to the tourism, promote the county The domain of economic development, building a harmonious society.

Pay attention to ecological protection, development of green economy, orderly and moderate tourism resources, scientific development, promote the balanced development of the economic benefits, ecological benefits and social benefits, tourism sustainable development and utilization of resources in the area.

Qiankun proposed Yellow River bay tourism project is a match made in heaven taiji diagram, is the Yellow River road jinshaan gorge on a big natural landscape, is the mostbeautiful, the most spectacular shanbei YanChuan a scenic spots, with mechanical ploughing through roads, traffic conditions, the scenic area the ground engineering content for qiankun fuxi culture center, qiankun bay scenic area, the southern gateway service area, at present, the qiankun scenic slope engineering machine has started the construction.

2. The Scenic Spot Design Plan

2.1. *Some slope protection scheme design unit*

Slope protection design concept to assume things around the scenic area of the Yellow River on the left bank slope for the sliding body, with anti-slide pile, anchor wire and soil measures to prevent the slope sliding plate structure, the pile length of 25.0 m, embedded 9.0 m for the foundation, soil layer section plus 2 Vom15 anchor cable, 35.0 m in length, along the eight-lane cliff elevation of 10.0 m, graphic (1) are arranged on axis 1.0 x 1.5 ten anti-slide pile, (2) set up 14 1 x axis 7.05 m, 0.9 m anti-slide pile, the minimum + / - 0.00 in the arrangement of the micromorphological 52 200 mm diameter pile, the length of 16.0 m, pile top have long penetration beam, slope 400 thick concrete site casting, all for the C30 concrete level [1].

2.2. *Revetment scheme defects*

Assume that landslides, the design idea of retaining wall with micro anti-slide pile shape against scientific principle, field investigation for the early stage of the design is not careful, caused by subjective inference. According to an engineering investigation institute issued in March 2012, the Yellow River qiankun bay scenic qiankun machine and qiankun bay area service area construction engineering geological disaster risk assessment report have clear conclusion has four [2].

The project looks remarkably loess Liang Mao area landscape for the quaternary terraces and exposed bedrock group of yanchang formation mainly for paper fang, extend the set of sandstone and mudstone and quaternary upper pleistocene series eolian loess. Fault and fold in the area is not development, new tectonic movement is mainly characterized by intermittent uplift movement, species in geotechnical engineering is simple, the better engineering geological properties, human engineering activities on the geological environmental impact is small, complexity of geological environmental conditions of evaluation area, is a simple ~ medium types, determine the level of project evaluation for level 3.

The current situation of geologic disaster danger evaluation shows, found no geological disaster survey area, small risk present situation assessment [3].

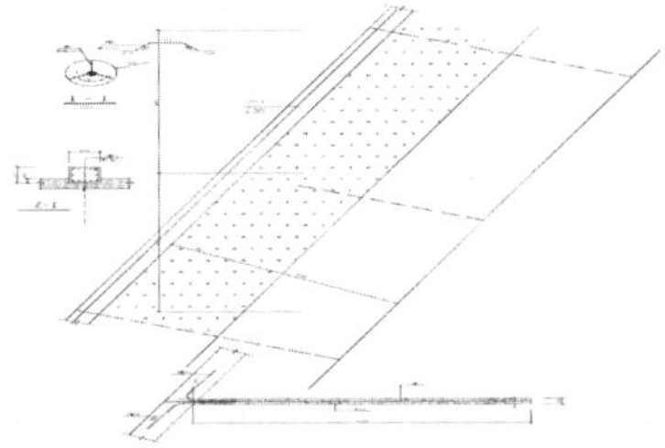

Fig. 1. Lattice column, anchor detail.

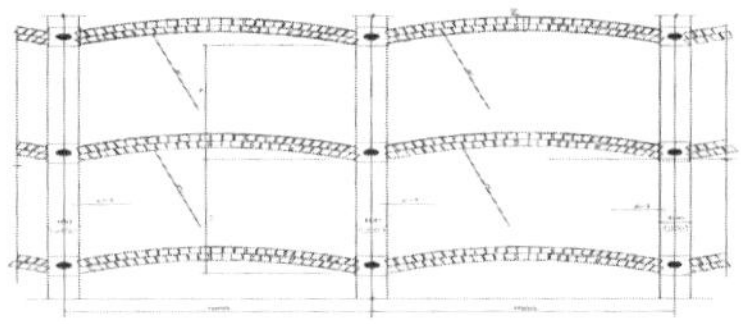

Fig. 2. Stone slope protection detail.

Design thinking for the assessment report have qualitative stability of the slopes to landslide treatment, with scientific principles, must increase investment, extend the time limit for a project, have money to waste, reducing investment performance, etc. This section of highway and railway, also do not have large vehicles disturbance, scenic also impossible to build a big project, small peripheral environment influence on the slope, slope governance can unload and and ecological measures is given priority to, the management of engineering measures are complementary.

No important scenic areas, buildings and structures, few people and decentralization, strong seasonality, actual situation, the slope at the top of the geomorphic features of residual beam width 80 ~ 150 m, the natural slope is too steep, the upper covering thickness is less than 20.0 m, according to the slope engineering technical specifications "GB50330-2002" regulation, the slope safety level should be level 2, should not be increased.

Slope governance should be given priority to with unloading and ecological measures, qiankun machine trailing edge of the scenic spot mountain ridge is wide, slope is relatively slow, covering layer of quaternary holocene respectively (Q4) tillage soil and quaternary pleistocene series (Q3) loess, paleosol and Triassic middle sandy shale argillaceous sandstone, no traces the ancient landslide, uninstall program can be used to eliminate the soil cliff fall of landslide hazard, which cut mountain, put slope, build order. Using artificial terrace slope is slower, planting trees grow grass, drainage, rainwater organized discharge, reduce the rain erosion on the slope surface and cutting, to achieve double effect of slope governance and beautify the environment.

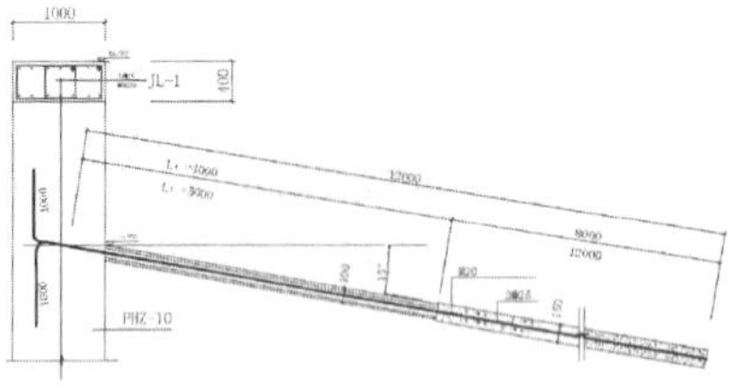

Fig. 3. Plate, anchor bolt grouting figure head, free package.

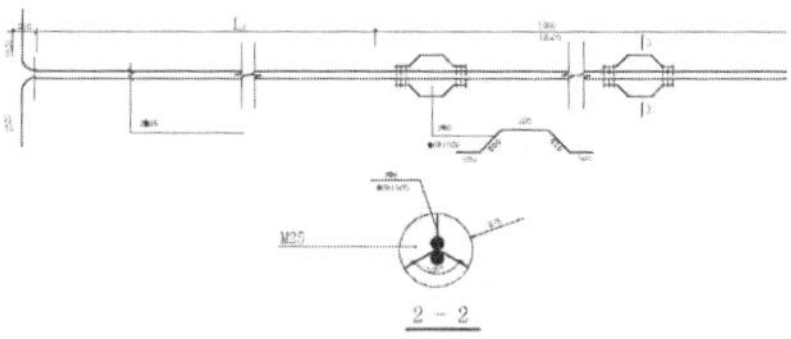

Fig. 4. Anchor reinforcement, positioning bracket drawing.

2.3. *Design change proposal*

Supplementary investigation: 1) the tide table a bid began construction of scenic spots, along the mountain scenic area square with earthwork excavation is in full swing, bareness mountain cliffs, prospecting units can be directly on the slope surface acquisition pattern, send laboratory analysis to determine the relevant data, than artificial exploratory well energy and time, also need not use drilling machinery, investigation personnel can be directly to watch the soil structure and vertical joint distribution. Supplementary investigation conclusion more in line with the actual, more scientific and reasonable proposals [4].

2) the supplementary investigation report after coming out, design personnel according to the investigation report as the basis, make a scientific, reasonable and economic, security, slope protection scheme.

3) slope protection engineering design for half experience and half theory, many practice will remain to be perfected, engineering experience rely on engineering measures also cannot completely eliminate the unsafe hidden trouble [5].

3. Conclusion

This project supporting the design of slope protection through detailed design and comparison, according to the soil and the surrounding topography and the surrounding site condition is different, use a variety of flexible revetment supporting technology, to ensure that the area safe and reliable, has obtained the good economic and social benefits. Therefore, in the mountain scenic area construction planning, especially in previous engineering investigation, should attach great importance to the investigation of geological disasters, take reasonable measures to prevent and control, on the one hand, want to be in planning work at the same time, strengthen the engineering geological conditions of the site survey, provides the prerequisite for planning and decision-making, prevent waste investment; On the other hand is to strengthen disaster regulation, to ensure the safety of the project.

References

1. GB50021-2001, geotechnical engineering specification [S]. Beijing: China building industry press, 2009.
2. GB50007-2010), the building foundation design specification [S]. Beijing: China building industry press, 2011.
3. GB50330-2010, building slope engineering specification [S]. Beijing: China building industry press, 2011.
4. season. The finite element simulation and stability analysis of slope excavation [J]. Journal of rock mechanics and engineering, 2002, 21 (6): 843-847.
5. DZ/T 0219-2006, the landslide prevention and control of engineering design and construction technical specification [S].

Shaanxi somewhere dangerous house static pressure pile in reinforcement engineering design and support

Xiao-Qiang Wu[1,a], Jian-Zhong Gao[2,b]

[1]Yan 'an university construction engineering school
Shaanxi yan 'an,China
[2]Yan 'an architectural design institute, China
Shaanxi yan 'an,China
E-mail: [a]825318170@qq.com, [b]nwwolf197724@163.com

By means of static pressure pile reinforcement technique of shanbei somewhere masonry structure reinforcement strip foundation, can improve the bearing capacity of foundation, the foundation settlement stabilized, meet the bearing capacity of normal use and use requirement. At the same time introduces the reinforcement principle of static pressure pile, construction method and matters needing attention, the construction quality control and its reinforcement effect, provides reference for similar regional static pressure pile reinforcement.

Keywords: Strip Foundation; Reinforcement; Static Pressure Pile; Design; The Supporting.

1. The Introduction

High bearing capacity, the static pressure pile quality has very good guarantee, and the construction noise and vibration are small, no slurry pollution environment, short construction period, the pile driving fast. Due to the advantages of the many, in the construction and expansion project, static pressure pile has been widely used. In the process of dangerous house foundation reinforcement, static pressure pile in the process of foundation reinforcement technology has also got very good application.

2. The Static Pressure Pile Penetration and Compaction Effect on Pile Driving Mechanism

2.1. *Static pile penetration of compaction mechanism*

When static pressure pile construction adopts the special frame weight and weight or the weight of the structures, through the whole pile pressure beam or column or the weight of the structures of balancing deadweight and reaction, and applied in hydraulic hoist pulley block or electric pump way at the top of the pile or pile body, when applied to the static pressure pile and soil resistance of pile of dynamic balance, pile on weight or under the action of static pressure Gradually into the foundation soil. Pressure during pile, pile rate generally remain within a certain value, so the pile penetration can be regarded as uniform motion in a straight line. Static pressure pile in the process of penetration caused by complex of movement, under the pile tip soil compression deformation. With the increase of pile penetration pressure, when the pile sharp edge pressure exceeds the shear strength of soil, sharp deformation and failure of soil, pile side soil plastic flow occurs (clay) or lateral compaction and towing sinking soil (sand), soil under the pile tip was downward and lateral compression is open. Surface area, mainly to the pile of viscous soil mass under weeks apart, keep close to the pile soil structure destroy completely, also affected by the bigger disturbance of soil around the pile body is a powerful method to soil resistance caused by the friction of pile and pile tip resistance of the resistance. At the same time, the saturated cohesive soil, due to the instantaneous drainage consolidation effect is not obvious, the pile penetration of pore water pressure, pore pressure dissipation, and consolidation and thixotropic recovery, in pile form shell layer. For static pressure pile, due to its nearly uniform in the process of penetration, therefore, in theory, the quasi static static equilibrium under the condition of appropriate to its simulation.

2.2. *Static pile pile mechanism*

Expansion theory is the main research methods for static pressure pile on international. Expansion theory first assume that the soil is the ideal elastic-plastic body, materials obey Treasa or Mohr - Coulomb yield criterion, according to the theory of elastic-plastic body give infinite soil body, has the initial radius of bore cylindrical or spherical, uniform distribution of internal pressure p is expansion of the general solution. Due to the pile penetration trendy, radius of certain depth in the soil gradually appear as the diameter of holes, a range of soil around into the plastic state, therefore, generally cylindrical cavity expansion is used to simulate remove near the deformation of the pile end and pile tip. Butterfield and Baneriggee prime minister under the condition of plane strain is proposed to cylindrical cavity expansion is used to solve the problem of pile penetration. They assumed that:

1) the soil is homogeneous and isotropic ideal elastic-plastic material;
2) soil saturation, incompressible;

3) the soil yield meet Mohr - Coulomb strength criterion;

4) the holes before the expansion, the soil with the effective stress to the equivalent. With the increase of internal pressure p, cylindrical hole around the bar area will by elastic state into plastic state. Rising in the plastic zone is increased and the p, outside the soil remains elastic state. For axisymmetric plane strain problem of the cylindrical cavity expansion, under the polar coordinate.

According to the principle of volume of soil, the solution of the differential equation, launch, and the radial displacement of expression, the pressure caused by pile instantaneous excess pore water pressure, pore pressure, the use of the henckel equation is derived.

Sagaseta application semi-infinite space spherical cavity expansion theory and found the surface horizontal displacement and vertical uplift quantity calculation formul.

Type: the radius of the said pile; Said the pile capacity; Said to calculate the distance from the pile axis; The horizontal displacement and uplift, respectively.

3. Design Reinforcement Scheme

3.1. *A project summary*

The project designed by the village committee and the construction unit, construction procedure is not standard, there is no formal drawings. Two-story brick structure, every three studio (3 x 3.6 x 10), a building area of 216 m2, hutch, defend, hall, a layer set indoor stair hall, brick strip foundation, the width of 500 mm, lower berth 300 thick 3:7 lime soil cushion, width 1.0 m, water, electricity, warm professional supporting is not complete, in front of the road and drainage, greening by village committee unified planning. Project started in July 2009, July 2011 subjects completed, no government quality, by a representative of the village committee to monitor the construction process, in December 2011 found the first few rows of houses in the west of uneven settlement of foundation, some outside longitudinal wall inclined cracks, west of the distance from the houses built having groove have multiple cracks, judging by the terrain, the cracks continue to grow [1].

3.2. *The foundation settlement and wall cracks circumstances*

Geomorphic units of construction sites with a gentle slope on the loess plateau gully, the terrain is open, the original state of farmland, site with two small gully, part of the bedrock outcrops, the crack line slope and to the mountain, covering the maximum thickness of 7.0 m, the west has a deep "V" shape use a ditch, rain erosion cutting for a long time, steep walls, human activities are natural at the bottom of the groove, as the expansion of construction land, the artificial digging filling regulation did not undertake unity when rolling, soil degeneration, rely on natural consolidation, with the filling thickness variation in different subsidence deformation, project performance is poor, potential harm is big, easy create housing inhomogeneous deformation of foundation. Rows at the end of 2011 found that the west is relatively close to gully edge building foundation uneven settlement, the longitudinal wall produces a few diagonal crack, Windows and doors hole in stress concentration phenomenon is obvious, crack wider, most of the crack width of 5 ~ 12 mm, subgrade settlement deformation, wall cracks to dig fill feature perfectly match with the original landform [2].

3.3. *Foundation reinforcement scheme*

On the basis of the geological report describes geological section, steel tube reinforced concrete pile foundation, pile end under the breeze bedrock 0.50 m, 100% of foundation subsoil underpinning, are expected to pile body is about 8.0 m. Construction method, in an attempt to dig under the existing foundation size gangway, form of space, jack will 168-6 x 1000 steel pipe piles in turn pressure into the foundation soil into rock. Single pile bearing capacity characteristic value of Quk = 250 kn, pressure pile to the foundation soil has certain role to form composite foundation compaction, underpinning, after building the original foundation bearing layer out of work, all the load borne by the static pressure pile, the deformation tends to be stable gradually[3].Static pressure pile is shown in figure 1.

3.4. *Static pressure pile construction*

1) track test pile: before the static pressure pile construction, about 6000 range from building pressure into the first three piles, jack not more than 45 mpa pressure readings, guarantee the pile tip into the layers of about 1000 mm, two days after the start loading test pile, and inform the design, the prospecting units to join. Temporarily consider the single pile vertical ultimate bearing capacity standard Quk = 500 kn, single pile vertical bearing capacity characteristic value of Ra = 250 kn. Test pile biggest load of 620 kn or less, the project can be instead of test pile. Pile length 10 m, finally determine when pile test report out. Static pressure pile construction process flow diagram is shown in figure 2.

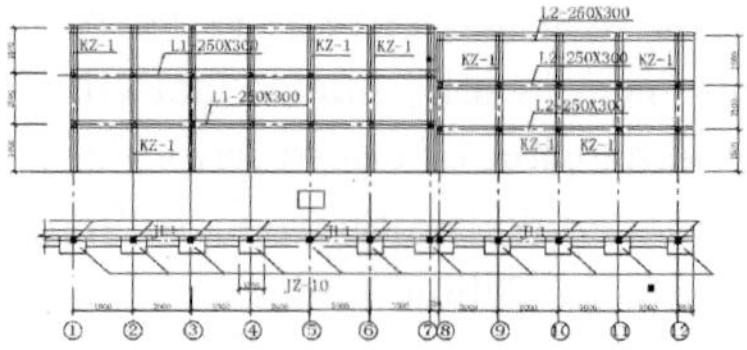

Fig. 1. detail of steel pipe pile

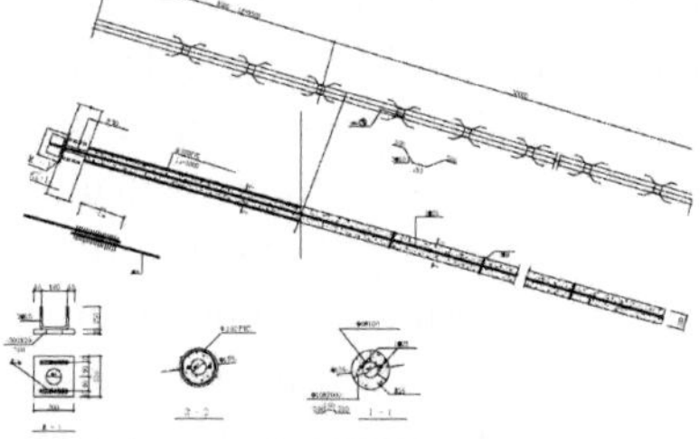

Fig. 2. static pressure pile construction process diagram

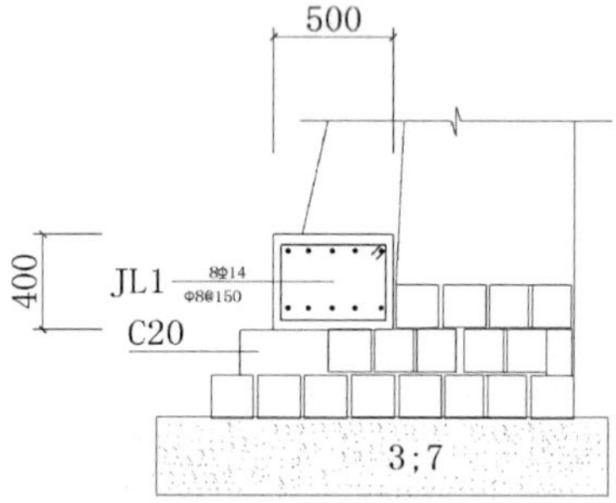

Fig. 3. steel frame detail

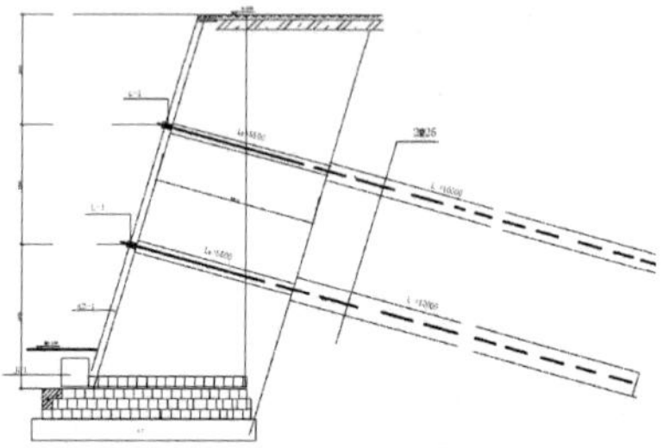

Fig. 4. steel column cap detail

2) material: steel tube concrete JYZ 168-6,1.2, pile cap closed for C30, pisolite concrete pile caps and ground beam bottom gap with dry hard C40 pisolite concrete fill in (expansive cement).Steel bar, steel plate, Angle of A3 stee.[1]

Rule 3.4.3 note: 1, the construction personnel should possess certain construction mechanics, foundation theory knowledge and practical experience, in strict accordance with the design requirements of the construction process. Familiar with the relevant provisions of the "building pile foundation specification JGJ". The order shall not alter the construction process.[2], piling of personnel to observe when the wall cracks, jack rises slowly, as far as possible the speed about 10/60 s, prevent the ground beam crack.[3], Windows and doors hole pressure pile, it must be in the ground beam on both sides of the strain gauge, a detailed record of shape. Wall vibration is found, counterweight in Windows and doors hole place, make sure that jack had enough reaction.[4], continuous pressure pile should be professional, do not allow the discontinuous (to prevent the pile friction recovery).Must ensure that the pile tip into the breeze bedrock. After the 5, piling, should be timely for size gangway 3:7 lime soil backfill, use wood of tamping compaction.(compaction coefficient is 0.98) to ground beam under the skin in 500.After bilateral symmetry in static pressure pile up top, packed column cap and ground beam cracks in concrete.6, piling, the detailed record jack readings, once every 200 height record table, in order to grasp the foundation of flooding and various soil thickness.

3) wall crack repair, waste recovery: 1, along the fracture direction with M10 mixed mortar paint.100 mm wide, along the height of about 100 mm hole, 24 h after water wet, high 2, all damage gutter press to restore the original design drawings. Groove outward looking for 2% slope 3.5.3 test pile: at 3% of the total number of pile extraction, and not less than three root. Can be used instead of test pile, but must choose under wall, ensure Nmax, beam and wall stability safety. Foundation beam or concrete base side stick strain gauge, observe whether the ground beam cracks or wall vibration timely adjust the test pile load.

4. Conclusion

The project according to the structure of the structuralcharacteristics, environmental conditions and geological conditions, using static pressure pile reinforcing foundation, has obtained the good economic and social benefits, the investment is saved, and meet the requirements of the use function, to guarantee the normal life;In a brick house, through the use of static pressure pile, can effectively solve the uneven settlement of foundation, strengthening effect is goodpasses elaborate design and calculation, in the dangerous house of masonry structure reinforcement engineering, the soft collapsible loess foundation under the walls of strip foundation settlement is too large, can use technology to reinforce the wall static pressure pile foundation.

References

1. Li Zengxuan, ying white.O - CELL method and its application in static load pile test [J]. Architectural technology, 2000, 31 (3) : 157-159.
2. GB50007-2010), the building foundation design specification [S]. Beijing: China building industry press, 2011.
3. The finite element simulation and stability analysis of slope excavation [J]. Journal of rock mechanics and engineering, 2002, 21 (6) : 843-847.

Simulation about speed-up character of split type reaming while drilling tool

Wen-Fei-Li, Zhi-Jun-Tang
Drilling Technology Research Institute Shengli Oil Service Corporation Sinopec, DongYing, ShanDong
E-mail: upclwf@163.com

Because low drill efficiency and high power of normal large bit, normal reaming while drilling is structural complex and drilling speed is slow, separate RWD technic is supplied, reaming bit and pilot bit are driven by two power sources. Based on numerical simulation, the character of drilling speed about speed up mechanism and drill parameter is researched, the analysis results show that, separate RWD technic can unload rock internal stress significantly, damage rock in less time; drill speed of separate RWD technic is decreasing with pressure increasing, increasing with pilot bit rotation speed increasing, increasing first and then decreasing with difference in height between reaming bit and pilot bit increasing, increasing with diameter ratio about two bit, and can reduce lateral vibration significantly and keep bit stable comparing to normal large bit, the result is the basis on optimum design of separate RWD tool and optimum select of drill parameter.

Keyword: Split Type; Reaming While Drilling; Rock Breaking; Speed Up; Simulation.

1. Technical Present Situation of Reaming While Drilling(RWD)

During the course of drilling deep well, super deep well, complex structure well, deep water well, low drilling speed and high energy of large bit is the trouble. RWD can drill large well, and solve drilling stick, improve casing structure, cementing, and decrease drilling cost [1-2], so RWD and tools are supplied [3-7].

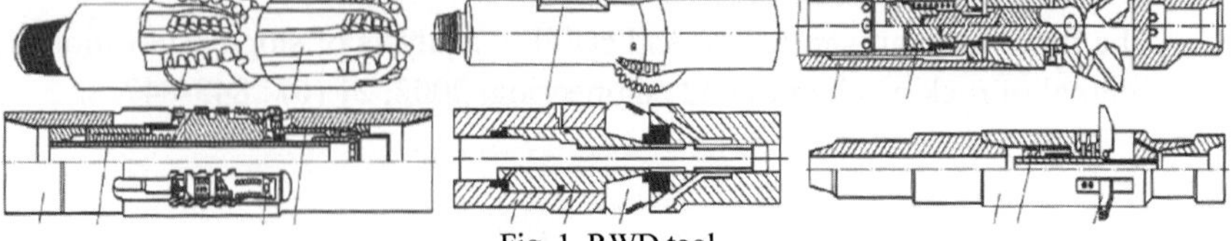

Fig. 1. RWD tool.

The RWD tool above can meet with large well drilling, but some disadvantage is found, (1) stability of BHA is poor, that can affect reliability of RWD tool; (2) decreasing pilot bit torque and drilling efficiency; (3) complex structure and high precision. On the basis of normal RWD tool [8-11], a new split type RWD method is supplied, in which reaming bit and pilot bit are driven by dual power, and can increasing drilling efficiency significantly. Because of reaming bit and pilot bit rotating and breaking rock independently, the ratio of diameter, height, rotary speed must be researched, finite element model method can analyze character quickly, that is significant for optimum design of structure, selecting of drilling parameter, decreasing research period and cost.

2. Finite Element Model

3D finite element model about normal bit and split type RWD bit is built, based on rock mechanism theory, speed up character of split type RWD bit is studied.

2.1. *Geometric and material model*

Assumption

(1) the bit is steel
(2) element is delete after failure
(3) water jet breaking rock is not considered

Normal bit RWD bit
Fig. 2. 3D model of bit.

2.2. *Meshing*

Based on Drucker-Prager and plastic failure rude, ock element is C3D8R and bit element is R3D4. Relationship between cutter and rock is nonlinear dynamic contact, nd bit is contact surface, ock is contact target, ontact property is surface-surface.

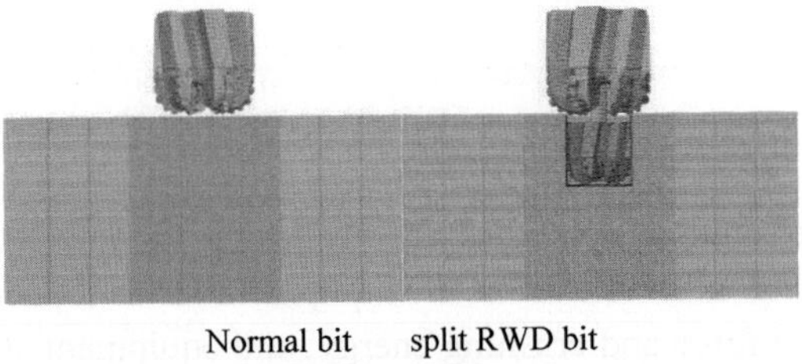

Normal bit split RWD bit
Fig. 3. Meshing.

2.3. *Boundary condition*

Bit freedom is no limit, ressure and rotation is applied on bit, onfining pressure is applied on rock, ydrostatic pressure is applied on bit and rock, ther area is applied on.

3. Simulation of Rock Breaking Character

Using 25s explicit dynamic analysis, tress cloud of bottom rock can be calculated. On the condition of bit cutting, ock in well has plastic deformation and failure, amage factor is 1, rock surround well has plastic deformation and not failure, ock far away well has not plastic deformation.

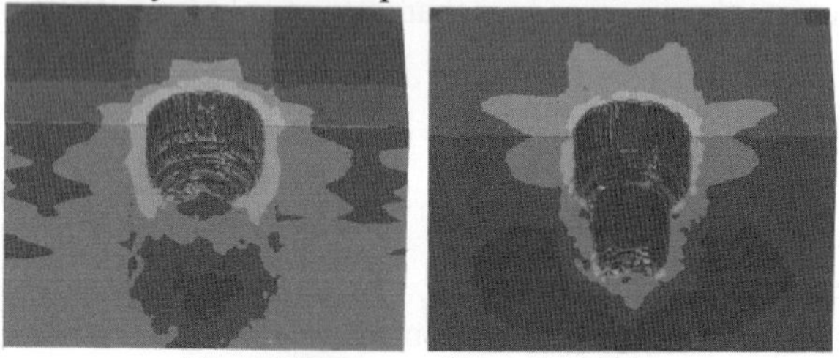

Normal bit lit type RWD bit

Fig. 4. Rock stress cloud of bottom hole.

3.1. *Mechanism of speed up*

Split type RWD technic is benefit for stress unload of bottom hole rock, and can make rock broken quickly, that can be analyzed in the term of maximum principal stress and damage class.

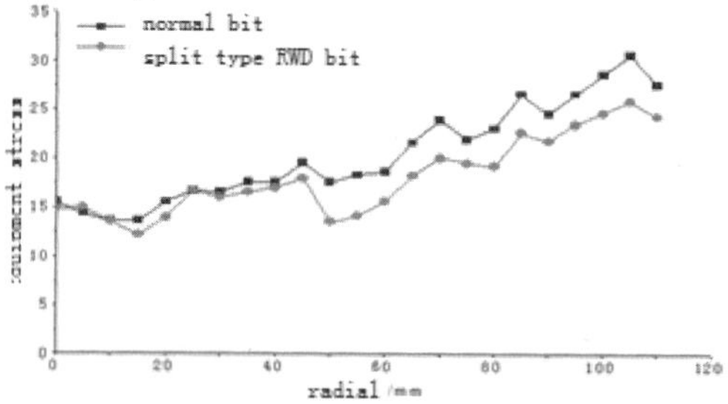

Fig. 5. Equipment stress of bottom hole rock.

In fig. 5, stress unloads obviously in well wall, equipment stress decreasing, and then increasing with well diameter. Equipment stress reducing is benefit for decreasing cutting force and crushing energy, and equipment stress of split type RWD is 88.5% of normal drilling technic, the speed up potential is significant.

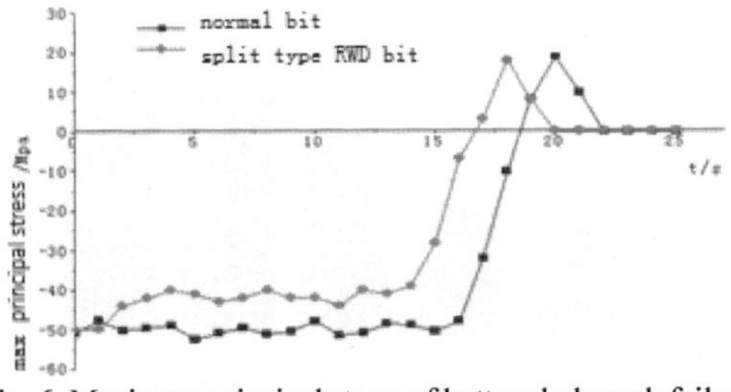

Fig. 6. Maximum principal stress of bottom hole rock failure.

In fig. 6, rock in bottom hole on the condition of split type RWD is failure quickly relative to normal bit, because the effect of pilot bit drilling, internal stress is decreased, and then balancing by hydraulic press and confining pressure, formation stress is released, that decrease the rock breaking difficulty of RWD bit.

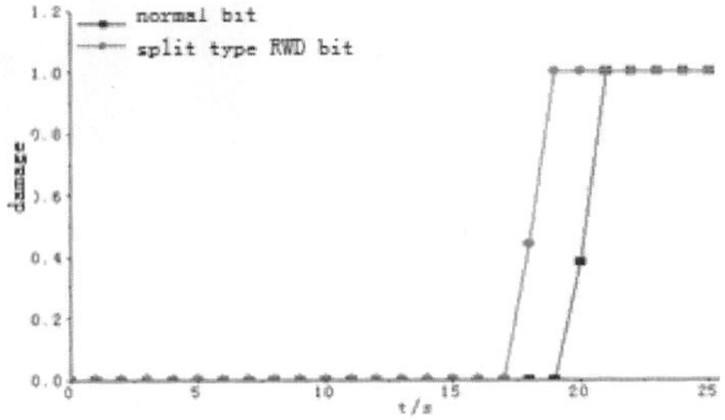

Fig. 7. Damage of bottom hole rock.

In fig. 7, rock is not damaged, damage factor is 0, rock is damaged completely, damage factor is 1, so we can see that period of rock damaging completely by RWD is short.

3.2. *Analysis of parameter influence character*

Pressure,rotary, tool size are key parameter,and artificial adjustable parameter,it is significant for optimum of drilling paramcter and tool sizc by researching character of split type RWD about speed up and vibration.

(1) pressure

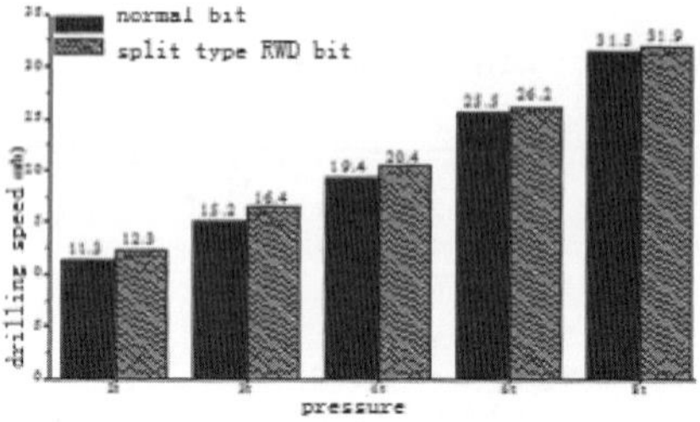

Fig. 8. Pressure influence character.

In fig. 8, split type RWD drilling speed is decreasing with pressure increasing, it means that higher pressure is not meet with split type RWD.

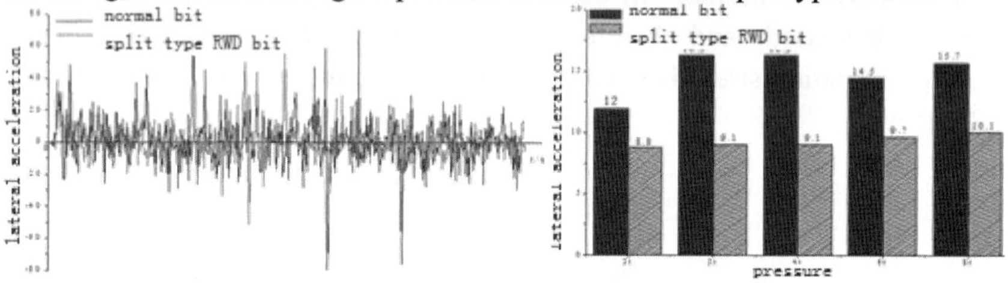

Fig. 9. Bit lateral acceleration.

In fig. 9, we can see that, lateral acceleration of split type RWD bit is lower than normal bit significantly, that is to say lower lateral acceleration, more reliability of bit, it is important for rock breaking quickly.

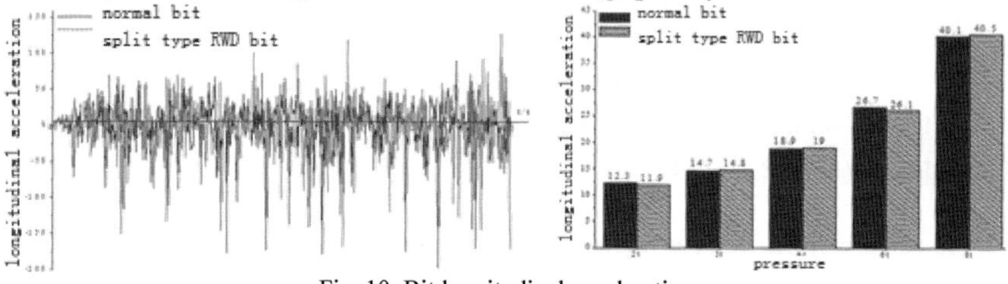

Fig. 10. Bit longitudinal acceleration.

In fig. 10, longitudinal acceleration of split type RWD bit is the same of normal bit, split type RWD technic can decrease longitudinal vibration and not enlarge vibration degree.

(2) Rotary

Simulation parameter, rotation speed of reaming bit of split type RWD is 50 r/min, rotation speed of pilot bit is A-60r/min, B-90r/min, C-120r/min, D-150r/min, E-180r/min, F-210r/min, rotation of normal bit is 50 r/min, the other parameter is same.

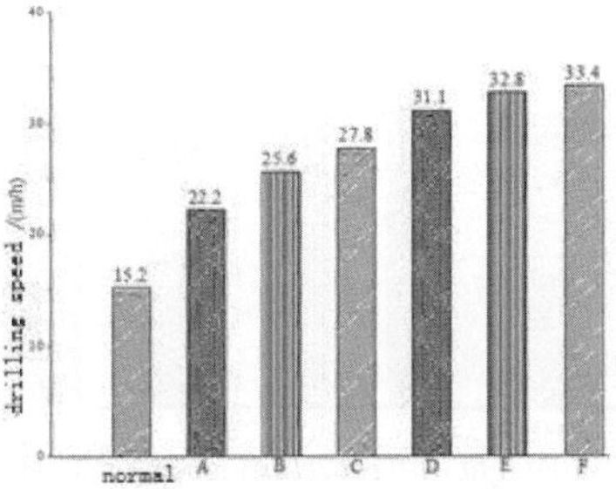

Fig. 11. Drilling speed of bit.

In fig. 11, drilling speed of split type RWD is increasing 46.1%-120% with rotation speed of pilot bit increasing, relative to normal bit, but drilling speed up degree reducing when rotation speed of pilot bit is more than 180r/min.

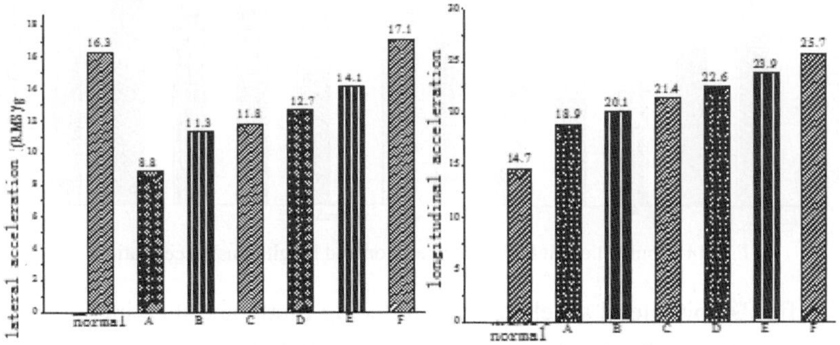

Fig. 12. Contrast of bit lateral acceleration and longitudinal acceleration.

In fig. 12, when rotation speed of pilot bit of split type RWD is lower than 180r/min, its lateral acceleration is reducing significantly, and when rotation speed of pilot bit of split type RWD is more than 180r/min, its lateral acceleration is severe, longitudinal acceleration is increasing with rotation speed of pilot bit of split type RWD increasing, so we know rotation speed of pilot bit of split type RWD is not high.

(3) Height difference

Simulation condition, height difference between reaming bit and pilot bit of split type RWD is A-75mm, B-100mm, C-125mm, D-150mm, E-200mm, the other parameter is same.

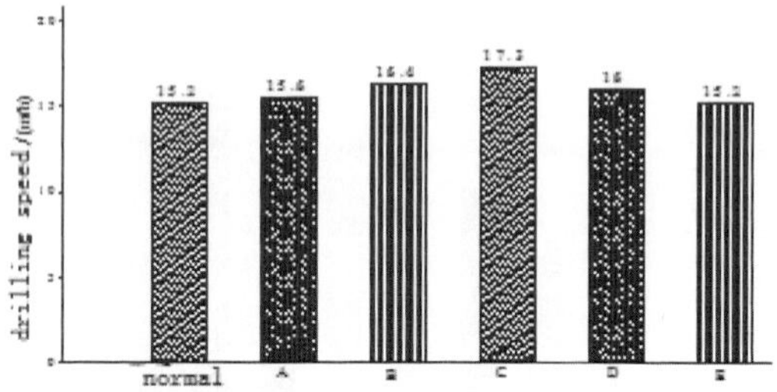

Fig. 13. Height difference influence drilling speed.

In Fig. 13, drilling speed of split type RWD is increasing firstly and then reducing with height difference increasing, drilling speed is faster as height difference is 125mm.

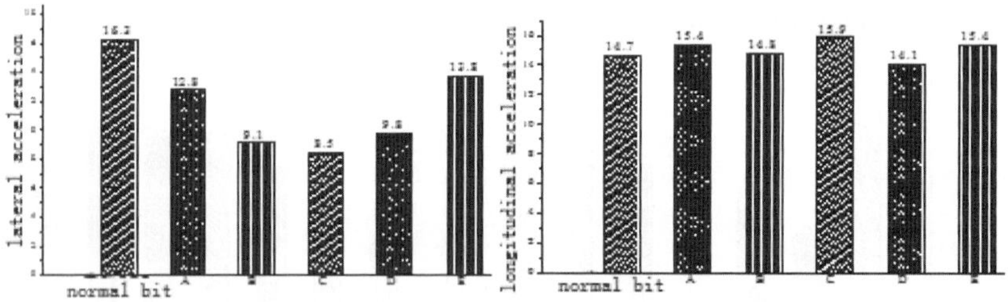

Fig. 14. Contrast of bit lateral acceleration and longitudinal acceleration.

In fig. 14, bit lateral acceleration is reducing firstly and then increasing with height difference increasing, bit is the most stable as height difference is 125mm,and has no affect longitudinal acceleration apparently yet.

(4)Diameter ratio

Simulation condition, diameter ratio between reaming bit and pilot bit of split type RWD is A-0.4, B-0.5, C-0.6, D-0.7, E-0.8, the other parameter is same.

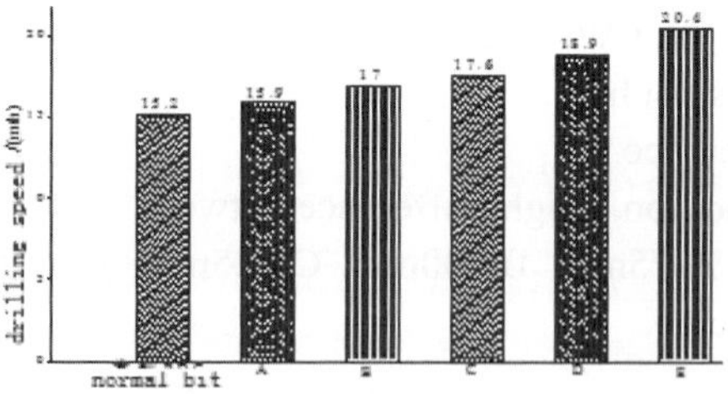

Fig. 15. Drilling speed on the condition of different diameter ratio.

In fig. 15, drilling speed of split type RWD technic is increasing with diameter ratio increasing, as diameter ratio is 0.8, speed up degree is the most apparent.

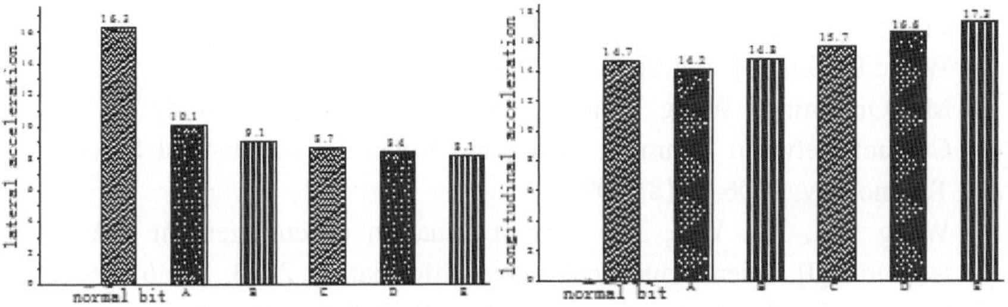

Fig. 16. Contrast of bit lateral acceleration and longitudinal acceleration.

In fig. 16, with diameter ratio increasing, bit lateral acceleration is reducing, that is to say bit lateral acceleration is stable, and longitudinal acceleration is increasing but magnitude is little, so as diameter ratio is 0.8, drilling speed is highest.

Bit stability and drilling speed character is analyzed on the condition of different drilling parameter and tool size parameter, the result shows that when parameter is the best, drilling speed of split type RWD is higher than normal large bit, and provides reference for down hole tool design and bit style optimum.

4. Conclusion and Suggestion

(1) on the basis of normal RWD technology analysis, a new split type RWD technic is provided, that can overcome normal large bit disadvantage, and increase drilling speed of large well further.

(2) according to finite element method, rock breaking mechanism of split type RWD technic and drilling parameter optimum are analyzed, the result shows that stress unloads obviously in bottom hole rock and damaging rock in short time; drilling speed of split type RWD is reducing with pressure increasing, increasing with rotary speed of pilot bit increasing, increasing first and then reducing with height difference between two bits increasing, increasing with diameter ratio adding, and can cut down lateral acceleration significantly, keep bit in stability.

(3) it is suggested that according to simulation result and drilling condition, making split type RWD tool and experiment.

References

1. Ma Rutao, Ji Youzhe, WangHui. The current situation and development trend of the research on hole reamer for oil- gas drilling[J]. china petroleum machinery, 2010, 38(8): 8-12.

2. Ma Qingming, Wang Ruihe. Study on Tools and Techniques of Reaming While Drilling[J]. Nature Gas Industry, 2006, 26(3): 71-74.
3. Ma Qingming, Wang Ruihe. Numerical Simulation Study of Friction Contact Between Reaming Tools and Rocks[J], Mechanical Science and Technology, 2006, 25(8):909-912.
4. Wang Hui, Xie Wei. The current situation of enlargement RWD tools technique[J]. Science and Technology Information, 2013, 12(36): 79-80.
5. Sun Wei-jia, Fu Jian-hong, Zhou Wei. Drill String Dynamics Analysis of DeepWell Reaming While drilling and Drilling Parameter Optimizing[J]. west drilling engineering, 2015, 20(2): 20-23.
6. Song Yi, Xia Hong-nan, Xu Chao. Reaming Tool Research and Optimization[J]. Equipment Manufactring Technology, 012, 03(2):282-284.
7. J. McCarthy. Study of Vertical Drilling Technology in High Dip and Applications of BHA with Offset Axes Sub, SPE paper 142431.
8. DAVE RODMAN. Near bit reamer allows under reaming while drilling with steerable systems[J] . JPT, 2001, 53(1): 20-21.
9. FORD Brett J, WARREN Thomas M, BEHR Suzanne M. Bit whirl — A new theory of PDC bit failure [R]. SPE 19571.
10. SH ESH TAWY A, H OWELL M. Cut costs with new BHA for reduced clearance casing programs[J]. World Oil, 1999, 220(12) : 59-62.
11. MORRISON W. Optimizing wellbore design using solid expandable tubular and bi center bit technologies[R]. SPE 92886, 2005.

Study on the criterion of time splitting of surrounding rock in underground chamber

Xiao-Jing Li[1,2,a,*], Xue-Rui Yang[3], Liang-Liang Liu[3]
[1]*School of Civil Engineering, Shandong Jianzhu University, Jinan, 250101(China)*
[2]*State Key Laboratory of Mining Disaster Prevention and Control, Shandong University of Science and Technology, Qingdao, (China)*
[3]*School of Civil Engineering, Shandong Jianzhu University, Jinan, 250101(China)*
E-mail: [a]li8021@163.com, [b]136699280@qq.com, [b]826687265@qq.com
www.sdjzu.edu.cn

Along with the development and application of underground cavern group in the southwest of China, a Engineering phenomenon occured frequently in the process of engineering development :Rock mass splitting in high side wall of underground cavern. This paper analyzes the mechanical characteristics of defective rock. It elaborate defective rock stress concentration should force distribution principle, and in the research process lieutenant general linear sliding mode crack for a specific object of study. At the same time, it establish timeliness splitting criterion, in order to analyze and judge the rock splitting. By using this method, the criterion of the age splitting of wall rock is used to provide reliable basis and scientific guidance for the engineering construction, design and optimization of large underground cavern.

Keywords: Timeliness; Splitting Failure; Uniaxial Compression.

1. Introduction

Engineering geological body in deep high ground stress area. When the strength of the surrounding rock mass and the stress difference are small, and the magnitude of the principal stress difference increases, Cavern surrounding rock into the yield condition, rock mass of the brittle turn for ductility and micro cracks in rock mass began along the direction of maximum principal should force direction of initiation and propagation, and along the direction perpendicular to the axial compression and expansion stretch, eventually lead to reduce the surrounding rock bearing capacity, higher stress to rock deep transfer until splitting. The process of stress adjustment and transfer will be accompanied by a rock aging rupture and large deformation [1-3].

In this paper, the analysis and study of the physical properties of rock and its energy dissipation principle are completed by considering the rheological properties of the rock and the energy dissipation principle.

2. Mechanism and Criterion of Splitting Failure of Rock Mass

2.1. *Selection of mechanical analysis model for splitting mechanism*

The following chart 1 shows a linear sliding crack as the main research object in figure1 (1), to study the rock splitting timeliness crack evolution. The crack form for the dilatancy phenomenon originated in the research process of rock. S.Hibino and others have carried out a detailed study on the issue: The crack of this model is composed of two parts: Initial crack and wing crack. The wing crack of the model is also originated from the initial crack. Initial crack in the far field compressive stress under the action of relative slip, which caused the wing crack.

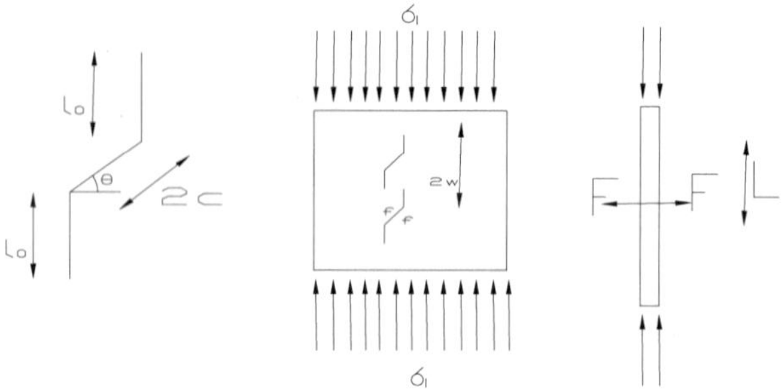

(1) (2) (3)

Fig. 1. Rock fracture model [4-5]

In figure 1 straight line sliding crack formation based on lateral should stress σ3 which is much smaller than σ1, wing crack along the parallel to the direction of σ1 for expansion. The final multiple cracks mutually communicated, formation rock splitting crack damage.

As shown in figure 1(2) initial crack 2c, wing crack 2L0, adjacent crack apart 2W, initial crack angle and horizontal direction θ) With the time change, when L0=w, two adjacent cracks occur, which leads to the formation of the split crack. The length of the crack is defined as L.

As shown in figure 1(3), the crack configuration, which is formed by theinteraction of adjacent cracks, is mainly controlled by the force F and the far field stress.

2.2. *Stress assumption of split model*

From the graph 1 and the research of S.Hibino [6] et al, we can know that such crack mechanism is wing cracks in the lateral force σ3 small state, along the parallel to the direction of the σ1 expansion and many engineering examples occurred to the lateral stress is much less than the maximum compressive stress position, so in this paper, we assume that in the event of a split in the process σ1 is the most important factor, that is assumed that the model stress is only for compressive stress σ1, since the rock complex is we simply rock uniaxial compression load as shown in figure 2, namely the focus on rock in uniaxial compression under the timeliness of splitting crack

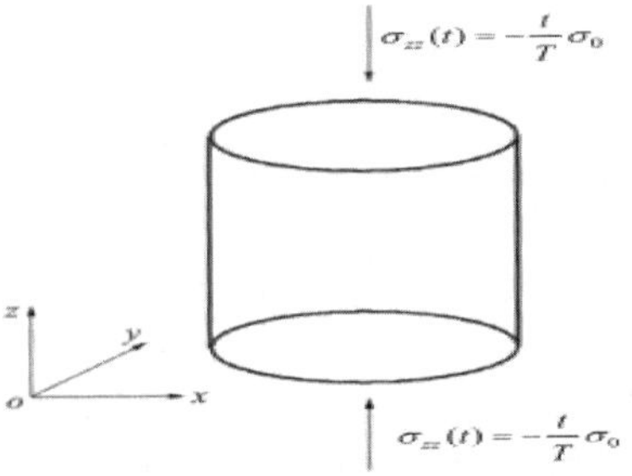

Fig. 2. schematic diagram of rock stress 2

2.3. *Establish criterion of splitting*

It is assumed that the rock mass is subjected to axial pressure only and the force increases linearly, and the lateral pressure is not affected by the pressure as shown in Figure 2. The expression of the axial pressure is shown in e.q.(1):

$$\sigma_{zz}(t) = -\sigma_0 \frac{t}{T} \tag{1}$$

σ0 for loading stress;T for rock loading time

(1)Rock splitting factor method [7-8].

According to Fig.1 the axial compressive stress under the condition of linear sliding crack will produce effective shear stress, the mechanical relationship formula as follows e.q. (2-4):

$$\tau^* = \tau - \mu\sigma_n \tag{2}$$

$$\sigma_n = \frac{1}{2}\sigma_1(1 - \cos 2\theta) \tag{3}$$

$$\tau = \frac{1}{2}\sin 2\theta \tag{4}$$

$\tau^{\cdot}$ —— Crack surface shear stress ; σn——Stress of crack surface method
θ——Straight crack and the angle between the horizontal plane
The linear slip crack stress intensity factor KI is shown in e.q. (5):

$$K_1 = \frac{2c\tau^* \cos\theta}{\sqrt{\pi l_0}} \tag{5}$$

For any cracks, assuming the shear stress component in the horizontal direction by the flank crack stress is f, so f is obtained in e.q. (6):

$$f = \tau^* \cos\theta \tag{6}$$

Combining equations e.q.(1) - e.q. (6), e.q. (7) is available:

$$F = \int_A f\mathrm{d}A = \tau^* \cos\theta \cdot L = \frac{1}{2} L \sin 2\theta \cdot \sigma_1 \left(\cos\theta - \mu \sin\theta\right) \tag{7}$$

According to the superposition principle, the Fig.1 (3) is shown as a set of splitting cracks, the stress intensity factor of the crack is shown in e.q. (8):

$$K_1 = \frac{F}{\sqrt{\pi L}} = \frac{L \sin 2\theta \cdot \sigma_1 \left(\cos\theta - \mu \sin\theta\right)}{2\sqrt{\pi L}} \tag{8}$$

Because of not considering the friction factors existing in the whole process, II type crack wing crack tip caused by stress concentration on the splitting process has little effect, so neglected[9]. From the basic knowledge of the split, the realization of splitting, to meet $K_1 \geq K_{IC}$: the end of the inequality can be obtained in e.q. (9):

$$\sigma_1 \geq \frac{K_{IC} 2\sqrt{\pi L}}{L \cdot \sin 2\theta \left(\cos\theta - \mu \sin\theta\right)} \tag{9}$$

(2) Maximum additional tensile stress method.[10]

As shown in figure3. When $t=t_1$, the additional defects caused by the stress reached the ultimate tensile strength of σp rock cracks are formed and began to expand and eventually reach the maximum bearing capacity of damaged rock, while the load reached σa, near the rock split state.

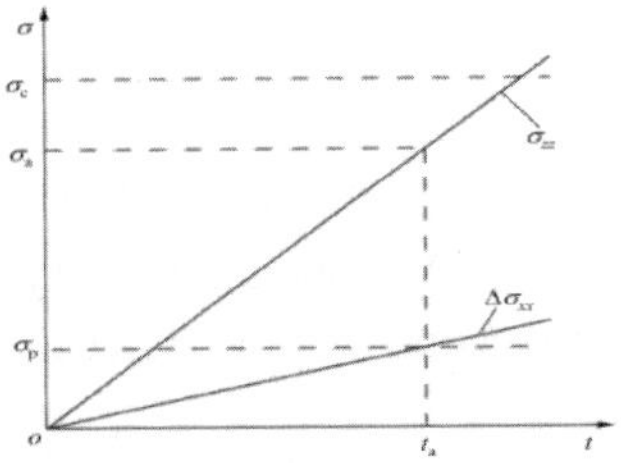

Fig. 3. stress state of rock under uniaxial compression

Analysis of stress concentration in the defect can be shown in e.q.(10-11):

$$\frac{d\Delta\sigma_{zz}}{dt} = Kpc_s^2 \dot{e}_{zz} - \eta\frac{\Delta\sigma_{zz}}{l} \tag{10}$$

$$\dot{e}_{zz} = -\frac{1}{3}\frac{\sigma_0}{pc_s^2}\frac{1}{T} \tag{11}$$

Analysis(e.q.10-11) can be known:

$$\Delta\sigma_{xx}\Big|_{t=t_a} = -\frac{1}{6}K\sigma_a = \sigma_p \tag{12}$$

$$\sigma_a = -\frac{6}{K}\sigma_p \tag{13}$$

$$K = -6\frac{\sigma_p}{\sigma_a} = 6\frac{\sigma_p}{-\sigma_c}\frac{\sigma_c}{\sigma_a} \tag{14}$$

Among e.q. (12-14), σa/σc=0.2-0.8, σc/σp=10-15. So, additional stress concentration factor K=0.5-3.0.

3. Conlusion

When rock mainly consider shear damage, such as high stress condition of deep rock stress destruction, the use of rock stress intensity factor method is more accurate. When using the criterion (e.q.9), initial crack and the horizontal direction of the angle theta to try to take large angle. At this time, the micro crack to along the direction of maximum compression were extended through, if little angle with the horizontal direction, even mainly subjected to shear failure. The criterion will no longer apply. If the rock mainly consider the tension failure, can give priority to the maximum tensile stress method by using e.q.14.

Acknowledgement

The study is jointly supported by grants from the National Natural Science Foundation of China (Grant No. 41372294 and 51579142), Opening Project Fund of State Key Laboratory of Mining Disaster Prevention and Control by Shandong Province and the Ministry of Science and Technology (No. MDPC201605) .

References

1. Lin QX, Liu YM, Tham LG, Tang CA, Lee PKK, Wang J. Time-dependent strength degradation of granite. Int J Rock MechMin Sci. 2009, Vol. 46, pp. 1103-14.
2. Heap MJ, Baud P, Meredith PG, Bell AF, Main IG. Time-dependent brittle creep in Darley Dale sandstone. Journal ofGeophysical Research. 2009, Vol. 114, pp.1-2.2.
3. Heap MJ, Baud P, Meredith PG, Vinciguerra S, Bell AF, Main IG. Brittle creep in basalt and its application to time-dependent volcano deformation. Earth Planet Sci Lett. 2011; Vol. 307, pp.71-82.
4. LI Xiaojing. The study onexperiment and theory of splitting failure in great depth openings[Ph.D. Thesis][D]. Jinan: Shandong University, 2007.(in Chinese).
5. X.J. Li; W.M. Yang; L.G. Wang. Displacement Forecasting Method in Brittle Crack Surrounding Rock under Excavation Unloading Incorporating Opening Deformation[J]. Rock Mechanics and Rock Engineering. 2013, Vol. 47(6) , pp.2211-2223.
6. LAJTAI E Z, CARTER B J, DUNCAN S. Mapping the state offracture around cavities[J]. Engineering Geology, 1991, Vol. 31(3/4) , pp. 277–289.
7. Guan Z, Jiang Y, Tanabashi Y, et al. A new rheological model and its application in mountain tunnelling[J]. Tunnelling and Underground Space Technology. 2008, Vol. 23(3) , pp. 292-299.
8. Hajiabdolmajid V. Mobilization of strength in brittle failure of rock[D]. Kingston, Canada: Queens University, 2001.
9. Hajiabdolmajid V, Kaiser P K, Martin C D. Modelling brittle failure of rock[J]. International Journal of Rock Mechanics and Mining Sciences. 2002, Vol. 39(6), pp. 731-741.
10. Shao J F, Zhu Q Z, Su K. Modeling of creep in rock materials in terms of material degradation[J].Computers and Geotechnics. 2003, Vol. 30(7) , pp. 549-555.

The research on the stability of the steel tube filled with steel-reinforce concrete subjected to mechanical after high temperature

Bai-Dong Zhao*, Peng Yu, Tian-Cheng Xu

Architectural and civil engineering institute, Shenyang university, Shenyang, China

*E-mail: 624865067@qq.com, *zhaibaidong@163.com*

In this paper, we do the research on the steel tube filled with steel-reinforced concrete subjected to Mechanical after high temperature by changing the temperature and the slender ratio. Influence of the temperature and slender ratio on the composite column was gotten by the research. It the stability formula of the composite column after high temperature was derived.

Keywords: ABAQUS; Slenderness Ratio; Stability; High Temperature.

1. Introduction

The stability of the column is the important problem in the design and construction of the building. The concept of structural instability is that under the action of external forces, the structure began to lose equilibrium state, and the little disturbance will suddenly increase the components distortion and finally lead to structural damage. Therefore, in the process of the use of the structure, the structural instability problems often cause structural collapse and suddenly extreme harm. At present, there is little research on the stability bearing capacity of the steel tube filled with steel-reinforced concrete after high temperature. So that, it is necessary to do some research on the stability of the composite column.

2. Temperature on the Bearing Capacity of Different Slender Ratio Composite Column

In order to study the temperature effect on the long composite column, this paper uses ABAQUS software to analysis the composite column's mechanical behavior after high temperature including $20\,^{\circ}C$, $300\,^{\circ}C$, $700\,^{\circ}C$ and $900\,^{\circ}C$. The component parameter is shown in table 1 [1].

Table 1. Component parameter.

Component number	$D \times t \times l_0$ (mm × mm × mm)	Slender ratio λ	Structural steel index	Confining index	f_{yt}	f_{ys}	f_{cu}
HM-1	166×3.5×1170	26	0.776	0.705	316	285	68.6
HL-1	166×3.5×1670	35	0.776	0.705	316	285	68.6

By simulating the component in table 1, get the curves of bearing capacity-deflection shown in Fig. 1.

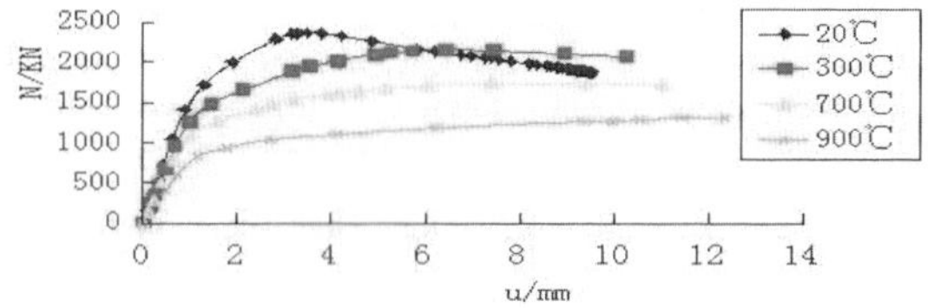

(a) the long column

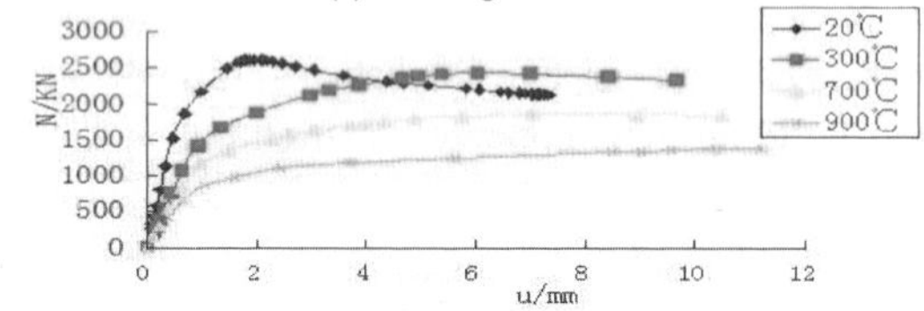

(b) the medium long column

Fig. 1. The bearing capacity-deflection curve of the column.

From the compare between (a) and (b), we can conclude that the bearing capacity and the slope of the elastic stage decrease with the increase of the temperature. This phenomenon indicates that the higher temperature the smaller modulus of elasticity and the bearing capacity. The other result is that when reach ultimate bearing capacity, the corresponding deflection increases with the increase of the temperature. From the compare between 20 curve and 300 $^\circ C$ curve in (a) and (b), we can conclude that the decrease of the bearing capacity is smaller; and this phenomenon is consistent with the research of scholars both in China and abroad. The relevant references shows that the compressive strength of the core concrete decreases with the increase of the temperature, but little change in less than 300 $^\circ C$. When the temperature exceeds 300 $^\circ C$, the heating process will damage the inner surface of the concrete. During the cooling process, due to the materials start cooling from the external surface and the inner surface is still maintained a higher temperature, so that there is a non uniform temperature field between the inner surface and the external surface to make the core concrete damaged again. Above all, the core concrete experienced two injuries after high temperature. The whole process will make the concrete volume expansion and intensify internal damage of core concrete. In Fig. 1, we can conclude another result. The result is that the decrease of the composite column slender ratio 35 is bigger than the slender ratio 26. This phenomenon indicates that the effect of slender ratio on the composite column bearing capacity is increasing with the increase of the temperature.

3. Deriving Stability Formula Subjected to Axial Compression

For the big slender ratio column, the bearing capacity of the column is decided by Euler formula, and the damage reason is Elastic Instability. But for the small slender ratio column, the damage reason is the crushed core concrete, external tube reaches the yield limit and the steel-reinforced reaches the critical load.

On the aspect of calculating the bearing capacity of the long column, this paper introduces the stability coefficient φ to calculate the bearing capacity of the big slender ratio column on the basis of short column bearing capacity $N_0(T)$.

In determining the value of φ, the domestic and foreign scholars have put forward different calculation methods. Virid and Dowling introduce the concept of relative slenderness ratio and make the steel-concrete composite column buckling curve and the steel column buckling curve reach agreement.

Based on the method of Virdi and Dowling, the formula to calculate the steel tube filled with steel-reinforced concrete short column subjected to axial compression was used to derive the relative slenderness ratio $\bar{\lambda}$. According to the φ - $\bar{\lambda}$ relationship curve, the stability coefficient φ was determined; and the deriving process is shown below.

The calculation formula of the slenderness ratio is shown in formula 1.

$$\bar{\lambda}=\frac{\lambda}{\lambda_c}=\frac{L_0}{L_c} \tag{1}$$

In formula 1, λ is the column slenderness ratio, L_0 is the effective length of the column, and λ_c is the slenderness ratio when the Euler critical load is equal to N_0. L_c is the critical length corresponding to the Euler critical load.

According to the relevant information, the strength of the steel tube filled with steel-reinforced concrete to resist bending is calculated by formula2.

$$EI = E_t I_t + E_s I_s + \beta E_c I_c \tag{2}$$

In formula 2, the symbol β is the stiffness reduction coefficient of the core concrete and the value of β equals to 0.6. The Euler critical load N_E is calculated by formula3.

$$N_E = \frac{\pi^2 (E_t I_t + E_s I_s + 0.6 E_c I_c)}{L_0^2} \tag{3}$$

The combination of formula 3 and N_0 can derive the value of L_c shown in formula4.

$$L_c = \pi \sqrt{\frac{E_t I_t + E_s I_s + 0.6 E_c I_c}{N_0}} \tag{4}$$

$$N_0(T) = A_{scs} f_{scs}^{y}(T) \tag{5}$$

The combination of formula 4 and formula 1 can derive the value of relative slenderness ratio, and the results of the slenderness ratio is shown in formula 6.In formula 4, the symbol E_t, E_s, E_c is the elastic modulus of tube, steel-reinforced and the core concrete; the symbol I_t represents moment of inertia, and the formula to calculate the moment of inertia is $I_t = \frac{\pi D^4}{64}(1-\alpha^4), \alpha = \frac{d}{D}$; the symbol I_s represents the moment of inertia of the steel-reinforced, and the moment of inertia of the core concrete is $I_c = \frac{\pi D^4}{64} - I_s$.

In determining the stability coefficient, the scholars in China and abroad

$$\bar{\lambda} = \frac{L_0}{\pi} \sqrt{\frac{N_0}{E_t I_t + E_s I_s + 0.6 E_c I_c}} \tag{6}$$

divide the column into three types according to the type of the column section. The section from of steel tube filled with steel-reinforced concrete can be considered in accordance with section type b. On the basis of the b-type steel column φ - $\bar{\lambda}$ curve, the stability coefficient can be calculated. Above all, the bearing capacity of the long composite column is determined in formula 7.

$$N_l = \varphi N_0 \tag{7}$$

Table 2. Contrast of calculated and simulation results.

Component number	$D \times t \times l_0$ (mm × mm)	Slenderness ratio	Stability coefficient	Temperature /°C	N_0	N_j/KN	N_m/KN	$\frac{N_j}{N_m}$
HM-1	166×3.5×1170	26	0.867	20	2850	2540	2617	1.00
HM-1	166×3.5×1170	26	0.867	300	2490	2220	2250	0.94
HM-1	166×3.5×1170	26	0.867	700	2150	1940	1750	1.068
HM-1	166×3.5×1170	26	0.867	900	1280	1250	1280	0.92
HL-1	166×3.5×1670	35	0.825	20	2870	2390	2260	1.028
HL-1	166×3.5×1670	35	0.825	300	2500	2080	2060	1.001
HL-1	166×3.5×1670	35	0.825	700	2100	1650	1690	1.044
HL-1	166×3.5×1670	35	0.825	900	1380	1090	1260	0.918

The average of $\frac{N_j}{N_m}$ is 1.033, the discrete coefficient is 0.1. So that, the simulation results coincide with the calculated results well.

4. Conclusion

(1) The effect of slender ratio on the composite column bearing capacity is increasing with the increase of the temperature.

(2) The decrease of the bearing capacity is smaller when the temperature below 700°C, and the bearing capacity decreases with the increase of the temperature.

References

1. Wang Qingxiang, Research on the mechanical behavior of axially loaded circular steel tube column filled with steel-reinforced high-strength concrete [J]. Journal of Southeast University, 2002, 32(5):711-714. (In Chinese)
2. Han Linhai, Concrete Filled With Steel Tube, [M] Beijing: Science Press, 2007.

Research on doubly-fed induction generator with adaptive rotor flux observer

Xiang-Hong Xu*, Ming Yin
Department of Mechanical Engineering, Shazhou Professional Institute of Technology, Zhangjiagang 215600, China
**E-mail: xxh602@126.com*
www.szit.edu.cn

By analyzing the mathematical model of a doubly-fed induction generator (DFIG) and vector control scheme, an adaptive rotor flux estimator and resistance and speed estimators for DFIG which can realizes the independent regulation of the active power and reactive power of DFIG are put forward. The method is based on model reference adaptive principle. The adaptive law has been obtained from the Lyapunov stability conduction. It has been proved that the estimating error caused by parameters mismatch is convergent. Results of simulation show that the estimator presented above is effective and the accuracy and robustness of control system are improved at the same time.

Keywords: DFIG; Model Reference Adaptive Principle; Rotor Flux Observer; Speed Sensorless; Control System; Robustness.

1. Introduction

Doubly-fed induction generator (DFIG) structure is simple, solid and reliable, which is the core part of wind power generation system. The running status of the control system, on the wind of the whole system performance, efficiency and output power quality has important influence. The operating characteristics of a generator and the control technology, which can be divided into constant speed constant frequency wind power generation technology VSCF are analyzed [1, 2]. Due to constant speed constant frequency wind turbine speed cannot change with wind speed, reducing the utilization of wind energy, so much the variable speed constant frequency wind turbine. VSCF are analyzed in the use of DFIG system, the application of the vector transformation control technology, through the power electronic converter of DFIG rotor amplitude, phase and frequency adjustable excitation, it can realize variable speed constant frequency generator run time of active and reactive power decoupling. In order to improve the performance of excitation on the electricity system, generator accurate observation of magnetic chain, is the generator of active and reactive power decoupling factor. Indirect flux observation method of practical, namely, by easy to measure physical quantities such as voltage, current to real-time calculation, the method of amplitude and phase of flux, parameters of the generator dependencies, and when in the operation of the generator parameters have certain degeneration, especially the rotor time constant change significantly with temperature and magnetic saturation degree, makes the actual control effect, serious when may spiral out of control. Flux observation error will make the generator control through the magnetic chain small or large, leading to the generator efficiency is lower, error of flux observation also affect the calculation of the slip frequency, which affects the calculation precision of speed.

Generator indirect usually with open-loop and closed-loop flux observing two observation models [3]. Open loop observation model is divided into current and voltage model. Contains the generator rotor resistance current model method, model accuracy is not high, the voltage has the potential of pure integral element model method, the observation accuracy is low. Closed-loop model by introducing a state feedback, and can effectively improve the stability of the flux observation model, the dynamic and static convergence properties, and improve the system anti-interference performance of parameter changes.

In order to make the generator of active and reactive power decoupling, get accurate observations accurately the rotor flux is very important. The paper proposes a doubly-fed generator algorithm of the adaptive rotor flux observation. This method was introduced to the observer in the stator current error feedback, belongs to the closed loop observation, has very strong robustness. Application by using Lyapunov stability condition, deduced the rotor flux and rotor resistance and speed of adaptive law and Matlab/Simulink simulation experiment, verify the feasibility of the algorithm. Simulation results show that this method constructs the flux observer and can accurately track the flux linkage, real-time identification of rotor resistance and rotating speed at the same time. This method is easy to implement, also does not depend on any prior knowledge of generator parameters, which study the high performance of doubly-fed generator active and reactive power decoupling has practical significance.

2. Theoretical Analysis

If the selection of the stator current, rotor flux as state variables, doubly-fed generator of α , β , the equation of static coordinate system is (1):

$$
\begin{cases}
\dfrac{d\psi_{r\alpha}}{dt} = -\dfrac{R_r}{L_r}\psi_{r\alpha} - \omega\psi_{r\beta} - \dfrac{R_r}{L_r}mi_{s\alpha} + U_{r\alpha} \\
\dfrac{d\psi_{r\beta}}{dt} = -\dfrac{R_r}{L_r}\psi_{r\beta} + \omega\psi_{r\alpha} - \dfrac{R_r}{L_r}mi_{s\beta} + U_{r\beta} \\
\dfrac{di_{s\alpha}}{dt} = -\dfrac{mR_r}{\sigma L_s L_r^2}\psi_{r\alpha} - \dfrac{m}{\sigma L_s L_r}\omega\psi_{r\beta} \\
\qquad -\dfrac{m^2 R_r + L_r^2 R_s}{\sigma L_s L_r^2} i_{s\alpha} - \dfrac{1}{\sigma L_s} U_{s\alpha} \\
\qquad + \dfrac{m}{\sigma L_s L_r} U_{r\alpha} \\
\dfrac{di_{s\beta}}{dt} = -\dfrac{mR_r}{\sigma L_s L_r^2}\psi_{r\beta} + \dfrac{m}{\sigma L_s L_r}\omega\psi_{r\alpha} \\
\qquad -\dfrac{m^2 R_r + L_r^2 R_s}{\sigma L_s L_r^2} i_{s\beta} - \dfrac{1}{\sigma L_s} U_{s\beta} \\
\qquad + \dfrac{m}{\sigma L_s L_r} U_{r\beta}
\end{cases} \tag{1}
$$

Take flux linkage, current, rotor resistance and estimate the rotational speed of respectively ($\hat{\psi}_{r\alpha}$, $\hat{\psi}_{r\beta}$, $\hat{i}_{s\alpha}$, $\hat{i}_{s\beta}$, $\hat{R}_r$, $\hat{\omega}$) construct observer state equation as (2) :

$$
\begin{cases}
\dfrac{d\hat{\psi}_{r\alpha}}{dt} = -\dfrac{\hat{R}_r}{L_r}\hat{\psi}_{r\alpha} - \hat{\omega}\hat{\psi}_{r\beta} - \dfrac{\hat{R}_r}{L_r}mi_{s\alpha} + U_{r\alpha} \\
\dfrac{d\hat{\psi}_{r\beta}}{dt} = -\dfrac{\hat{R}_r}{L_r}\hat{\psi}_{r\beta} + \hat{\omega}\hat{\psi}_{r\alpha} - \dfrac{\hat{R}_r}{L_r}mi_{s\beta} + U_{r\beta} \\
\dfrac{d\hat{i}_{s\alpha}}{dt} = -\dfrac{m\hat{R}_r}{\sigma L_s L_r^2}\hat{\psi}_{r\alpha} - \dfrac{m}{\sigma L_s L_r}\hat{\omega}\hat{\psi}_{r\beta} - \dfrac{m^2\hat{R}_r}{\sigma L_s L_r^2} i_{s\alpha} \\
+\dfrac{L_r^2 R_s}{\sigma L_s L_r^2} i_{s\alpha} - \dfrac{1}{\sigma L_s} U_{s\alpha} + k(i_{s\alpha} - \hat{i}_{s\alpha}) + \dfrac{m}{\sigma L_s L_r} U_{r\alpha} \\
\dfrac{d\hat{i}_{s\beta}}{dt} = -\dfrac{m\hat{R}_r}{\sigma L_s L_r^2}\hat{\psi}_{r\beta} + \dfrac{m}{\sigma L_s L_r}\hat{\omega}\hat{\psi}_{r\alpha} - \dfrac{m^2\hat{R}_r}{\sigma L_s L_r^2} i_{s\beta} \\
+\dfrac{L_r^2 R_s}{\sigma L_s L_r^2} i_{s\beta} - \dfrac{1}{\sigma L_s} U_{s\beta} + k(i_{s\beta} - \hat{i}_{s\beta}) + \dfrac{m}{\sigma L_s L_r} U_{r\beta}
\end{cases} \tag{2}
$$

Among them, the positive constant k. Set observation error for (3):

$$\begin{cases}\Delta i_{s\alpha} = i_{s\alpha} - \hat{i}_{s\alpha} \\ \Delta i_{s\beta} = i_{s\beta} - \hat{i}_{s\beta} \\ \Delta \psi_{r\alpha} = \psi_{r\alpha} - \hat{\psi}_{r\alpha} \\ \Delta \psi_{s\beta} = \psi_{r\beta} - \hat{\psi}_{r\beta} \\ \Delta \sigma = R_r - \hat{R}_r \\ \Delta \omega = \omega - \hat{\omega} \end{cases} \tag{3}$$

Generator state equation (1) minus the observer equation (2) get the error equation (4):

$$\begin{cases}\dfrac{d\Delta\psi_{r\alpha}}{dt} = -\dfrac{\hat{R}_r}{L_r}\Delta\psi_{r\alpha} - \hat{\omega}\Delta\psi_{r\beta} - \hat{\psi}_{r\beta}\Delta\omega \\ \qquad + \Delta\sigma(-\dfrac{\hat{\psi}_{r\alpha}}{L_r} - \dfrac{m}{L_r} i_{s\alpha}) \\ \dfrac{d\Delta\psi_{r\beta}}{dt} = -\dfrac{\hat{R}_r}{L_r}\Delta\psi_{r\beta} + \hat{\omega}\Delta\psi_{r\alpha} + \hat{\psi}_{r\alpha}\Delta\omega \\ \qquad + \Delta\sigma(-\dfrac{\hat{\psi}_{r\beta}}{L_r} - \dfrac{m}{L_r} i_{s\beta}) \\ \dfrac{d\Delta i_{s\alpha}}{dt} = -k\Delta i_{s\alpha} + \beta\dfrac{d\Delta\psi_{r\alpha}}{dt} \\ \dfrac{d\Delta i_{s\beta}}{dt} = -k\Delta i_{s\beta} + \beta\dfrac{d\Delta\psi_{r\beta}}{dt} \end{cases} \tag{4}$$

3. Result and Discussion

In doubly-fed generator double PWM inverter power, the observer in the adaptive law is under the control of doubly-fed generator for observation and identification. Based on Matlab simulation model is established, including: the data of doubly-fed generator rated power 1500 kw / 220 V, 2 to the pole, the R_s = 0.435 Ω, R_r = 0.816 Ω, L_{1s} = 0.3042 H, L_{1r} = 0.3107 H, J = 0.089 kg·m^2, L_m = 0.292 H. Parameter convergence rate depends on r_1, r_2, r_3 and r_4 choice, here the selection for r_1 = 20, r_2 = 30, r_3 = 40 and r_4 = 50, feedback coefficient selection for k = 100.

Fig.1 is expressed as the rotor resistance estimation process, observation resistance after t = 1.5 s tend to be resistance real value. Fig. 2 for generator's actual flux and flux amplitude waveform comparison chart, and the observation of magnetic chain value at t = 0.5 s can rapidly track after the actual flux value, fig. 3 for the observations of two phase flux linkage, said it can be seen that it can quickly converge to become a round. Fig. 4 for the estimation of the rotational speed process, estimate to stabilize after t = 0.6 s. Can be seen from the simulation experiments, the flux observer is able to accurately observe the rotor flux, but also to identification of rotor resistance and the rotor speed can achieve good effect. Further show that the designed adaptive flux observer is correct, so can use the observed value composition, rotor resistance and rotor speed is the vector control system.

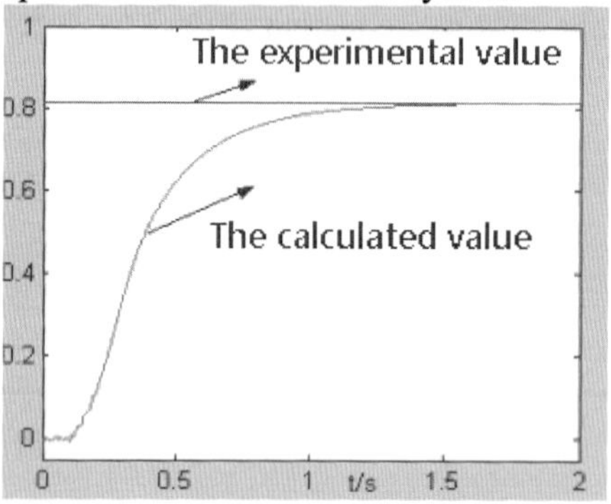

Fig. 1. Rotor resistance estimated.

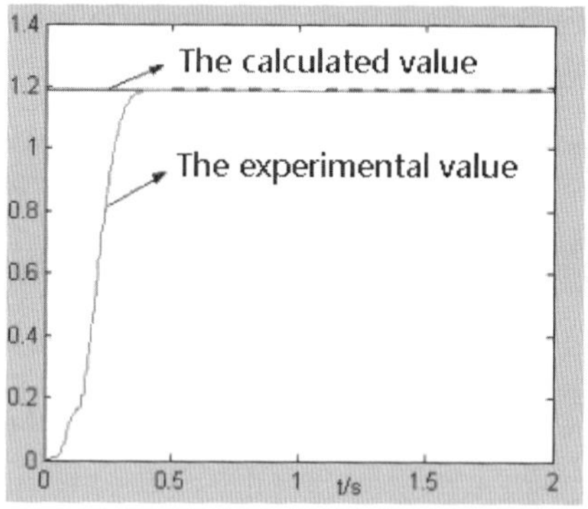

Fig. 2. Estimate rotor flux.

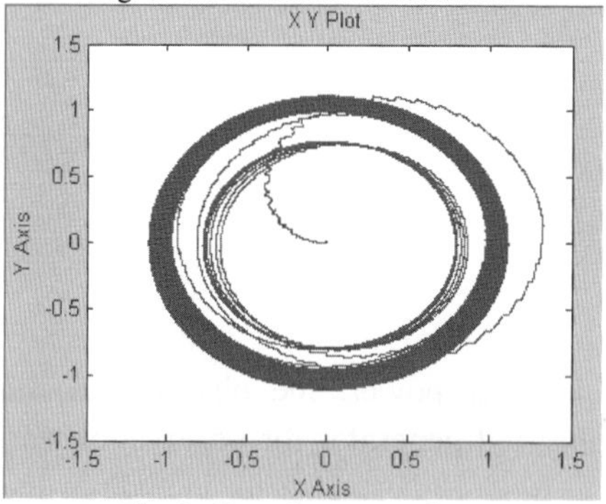

Fig. 3. Two phase-rotor flux.

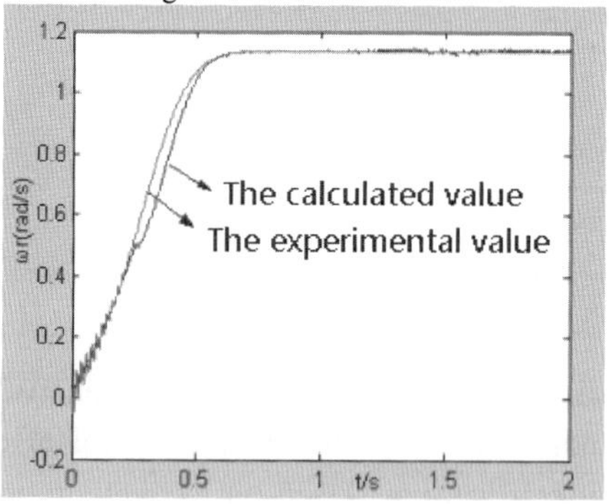

Fig. 4. Speed estimated figure.

Through the above analysis and calculation and simulation experiment, proves that to adopt the adaptive algorithm of rotor flux observer flux observation well, to estimate the rotor resistance, rotational speed and has good performance. The study high-performance doubly-fed generator active and reactive power decoupling has practical significance.

Acknowledgments

This work was financially supported by Jiangsu university brand professional construction engineering projects in 2015 (PPZY2015B184).

References

1. Hossein Madadi Kojabadi, Liuchen Chang, and R. Doraiswami. A MRAS Based Adaptive Pseudo reduced Order Flux Observer for Sensorless Induction Motor Drives. IEEE Transactions on Power Electronics, 2005, 20: 930-938.
2. Liuchen Chang. Wind Energy Conversion Systems. IEEE Canadian Review, 2002: 12-16.
3. Quincy Wang and Liuchen Chang. PWM Control Strategies for Wind Turbine Inverters. Journal of Wind Engineering, 2001, 25: 33-40.

The application of IGA_BP model in GPS elevation fitting

Ji-Xian Wu[1], Hai-Ying Cao[1], Hai-Yan Du[2]
[1]*College of Civil Engneering & Mechanics, YanShan University, Qinhuangdao 066004, China*
[2]*Fugro Pacifica Qinhuangdao Co.Ltd, Qinhuangdao 066004,china*
E-mail: wjx44@163.com, 16543755@qq.com, danger_smith@126.com

Based on the advantages and disadvantages of genetic algorithm(GA)and BP neural network theory, this paper builds a new mixed model-IGA_BP model, which is applied to GPS elevation fitting. In GPS elevation fitting example, this paper compares the accuracy in IGA_BP model, BP network and the quadratic polynomial. The result shows that IGA_BP model is better in accuracy.

KeyWords: IGA_BP Model, BP Neural Network, GPS Elevation Fitting.

1. Introduction

With the development of the global positioning system technology, GPS is widely used in various engineering fields, but only in the plane coordinate system. Because of the normal height in our country, the geodetic height of the WGS84 ellipsoid which is obtained by the GPS measurement cannot be directly used. Elevation fitting model are commonly used as follows: polynomial fitting, multi-surface function, surface interpolation approximation, regression approximation method, fast fourier transform [1-4]. All the conventional methods are based on statistics, needs a sufficient number of samples, but in actual engineering application, the number of samples is usually limited, which prevent the fitting model to obtain the best solution. Artificial neural network is a distributed information storage, and has a good fault tolerance. Many scholars have studied the GPS height fitting model of artificial neural network [5-7]. All papers have shown that BP neural network can be used in elevation fitting, but there are also some disadvantages: (1) local extremum problem; (2) network topology structure needs to be empirically determined, with larger randomness; (3) the initial weights determination seriously affect the results. In order to solve the above problems, this paper introduces genetic algorithm, builds new height fitting model-IGA_BP. The model gives full play to the advantages of GA and

BP, to make up for their deficiencies. In this paper, the IGA_BP model is used in practical engineering example, the results is better than traditional models.

2. IGA_BP Model Process

BP neural network is based on the gradient descent learning algorithm, which is easy to fall into the local minimum value of the solution. Genetic algorithm is a general algorithm framework, types of non-adherence to any problem, expansion of strong, Easy integration with other algorithms. BP neural network optimization accuracy, fault tolerance is strong, but Seriously depends on initial weight which is obtained by experience or randomly, if the initial weight is unreasonable, the BP network does not converge, or limited to local optimum, which led to obtain the optimal value failure. The genetic algorithm is a global optimization algorithm, which can be applied to the global optimal solution, but the convergence rate of the algorithm is slow, which affects its efficiency. Based on this point, we will combine the two together to form a hybrid algorithm IGA_BP algorithm. Basic idea: in the beginning of the algorithm, with a larger number of populations, with intensive algorithms, a corresponding reduction in number of population, so as to ensure the diversity of individuals, avoid algorithm falling into local minimum, and improves the convergence speed of the algorithm, allows the algorithm to find the optimal solution quickly. The specific steps are as follows:

Step 1: Randomly generated population, each individual represents an initial value of network, the value of each gene is a connection weight.

Step 2: According to the fitness function value, the evaluation of individuals is obtained. By decoding each individual, a BP neural network input samples is built, By calculating the output error value E, the fitness function is adapted:

$$f = 1/(1+E) \tag{1}$$

Calculate the value of each individual, The individuals that have maximum values directly join into sub populations.

Step 3: Genetic operation

(1) selection operator. Using the roulette principle. The following function relationship between individual value and probability:

$$p_i = f_i / \sum_{i=1}^{m} f_i \tag{2}$$

(2) crossover operator. The cross rate is Pc.

$$\begin{cases} X_i(k+1) = \alpha X_i(k) + \beta X_{i+1}(k) \\ X_{i+1}(k+1) = \alpha X_{i+1}(k) + \beta X_i(k) \end{cases} \quad \begin{cases} \alpha = 0.5 + \lambda \\ \beta = 0.5 - \lambda \end{cases} \quad \lambda \in [0,1] \tag{3}$$

(3) mutation operator. The mutation rate is Pm.

$$X_i = X_i(p) + r \times \varepsilon + X_i(n-p-1) \qquad r \in [0,1] \tag{4}$$

ε is the domain of individual p+1.

Step 4: To determine whether the maximum number of iterations or meet the accuracy requirements, return Step 2.

Step 5: Taking the best individual in genetic operation as the initial weights of the neural network, through the given sample data, according to the BP algorithm for training the neural network to obtain the optimal solution.

Step 6: Inputting new sample data, the corresponding results are calculated by BP algorithm.

3. Experimental

3.1. *Introduction*

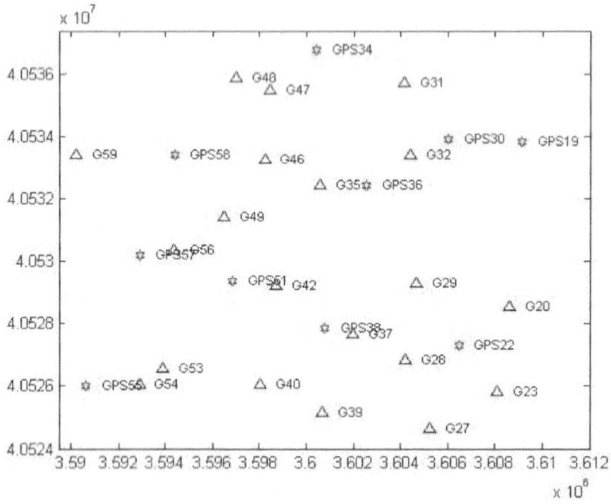

Fig. 1. View of GPS points.

Test area is located on the eastern coast of China, with a total area of 300km^2.The D grade GPS network has been established in the region, the average elevation is 7.587m, maximum elevation is 24.199m, minimum elevation is 2.404m, belongs to the plain terrain. The area has 30 GPS control points, each point has coordinate (X,Y), WGS84 height(H).and the normal height(h). GPS control points distribution diagram is shown in figure 1. In order to test the accuracy of the IGA_BP model, The traditional BP neural network and polynomial function are used. In the process of BP calculation, there are two plans. In Plan A, the learning set has 10 data while the working set has 20 data; On the contrary, In Plan B the learning set has 20 data while the working set has 10 data.

3.2. *Model parameter*

The paper constructs three layer BP neural network, the input layer is coordinates (x, y) and geodetic height (H), the hidden layer is provided with four nodes and the output layer is normal height (h). Before calculating, data is normalized [8]. Excitation function is Sigmoid [9], learning rate is 0.01, Minimum expected error is 1×10^{-5}, impulse rate is 0.9, the maximum training number is 10000. Individual gene length is 21, population size is 20, Pc=0.6, Pm=0.2, the maximum iterations is 1000.

3.3. *The GPS elevation fitting*

In the example, the residual value and the error is selected to test models. The comparison results are shown in table 1, Learning and working set of control points residuals after fitting is shown in Figure 2.

Table 1. Comparison of GPS elevation fitting accuracy in Three Models.

(1) the error of learning

		Plan A			Plan B		
		polynomial	BP	IGA_BP	polynomial	BP	IGA_BP
Residual	Max	-0.044	0.056	-0.038	0.030	0.036	-0.023
	Min	0.002	-0.003	0.001	-0.003	0.005	0.001
error		0.034	0.031	0.019	0.027	0.026	0.014

(2) the error of Working

		Plan A			Plan B		
		polynomial	BP	polynomial	BP	polynomial	BP
Residual	Max	-0.132	0.108	-0.064	-0.048	0.063	0.035
	Min	0.006	-0.001	0.001	0.002	0.000	-0.001
error		0.043	0.045	0.027	0.023	0.021	0.015

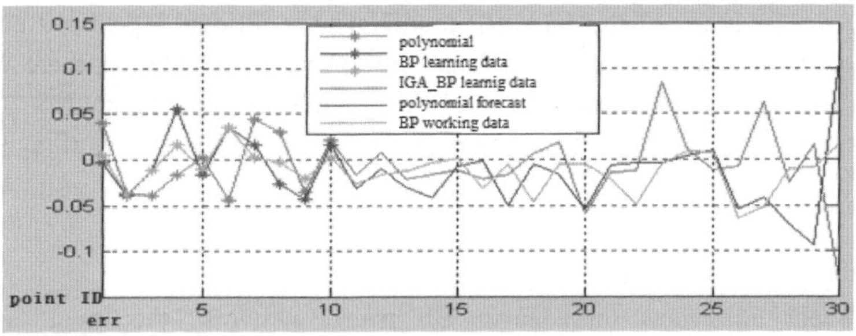

(1)Plan A

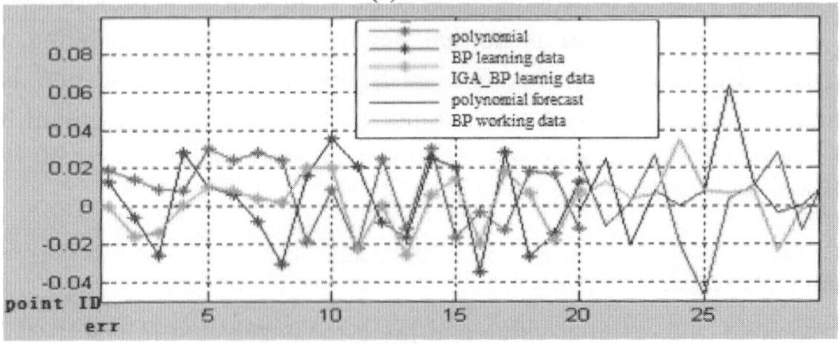

(2)Plan B

Fig. 2. Comparision of GPS elevation fitting.

Table 1 and Figure 2 shows that the fitting accuracy of the three models in the plan B is better than plan A. It indicates that more learning samples to establish the network, the network is more stable, and the fitting precision is also relatively higher. Comparison of three models, the traditional BP neural network is slightly higher than the polynomial fitting accuracy, and the accuracy of IGA_BP is highest, which shows that the new model is feasible, and has certain application prospect. The maximum residual of IGA_BP hybrid model in the plan B is 0.035m, error is 0.015.The precision can reach four grade leveling surveying.

4. Conclusion

In this paper, an IGA_BP model is established and tested by a specific example. The result shows that IGA_BP model is feasible and has high precision. We also draw the following conclusion:

(1) the precision of IGA_BP model in GPS height fitting is higher than that of the traditional BP neural network and polynomial algorithm, and the number of iterations is significantly less than the traditional BP neural network;

(2) IGA_BP model can be used with less control points, while quadratic polynomial algorithm requires more control points.

(3) the IGA_BP model has strong generalization capability, thus, The fitting precision of intermediate control points and edge control points are equal.

References

1. Xu SQ, Zhang HH, Yang ZQ, Wang ZM. GPS measurement principle and Application[M] (In Chinese).
2. Wang SW, Chen ZY. GPS Height Collocation Based on Moving Surface Model[J]. Hydrographic Surveying and Charting, 2005, 25(5):13~14 (In Chinese).
3. Li JW, Yang YK, Gao J, Zhou GY, et al. Compare and Analysis of GPS Height Fit by Polyhedral Function and Conicoid Fuction[J]. Shandong Metallurgy, 2006, 28(3):42~43 (In Chinese).
4. Jiao ML, Jiang TC. A Conversion Method of GPS Height Based on Kriging Algorithm and Wavelet[J]. JOURAL OF GEODESY AND GEODYNAMICS, 2008, 28(2):100~103 (In Chinese).
5. Yang MQ, etc. Conversion of GPS Height by neural network method.[J]. Acta Geodaetica et Cartographica Sinica, 1999, 28(4):301-307 (In Chinese).
6. Hu WS, Hua XS, Zhang ZW. Conversion of GPS Height in plainness Area by the CF&NNM Method[j]. Acta Geodaetica et Cartographica Sinica, 2002, 31(2):128-133(In Chinese).
7. Kavzoglu T and Saka M H. Modeling local GPS/leveling geoid undulations using artificial neuralnet works[J]. Journal of Geodesy, 2004, 78(9): 520-527.
8. Zhang HW, Wang QJ, Zhu JJ, Zhang XH. Influence of Sample data Preprocessing on BP Neural Network-Based GPS Elevation Fitting[J]. JOURAL OF GEODESY AND GEODYNAMICS, 2011, 31(2):125~128 (In Chinese).
9. Zhang GY, Hu Z. Improved BP neural network model and its stability analysis[J]. Journal of Central South University (Science and Technology), 2011, 42(1):115~124 (In Chinese).

Preliminary study on the soft ceramic tile in building exterior wall---case study on Zunyi Normal University campus Xinpu project as example

Zhen-Yu Song[1], Xiao-Shan Fang[2,3*], Ya-Nan Qin[3]

[1]*Architectural Design & Research Institute of SCUT, Guangzhou, Guangdong*

[2]*State Key Lab of Subtropical Building Science, Guangzhou, China*

[3]*School of Architecture, South China University of Technology, Guangzhou, China*

E-mail: xshfang@scut.edu.cn

* *Corresponding author*

Soft ceramic is a new type of building decoration material, which has the characteristics of thin material, light weight, strong performance, energy saving, environmental protection, etc. According by the requirements of the design idea, we finally chose the soft ceramic as the external wall material in the new campus of Zunyi Normal University. Based on the introduction of the material selection and the scene effect of using soft ceramic, this paper put forward reasonable suggestions on the application of the soft ceramic tile in the building external wall.

Keywords: Soft Ceramic Tile; External Wall; Application Program; The New Campus of Zunyi Normal University.

1. Introduction

External wall decoration is one of the important contents of building decoration. Its purpose is to improve the capability of exterior wall to resistance nature of various factors such as dust, rain and snow, ice, sun invasion and damage, and together with the wall structure to meet the requirements of heat preservation, heat insulation, sound insulation, waterproof, landscaping and other functions. Therefore, good external wall decoration materials should balance the dual functions of protecting wall and landscaping building. In recent years, due to the rapid development of the construction industry as well as people's increasing demand for material and spiritual life, the development of modern decorative materials improved quickly and many new exterior wall decoration materials constantly appeared.

The soft ceramic is one kind of new decorative surface material in recent years. The soft ceramic as an external wall decoration materials was used in the first-stage project of Zunyi Normal College Campus in Xinpu. This paper introduces the material selection, design and application effect of soft ceramic in this project and summarizes some application suggestions of the soft ceramic.

Soft ceramic is composed of modified clay material (MCM) which is the main raw material, a small amount of water soluble polymer. It is made under dynamic temperature curve, in the light of irradiation crosslinking, roasting and formed into a kind of building decorative surface materials which is flexible, thin and light ,short soft ceramic tile . [1] (Fig.1) The soft ceramic with inorganic materials can achieve the flexibility of organic materials and in the same time, retain the durability of inorganic materials. The soft ceramic is made at low temperatures with electric roasting, which is one kind of recycling, energy-saving and ecological building materials. In addition, the soft ceramics decoration materials expressive force is so strong that it not only can perform the color and texture of the stone, ceramic, wood, leather, knitting, metal plate, weaving, water board of the existing material, but also can create the personalized forms according to the architects' and designers' design requirements. [2]The soft ceramics tile decoration materials are divided into four types such as flexible ceramics external wall decoration materials, flexible ceramics interior wall decoration materials, flexible ceramics indoor and outdoor floor, and bathroom flexible brick [3] .

The soft ceramics decoration material is a new type of environmental protection decoration materials with excellent physical properties. Used as wall decoration materials, it has the characteristics of light quality, good flexibility, modeling the appearance of diverse and good weather resistance. Used as a decoration materials on the ground, it has the characteristics of wear-resistant, anti-slip and comfortable feeling. The product construction is simple and quick, not easy to fall off. It applies as building decoration to the outer wall, the inner wall and ground , especially in the projects of high-rise building exterior decoration, building facade decoration renovation exterior wall insulation system and curved walls or arc columns. Soft ceramics decoration material can be directly pasted on the surface layer of ceramic tile, mosaic, aluminum-plastic plate etc., not need to knock out the original base. Therefore, it is not only suitable for new building decoration but also is very applicable to the renovation of existing buildings. [4] The soft ceramic has been widely used in the transformation of the old city of Chongqing, the Guangzhou Asian Games, the Shenzhen Universiade and other major national projects.

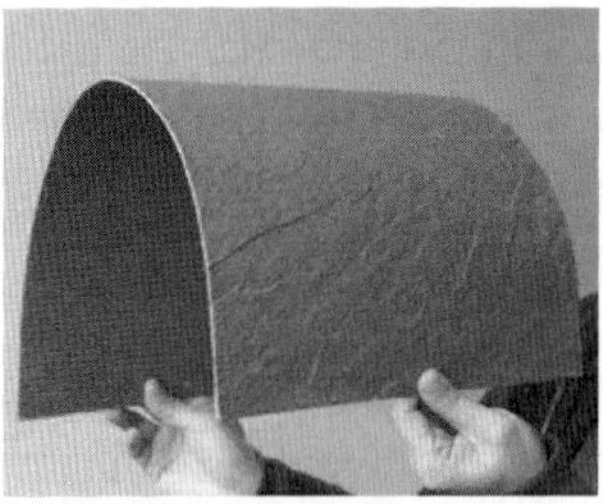

Fig. 1. Picture of Soft ceramic

Fig. 2. Aerial view of the Zunyi Normal University

2. Case Study

2.1. *Case introduction*

Zunyi Normal University is located in Zunyi City ,Guizhou Province. Zunyi is a famous city known as the historical and cultural city of revolution and named as" turning point in history" because there were some famous conferences held in the city and that was very important to Chinese history. The Zunyi Normal University is a university with more than one century history of traditional institutions of higher learning. The new campus of the college is located in Xinpu District University City of Zunyi City. The first-stage project is 925 acres, with a total construction area of 500,000 square meters, which would serve for about 15000 students. There are 30 buildings in the whole new campus and 27 of them are used soft ceramic as building exterior wall materials. (Fig.2).

In the campus planning, the design concept is "harmony with nature, source from the regions, based on the southwest district, recurrence to the humanities". [5,6] "Based on the southwest" means based on the current situation of economic underdevelopment in the southwest region. The architecture were designed as high efficient buildings and without too much decoration. Local materials and traditional techniques were used in the design to create the regional characteristics of the buildings. The colour tone of the whole campus is red and combination with white, meaning the school is from history and facing the future. The architects expressed their respect to the original landform and regional characteristics of the environment by simulating the material and texture of red sandstone. (Fig.3,Fig.4,Fig.5).

Fig. 3. The effects of public school building

Fig. 4. The effects of Public Conference Center

Fig. 5. The effects of art college building

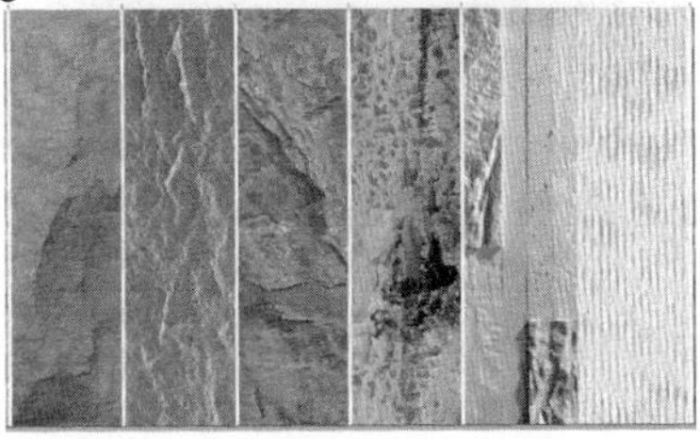

Fig. 6. rich expressive form of soft ceramic

2.2. *Application programs*

Under the premise of project investment and time limit for this project , the principles of selecting materials are appropriate cost and convenient construction . In order to make the red part of the wall to create the effect of the local red sandstone, the design team selected a variety of external wall materials for comparison.

Red sandstone: This type material is with low strength and high bibulous rate . It is not an ideal material for stone curtain wall panels for its poor weather resistance. But its unique sandstone texture, color and style, always make the architects like to use them as a stone curtain wall. [7] From the completion of the effect, the direct use of dry hanging sandstone curtain wall would be the best choice. But the cost of dry-hang sandstone is too much expensive. The cost of wet paste red sandstone would be lower, but the buildings in the Zunyi Normal University are too high to use them. So based on the point of view of investment control, it had to select other close effects materials.

Facing tile: This material is with strong tile texture and rich decorative effect,. It's durable, easy to clean, fire resistance, water resistance, corrosion resistance and other characteristics, suitable for polluted environment and the dusty air. But the process of external wall tile is complicated and the demand of the construction technology control is higher. It's easy to cause the shedding phenomenon of security risks. In view of the existence of security issues, some cities have issued a relevant building regulations and forbidden the use of tiles. The main reason for restricting the external wall tile use is tiles are easy off to hurt people [8] .

Coating: coating is more economical and convenient to update. Due to its light weight, it is an ideal thermal insulation facing layer from security. But in China, dusty environment, acid rain, will seriously affect the paint effect. It need to brush paint again to protect the visual effect of the facade every few years. [9] And from a decorative point of view, the texture of coating is poor. Although now there are a variety of texture to choose, but that greatly enhance the price of coating and the complexity of the construction process.

Soft ceramic: This material has the following features: waterproof breathable, light weight (weight per square meter between the 3-6kg); good fire-retardant performance; good compatibility with the flexible insulation materials and surface mortar ; good ageing resistance and artificial accelerated resistance aging test reached 3500h, which is 14 times of the ordinary latex paint for exterior wall; strong stain resistance. MCM has the "Lotus" effect, dirty water and dirt is hard to invade the surface, and it has rich expressive power. Compared with ceramic wall tiles and coating, the soft ceramic gets more rich expressive form. [9] (Fig.6).

Integrated the specific circumstances of the Zunyi Normal University project, the construction party chose the soft ceramic as the external wall material. The specific colors and materials scheme are that the campus overall color according to the partition function and each building monomer has the contrast of color and texture, such as public teaching building with yellow and white combination. The yellow is a concave convex sandstone while the white is flat sandstone. The College of liberal arts with red and white combination. The red color is like Mara slate while the white is flat sandstone. Because the soft ceramic tile relatively thin, the block size can be larger, such as the maximum size can be 2440mm*1220mm.The size of the plate material of sandstone maximum can be 1200mm*600mm. The size of the south wall rock material maximum can be 1800mm*600mm. The bump material of sandstone maximum size can be 600mm*300mm. In order to let the red and white exterior wall texture contrast more strongly, we use two lattice such as 1200mm*600mm and 600mm*600mm in the area of the white walls, while another two lattice such as 600mm*300mm and 600mm*150mm in the red wall area. However, because of its thin thickness and soft texture, it is needed to choose a combination of the block to use in the corner of the wall and other parts of the body.

After choosing the external wall material, the construction can refer to the "Technical Specification for Flexible Ceramics Decoration Work DBJ/T 15-68-2009 ". There are a few points will impact greatly on the appearance: ①It should use waterproof and drainage construction on the windowsill, cornice, decorative line, awning, balcony and water hole wall of concavo convex parts.

②To windowsill, cornice, decorative line, awning, balcony coping and component, the top surface drainage slope should not less than 6%, and should be set to drip structure. ③Parts of the closing of the basement and windowsill edge must be pointed agent by pointing dense, avoid warping, the wall can also be designed according to requirements by stone feet. ④It should select with the product color similar matching binder, choose matching sealant or silicone rubber. In order to avoid the occurrence of efforescence albescent phenomenon, it need to use low alkaline or alkaline non special adhesive and sealant.

3. Completion Effect

In September 2015, more than half of this project's construction has been completed and the rest of the building is capped. The whole campus is planned to put into use in September 2016. From the building effect, the result of flexible ceramics decoration material to create the texture of sandstone is ideal. Through a variety of different textures, different color material selection and combination, the campus succeeded to create a rich visual image. It is closer to the design intention of the original idea of the southwest region of the college. (Fig.7-12).

Fig. 7-12. scene photo of school building

4. Conclusion

As a new type of building decoration materials, soft ceramic has been vigorously promoted in recent years, and related applied research also has been gradually expanded. But among the published research results, there is still relatively little combined with practical engineering case research.

The majority of architects with practical engineering need to sum up the experience of this meterial use in order to obtain more appropriate use effect. Soft ceramic has the features of soft, thin, light, safety. Because of its forming mechanism, performance and construction advantages, it has been widely used. However, the actual application of this material is not long, the adaptation to the climate and the effects of completion still need time to test.

Acknowledgements

This article was financially supported by the Natural Science Foundation of Guangdong Province, China (No: 2015A030313218), supported by the independent research project (No: 2016KB09) from State Key Lab of Subtropical Building Science, South China University Of Technology. Special thanks to the chief engineer of Architecture Design Research Institute of SCUT——Mr. Yongxiang LIN. Special thanks to all the participation of these research projects.

References

1. DBJ/T15-68-2009, *Technical Specification for Flexible Ceramics Decoration Work*, Construction Department of Guangdong Province, Guangdong, 2009, pp. 2.
2. DBJ/T15-68-2009, *Technical Specification for Flexible Ceramics Decoration Work*, Construction Department of Guangdong Province, Guangdong, 2009, pp. 24.
3. DBJ/T15-68-2009, *Technical Specification for Flexible Ceramics Decoration Work*, Construction Department of Guangdong Province, Guangdong, 2009, pp.3.
4. DBJ/T15-68-2009, *Technical Specification for Flexible Ceramics Decoration Work*, Construction Department of Guangdong Province, Guangdong, 2009, pp. 23.
5. Song, Zhen Yu; Lin, Yong Xiang; Qiao, De Bao; Fang, Xiao Shan, *The exploration of design strategies on regional characteristics of southwest campus architecture-- case study on art teaching building group of Zunyi Normal College in Xinpu new campus,* Applied Mechanics and Materials, v 584-586, 2014, pp 247-252
6. Song, Zhen Yu; Fang, Xiao Shan, *Harmony with site and assimilated into nature - The regional exploration of southwest mountain campus planning in China*, 2011 International Conference on Multimedia Technology, ICMT 2011, pp 3979-3982
7. LeiZibin, *How to ensure the quality of dry red sandstone curtain wall decoration*, Stone, 2009, 2009(5), pp. 31-33.

8. MaJunhui, WangJun, ZhangHaimei, *Application and existing problems of building facade decoration materials in China*, Master, 2012, 2012(9), pp. 107.
9. WangYungang, SuXiange, *Application and engineering demonstration of modified inorganic powder composite building facing sheet in external thermal insulation system*, Construction Science and Technology, 2013, 2013(11), pp. 55-57.

Effect of rolling-tension treatment on the vibration characteristics of circular saw blade

Yan-Chun Wang
College of Mechanical and Electronic Engineering, Qingdao University of Science and Technology,
Qingdao, China
E-mail: wangyanchun0603@126.com

In this paper the effect of rolling-tension treatment on the vibration characteristic of circular saw blade is analyzed and calculated with nonlinear theory and perturbation method and finite element analysis, the results show that : cutting temperature can reduce frequency of circular saw blade, rotation speed can increase frequency; Rolling-tension treatment can increase frequency of (0,2) mode, therefore, the smallest critical rotation speed increases and there is the optimal position; Ratio of outside radius and clamping radius (Rb/Ra) can cause very big effect. Thence carrying out studies have provide a reference for designing the new type of saw blade and quickly confirming the optimal position of rolling-tension treatment of the circular saw blade.

Keywords: Circular Saw Blade; Non-linear Vibration; Tensioning Treatment; Stability.

1. Topic Background

Circular saw blade is the main equipment in timber-process field. According to statistics [1]: the thickness of the saw blade or saw road deviation reduces to 0.8mm, which enables the material use rate to increase (1-3)%. In America, a medium-scale timber-process plant, timber losses amount to as much as 800,000 dollars in every year because thickness of saw is too thick. Use of high-speed cutting of thin blade, we must study effective measures to ensure the smooth cutting. Currently, there are many ways to improve the stability of the saw blade. For example, 1) active control of cutting process; 2) on-line cooling or heating; 3) change the structure of the saw blade; 4) tension of the saw blade and so on. Adopting rolling-tension treatment approach is the effective way to control vibration and improve the stability of cutting. Theory and practice shows: in various methods of tension treatment, rolling-tension approach is recognized as an effective method, which is a simple, high efficient, reliable, repeatable and consistent method. This method can improve effectively stability and precision of cutting.

2. Theory Analysis

2.1. *The nonlinear vibration equation of the saw blade*

Outside radius of circular saw blade is Rb, clamping radius is Ra, uniform and equal thickness saw blade will be simplified as mechanical model and will be shown in Fig. 1.

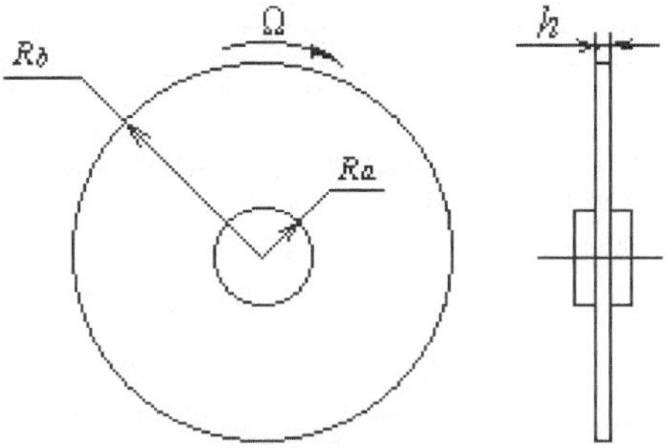

Fig. 1. Mechanical model.

According to the nonlinear vibration theory[2]-[4], dynamic equation of the saw blade is:

$$M\ddot{U}(t) + C\dot{U}(t) + K_T U(t) = P(t) \tag{1}$$

In this formula, M—quality matrix of the system; C—damp matrix of the system; KT—tangent stiffness matrix of the system; —vibration displacement, velocity and acceleration arrays of the system; P(t)—external load arrays. Therefore, if quality and damping of saw blade is the constant, saw rotating, tension treatment, cutting temperature and tension of band saw and so on, which will make change of tangent stiffness matrix KT, so that will cause effect of dynamic characteristic system.

In process of cutting, axisymmetric temperature distribution will be cause by cutting heat and friction heat. According to reference documents[5], the actual measured temperature distribution equation T(r) when normal cutting, which will be simplified as normal distribution equation:

$$T(r) = \frac{T_m}{\sigma\sqrt{2\pi}} \exp\left[-\frac{1}{2}\left(\frac{r - r_m}{\sigma}\right)^2\right] \tag{2}$$

In this equation, σ—temperature distribution curve variance; rm—maximum temperature of radial position; Tm—temperature coefficient; In normal cutting, σ=0.3, Tm=26.32, rm=(Rb-0.02)m.

3. Critical Rotation Speed

Test proves that critical rotation speed is main parameter of controlling the stability of circular saw blade. As for disk of middle clamping, around free, formula of critical rotation frequency as following [6]:

$$(\omega_{mn} - n\omega)^2 = \beta^2 \tag{3}$$

In this formula, ω_{mn}—natural frequency of vibration mode;

ω— rotation frequency;

β— excitation frequency;

n— the nodal diameter number of vibration mode.

Critical rotation frequency ω_{crit} is the minimum rotation resonance frequency of saw blade under constant horizontal force (β=0), which is the fourth formula:

$$\omega_{crit} = \min\left(\frac{\omega_{mn}}{n}\right) \quad (n \geq 2) \tag{4}$$

4. Example Analysis

Using the above theory and perturbation method [7]-[9], we give an example analysis for saw blade structure of Fig . 2, Rc is rolling radius.

Parameters as following: Rb=0.225m, Ra=0.060m, h=0.0016m, elastic modulus E=2.1×1011N/m2, mass density ρ=7800kg/m3, Poisson's ratio μ=0.3, coefficient of thermal expansion α=1.1×10-8/°C. Results data, (m, n) mode is that mode of m pitch and n pitch diameter, (m,n:C) is cosine mode; (m,n:S) is sine mode.

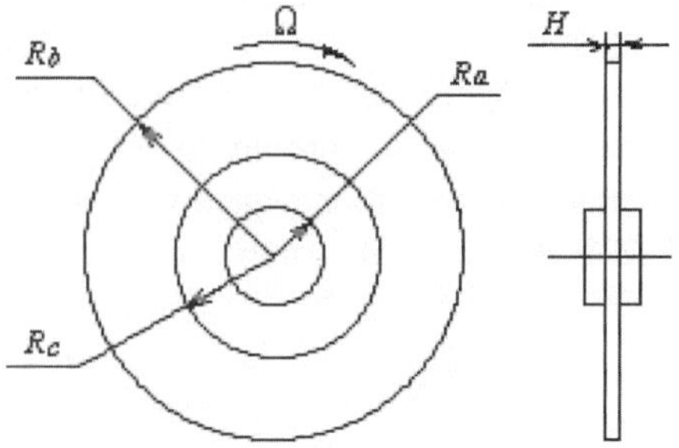

Fig. 2. Rolling-tension treatment position R_c.

Table 1 is that without rolling-tension, centrifugal force field, normal cutting temperature field, and clamping-radius have effect on saw frequency.

Table 1. centrifugal force field, normal cutting temperature field, and clamping-radius have effect on saw frequency (Hz).

$\Omega=0$, $T_m=0$, $R_b=0.225m$, $R_a=0.060m$				
(0,1:C)mode	(0,1:S)mode	(0,0)mode	(0,2:C)mode	(0,2:S)mode
46.552317	46.552338	47.990962	57.928004	57.928223
$\Omega=0$, $T_m=26.32$, $R_b=0.225m$, $R_a=0.060m$, $r_m=0.205m$, $\sigma=0.3$				
(0,1:C)mode	(0,1:S)mode	(0,0)mode	(0,2:C)mode	(0,2:S)mode
41.836862	41.836883	44.514783	51.394683	51.394902
$\Omega=198r/s$, $T_m=26.32$, $R_b=0.225m$, $R_a=0.060m$, $r_m=R_b$, $\sigma=0.3$				
(0,0)mode	(0,1:C)mode	(0,1:S)mode	(0,2:C)mode	(0,2:S)mode
54.825080	56.547973	56.547990	73.115507	73.115740
$\Omega=198r/s$, $T_m=26.32$, $R_b=0.225m$, $R_a=0.04m$, $r_m=0.205m$, $\sigma=0.3$				
(0,0)mode	(0,1:C)mode	(0,1:S)mode	(0,2:C)mode	(0,2:S)mode
45.396646	46.627161	46.627196	65.931308	65.931468

With rolling-tension treatment radius Rc have effect on various mode frequency as following Fig. 3. The following is parameters of all curves.

Parameters of curve B: $\Omega=0, Tm=0$;

Parameters of curve C: $\Omega=198.0r/s$, $Tm=26.32$, $\sigma=0.3$, $Rb=0.225m$, $Ra=0.060m$, $rm=0.205m$;

Parameters of curve D: $\Omega=198.0r/s$, $Tm=26.32$, $\sigma=0.3$, $Rb=0.225m$, $Ra=0.060m$, $rm=0.225m$;

Parameters of curve E: $\Omega=198.0r/s$, $Tm=26.32$, $\sigma=0.5$, $Rb=0.225m$, $Ra=0.060m$, $rm=0.225m$;

Parameters of curve F: $\Omega=198.0r/s$, $Tm=26.32$, $\sigma=0.3$, $Rb=0.225m$, $Ra=0.040m$, $r_m=0.225m$.

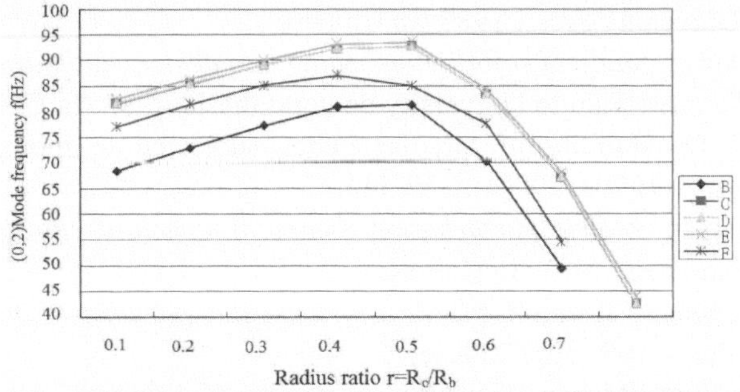

Fig. 3. Hange curve of radius ratio r with (0,2) mode frequency f.

5. Conclusions

(1) From table 1, we can see, for middle clamping saw, (0,0) mode frequency is single frequency. (0,1) mode frequency and (0,2) mode frequency are overlapping frequency. Cutting temperature can reduce frequency of circular saw blade, rotation speed can increase frequency of saw blade.

(2) From table 2, We can see from all results of various curves change, rolling-tension treatment can increase frequency of (0,2) mode. Therefore, the smallest critical rotation speed increases and there is the optimal position.

(3) With a little change of parameter (rm,σ) of centrifugal force field and cutting temperature field, which can cause a little effect on optimal position of rolling-tension treatment, but, ratio of outside radius and clamping radius (Rb/Ra) can cause very big effect, which we can be not ignored. We should take care in actual application.

Acknowledgments

This work is supported by Doctoral Found of QUST(0022655), where giving many acknowledges.

References

1. Szymani, R & Mote, C D. *Theoretical and experimental analysis of circular*. Wood Sci. Technol., 1979, 13(2):211-237.
2. Guo Zhongheng, *Nonlinear theory of elasticity.* Science Press, 1980 (In Chinese).
3. Xu Zhilun, *Theory of elasticity*. People's Education Press, 1982 (In Chinese).
4. Yin Youquan, *Nonlinear Finite Element Introduction of Solid Mechanics*. Beijing University Press, 1986 (In Chinese).
5. Zhang Dawei, *Study of stability of sawing and design methods of tensioning of circular saw*. Tianjin University doctoral thesis, 1995 (In Chinese).
6. Mote, C D. Stability of circular plates subjected to moving loads. J. Franklin Inst.,1970a, 290(4): 329-344.
7. Chen Suhuan, *Matrix perturbation theory of structure vibration analysis*. Chongqing Press, 1991 (In Chinese).
8. Chen Suhuan, *Vibration theory of stochastic parameter structure*. Jilin Science Press, 1992 (In Chinese).
9. Chen Suhuan, *Matrix perturbation theory of Dynamic design*. Science Press, 1999 (In Chinese).
10. Chen Suhuan, *Numerical Methods for Structural Vibration Analysis*. JiLin Science Press, 1996 (In Chinese).

Author Index